AF413036

Cellular Biology of Endotoxin

Handbook of Endotoxin

Series Editor: R.A. PROCTOR
University of Wisconsin Medical School, Madison, Wisconsin, U.S.A.

VOLUME 3

Cellular Biology of Endotoxin

Editor:

L.J. BERRY

Department of Microbiology,
University of Texas,
Austin, Texas, U.S.A.

1985

ELSEVIER

Amsterdam – New York – Oxford

ISBN 0 444 90389 5

Library of Congress Cataloging in Publication Data
Main entry under title:
Cellular biology of endotoxin.
(Handbook of endotoxin ; v. 3)
Includes bibliographies and index.
1. Endotoxins—Physiological effect.
I. Berry, L.J. (L. Joe) II. Series. [DNLM: 1. Endotoxins–pharmacodynamics. QW 630 C393]
QP632.E4C45 1985   616'.014   85-7060
ISBN 0-444-90389-5 (U.S.)

Published by:
Elsevier Science Publishers B.V.
P.O. Box 1126
1000 BC Amsterdam

Sole distributors for the USA and Canada:
Elsevier Science Publishing Co. Inc.
52 Vanderbilt Avenue
New York, NY 10017

Typeset in The Netherlands by Pecasse Intercontinental B.V., Maastricht
Printed in The Netherlands by Casparie, Amsterdam

Contributors

DR. L.J. BERRY
Department of Microbiology
University of Texas
Austin, TX 78712
U.S.A.

DR. S.G. BRADLEY
Department of Microbiology
  and Immunology
Virginia Commonwealth University
MCV Station, Box 110
Richmond, VA 23298
U.S.A.

DR. C.E. BURROWES
Division of Experimental Pathology
Department of Pathology
University of Toronto
Medical Sciences Building
Toronto
Canada M5S 1A8

DR. L.C. CASEY
Naval Medical Research Institute
Bethesda, MD 20814
U.S.A.

DR. J.-M. CAVAILLON
Unité d'Immuno-Allergie
Institut Pasteur
28 rue du Dr. Roux
75015 Paris
France

DR. J.A. COOK
Department of Physiology
Medical University of South Carolina
171 Ashley Avenue
Charleston, SC 29425
U.S.A.

DR. K. EGAWA
Department of Microbial Chemistry
Showa University School of
  Pharmaceutical Sciences
Hatanodai, Shinagawa-ku
Tokyo
Japan

DR. N. HAEFFNER-CAVAILLON
Institut de Biochimie
Université de Paris XI
91405 Orsay
France

DR. P.V. HALUSHKA
Departments of Pharmacology
  and Medicine
Medical University of South Carolina
171 Ashley Avenue
Charleston, SC 29425,
U.S.A.

DR. A.G. JOHNSON
Department of Microbiology and
  Immunology
University of Minnesota
  School of Medicine
Duluth, MN 55812
U.S.A.

DR. N. KASAI
Department of Microbial Chemistry
Showa University School of
  Pharmaceutical Sciences
Hatanodai, Shinagawa-ku
Tokyo
Japan

Contributors

DR. L. MELA-RIKER
Departments of Surgery and
  Biochemistry
Oregon Health Sciences University
Portland, OR 97201
U.S.A.

DR. R.J. MOON
Department of Basic Sciences
Mercer University School of Medicine
1400 Coleman Avenue
Macon, GA 31207
U.S.A.

DR. R.N. MOORE
Department of Microbiology
University of Tennessee
Knoxville, TN 37996
U.S.A.

DR. D.C. MORRISON
Department of Microbiology
Kansas University Medical
  Center
39th and Rainbow Boulevard
Kansas City, KS 66103
U.S.A.

DR. H.Z. MOVAT
Division of Experimental Pathology
Department of Pathology
University of Toronto
Medical Sciences Building
1 Kings College Circle
Toronto
Canada M5S 1A8

DR. A. NOWOTNY
Center for Oral Health Research
University of Pennsylvania
4001 Spruce Street
Philadelphia, PA 19104
U.S.A.

DR. R.A. PROCTOR
Department of Medical Microbiology
University of Wisconsin
  Medical School
Medical Sciences Center
Madison, WI 53706
U.S.A.

DR. T.S. ROGERS
Department of Pharmacology
  and Medicine
Medical University of South Carolina
171 Ashley Avenue
Charleston, SC 29425
U.S.A.

DR. D.L. ROSENSTREICH
Departments of Medicine
  and Microbiology and
  Immunology
Albert Einstein College of Medicine
1300 Morris Park Avenue
Bronx, NY 10461
U.S.A.

DR. G.M. SHACKLEFORD
Department of Microbiology
  and Immunology
University of California
  School of Medicine
San Francisco, CA 94143
U.S.A.

DR. R.C. SKARNES
Worcester Foundation for
  Experimental Biology
Shrewsbury, MA 01545
U.S.A.

DR. J.A. SPITZER
Departments of Physiology and
  Medicine
Louisiana State University
  Medical Center
New Orleans, LA 70112
U.S.A.

DR. L. SZABÓ
Institut de Biochimie
Université de Paris XI
91405 Orsay
France

DR. H. TAVAKOLI
Departments of Physiology
   and Surgery
Michigan State University
East Lansing, MI 48824-1101
U.S.A.

DR. R.J. ULEVITCH
Department of Immunology
Scripps Clinical and Research
   Foundation
10666 North Torrey Pines Road
La Jolla, CA 92037
U.S.A.

DR. R.I. WALKER
Naval Medical Research Institute
Bethesda, MD 20814
U.S.A.

DR. W.C. WISE
Department of Physiology
Medical University of
   South Carolina
171 Ashley Avenue
Charleston, SC 29425
U.S.A.

Contents

Contents

Chapter 3
In vivo distribution and detoxification of endotoxins
R.C. Skarnes

Chapter 4
Genetic control of endotoxin response: C3H/HeJ mice
D.L. Rosenstreich

Chapter 5
Glucocorticoid-antagonizing factor
R.N. Moore, G.M. Shackleford and L.J. Berry

Chapter 6
Effect of endotoxin on lipid metabolism and on adipocytes
J.A. Spitzer

Chapter 7
Effect of endotoxin on mitochondrial function
L. Mela-Riker and H. Tavakoli

Chapter 8
Effect of endotoxin on cytochrome P-450 activity
N. Kasai and K. Egawa

Chapter 9
Effect of endotoxin on tryptophan metabolism
R.J. Moon

Chapter 10
Modulation of antibody synthesis by bacterial endotoxins
A.G. Johnson

Chapter 11
Endotoxin interactions with platelets
R.I. Walker and L.C. Casey

Chapter 12
Effects of endotoxins on neutrophils
R.A. Proctor

Chapter 13
Local Shwartzman reaction: endotoxin-mediated inflammatory and thrombo-hemorrhagic lesions
H.Z. Movat and C.E. Burrowes

Chapter 14
Agents which modulate reticuloendothelial function and affect endotoxin sensitivity
J.A. Cook, T.S. Rogers, W.C. Wise and P.V. Halushka

Contents

Chapter 17
Antitumor effects of endotoxins
A. Nowotny

Introduction

Few, if any, biological materials have been the object of as many scientific investigations as bacterial endotoxin or lipopolysaccharide. Starting soon after World War II, when its antitumor properties caught the imagination of a variety of specialists, the initial gush of publications that lasted for more than two decades has now slowed to a steady but apparently undiminishing flow. An important reason why endotoxin has been the object of so much research is its diversity of action. It poses challenges to chemists, physiologists, immunologists, microbiologists, pharmacologists, endocrinologists, biochemists, anatomists or structural biologists, pathologists, internists, surgeons, infectious disease specialists and others. A substance with an appeal as broad as this is rare in science.

Volume 3 of this Handbook focuses on the cellular biology of endotoxin. The chapter titles confirm the diversity of the responses endotoxin elicits in cells and organs and so, of necessity, the authors represent a variety of disciplines. They are authorities in their specialty and come from several different countries. The background of the scientists who have become involved in research on endotoxin covers a broad range of disciplines and hence their training has differed. It is not surprising, therefore, that the approach used by some may be called into question by others.

In these introductory remarks, I want to address three issues (a) the question of dosages that investigators of endotoxin have used, (b) the importance of route of endotoxin administration in natural and experimental situations, and (c) the importance of variation of species in the type of responses elicited by endotoxin. No references will be cited in order to avoid 'finger-pointing' at scientists who have done their work from different perspectives. After all, there are many reasons for studying endotoxin and each investigator must decide on the experimental design that best serves his purposes. He must choose the animal or cell type to be studied, the dosage and route of injection of test materials and the times when observations are to be made. All of these parameters may influence the outcome of the experiment, at least potentially. When the focus is only on endotoxin, the generalizations that have been drawn as a result of ignoring the experimental circumstances involved have resulted in erroneous conclusions. One of the most serious offenses in endotoxin research has been the use of excessively high doses. Everyone should be cognizant of the points made here before assuming that observations made under one set of conditions are typical of the biological effects of endotoxin under all circumstances.

To illustrate this point, the range of dosages that have been employed, especially those on the high side, leaves room for criticism. I once heard a distinguished investigator say in defense of the large dose he had administered to dogs: 'I am studying shock and I will use the amount of endotoxin required to produce it'. He did not say that the dose was also large enough to cause death of some of the animals within a period of six or seven hours, that is, within the usual work-day of the technicians. This is a matter of convenience unrelated to any other consideration. His approach represents a justifiable experimental design (a) as long as everyone is aware of what is being done and (b) no one assumes that the ensuing biological effects are typical of endotoxemia, regardless of dose. Unfortunately, this is rarely the case. In fact, I do not recall any statement in writing (other than my own) in which large doses have been called into question. On the other hand, there are many examples where the biological or cellular responses to large doses of endotoxin are cited as typical effects of endotoxin.

Why do people of differing scientific interests use endotoxin in their research? Physiologists have focused on endotoxin primarily because it provides a convenient means of producing shock. As long as this is an end in itself and not an attempt to explain the shock that often accompanies gram-negative sepsis, then there is no argument about dosage. Shock can be produced also by other techniques that are used from time to time but none is as simple as giving an injection of endotoxin. For a given dose a well-timed response usually follows. Physiologists are used as an example and not as the only ones who inject large quantities of endotoxin into animals or add them to cells in culture. When this is done, they rarely, if ever, are attempting to understand the pathogenesis of gram-negative infections. They know that endotoxin elicits a response they hope to analyze and they use whatever amount is needed to produce the effect of interest to them.

Textbooks of medical microbiology in discussing the pathology of some diseases caused by gram-negative bacteria contain statements that endotoxin is, or probably is, responsible for some, or all, of the symptoms. I have seen no references cited to substantiate these claims and yet there may be some justification for making them since endotoxin is the only known toxic moiety that has been identified in the organisms referred to. Examples are meningococci, gonococci, the treponemes of syphilis, and some of the opportunistic bacteria responsible for nosocomial infections (*Serratia marcescens*, *Proteus vulgaris*, etc.). Careful investigations designed to determine whether or not endotoxin duplicates the symptoms of the infection have not always been successful. In fact, inbred strains of mice known to be unusually susceptible or resistant to endotoxin fail to respond to gram-negative infections in a similar manner. Actually, some of the strains most resistant to endotoxin are unusually susceptible to infection and vice versa. This has naturally led to the conclusion that endotoxin makes little, if any, contribution to the pathogenic activity of these gram-negative organisms. Since the biological effects of endotoxin are typically assessed in the laboratory following the administration of a bolus via the intraperitoneal route (most commonly) or the intravenous route in some studies, any endotoxin released from bacteria during the course of an infection would be expected to occur more or less continously over a matter of hours or days, and the site (or sites) of the release would then depend on the in vivo distribution of the bacteria. In the absence of definitive studies comparing host response to a bolus of endotoxin (regardless of size) with what occurs after the infusion of the same amount of toxin over a period of time, the ability to draw valid conclusions is impossible. It may be that the inferences drawn from studies of inbred mice are misleading because of the way in which the endotoxin is presented to the animal. Since it has been found that the extensively studied C3H/HeJ mouse responds normally to endotoxin prepared by the Boivin technique but is unresponsive to endotoxin extracted from cells with phenol-water, it would be a mistake to assume that only a phenol-water-type of material is present in an active gram-negative infection. It is more likely that Boivin-type endotoxin is present since it contains both lipopolysaccharides and proteins. The very uncertainty about the endotoxic moiety released by the pathogens during an infection is warning enough to wary investigators who might otherwise be misled by experiments of this type. There is a related point. In many of the reports in the literature doses of 100 µg/ml of endotoxin have been used in in vitro experiments. A few simple calculations make it possible to put these amounts in terms of the number of bacteria required to yield this quantity of endotoxin. A liter of culture has about 1 g wet weight of bacteria (*Escherichia coli* or *Salmonella typhimurium*). The same culture contains about 10^9 cells/ml or a total of 10^{12} bacteria per liter. It takes about 5×10^9 living bacteria to yield 1 mg of dry weight. The amount of endotoxin that can be extracted is at most about 5% of the dry weight so 1 mg yields about 50 µg. The number of bacteria needed to provide 100 µg/ml would be at least 10^{10}. To put this in a little better perspective, this is the number obtained when the bacteria present in 10 ml of a full-grown broth culture are concentrated into 1 ml. In fact, it has been shown in several laboratories that mice die of an infection when the recoverable number of pathogens is about 10^9. A challenge of 10^{10} would be a 10-fold overkill for a mouse. Dogs have often been given injections of 1 mg/kg of endotoxin. Sometimes even larger doses are used. A comparatively small dog would receive 10 000 µg, the quantity that would be obtained from 10^{12} bacteria. This number would be found in one liter of rich broth culture and were they concentrated into one ml, their consistency would be almost caseous.

Even though endotoxin does not lend itself to good dose–response curves, the use of excessive amounts is reason enough to question the validity of relating the observations obtained to natural phenomena. There are, unfortunately, references in the literature of animals with a gram-negative infection that fail to manifest the same symptoms that have been reported in publications where the equivalent of 50 µg to 100 µg/ml of endotoxin was administered. After all, one does not study an animal's response to burns by using a blow torch to injure it. The analogy does not miss the mark. Doses of the magnitude used in numerous publications should be related to the bacterial numbers required to provide that amount of endotoxin and then they can decide whether a similar quantity of toxin is likely to occur during an infection.

Just as unusually large doses have been used by many investigators, it is equally important to realize that studies with nanogram amounts have been successful. These experiments were done with cells in culture where minute amounts of endotoxin were found to stimulate the release of such mediators as colony stimulating factor, tumor necrosis factor, and interleukins. When sensitive cells such as macrophages and lymphocytes are the target, it is important to recognize that responses are obtained with quantities of endotoxin that might be expected in the course of an infection. Certainly, no more than 10^6 bacteria would be able to provide sufficient endotoxin to produce detectable effects. When massive doses are required for a measurable change, it can be questioned whether this would occur naturally rather than under the artifical conditions of the experiment.

The question of dosage can be summarized with a plea. No one should generalize about the effects of endotoxin without taking into account the amount administered, the route of injection and at least some mental calculation of the number of bacteria this represents. As long as this is done, it would be inexcusable to maintain that endotoxin does this or that independent of other considerations. Moreover, the species of animal being employed makes a difference in the type of response that is obtainable. In the first place, mice and rats are far more resistant to endotoxin, on a weight basis, than rabbits and dogs. Man is also exquisitely sensitive. A dose per kilogram in a rabbit that yields a 2–3 °C rise in body temperature and is life-threatening would produce minimal changes in mice. Dogs, for example, show hemorrhages in the intestinal wall following a sublethal but shocking dose of endotoxin while other commonly used laboratory animals do not. Mice, housed at usual room temperatures (20–25 °C) become hypothermic after an injection of endotoxin but when they are housed at 30 °C, they develop fever. Guinea pigs are unique insofar as the literature has revealed since they cannot be protected against the effects of endotoxin by prompt treatment with adrenal steroids. Before assuming that an observation made following an injection of endotoxin can be extended to other species, one should be sure that the relative responsiveness of the animals being compared is known and the dosage employed is relevant in the two situations. Had this simple concept been followed through the years, much confusion and erroneous generalizations would have been avoided.

One of the intriguing aspects about the effects critically evaluated in this volume is how a molecule as large as endotoxin exerts effects on cells. From what has been reported in the literature, little if any endotoxin actually penetrates the cells of most tissues. This does not keep it from eliciting responses, however. The hypothalamus is believed to be the center of regulatory control over body temperature, the so-called thermostat of the body, and yet it has not been possible to detect by means of several techniques the entry of endotoxin into its cells. This was one reason why it seemed reasonable to postulate the formation of an endogenous pyrogen outside the central nervous system that could then be transported to the hypothalamus via the bloodstream. This is what has been found to occur. There are a number of other examples of an intermediary substance serving as the transmitter of a response from a primary target site to the effector site. The action of these mediators is now well known even though much remains to be done. It will be a challenge for future investigators to work out the origin of the mediators along with the details of how they act. Moreover, there is reason to believe that some of the cellular and biological effects of endotoxin are primary responses. These and other questions about this bacterial substance are sufficient to keep the flow of manuscripts in a steady state for a number of years ahead.

L.J. BERRY

Handbook of Endotoxin, Vol. 3: Cellular Biology of Endotoxin
L.J. Berry, editor
© Elsevier Science Publishers B.V., 1985

CHAPTER 1

Cellular receptors for endotoxin

NICOLE HAEFFNER-CAVAILLON, JEAN-MARC CAVAILLON
AND LADISLAS SZABÓ

1. INTRODUCTION

Endotoxins of gram-negative bacteria possess a very large number of biological
activities. While it is easy to conceive that the different activities result from their
interactions with host cells and tissues, characterization of this interaction has been
more difficult to achieve, and in most cases the very existence of a receptor remains
open to question. At the molecular level, the various biochemical and physiological
responses to endotoxin stimulation might be the consequence of its specific binding
to a receptor or to nonspecific interaction with a membrane component, for example the lipid bilayer. Insertion of endotoxins into the cell membrane bilayer (reviewed in Chapter 2) has been demonstrated. Analogous observations were made
during the study of endotoxins with artificial membranes by measuring modifications of surface pressure, electrical resistance or permeability of artificial membranes such as thin lipid films in aqueous medium (Benedetto et al 1973; Romeo
et al 1970; Schuster et al 1970) or liposomes (Davies et al 1978; Onji et al 1979).
It has also been established that nonspecific interactions due to hydrophobic and
electrostatic forces exist between endotoxins and lipidic membrane components.

As regards specific interactions, the involvement of endotoxin receptors in
biological responses following endotoxin action has been often suggested and
sometimes demonstrated. Before describing endotoxin-receptors, it is important
to define the term 'receptor': following Dorrington (1976), who gave the definition
for the receptor for the Fc portion of immunoglobulins, we shall refer to a binding
site as a receptor if, and only if, the binding site has been shown to be a distinct
molecular entity present on the plasma membrane of the cells and has the following
properties: '(i) *Saturability*. There are a finite number of receptors per cell, all of
which can be occupied at high free-ligand concentrations. (ii) *Specificity*. The receptor shows specificity for only a certain type of ligand' (e.g. immunoglobulins,
complement components and hormones). '(iii) *High affinity*. The specific ligands
are bound with high energy implying a close stereochemical relationship to the
receptor.' We should also emphasize that the expression 'specific binding' will be
used according to Cuatrecasas' definition (1973): 'the amount of labeled ligand
bound to the cells or membranes which can be specifically "displaced" by adding
an excess of unlabeled ligand to the cells or membranes before addition of the

labeled material'; it thus represents the amount of specifically bound ligand. In our view, however, the notion of 'receptor' comes to be complete only when a correlation between ligand binding and cellular response is established.

As difficulties inherent in the precise characterization of macromolecules in general, and those of amphipathic endotoxins in particular, led to inconsequential terminology, and, as a consequence, sometimes to apparently conflicting biological results, it seems appropriate to define the terms endotoxin, lipopolysaccharide (LPS), and lipid A as used in this chapter. Endotoxins are biochemical preparations which possess specific biological properties: they are defined by the microorganism from which they originate, the culture conditions, and the methods of extraction and purification; unless otherwise specified, we refer to phenol-extracted material (Westphal et al 1952). Lipopolysaccharides are physically and chemically defined molecular species that make up endotoxins; few have been fully characterized chemically. Lipid A is a heterogeneous precipitate formed when an endotoxin or lipopolysaccharide is treated with aqueous acids; its composition varies with the parameters of the hydrolytic conditions used.

It would appear that not enough attention has been paid to differences that exist between endotoxin preparations from different bacterial species, in particular in such exacting studies as binding or stimulation assays. Indeed, although similarities of the main features of endotoxins of all gram-negative bacteria undoubtedly exist, structures present in both the hydrophobic and the hydrophilic regions may differ considerably from those established for enterobacterial lipopolysaccharide (Lüderitz et al 1973). The influence of these upon any given biological activity is far from being established. Peculiar endotoxin structures appearing either in the hydrophobic or hydrophilic regions have been reported for *Chromobacterium violaceum* (Hase and Rietschel 1977), *Rhodopseudomonas viridis* (Roppel et al 1975), *Rhodospirillum tenue* (Tharanathan et al 1978), *Bordetella pertussis* (Chaby and Szabo 1975; Haeffner-Cavaillon et al 1977; LeDur et al 1980), *Brucella abortus* (Leong et al 1970; Moreno et al 1979), *Bacteroides fragilis* (Kasper 1976; Wollenweber et al 1980), and *Pseudomonas aeruginosa* (Drewry et al 1973; Pier et al 1978). Endotoxins from different gram-negative bacteria may also vary in their immunological properties; for example, enterobacterial endotoxins have been shown to be mitogenic for B cells of the so-called 'endotoxin-resistant' C3H/HeJ mouse strain, whereas endotoxins from *B. abortus* (Moreno and Berman 1979), *P. aeruginosa* (Pier et al 1981), *B. pertussis* (Girard et al 1981) and *B. fragilis* (Joiner et al 1982) were found to be good mitogens for B cells of the same mouse strain. Attribution, to endotoxins in general, of biological properties observed with enterobacterial endotoxins only, can lead to misinterpretations of experimental results. Thus, in a study bearing on the effects of *B. pertussis* on the 'endotoxin-resistant' C3H/HeJ mouse strain, Wirsing et al (1981) observed that killed *B. pertussis* cells were able to produce an adjuvant effect, hypersplenia, leukocytosis and increase of nonspecific resistance; it was concluded that the biological activities induced by *B. pertussis* were due to structural components other than the endotoxin. Since Girard et al (1981) demonstrated that the C3H/HeJ mouse strain does respond to stimulation by *B. pertussis* lipopolysaccharide, the above conclusions may be ill-founded.

Before we discuss the endotoxin receptors, it is interesting to note that lipopolysaccharides may be themselves receptors for other ligands. Thus it was demonstrated that lipopolysaccharides constitute a part of the receptor for some bacteriophages (Castillo 1980; Jarrel and Kropinski 1981), and that the hydrophobic and the hydrophilic regions of the lipopolysaccharides participate in the recognition process (Yamada et al 1981, Yu et al 1981).

2. ENDOTOXIN-BINDING SITES ON DIFFERENT CELLS

In early attempts to characterize the target cells of endotoxin action, studies of uptake and distribution of radiolabeled endotoxin have been performed in vivo after intravenous injection of the labeled material. Thus, in 1955, Braude et al reported that ^{51}Cr-labeled *Escherichia coli* endotoxin injected intravenously into the rabbit remained in the plasma; it was not associated with erythrocytes and rapidly passed to the liver. Similar observations were made in mice and in guinea pigs (Noyes et al 1959; Rowley et al 1956). Injected ^{32}P-labeled *E. coli* endotoxin was also detected in spleen (Rowley et al 1956) and after intravenous injection of *Salmonella enteritidis* endotoxin, degraded material was found in the urine (Chedid et al 1963). Herring et al (1963) established that circulating *E. coli* endotoxin was distributed between plasma and platelets.

Following these preliminary experiments, in vitro studies have been undertaken to characterize the binding of endotoxins to individual cell populations such as platelets, granulocytes, liver cells, lymphocytes, erythrocytes, macrophages and monocytes. These studies will be analyzed below with particular attention to the specificity of binding suggesting the presence of a receptor.

2.1. Platelets

Interaction of endotoxin with platelets is a well-known phenomenon: it leads to the aggregation of platelets, causes release of vasoactive amines and adenine nucleotides, unmasks the phospholipid procoagulant (PF3), and initiates platelet lysis mediated by complement. In the reports dealing with the attachment of endotoxin to platelets, very little information is available on the specific binding of endotoxin through receptor sites. Washida (1978a), using labeled *E. coli* O111:B4 endotoxin, concluded from the binding sequence: platelets = granulocytes > lymphocytes > erythrocytes observed upon diffusion analysis, that endotoxin was bound preferentially to platelets. Pretreatment of platelets with unlabeled endotoxin altered the equilibrium as measured by diffusion analysis (Washida 1978b). Modification of certain physiologically important activities of human platelets by *S. enteritidis* endotoxin (Boivin-type, alkali-treated) led Hawiger et al (1975, 1977) to postulate the existence of an endotoxin binding component or receptor on the platelet membrane. In these studies, attachment of endotoxin to platelets was not characterized by the criteria that define specific binding to a receptor. Morrison and Oades (1979) concluded that the hydrophobic region of lipopolysaccharides was responsible for their attachment to platelets, because alkali-treated endotoxin (*E. coli* O111:B4) and endotoxin treated with polymyxin B – lipopolysaccharide preparations in which the hydrophobic region is modified or rendered inaccessible – were ineffective; moreover, polysaccharide-free lipopolysaccharide from *Salmonella minnesota* R595 – which cannot activate the alternative pathway of the complement system and thus does not initiate lysis – inhibited lysis by *E. coli* endotoxin. Although these results unambiguously demonstrate that the lipophilic region of lipopolysaccharides modulates their binding to platelets, they do not disclose whether endotoxin intercalates with the phospholipid bilayer of the cellular membrane or binds through a receptor site. In a study on the nature of human platelet membrane components interacting with endotoxin, Hawiger et al (1977) reported that no significant change in the sensitivity of human platelets to endotoxin was noted after trypsin or neuraminidase treatment. This correlates well with the findings of Springer and Adye (1975), who demonstrated that the binding of endotoxin to components of a crude platelet extract was due to the lipidic constituents of the extract; neither proteins, carbohydrates nor nucleic acids were involved. The most reactive substance showed, on polyacrylamide gel electrophoresis, only one band: it was a lipid which migrated like phosphatidyl lipids. Taken together, the different data suggest lipid–lipid interactions between lipids of the platelet cell membrane and the hydrophobic region of endotoxin.

2.2. Granulocytes

The interaction of endotoxin with granulocytes was investigated by Gimber and Rafter (1969), who showed that rabbit granulocytes, recovered from induced acute peritoneal exudate, bind ^{32}P-labeled *E. coli* endotoxin. The amount of endotoxin bound was, in the range of 2 to 12 µg, proportional to the amount of endotoxin added; 50–60% of the radioactivity added was taken up by the cells; the maximum of binding was reached within 10 minutes and an approximately 30% inhibition of the binding reaction could be obtained by addition of an excess of unlabeled endotoxin. It appears from these results that rabbit granulocytes do possess receptor sites for the endotoxin used. Binding of endotoxin to granulocytes was also demonstrated by Washida (1978a), who reported that granulocytes, like platelets, had binding affinities for endotoxin (^{125}I-labeled endotoxin from *E. coli*) whereas lymphocytes and erythrocytes had little or none; from the published data no conclusion can be reached whether binding occurs through a specific receptor or not. Springer et Adye (1975), who demonstrated the existence of a specific receptor for endotoxin on erythrocytes (see Section 2.5.), established that binding to granulocytes and platelets occurred in the phospholipid bilayer and was due to hydrophobic interactions between this bilayer and the lipid moiety of endotoxin.

2.3. Liver cells

The major role played by the liver in the clearance of endotoxin from the blood was clearly established in 1955 by experiments performed in vivo with radiolabeled endotoxin from *E. coli* (Braude et al 1955; Noyes et al 1959; Rowley et al 1956). More recent experiments by Willerson et al (1970) and by Zlydaszyk and Moon (1976) have indicated that radiolabeled endotoxins administered by intravenous injection were internalized by liver cells and mainly associated to cell nuclei and mitochondria. It is noteworthy that within 15 minutes over 80% of ^{51}Cr-labeled endotoxin from *S. typhimurium* injected intravenously was trapped in the liver. It is generally accepted that Kupffer cells are responsible for the hepatic clearance and respond to endotoxin stimulation (Maier and Ulevitch 1981). Thus, Mathison and Ulevitch (1979) have demonstrated that fixed macrophages of liver and spleen are the principal cellular targets of the endotoxins of *E. coli* O111:B4 and *S. minnesota* injected intravenously into the rabbit. More than 40% of the endotoxin injected was found to be sequestered in the liver and located by electron microscopic autoradiography in the phagocytic vacuoles of hepatic macrophages (Kupffer cells); no endotoxin could be detected in hepatocytes. According to Ramadori et al (1979), however, parenchymal liver cells are also involved in the hepatic clearance of endotoxin. In their study, the binding sites for *E. coli* endotoxin on the plasma membrane of isolated rabbit hepatocytes were located by immunofluorescence: endotoxin-treated hepatocytes were detected either with fluorescein-isothiocyanate(FITC)-conjugated rabbit anti-LPS IgM or by specific rabbit anti-LPS IgM (anti-*E. coli* O26:B6 and anti-*E. coli* O111:B4 LPS) and FITC-conjugated anti-rabbit IgM from goat. Binding sites for endotoxin could be detected on the plasma membrane of hepatocytes but not on Kupffer cells; specificity of the binding and saturability of the receptor site have not been demonstrated. Since the anti-LPS O26:B6 and anti-LPS O111:B4 were specific for the *E. coli* strains used, and since neither of the antisera inhibited the binding of endotoxin to the plasma membrane, the binding may have occurred through the hydrophobic region of the lipopolysaccharide, the O-specific side chain being probably embedded in the antibodies directed against the O-specific determinants.

2.4. Lymphocytes

Data reported on binding of endotoxin to lymphocytes originating from blood, from various organs and animals species, are summarized in Table 1. While it is generally agreed that endotoxin does bind to lymphocytes, from the disparate and sometimes contradictory data no particular cell type or cell subpopulation emerges as the specific target for endotoxin. Even within the same species, the percentage of cells able to bind endotoxin varies greatly from one report to another. Thus, among mouse splenocytes, only 5–10% of the cells bind endotoxin according to Nygren et al (1979), whereas Bona et al (1976) detected up to 50%. Since Gregory et al (1980a) have observed that B lymphocytes of CBA mice bind 2.4 times more

Table 1 *Lipopolysaccharide binding to lymphocytes*

Ligand	Binding detection	Time necessary for maximal binding	Species	LPS-binding cells (% positive cells)	Reference
^{3}H-LPS *S. enteritidis*	autoradiography and cpm	12 hr (37 °C) 3 hr (4 °C)	nu/+	splenocytes > thymocytes (47% *vs* 17% after 1 hr at 4 °C)	Bona et al 1976
^{14}C-LPS *S. minnesota*	ng ^{14}C-LPS bound	4 hr (37 °C)	C3H/HeN nu/nu	macrophage-depleted spleen cells < normal spleen cells; B > T (2 ×)	Kabir and Rosenstreich 1977
^{125}I-lipid A *S. abortus equi*	cpm	1 hr (37 °C)	rabbit, rat C3H/HeN Balb/c	splenocytes depleted of adherent cells ≃ normal splenocytes; splenocytes ≫ thymocytes; B ≫ T	Zimmerman et al 1977
^{3}H-LPS *E. coli*	cpm	5 min (37 °C)	mouse, pig	peripheral blood lymphocytes = mesenteric lymph node cells = thymocytes	Symons and Clarkson 1979
LPS-peroxidase *E. coli*	electron microscopy	ND	C3H/Tif	5–10% splenocytes	Nygren and Dahlen 1979
^{125}I-lipid A *S. abortus equi*	autoradiography	ND	C3H/HeN CBA rabbit	B (23%) > NULL (10%) > T (< 1%) B(55%) B (14%) > T = thymocytes (< 1%)	Gregory et al 1980b
Mutants from *Salmonella*	rosettes	15 min (4 °C)	AKR; SJL C57Bl; C3H Balb/c; A/J	20–25% activated cells 0%	Lehmann et al 1980
LPS *E. coli*	fluorescence	ND	C3H/St	splenocytes (50%) > lymph node cells (15%); spleen Thyl, 2$^+$ (25%) ≫ thymocytes (4%)	Jacobs and Eldridge 1984
Rhodamine-LPS *S. choleraesuis*	fluorescence	ND	C3H/He	B (26%) > T (2%)	Chaby et al 1984
^{125}I-lipid A	autoradiography	ND	C3H/He C57BP	B (25–30%) > thymocytes (1%)	Kern 1984

ND = not determined

endotoxin than those of C3H/HeN mice, the discrepancy between the results reported by Bona et al (1976) and Nygren and Dahlen (1979) could be due to the different mouse strains used in the assays. It can be seen from the data reported in Table 1 that the question of contribution of adherent cells in the binding of endotoxin remains open, since Kabir and Rosenstreich (1977) reported a decrease in binding of *S. minnesota* endotoxin after removal of macrophages from the spleen cell populations, whereas Zimmerman et al (1977), who used the lipid A fragment of *E. coli* endotoxin, observed that the spleen cell populations depleted of adherent cells were unaltered with regard to their lipid A-binding capacity as compared to the original population.

As far as purified populations of T lymphocytes are concerned, it appears from the data reported in Table 1 that unfragmented lipopolysaccharide binds to T cells whereas the lipid A fragment does not. The proportion of T cells which binds endotoxins varies from one report to another. For instance, according to Bona et al (1976), 17% of thymus cells of the mouse bind endotoxin, whereas Jacobs and Eldridge (1984), who reported that 25% of Thy $1,2^+$ splenic lymphocytes were labeled with fluorescent LPS, found only 4% labeling of thymocytes. Working with whole bacteria, Lehmann et al (1980), who used a number of rough and smooth mutants derived from *Salmonella* in a rosette assay, and purified lymphocyte populations, reported attachment of *Salmonella* mutants to T lymphocytes, activated allogeneically or by concanavalin A; neither unstimulated T or B lymphocytes nor B cell blasts could bind the same bacteria. Bacterial attachment was specific for the mouse strain used, since the binding could be observed with cells from AKR, C57Bl/6 and C3H but not with cells from Balb/c and A/J mice. Binding of the *Salmonella* mutants was not abolished by the deep rough Re mutant but was completely inhibited by mutants whose lipopolysaccharide contained a polysaccharide chain. It was concluded that lipopolysaccharide binding was mediated by the polysaccharide moiety of the macromolecule through a protein receptor since proteolytic treatment of T cells with trypsin abolished rosette formation. Attachment of various bacteria due to endotoxin recognition by unstimulated human B and T cells was reported by Teodorescu and Mayer (1982).

It is hard to draw definitive conclusions from these reports; since different techniques were employed to evaluate binding, different endotoxins or endotoxin fragments were used. However, a few comments can be made as regards the incidence of the techniques used on the results obtained in the various binding assays. Thus, only few authors have ascertained that the procedure applied for radiolabeling the endotoxin did not modify its biological properties. Moreover, in most studies it was not clearly established whether the data reported referred to specific binding through a receptor or nonspecific binding by, for example, hydrophobic lipid–lipid interactions, or both. It is quite probable that most of the data reported deal with nonspecific binding, because after incubation with labeled endotoxin, the cells were usually washed 2 to 3 times over long periods of time (from 10 minutes to 18 hours) or even until no radioactivity was observed in the supernatant. The behavior of endotoxin receptors present on human erythrocytes

(Springer et al 1974; Yokoyama et al 1978) on mouse and pig lymphocytes (Symons and Clarkson 1979), and rabbit peritoneal macrophages (Haeffner-Cavaillon et al 1982a) suggests that endotoxin binding is not accompanied by formation of covalent bonds between the ligands and the receptor. If this is indeed the case, long washing procedures are inappropriate for demonstrating the existence of specific endotoxin receptors since during the extended washing the equilibrium of the equation: LIGAND + RECEPTOR $\leftrightarrows$ (LIGAND–RECEPTOR) is displaced to the left. In the case of human erythrocytes, the order of magnitude of the association constant of the endotoxin/receptor complex was estimated to be 10^6 M^{-1}, which suggests that washing for extended periods of time would lead to complete dissociation of the complex. The technique employed by Symons and Clarkson (1979) did not lead to desorption of bound endotoxin from cells during the washing procedure; their subsequent observation of unloading of radiolabeled endotoxin bears witness of a reversible reaction between endotoxin and cell surface components. Reversibility of endotoxin binding has also been reported by Bona et al (1976), whose chasing experiments with an excess of unlabeled endotoxin led to release of radiolabeled material from spleen cells of nu/nu mice, whereas up to 12 hours the binding to thymocytes of nu/+ mice was not affected; the same authors also observed initiation of patching and capping suggesting ligand binding on receptor sites.

Another attempt to demonstrate an endotoxin receptor on murine B lymphocytes was made by Coutinho and his colleagues (Coutinho et al 1978; Forni and Coutinho 1978). These authors have prepared a rabbit antiserum against C3H/tif mouse splenocytes which was then absorbed with pooled liver, thymus and spleen cells of mice not responding to enterobacterial endotoxin (C3H/HeJ). This antiserum reacted with a fraction of B cells (endotoxin-reactive precursors and endotoxin-activated blasts) of mice responding to endotoxin but did not react with any cell from mice not responding to enterobacterial endotoxin (C57Bl/10ScCr). It combined with cells from (C3H/Tif × C3H/HeJ)F_1 mice and backcross generations in a proportion that reflected the high-, intermediate-, or non-responder qualities of these mice; it also competed with endotoxin for binding on the B cell surface and inhibited endotoxin-initiated mitogenic cell-responses. It was concluded that this antiserum contained antibodies specific for the endotoxin receptor on B cells. Taken together, some of these data might suggest the presence of a receptor site for endotoxin on the lymphocyte.

The existence of an endotoxin receptor on lymphocytes needs, nevertheless, to be confirmed. Indeed, Watson et al (1980), who introduced minor modifications in Coutinho's procedure (other C3H strains, other rabbit populations, different immunofluorescent agents), could not obtain any antiserum bearing anti-endotoxin-receptor activity, and a number of authors (Eldridge and Jacobs 1981; Goodman et al 1984; Gregory et al 1980a; Kabir and Rosenstreich 1977; Kern 1984; Symons and Clarkson 1979; Watson and Riblet 1975) failed to observe any difference between the extent of endotoxin binding by splenocytes of C3H/HeJ and splenocytes of other C3H-derived mouse strains. A difference was reported only in two studies: Nygren and Dahlen (1979) found that at high doses (70 µg/ml) a large proportion of lymphocytes of the low-responder strain bound peroxidase-labeled endotoxin, but at lower doses (0.7 and 7 µg/ml), C3H/HeJ splenocytes did not bind endotoxin, and Chaby et al (1984), using a 40 µg/ml final concentration of rhodamine-labeled lipopolysaccharide, reported that B cells from C3H/HePas mice were significantly more labeled than were B cells from C3H/HeJ. One could argue that most authors have described nonspecific endotoxin binding to spleno-cytes, probably mediated by the lipid A moiety of lipopolysaccharide (Kern 1984), but the fact remains that Symons and Clarkson (1979) observed a similar extent of endotoxin-binding and -unloading for splenocytes of C3H/HeJ and C3H/He mice, in experimental conditions compatible with the measurement of specific binding.

In this context, experiments reported by Niederhuber et al (1975) should be recalled which showed that in the presence of complement, endotoxin-induced mitogenesis can be abolished by specific killing of Ia$^+$ cells with specific anti-Ia antiserum; mitogenesis was inhibited by addition of anti-Ia antiserum, but not by anti-H2-K antibodies. It is unfortunate that the target cell of the endotoxin has not been determined in this study: indeed, it is now known that Ia molecules are present not only on B cells, but also on T cells – which were shown to play a 'helper' role in endotoxin stimulation (Yokoyama and Osawa 1979) – and on macrophages – which were shown to cooperate with lymphocytes (Haeffner-Cavaillon et al 1982b). Other antibodies, such as anti-immunoglobulins, can also inhibit endotoxin-induced stimulation of B cells (Sidman and Unanue 1976); whether surface immunoglobulins combine with endotoxin – like aggregated Igs (Ginsberg and Morrison 1978) – remains to be studied.

In an attempt to identify the entity to which endotoxins are bound, Yokoyama et al (1979), who showed that the binding of radiolabeled endotoxin to B and T cells, and the mitogenesis induced by endotoxin, can be inhibited by both anti-Ia and anti-H2 antibodies, have isolated components from solubilized membranes of mouse B and T cells, using an affinity adsorbent which was prepared by coupling *S. minnesota* R595 endotoxin to activated Sepharose 4B. Although, after elution, the retained membrane proteins were rather heterogeneous, they consisted of immunoglobulins (B cells: 30%) and proteins of the histocompatibility H2 com-plex: H2-D, H2-K, Ia (B cells: 40%; T cells: 65%).

A simultaneous and parallel study of specific endotoxin-binding and endotoxin-induced stimulation of lymphocytes could be an approach to uncover the existence of the putative endotoxin receptor. To elucidate the role of the lymphocyte plasma membrane in regulating the response to mitogens, Jakobovits et al (1982) have used a new technique: they fused B-cell membrane components from the C3H/eb strain with B splenocytes derived from C3H/HeJ strain. They demonstrated that upon insertion of membrane components from lymphocytes responding to endo-toxin into the membranes of nonresponding cells, the latter could be stimulated by this mitogen. They concluded that the response obtained depends on the pre-sence of a 'triggering unit' which consists of appropriate receptors and the transducer of the mitogenic signal. They have not yet isolated the specific membrane constituents which act either as a receptor or as a transducer in cell activation. A similar hypothesis was also reported by Kern (1984) who postulated that LPS-mediated reactions result from two phenomena: binding and cell triggering, the triggering activity being lost in low-responder C3H/HeJ mice.

It is now well established that in mice only a subpopulation of B cells is endotoxin-responsive. This Ia^+ subpopulation has been shown to be different from the one responding to *Nocardia* (Bona et al 1978); it is a 14G8-positive population (DeFranco et al 1982) and it produces predominantly the IgG3 subclass (McKearn et al 1982). The proportion of endotoxin-responding cells is higher in CBA mice than in other strains (Andersson et al 1977). The latter observation correlates with the finding of Gregory et al (1980a), who stressed that only a subpopulation of B cells, more abundant in CBA than in C3H/HeN mice, can bind radiolabeled lipid A. As far as T cells are concerned, no clear-cut information seems to be available. Thus, while some authors have reported a similar degree of binding of endotoxin to B and T cells, others concluded that endotoxin was not bound to T cells.

2.5. Erythrocytes

Adsorption of endotoxin to the surface of red blood cells was described in 1956 by Neter (Neter 1956; Neter et al 1956): it was detected by specific agglutination of endotoxin-coated erythrocytes with homologous bacterial antibodies; agglutination augmented significantly following heat and sodium hydroxide treatment of the endotoxin. As isolated lipid A inhibited binding of alkali-treated lipopolysaccharides to erythrocytes, Hämmerling and Westphal (1967) concluded that the hydrophobic moiety of lipopolysaccharides was operative in the binding. Inasmuch as the alkaline treatment used by Neter (0.25 M sodium hydroxide at 56°C for 60 min) probably removed a substantial proportion of the ester-bound fatty acids, constituents of the hydrophobic moiety of lipopolysaccharide are presumably responsible for the binding. According to Springer and Adye (1975), the amide-bound 3-hydroxytetradecanoic acid residues are involved. Attachment of endotoxin to erythrocytes, satisfying the criteria of specific binding of a ligand to its receptor, was first reported by Springer et al (1970). From human erythrocyte ghosts they isolated a fraction which specifically inhibited endotoxin coating of red cells. Since the components of that fraction had no significant inhibitory activity against antigens not related to endotoxin (V_i antigens of gram-negative bacteria, common antigens of gram-positive bacteria), ligand specificity was demonstrated; accordingly, the authors named the components of that fraction 'LPS-receptor'. Since interaction of the receptor with endotoxin was shown to be reversible and since it was established that the isolated 'LPS-receptor' could displace the endotoxin from the red cell membrane, the three most important criteria defining a 'specific receptor' were satisfied. However, if the opinion generally accepted that the lipophilic moiety of endotoxins is the one that binds to erythrocytes is correct, it remains difficult to explain the observation that seven times as much receptor material was needed to prevent attachment of the endotoxin produced by a rough mutant of *S. minnesota* than that of endotoxin produced by a smooth strain of the same organism.

The binding of endotoxins to red cells was also investigated by Davies et al (1978), who used ^{32}P- or ^{14}C-labeled *E. coli* endotoxin; they reported that saturation of the receptor sites on the cell membrane could be achieved under various conditions. These authors also observed a consistent pattern of cyclic fluctuation between adsorption and desorption of endotoxin molecules; a hypothesis explaining this phenomenon was given. Specific binding of endotoxin on the surface of red cells has been observed by electron microscopy using *Fusobacterium nucleatum* peroxidase-conjugated lipopolysaccharides (Nygren and Dahlen 1979).

Detailed investigations performed by Springer et al (1973; 1974) led to the physicochemical characterizations of the lipopolysaccharide-receptor from human erythrocytes. The receptor, isolated in an apparently homogeneous form, is a lipoglycoprotein (mol. wt. = 256 000; pI = 3.0) which, after incubation with sodium dodecylsulfate (SDS) for 2 hours at 37°C, dissociated into two bands of apparent molecular weights of 95 000 and 30 000 daltons. Sixty-eight percent of the activity of the whole receptor was due to the large fragment, the small fragment was inactive. If the incubation with SDS lasted overnight, three additional bands (20 000, 10 000 and 7000 daltons) could be detected. When kept at 56°C for 6 hours, the receptor lost about 40% of its activity and more than 80% when exposed to pH below 6 and above 8. Treatment with hexosaminidase or galactosidase had no effect, and extraction with chloroform-methanol led to only a 7% decrease of receptor activity. In contrast, after papain or trypsin treatment, less than 30% of the activity was recovered. These experiments suggested that lipid and carbohydrate were not part of the receptor's binding site.

More recently, Yokoyama et al (1978), who studied the binding capacity of human erythrocytes for ^{3}H-labeled *E. coli* and *S. minnesota* R595 endotoxins, observed that binding reached a plateau after 30 to 60 minutes. The formation of the endotoxin–erythrocyte complex was found to be specifically reversed when unlabeled endotoxin (90% binding inhibition with the homologous material) or isolated lipid A (70% binding inhibition) were added. The apparent association constants were $K_A = 1.4 \times 10^6$ M^{-1} and 7×10^5 M^{-1} for lipopolysaccharide and lipid A, respectively; the average number of major receptor sites were n = 4.5 × 10^6 and 1.5 × 10^7 per cell for lipopolysaccharide and lipid A, respectively. It should be noted that the association constants for both lipopolysaccharide and lipid A were determined according to the method described by Steck and Wallach (1965) and thus calculated according to Scatchard. The values reported above are therefore only approximate since neither the lipopolysaccharides nor the lipid A are in monomeric form in aqueous dispersion. When treated with SDS, the receptor, isolated by affinity chromatography on Sepharose-bound *S. minnesota* R595 endotoxin, dissociated into two subunits: Band 3 and PASS 1; their properties are given in Table 2. Because of the similarity of molecular weights, Yokoyama et al (1978) suggested that the receptor defined by Springer et al (1974) was an aggre-

Table 2   *Characteristics of lipopolysaccharide–receptor on human erythrocytes*

Biochemical nature	Molecular weight	Inhibition of LPS binding by receptor material	Reversibility Saturability Specificity	K_A (M^{-1}) (LPS)	No. of sites	Reference
Lipoglycoprotein – large fragment – small fragment	256 000 95 000 30 000	80% 55% 5%	yes	ND	ND	Springer et al 1970, 1973, 1974
Glycoprotein – Band 3 – PASS 1 (dimeric form of PASS 2)	ND 90 000 83 000	ND 65–70% 45–52%	yes	1.4×10^6	4.5×10^6	Yokoyama et al 1978

ND = not determined

gate of PASS 2 and Band 3 glycoproteins and contained a small amount of membrane lipids. Band 3 and PASS 1 (a dimeric form of PASS 2) are the most prominent red cell membrane proteins: Band 3 is the subunit of a presumed anion channel, and PASS 1 is a transmembrane sialoglycopeptide (Marchesi et al 1976).

2.6. Macrophages

'Activation' of macrophages by endotoxins, expressed by morphological changes, increased rate of metabolism and phagocytosis, product secretion, etc, has been observed long ago. The observed mitogenic activation of B cells by endotoxin led to postulating the existence of endotoxin receptors on this type of cell and to the hypothesis that at least some features of macrophage activation were due, not to the direct action of endotoxins on the phagocytic cell, but to factor(s) released by B cells under endotoxin stimulation. For instance, Yoshida et al (1973) observed that endotoxin-induced inhibition of macrophage migration was due to migration inhibiting factor (MIF) produced by lymphocytes stimulated by endotoxin; Wilton et al (1975) have shown that the enhancement of glucosamine uptake by endotoxin-stimulated macrophages was mediated by B cell-derived products. Evidence that endotoxin also acted directly upon macrophages, has been offered by others who established that lymphocyte activating factor (IL-1) was produced by lymphocyte-depleted adherent cells or macrophage-like cell lines under the influence of endotoxins (Gery and Waksman 1972). Schubert and David (1980) reported enhancement of pinocytosis of horseradish peroxidase by guinea pig macrophages as the result of direct stimulation of these cells by endotoxin. Esser and Russel (1979) have shown by electron spin resonance that exposure to endotoxin of cultured tumor macrophages causes plasma membrane perturbation. Bona et al (1973) concluded from morphologic observation that endotoxin was taken up by pinocytosis by the macrophages.

In the absence of antibody or complement, macrophages ingest a wide variety of particles, for example, latex beads, red cells, tumor cells, and bacteria. From experiments in which different test particles were used, Benoliel et al (1980) concluded that a number of different mechanisms, different physical parameters and chemical structures were involved in the process of adhesion preceding the ingestion; no attempts were made, however, to characterize membrane structures responsible for the binding. Specific receptors do, nevertheless, exist on the plasma membrane of the macrophages: discovery of the Fc receptor was followed by those of components of the complement system, and today more than 20 different specific receptors have been shown to be present on the surface of this type of cell.

In studies using whole, gram-negative (*E. coli, S. typhimurium*, including 'rough'-type mutants, *Klebsiella aerogenes*) and gram-positive (*Staphylococci* and *Streptococci*) bacteria, Freimer et al (1978) observed that almost all microorganisms tested were bound to mouse peritoneal macrophages; at concentrations of 10^8–10^{11} bacteria/ml, 20-30% binding could be achieved. Binding could be inhibited by preexposure of the macrophage monolayers to a variety of monosaccharides; glucose and galactose inhibited binding of a large majority of the organisms tested. These observations were interpreted as indicating the presence of a lectin-like receptor on macrophages and it was suggested that, as far as gram-negative bacteria are concerned, the carbohydrate regions of the LPS molecules could provide the ligand for these receptors. Involvement of the hydrophobic region of lipopolysaccharide in the binding of gram-negative bacteria to macrophages was, however, not ruled out completely, since the lipid A fragment of the *Salmonella typhi* lipopolysaccharide also inhibited binding to the cells of a number of rough-type mutants. The authors suggested that inhibition by the lipid A preparation was due to its galactose content; the presence of this hexose was, however, not demonstrated experimentally. Moreover, as Gregory et al (1980b) have shown that the lipid A fragment derived from the *Salmonella abortus equi* lipopolysaccharide does bind to mouse peritoneal macrophages, it can be suggested that the wild type of *Salmonella* strains bind preferentially to a lectin-like receptor present on macrophages through the polysaccharide region of their lipopolysaccharide, whereas the rough types bind through the hydrophobic region, perhaps by nonspecific interactions with constituents of the lipid bilayer. The latter restriction must be made because the data given by Gregory et al (1980b) do not establish that the binding of radiolabeled lipid A to adherent cells was specific, saturable and reversible.

Recent studies with *B. pertussis* and *Neisseria meningitidis* lipopolysaccharides and their isolated fragments (Haeffner-Cavaillon et al 1982a; 1985) demonstrated, on certain types of adherent cells, the presence of lectin-like receptors, specific for molecular structures present in the carbohydrate region of some bacterial endotoxins. Labeled *B. pertussis* endotoxin was found to bind to lung and peritoneal macrophages of the rabbit; on peritoneal macrophages 40-60% of the binding was specific whereas on lung macrophages only nonspecific binding could be detected. The specific binding of *B. pertussis* endotoxin to peritoneal macrophages (resident and elicited) was shown to be time-dependent, dose-dependent, saturable and reversible. In an attempt to obtain some information about the ligand-specificity of this receptor, inhibition studies were carried out with endotoxins of other bacterial species. The specific binding was completely inhibited by the homologous unlabeled endotoxin, but not at all by that from *Proteus mirabilis*; endotoxins from *Klebsiella pneumoniae*, *Shigella flexneri*, *Salmonella johannesburg*, *Salmonella senftenberg*, *S. typhimurium* and *E. coli* gave intermediate inhibition values (33–90%); ligand specificity of the binding site was thus established. High binding inhibition (90%) was also observed with the polysaccharide chain (PS-1) isolated from *B. pertussis* endotoxin, showing that the attachment of this endotoxin to the macrophage membrane receptor was probably mediated through structures present in the polysaccharide chain. The lipid A fragment did not inhibit binding of labeled *B. pertussis* lipopolysaccharide to the membrane receptor. The conclusion that this part of the lipopolysaccharide does not participate at all in the attachment of the endotoxin to the macrophage membrane is, however, not warranted because it is known (M. Caroff, unpublished observation) that the chemical structures of the

14

Table 3   *Bordetella pertussis endotoxin binding to different macrophage populations*

Species and strains	Cell sources	Binding (cpm $\pm$ SD)*		Specific binding (%)
		0.1 µg ^{3}H-LPS	0.1 µg ^{3}H-LPS + 10 µg unlabeled LPS	
Mice: Balb/c	elicited peritoneal macrophages	3307 $\pm$ 353	1270 $\pm$ 278	62%
DBA 2	elicited peritoneal macrophages	2110 $\pm$ 568	1081 $\pm$ 72	49%
C3H/He Pas	elicited peritoneal macrophages	3057 $\pm$ 307	1595 $\pm$ 303	48%
C3H/HeJ	elicited peritoneal macrophages	3847 $\pm$ 1237	1658 $\pm$ 196	57%
Rabbit	resident peritoneal macrophages	938 $\pm$ 134	436 $\pm$ 75	53%
(New Zealand type)	elicited peritoneal macrophages	1307 $\pm$ 94	736 $\pm$ 32	44%
	alveolar macrophages	206 $\pm$ 24	196 $\pm$ 3	5%
Man	monocytes	3602 $\pm$ 511	429 $\pm$ 117	88%

* 0.8×10^6 cells/well

B. pertussis lipid A preparation have undergone considerable destruction during its isolation. A specific binding of *B. pertussis* and *N. meningitidis* lipopolysaccharides mediated by the polysaccharide moiety has also been shown to occur on peritoneal macrophages isolated from different mouse strains, including C3H/HeJ, and on human monocytes (Table 3; Haeffner-Cavaillon et al 1985); the specific binding was only observed in the presence of serum. The binding was not mediated by natural anti-LPS antibodies present in normal sera, but the data suggest a role for complement components (Haeffner-Cavaillon et al 1985). Although the polysaccharide chains of *B. pertussis* and *N. meningitidis* lipopolysaccharides are small oligosaccharides with no repetitive units, only partial structures have been established (Chaby et al 1977; Jennings et al 1984; Moreau et al 1982), and for the time being, the molecular structure responsible for the binding of LPS to mononuclear phagocytes have not been identified.

As regards the receptor itself, it is most probably different from that described by Ezekowitz et al (1981) which mediates D-mannose-specific endocytosis, and also from that discovered by Nagamura and Kolb (1980) which is specific for D-galactose, since, even at high concentrations, neither D-mannose nor D-galactose did inhibit the binding of the *B. pertussis* endotoxin to peritoneal macrophages. Inasmuch as Weir (1980) found that binding of whole bacterial cells to peritoneal macrophages of the mouse was inhibited by D-galactose, it seems that this receptor is also different from that which binds the PS-1 of the *B. pertussis* endotoxin.

In a first attempt to establish a relationship between a cellular function and the endotoxin receptor, the role of rabbit macrophages in the endotoxin-induced polyclonal activation of immunoglobulin-secreting cells was investigated (Haeffner-Cavaillon et al 1982a). It was found that whole *B. pertussis* lipopolysaccharide and both isolated PS-1 and lipid A fragments initiated polyclonal activation of rabbit splenic B cells; the presence of macrophages was an absolute requirement. Inasmuch as on the macrophage there is a specific receptor for the polysaccharide PS-1 only, it would appear that the polyclonal activation induced by this fragment and by lipid A is mediated by two different mechanisms. Moreover, it was observed that isolated polysaccharides from *B. pertussis* as well as from *N. meningitidis* induced secretion of interleukin 1 (IL-1) by human monocytes to a similar extent as native lipopolysaccharide (Haeffner-Cavaillon et al 1984). IL-1 induction by both isolated polysaccharides and lipopolysaccharides was increased in the presence of serum (Cavaillon and Haeffner-Cavaillon 1985). Recent data demonstrated that the determinants involved in the IL-1 induction by polysaccharides are located in the heptosyl (1-5)3-deoxyl-D-*manno*-2-octulosonic acid disaccharide present in almost all lipopolysaccharides from gram-negative bacteria. The hypothesis that the same structures are responsible for the specific binding of lipopolysaccharides to monocytes remains to be proven.

3. CONCLUSION

In an attempt to explain immunological activities of endotoxins, a fairly large body of observations bearing on the mechanism of action, especially on the nature of the link by which lipopolysaccharide is bound to cells of the hematopoietic system, has been collected during the last fifteen years. Many methods have been developed for the purpose, but they are not equivalent in their capacity to detect binding of lipopolysaccharide to various cell populations. While differences in sensitivity could, in part, account for nonconcording observations, other factors, such as variations in affinity of lipopolysaccharides for different cell populations, the origin, the method of isolation, and the purity of the lipopolysaccharides employed, or even parameters as yet unidentified, may also be involved.

It is clear that the vast majority of lymphoid cells bind endotoxin. While in some instances binding through a specific receptor has been established, in the large majority of cases examined the binding seems to be mediated through nonspecific hydrophobic interactions which involve the lipidic region of the lipopolysaccharide molecule. In our opinion, the conflicting conclusions reached when demonstration of the existence of specific endotoxin receptors (on lymphocytes, on liver cells) was attempted could be due to technical factors. The use of high lipopolysaccharide concentrations favors nonspecific binding and may lead to insertion of the lipid region into the phospholipid bilayer of the cellular membrane (Benedetto et al 1973); therefore, above certain ligand concentrations, the specific binding can be completely obscured by nonspecific interactions (Haeffner-Cavaillon et al 1982a). Furthermore, dissociation of the specifically bound lipopolysaccharide from macrophages being a very fast process (Haeffner-Cavaillon et al 1982a), the duration of the washing stage used to remove unbound ligand may lead to conflicting results.

It has been established that the molecular structures present in endotoxin specifically 'recognized' by different cells may be located in the hydrophobic or in the hydrophilic regions of the macromolecule: thus a receptor on erythrocytes recognizing the lipid A region has been described (Springer et al 1974; Yokoyama et al 1978), whereas a lectin-like binding site for endotoxin has been identified on activated T cells (Lehmann et al 1980), and on macrophages (Haeffner-Cavaillon et al 1982a; 1985). Participation of both the hydrophilic and the hydrophobic regions of endotoxins in the immunological phenomena induced by these substances has been reported and can now be considered as firmly established. Thus the hydrophilic region can activate the alternative pathway of the complement system (Morrison and Kline 1977), whereas the hydrophobic region activates the classical complement pathway (Loos et al 1974). Both regions have been found to induce an adjuvant effect (Scibiensky 1980). Isolated polysaccharide chains of the *B. pertussis* endotoxin can trigger polyclonal activation of B cells like homologous lipid A (Haeffner-Cavaillon et al 1982b); both isolated polysaccharide chains of the same endotoxin induce B cell mitogenesis in C3H/HeJ mice whereas the corresponding lipid A triggers B cells in C3H/HeN mice (Girard et al 1981). In rabbits immunized with cells of *S. johannesburg*, specific blast response, induced by the isolated polysaccharide of the corresponding endotoxin, reaches a higher level than that induced by homologous lipid A (Cavaillon et al 1978), while in man, isolated polysaccharide chains from *B. pertussis* and *N. meningitidis* induce IL-1 secretion by monocytes (Cavaillon and Haeffner-Cavaillon 1985; Haeffner-Cavaillon et al 1984). Use of endotoxin fragments to study immunological phenomena should, therefore, not be limited to or focused on the hydrophobic region alone, nor should cellular responses elicited with one isolated region (hydrophobic or hydrophilic) be considered to reflect all properties of the unfragmented endotoxin. Like whole lipopolysaccharides, isolated hydrophilic (and hydrophobic?) regions do have specific properties which vary, individually, with the molecular structure.

Information on the chemical composition of endotoxin-receptors is very scarce. According to data available, the endotoxin-receptor on erythrocytes is a lipo-glycoprotein (Springer et al 1974), the lipid and carbohydrate moieties of which are not involved in the activity of the receptor. Endotoxin-receptors appear to be located adjacent to glycoproteins coded for by genes of the major histocompatibility complex, since both binding of endotoxin and the proliferative response of lymphocytes are inhibited by anti-Ia and anti-H2 antibodies (Yokoyama et al 1979).

As regards the functions of the endotoxin-receptor, it is not clear whether the specific structures that bind endotoxins are required for cell activation. Receptors are part of the immune system and required to obtain an immune response: they recognize 'self'-mediators and surface components which elicit immunological responses and ensure cell communication; selection, during evolution, of cells developing structures 'recognizing' 'non-self'-endotoxin appears to be unlikely. It has been suggested by Teodorescu and Mayer (1982) that binding of bacteria to epithelial cells could be due to interaction of lectins present on their surface with the carbohydrate moiety of the lipopolysaccharide. It is also known that lectins or lectin-like molecules are present on lymphocytes (Lehmann et al 1980) and macrophages (Haeffner-Cavaillon et al 1982a). The interaction of these receptor-structures – which, if required for cell activation, should have their 'self'-ligand – to carbohydrates of 'non-self'-endotoxin could, therefore, be responsible for delivering signals that eventually lead to the stimulation of the host's immune system.

Attachment of endotoxin that is not mediated by specific receptors seems to occur to almost all mammalian cells. Such interaction may (platelets) or may not (lymphocytes of C3H/HeJ mice, T cells) lead to cell activation. As regards the mechanism of this type of action, Morrison and Rudbach (1981) reported that endotoxins may be internalized and envisaged an intracellular signal. Membrane perturbation – for instance by endotoxin insertion into the lipid bilayer, or by other, as yet unknown mechanism – may also be sufficient to trigger transmembrane signals to the cell. Cell activation by both specific and nonspecific binding has been substantiated by the observation that the lipid A and the polysaccharide chain of the *B. pertussis* endotoxin could induce polyclonal activation of rabbit spleen cells, macrophages being always required. Therefore, it would appear that macrophages are target cells for both the hydrophilic and hydrophobic regions of this endotoxin, and although specific binding of the endotoxin occurs via the polysaccharide, nonspecific binding, reasonably assignable to the hydrophobic region (Zimmerman et al 1977; Kern 1984), also occurs, and both interactions lead to cell activation.

REFERENCES

Andersson J, Coutinho A, Melchers F (1977) Frequencies of mitogen-reactive B cells in the mouse. I. Distribution in different lymphoid organs from different inbred strains of mice at different ages. *J. Exp. Med. 145*, 1511.

Benedetto DA, Shands JW Jr, Shah DO (1973) The interaction of bacterial lipopolysaccharide with phospholipid bilayers and monolayers. *Biochim. Biophys. Acta 298*, 145.

Benoliel AM, Capo C, Bongrand P, Ryter A, Depieds R (1980) Non-specific binding by macrophages: existence of different adhesive mechanisms and modulation by metabolic inhibitors. *Immunology 41*, 547.

Bona C, Robineaux R, Anteunis A, Heuclin C, Astesano A (1973) Transfer of antigen from macrophages to lymphocytes. II. Immunological significance of the transfer of lipopolysaccharide. *Immunology 24*, 831.

Bona C, Juy D, Truffa-Bacchi P, Kaplan GJ (1976) Binding, capping and internalization of lipopolysaccharide in thymic and non-thymic lymphocytes of the mouse. Biological and autoradiographic study. *J. Microsc. Biol. Cell. 25*, 47.

Bona C, Yano A, Dimitriu A, Miller RG (1978) Mitogenic analysis of murine B-cell heterogeneity. *J. Exp. Med. 148*, 136.

Braude AI, Carey FJ, Zalesky M (1955) Studies with radioactive endotoxin. II. Correlation of physiologic effects with distribution of radioactivity in rabbits injected with lethal doses of *E. coli* endotoxin labeled with radioactive sodium chromate. *J. Clin. Invest. 34*, 858.

Castillo FJ (1980) Partial characterization of *Pseudomonas* phage 2 receptor. *Can. J. Microbiol. 26*, 1015.

Cavaillon JM, Haeffner-Cavaillon N (1985) The role of serum in interleukin 1 production by human monocytes activated by endotoxins and their polysaccharide moieties. *Immunol. Lett.*, in press.

Cavaillon JM, Bona C, Staub AM (1978) Specific blast transformation of rabbit spleen cells induced by a bacterial polysaccharide and inhibitory effect of peripheral blood lymphocytes. *Ann. Immunol. (Paris) 129C*, 199.

Chaby R, Szabó L (1975) 3-Deoxy-2-octulosonic acid-5-phosphate: a component of the endotoxin of *Bordetella pertussis*. *Eur. J. Biochem. 59*, 277.

Chaby R, Metezeau P, Girard R (1984) Binding of a rhodamine-labeled lipopolysaccharide to lipopolysaccharide responder and nonresponder lymphocytes. *Cell. Immunol. 85*, 531.

Chaby R, Moreau M, Szabó L (1977) 2-O-β-(D-Glucuronyl)-7-O-(2-amino-2-deoxy-α-D-glucopyranosyl)-L-glycero-D-manno-heptose: a constituent of the *Bordetella pertussis* endotoxin. *Eur. J. Biochem. 76*, 453.

Chedid L, Skarnes RC, Parant M (1963) Characterization of a ^{51}Cr-labeled endotoxin and its identification in plasma and urine after parenteral administration. *J. Exp. Med. 117*, 561.

Coutinho A, Forni L, Watanabe T (1978) Genetic and functional characterization of an antiserum to the lipid A-specific triggering receptor on murine B lymphocytes. *Eur. J. Immunol. 8*, 63.

Cuatrecasas P (1973) Interaction of *Vibrio cholerae* enterotoxin with cell membranes. *Biochemistry 12*, 3547.

Davies M, Stewart-Tull DES, Jackson DM (1978) The binding of lipopolysaccharide from *Escherichia coli* to mammalian cell membranes and its effect on liposomes. *Biochim. Biophys. Acta 508*, 260.

DeFranco AL, Kung JT, Paul WE (1982) Regulation of growth and proliferation in B cell subpopulations. *Immunol. Rev. 64*, 161.

Dorrington KJ (1976) Properties of the Fc receptor on macrophages and monocytes. *Immunol. Commun. 5*, 263.

Drewry DT, Lomax JA, Gray GW, Wilkinson SG (1973) Studies of Lipid A fractions from the lipopolysaccharides of *Pseudomonas aeruginosa* and *Pseudomonas alcaligenes*. *Biochem. J. 133*, 563.

Esser AF, Russel SW (1979) Membrane perturbation of macrophages stimulated by bacterial lipopolysaccharide. *Biochem. Biophys. Res. Commun. 87*, 532.

Ezekowitz RAB, Austyn J, Stahl PD, Gordon J (1981) Surface properties of Bacillus Calmette-Guérin-activated mouse macrophages. Reduced expression of mannose-specific endocytosis, Fc receptors and antigen F4/80 accompanies induction of Ia. *J. Exp. Med. 154*, 60.

Forni L, Coutinho A (1978) An antiserum which recognizes lipopolysaccharide-reactive B cells in the mouse. *Eur. J. Immunol. 8*, 56.

Freimer NB, Ögmundsdottir HM, Blackwell CC, Sutherland IW, Graham L, Weir DM (1978) The role of cell wall carbohydrates in binding microorganisms to mouse peritoneal exudate macrophages. *Pathol. Microbiol. Scand. (Sect. B) 86*, 53.

Gery I, Waksman BH (1972) Potentiation of the T-lymphocyte response to mitogens. II. The cellular source of potentiating mediator(s). *J. Exp. Med. 136*, 143.

Gimber PE, Rafter GW (1969) The interaction of *Escherichia coli* endotoxin with leukocytes. *Arch. Biochem. Biophys. 135*, 14.

Ginsberg MH, Morrison DC (1978) The selective binding of aggregated IgG to Lipid A-rich bacterial LPS. *J. Immunol. 120*, 317.

Girard R, Chaby R, Bordenave G (1981) Mitogenic response of C3H/HeJ mouse lymphocytes to polyanionic polysaccharides obtained from *Bordetella pertussis* endotoxin and from other bacterial species. *Infect. Immun. 31*, 122.

Goodman SA, Vukajlovich SW, Mukenbeck P, Morrison DC (1984) Selective interaction between lymphocytes and lipid A subunits in lipopolysaccharide macromolecular aggregates. *Rev. Infect. Dis. 6*, 511.

Gregory SH, Zimmerman DH, Kern M (1980a) The lipid A moiety of lipopolysaccharide is specifically bound to B cell subpopulations of responder and non responder animals. *J. Immunol. 125*, 102.

Gregory SH, Zimmerman DH, Kern M (1980b) ^{125}I-lipid A binding by individual macrophage cells: cell purification culture, and autoradiographic analysis on a single surface. *Anal. Biochem. 107*, 246.

Haeffner-Cavaillon N, Chaby R, Szabó L (1977) Identification of 2-methyl-3-hydroxydecanoic and 2-methyl-3-hydroxytetradecanoic acids in the 'lipid x' fraction of the *Bordetella pertussis* endotoxin. *Eur. J. Biochem. 77*, 535.

Haeffner-Cavaillon N, Chaby R, Cavaillon JM, Szabó L (1982a) Lipopolysaccharide receptor on rabbit peritoneal macrophages. I. Binding characteristics. *J. Immunol. 128*, 1950.

20

Haeffner-Cavaillon N, Cavaillon JM, Szabó L (1982b) Macrophage dependent polyclonal activation of splenocytes by *Bordetella pertussis* endotoxin and its isolated polysaccharide and lipid A regions. *Cell. Immunol. 74*, 1.

Haeffner-Cavaillon N, Cavaillon JM, Moreau M, Szabó L (1984) Interleukin 1 secretion by human monocytes stimulated by the isolated polysaccharide region of *Bordetella pertussis* endotoxin. *Mol. Immunol. 21*, 389.

Haeffner-Cavaillon N, Cavaillon JM, Etiévant M, Lebbar S, Szabó L (1985) Specific binding of endotoxin to human monocytes and mouse macrophages: serum requirement. *Cell. Immunol. 91*, 119.

Hämmerling U, Westphal O (1967) Synthesis and use of O-stearoyl polysaccharides in passive hemagglutination and hemolysis. *Eur. J. Biochem. 75*, 23.

Hase S, Rietschel ET (1977) The chemical structure of the lipid A component of lipopolysaccharide from *Chromobacterium violaceum* NCTC 9694. *Eur. J. Biochem. 75*, 23.

Hawiger J, Hawiger A, Timmons S (1975) Endotoxin-sensitive membrane component of human platelets. *Nature (London) 256*, 125.

Hawiger J, Hawiger A, Steckley S, Timmons S, Cheng C (1977) Membrane changes in human platelets induced by lipopolysaccharide endotoxin. *Br. J. Haematol. 35*, 285.

Herring WB, Herion JC, Walker RI, Palmer JG (1963) Distribution and clearance of circulating endotoxin. *J. Clin. Invest. 42*, 79.

Jacobs DM, Eldridge JH (1984) Surface phenotype of LPS-binding murine lymphocytes. *Proc. Soc. Exp. Biol. Med. 175*, 458.

Jakobovits A, Sharon N, Zan-Bar I (1982) Acquisition of mitogenic responsiveness by non-responding lymphocytes upon insertion of appropriate membrane components. *J. Exp. Med. 156*, 1274.

Jarrel KF, Kropinski AM (1981) *Pseudomonas aeruginosa* bacteriophage ØPLS27-lipopolysaccharide interactions. *J. Virol. 40*, 411.

Jennings AJ, Lugowski C, Ashton FE (1984) Conjugation of meningococcal lipopolysaccharide R-type oligosaccharide to tetanus toxoid as route to a potential vaccine against group B *Neisseria meningitidis. Infect. Immun. 43*, 407.

Joiner KA, McAdam KPWJ, Kasper DL (1982) Lipopolysaccharides from *Bacteroides fragilis* are mitogenic for spleen cell from endotoxin responder and non-responder mice. *Infect. Immun. 36*, 1139.

Kabir S, Rosenstreich DL (1977) Binding of bacterial endotoxin to murine spleen lymphocytes. *Infect. Immun. 15*, 156.

Kasper DL (1976) Chemical and biological characterization of the lipopolysaccharide of *Bacteroides fragilis* subspecies fragilis. *J. Infect. Dis. 134*, 59.

Kern M (1984) Selective binding of lipid A to responder and non-responder B lymphocyte subpopulations. *Rev. Infect. Dis. 6*, 506.

LeDur A, Chaby R, Szabó L (1980) Isolation of two protein free and chemically different lipopolysaccharides from *Bordetella pertussis* phenol extracted endotoxin. *J. Bacteriol. 143*, 78.

Lehmann V, Streck H, Minner I, Krammer PH, Ruschmann E (1980) Selection of bacterial mutants from Salmonella specifically recognizing determinants on the cell surface of activated T lymphocytes. A novel system to define well surface structure. *Eur. J. Immunol. 10*, 685.

Leong D, Diaz R, Milner K, Rudbach J, Wilson JB (1970) Some structural and biological properties of Brucella endotoxin. *Infect. Immun. 1*, 174.

Loos M, Bitter-Suermann D, Dierich M (1974) Interaction of the first (C1), the second

Watson J, Kelly K, Whitlock C (1980) Genetic control of endotoxin sensitivity. In: Schlessinger D (Ed), *Microbiology – 1980*, p 4, American Society for Microbiology, Washington D.C.

Weir DM (1980) Surface carbohydrates and lectins in cellular recognition. *Immunol. Today 1*, 45.

Westphal O, Lüderitz O, Bister F (1952) Über die Extraktion von Bakterien mit Phenol-Wasser. *Z. Naturforsch. 7b*, 148.

Willerson JT, Trelstad RL, Pincus T, Levy SB, Wolff SM (1970) Subcellular localization of *Salmonella enteritidis* endotoxin in liver and spleen of mice and rats. *Infect. Immun. 1*, 440.

Wilton JM, Rosenstreich DL, Oppenheim JJ (1975) Activation of guinea pig macrophages by bacterial lipopolysaccharide requires bone marrow-derived lymphocytes. *J. Immunol. 114*, 388.

Wirsing von Koenig CH, Heyner B, Hof H, Finger H, Emmerling P (1981) Biological activity of *Bordetella pertussis* in lipopolysaccharide-resistant mice. *Infect. Immun. 33*, 223.

Wollenweber H-W, Rietschel ET, Hofstad T, Weintraub A, Lindberg AA (1980) Nature, type of linkage, quantity, and absolute configuration of (3-hydroxy) fatty acids in lipopolysaccharides from *Bacteroides fragilis* NCTC 9343 and related strains. *J. Bacteriol. 144*, 898.

Yamada H, Nogami T, Mizushima S (1981) Arrangement of bacteriophage lambda receptor protein (LamB) in the cell surface of *Escherichia coli*: a reconstitution study. *J. Bacteriol. 147*, 660.

Yokoyama K, Osawa T (1979) The collaboration of T-cell subsets in the mitogenic stimulation of purified B cell subpopulations. *Immunology 38*, 789.

Yokoyama K, Terao T, Osawa T (1978) Membrane receptors of human erythrocytes for bacterial lipopolysaccharide (LPS). *Jpn. J. Exp. Med. 48*, 511.

Yokoyama K, Mashimo J, Kasai N, Terao T, Osawa T (1979) Binding of bacterial lipopolysaccharide to histocompatibility-2-complex proteins of mouse lymphocytes. *Hoppe-Seyler's Z. Physiol. Chem. 360*, 587.

Yoshida T, Sonozaki H, Cohen S (1973) The production of migration inhibition factor by B and T cells of the guinea pig. *J. Exp. Med. 138*, 784.

Yu F, Yamada H, Mizushima S (1981) Role of lipopolysaccharide in the receptor function for bacteriophage TuIb in *Escherichia coli*. *J. Bacteriol. 148*, 712.

Zimmerman DH, Gregory S, Kern M (1977) Differentiation of lymphoid cells: the preferential binding of the lipid A moiety of lipopolysaccharide to B lymphocyte populations. *J. Immunol. 119*, 1018.

Zlydaszyk JC, Moon RJ (1976) Fate of ^{51}Cr-labeled lipopolysaccharide in tissue culture cells and livers of normal mice. *Infect. Immun. 14*, 100.

Handbook of Endotoxin, Vol. 3: Cellular Biology of Endotoxin
L.J. Berry, editor
© Elsevier Science Publishers B.V., 1985

CHAPTER 2

Nonspecific interactions of bacterial lipopolysaccharides with membranes and membrane components*

DAVID C. MORRISON

This review is dedicated to the memory of Dr. Erwin Neter, in acknowledgement of his pioneering contributions to the analysis of the mechanism of interaction of lipopolysaccharides with mammalian cell membranes. His definitive studies have provided the experimental focus for much of the research summarized in this review.

1. INTRODUCTION

The almost ubiquitous capacity of endotoxins to interact with cells and tissues of mammalian origin is, perhaps, one of the most intriguing properties of these unique bacterial macromolecules. As the various contributors to this volume will document, endotoxins profoundly affect the cellular responses of liver cells and adipocytes, lymphocytes and macrophages, polymorphonuclear leukocytes, platelets and endothelial cells. Central to the understanding of the effects of entodoxins on these diverse cell types is a delineation of the biochemical mechanism by which a given cell recognizes and is triggered by endotoxin. Indeed, the elucidation of the mechanism (or mechanisms) by which bacterial endotoxins interact with mammalian cell membranes may be considered the keystone to understanding the multiple immunologic and immunopathologic activities of these potent bacterial products.

From a conceptual point of view, it is useful to define two fundamentally distinct mechanisms by which endotoxins may initially interact with mammalian cell membranes, namely specific and nonspecific. Specific interactions would result from the binding of the endotoxin ligand to a defined endotoxin receptor localized on the cell membrane. Such interactions would be governed by all of the constraints generally attributable to receptor–ligand interactions, including ligand specificity,

*   Research support during the preparation of this manuscript was provided by the National Institute of Allergy and Infectious Disease Grants AI-17225 and 17304.

saturability and reversibility. Nonspecific interactions, on the other hand, result from the binding of endotoxin macromolecules to any/all membrane constituents other than endotoxin receptors. These latter interactions would, in general, not be saturable or ligand-specific. Research over the past two decades has provided experimental evidence for both of these mechanisms as potentially contributing to interactions of endotoxins with cell membranes. In Chapter 1 Drs. Cavaillon and colleagues review the topic of specific membrane-localized endotoxin receptors. In this chapter, I shall discuss the concept of nonspecific interactions of endotoxins with mammalian cell membranes.

It is readily apparent that nonspecific interactions of endotoxins with membranes can occur via a variety of membrane constituents, including glycoproteins, glycosphingolipids and/or phospholipids. There is, in fact, ample experimental evidence to support the concept that endotoxins will bind to each of these components (reviewed in Morrison and Rudbach 1981). For the purpose of this review, however, I shall consider primarily the interaction of endotoxin with the lipid component of cell membranes. The reasons for this are twofold. First, the concept of a nonspecific 'insertion' of endotoxin into the membrane lipid bilayer is a rather attractive hypothesis which has earlier been suggested by a number of investigators to contribute to endotoxin-initiated cell-triggering events (Bara et al 1973b; Kabir and Rosenstreich 1977; Morrison et al 1981; Shands 1973; Symons and Clarson 1979). Second, the interaction of endotoxin with phospholipid represents a true nonspecific interaction. In this respect, membrane-localized glycoproteins with a relatively high affinity for binding endotoxins, while not being endotoxin receptors per se, would nevertheless be difficult to distinguish functionally from true endotoxin receptors. While there is some evidence to suggest that endotoxins can interact with a variety of membrane-localized glycoproteins (Springer et al 1970, 1974; Yokoyama et al 1979), the majority of studies have dealt with binding of endotoxin to phospholipids, which will form the focus of this review.

Although endotoxins are classically defined as high molecular weight complexes of lipopolysaccharide and protein, it has been amply documented that a major contribution to the manifestation of endotoxin activity is provided by the lipopolysaccharide (LPS) component. In most instances, LPS activity has been shown to reside in the lipid A component of the LPS macromolecule (reviewed in Galanos et al 1977; Morrison and Ryan 1979; Morrison and Ulevitch 1978). Recent studies, however, have shown that the protein component of endotoxin will also elicit a variety of cellular responses, some of which are at least qualitatively similar to those previously characterized in response to the LPS component (reviewed in Volume 1 of the series (Chapter 11)). Nevertheless, it is probably fair to conclude that the majority of the endotoxic properties of endotoxin still occur as the result of lipid A-dependent cellular responses to the LPS. In this review, therefore, attention will center on the nonspecific mechanisms of interaction of LPS, and in particular lipid A, with membrane phospholipids.

Many investigators have underscored the need for caution in the analysis of experiments performed by different investigators using lipopolysaccharides from different bacterial sources prepared by different extraction procedures (see e.g. Nowotny 1969; Shands 1971b). In addition, as documented a number of years ago by Nowotny and colleagues (Tripodi and Nowotny 1966), extensive heterogeneity can exist in an LPS preparation derived from a single organism. Finally, recent experiments from a number of laboratories (Goldman and Leive 1980; Jann et al 1975; Palva and Mäkelä 1980) have provided evidence that LPS preparations which are homogeneous by macromolecular constraints still contain considerable heterogeneity in their subunit composition. These and other variables may prove to be of critical relevance in defining the effective nonspecific interaction of LPS macromolecules with membrane phospholipids.

Three such variables merit particular consideration in this review because of their apparent major contributory effects to LPS–membrane interactions. As emphasized by Galanos and Lüderitz (1975), virtually all LPS preparations contain low molecular weight cationic amines as well as inorganic cations. As elegantly demonstrated by Galanos (1975), removal of these low molecular weight constituents and subsequent conversion of LPS to a uniform salt form can have a profound influence on the subsequent physical-chemical properties and (vide infra) net electrical charge of the LPS macromolecule. Such alterations in LPS structure may be manifest in markedly different endotoxic activity. As I will discuss, the available evidence would suggest that ionic interactions contribute to the net association of LPS with phospholipids, and thus would depend upon the actual salt form of the LPS preparation employed.

It is also relevant to consider the bacterial source from which the LPS is derived, particularly with respect to rough (R) versus smooth (S) organisms. LPS from the former is deficient in O-antigen polysaccharide and, as a consequence, contains considerably more lipid on a per weight basis. While such LPS preparations have proven to be invaluable in the study of the biological activity of LPS and for the demonstration that activity resides in the lipid A region of the molecule, there exists no a priori reason why such preparations should behave, from a physical-chemical point of view, in a manner identical to more hydrophilic smooth LPS preparations. (Similar arguments could be employed with respect to analysis of isolated lipid A.) Indeed, the recently published experiments of Magnusson et al (1977) would suggest that both LPS polysaccharide composition and ion content influence selectively the capacity of LPS to partition into a hydrophobic or ionic environment. On the basis of these studies, these authors conclude that R-type LPS may have a much greater tendency to interact with mammalian cell membranes than S-type LPS.

Finally, it should be noted that some investigators have employed physical and/or chemical modification of the LPS as a technique for the subsequent analysis of LPS–membrane interactions. In particular, heating of LPS to 100 °C and treatment of LPS with 'mild' alkali have been rather extensively utilized. The use of such procedures is predicated upon the pioneering studies of Neter and his colleagues (Lüderitz et al 1958; Neter et al 1956) who demonstrated rather marked enhancement of the erythrocyte-modifying capacity of such LPS. As will be discussed, evidence has suggested both enhancement and inhibition of existing LPS activity following such LPS modification as well as the generation of 'new' biological activity. In most incidences, the precise correlation of activity with membrane affinity has not been addressed. Therefore, while the use of such modified LPS, and in particular alkali-modified LPS, may well provide a molecular tool with which to modify nonspecifically mammalian cell membranes, the relevance of such modified LPS interaction with mammalian cells to the actual biochemical mechanism(s) of endotoxin-initiated host responses is questionable. Parenthetically, the observation by Goodman and Sultzer (1977) that at least twelve different protocols to modify LPS by mild alkali have been described in the literature, 'all of which were more drastic than that originally described by Neter' would indicate an additional complicating variable in the analysis of such preparations. There is, in fact, rather compelling physical-chemical evidence to support the concept that the methodology employed to generate alkaline-hydrolyzed LPS can have a profound effect on its biological activity (Niwa et al 1969).

With these caveats in mind, in the following sections I will consider the interaction of LPS with mammalian cell membranes. It is readily apparent that the effective interaction of LPS with the membrane, whether by specific or nonspecific means, is but the initial step in the chain of biochemical events leading to the generation of a transmembrane signal and the stimulation of a cellular response. A consideration of putative events subsequent to the nonspecific interaction of LPS with the cell membrane, while beyond the scope of this review, has been discussed in an earlier review (Morrison and Rudbach 1981).

2. THE INTERACTION OF LIPOPOLYSACCHARIDE WITH MEMBRANE PHOSPHOLIPIDS

Lipopolysaccharide is one of the major constituents of the cell surface of gram-negative bacteria. Evidence accumulated during the last twenty years has provided convincing evidence that LPS is localized in the outer membrane component of the bacterial surface (reviewed in Nikaido and Nakae 1979). As suggested by numerous investigators, the outer membrane has the morphological characteristics of a typical bilayer structure (Freer and Salton 1971), and contains, in addition to LPS, the phospholipids and protein constituents characteristic of virtually all biological membranes. Of particular interest has been the demonstration, by Mühlradt and Golecki (1975), of the asymmetric distribution of LPS in the outer membrane. Using ferritin-conjugated antibody to LPS, these authors showed that LPS is restricted to the outer leaflet of the outer membrane. These conclusions were also reached by Funahara and Nikaido (1980) using a biochemical approach in which accessibility of bacterial LPS carbohydrate residues to enzyme modification was employed. Thus, the majority of the available evidence would suggest that LPS, in its most natural environment, exists in an asymmetric bilayer configuration in association with itself, with phospholipid and with outer membrane proteins.

Most investigators would agree that hydrophobic interactions between the lipid A component of LPS and outer membrane phospholipids contribute to the stability of LPS in the membrane bilayer even though there is evidence that outer membrane phospholipids can exist as monolayers and/or bilayers which do not enter into association with LPS (Nikaido et al 1977), and vice versa (Mühlradt et al 1974). In addition there exists good evidence for a major role of divalent cations in the interaction of LPS in the outer membrane (Gray and Wilkinson 1965; Leive et al 1968). Of particular note, however, is the convincing demonstration by Rottem and Leive (1977) that polysaccharide–polysaccharide interactions between neighboring LPS molecules also contribute to the stability of LPS. As shown by these investigators, the presence of long-chain O-antigen polysaccharide, as well as its surface density in the outer membrane, both restricted the mobility of LPS and/or associated phospholipid in the outer membrane. These rather surprising findings indicate that LPS–LPS interactions via both hydrophobic and (presumably) hydrogen bonding as well as hydrophobic LPS–phospholipid and ionic LPS–divalent cation interactions may be relevant considerations in the evaluation of LPS membrane interactions.

In any case, the evidence is convincing that LPS can 'integrate' with phospholipids on the bacterial surface in the formation of a membrane bilayer. Further, such bilayer structures would appear to be energetically favorable configurations. As demonstrated by Nakamura and Mizushima (1975), outer membranes reformed from purified outer membrane proteins, phospholipid and lipopolysaccharide displayed the characteristic morphology of outer membrane trilaminar structures. It is noteworthy that, in these experiments, the reformation of membranes required the presence of magnesium.

There is also considerable experimental support to indicate that the isolated LPS will itself form natural bilayer structures. Although the LPS monomer structural unit has a molecular weight in the order of 3000–5000 (reviewed in Shands and Chen 1980), its extraction from the bacterial surface leads to LPS macromolecular structures with molecular weights in the order of millions (reviewed in Shands 1971b). While a variety of morphological structures for these LPS macromolecular aggregates have been described (reviewed in Shands 1971a, 1971b), it has been a common finding that LPS can be described as consisting of ribbon-like bilayer strands and discs (see, e.g., Burge and Draper 1967; Goodman and Sultzer 1977; Rudbach et al 1966; Shands et al 1967; Shands 1971a). In the studies of Burge and Draper (1967), the results obtained by electron microscopy were also confirmed by analysis of X-ray diffraction of the LPS bilayers. In addition, the more recent experiments of Nikaido et al (1977) provide electron spin resonance data which are also consistent with the concept of LPS bilayer structures.

As proposed by many of these investigators, in these LPS ribbon-like structures, the lipid A region represents the most central region of the bilayer and the 'core' polysaccharides the two outer dense lines. The findings of a negative coefficient of linear expansion described by Burge and Draper (1967) would provide thermodynamic evidence for this interpretation. Although the O-antigen polysaccharides were proposed not to contribute to the trilaminar structure of isolated LPS (Shands et al 1967), such an interpretation might be an oversimplification considering the results of Rottem and Leive (1977). Nevertheless, the finding that LPS from semi-rough strains had an equivalent laminar structure provides good evidence for a prominent role for lipid A and core polysaccharide as structural determinants. The fact that LPS from Re mutants had a smaller and 'more delicate' trilaminar structure would also be consistent with this view. These combined studies clearly indicate that the intrinsic structure of the LPS monomer lends itself to the formation of bimolecular leaflets characteristic of biological membranes.

It is of relevance that the tertiary macromolecular structure of LPS is subject to considerable modification. As shown by Rudbach et al (1966), treatment of LPS with detergent bile salts, followed by removal of detergent, causes marked modification of morphology as assessed by electron microscopy. Shands et al (1967) demonstrated a conversion of LPS from ribbon-like structures to homogeneous discs by sonication or by treatment with aqueous ether. Finally, chemical procedures which have earlier been discussed with respect to their capacity to modify LPS interaction with membranes, in particular heating to $100\,°C$ or treatment with mild alkali, also have significant and often profound effects on LPS trilaminar morphology. As suggested by Burge and Draper (1967), heating LPS to $50\,°C$ or $100\,°C$ resulted in the appearance of a new X-ray reflection plane (at 14 nm) as well as a slight shift in the characteristic reflection from 9.6 to 9.1 nm. It is of importance that these changes in X-ray diffraction patterns appeared to be irreversible and were maintained following a return of LPS to ambient temperature.

Treatment of LPS with mild alkali has been shown by light scattering (Tripodi and Nowotny 1966) to result in a marked reduction in particle size and by electron microscopy to result in destruction of the ribbon-like lamellar structures (Goodman and Sultzer 1977; Shands 1971a). Alkaline hydrolysis of LPS is generally recognized as causing chemical modification via cleavage of ester-linked fatty acids contained within the lipid A (Nowotny 1969; Rietschel et al 1972), which might reduce the hydrophobic interactions required for maintenance of the trilaminar structure, and thus a breakdown of the tertiary LPS structure. In the experiments of Goodman and Sultzer, however, loss of bilayer structure was reported to have been achieved in the absence of detectable loss of fatty acids from the LPS preparation. It may well be, therefore, that additional alkaline-labile physical-chemical interactions (e.g. divalent cations, hydrogen bonding) not only contribute, but are in fact critical to the maintenance of LPS trilaminar structures.

Whatever the nature of the precise interaction between LPS subunits, the binding is apparently of relatively high affinity. Following dissociation of LPS by detergent and subsequent removal of detergent, reformation of filamentous rods (Rudbach et al 1966) or even distinct trilaminar structures (Nakamura and Mizushima 1975) has been reported. Of relevance, in the latter study, a requirement for magnesium was reported for the reformation of LPS bilayers.

In the absence of LPS-dissociating procedures, the available evidence would suggest that LPS (unlike phospholipids or alkaline-hydrolized LPS) does not dissociate to form monomolecular films in aqueous solution (Benedetto et al 1973; Romeo et al 1970a). This observation most probably reflects hydrophobic lipid A interactions between LPS subunits and hydrogen bonding between polysaccharide side chains of higher energy than is found in phospholipid dispersions. It is noteworthy in this respect that Nikaido et al (1977), in examining the thermal electron spin resonance (ESR) melting profiles, found the spectra of LPS and phospholipid to be approximately superimposable with a 30 °C temperature shift. Thus, the ESR of phospholipid at 15 °C was comparable to that of LPS at 47 °C. In fact, these authors reported that they were never able to achieve total 'melting' of the fatty acids in the region of the lipid A diglucosamine backbone. Thus, the energy required to separate the leaflets of the LPS bilayer may preclude their spontaneous dissociation and formation of monolayers on aqueous surfaces.

Since lipopolysaccharides are derived from bacterial membranes and, in an isolated state, have the capacity to align themselves in trilaminar membranous structures, it is logical to anticipate that LPS would have the capacity to interact with membranes of mammalian origin, and, in particular, with membrane phospholipids. It is constructive to consider such LPS–membrane phospholipid interactions as being either passive or active. Passive interactions involve the surface absorption of LPS to phospholipid micellar structures or membranes by surface charge and/or hydrophobic interactions. In general, such interactions do not result in extensive modification of either the LPS macromolecular structure or of the phospholipid bilayer or lamellar structure. The active interaction of LPS with phospholipids, on the other hand, may be considered as resulting from the extensive disruption of the LPS–tertiary macromolecular structure by the phospholipid (and vice versa), the net consequence of which is a new complex LPS-phospholipid micellar structure (or membrane). The active interaction of LPS with phospholipids would require that the LPS subunits have, in general, a greater affinity for phospholipid than for themselves. As will be discussed later in this review, there is a growing body of evidence to support this type of dynamic interaction between LPS and mammalian cell membranes, in spite of the apparent high affinity of LPS subunits for each other.

An extensive series of experiments published by Rothfield and his collaborators have explored the nature of the interaction of LPS derived from a rough (Rc) mutant of *Salmonella typhimurium* with phospholipid dispersions, monolayers and bilayers. While a major objective of these studies was the delineation of the role of outer membrane constituents in the biosynthesis of LPS mediated by specific monosaccharide transferases, these experiments also provided considerable insight into the physical-chemical interactions between LPS and phospholipid in general.

Perhaps the first indirect indication of extensive complex formation between LPS and phospholipid was the demonstration that LPS–phospholipid mixtures, when added to purified transferase, protected the enzyme against thermal inactivation (Rothfield and Takeshita 1965). Neither component alone afforded protection against inactivation. Of interest, protection required the appropriate (LPS) substrate and, in additional studies, demonstrated remarkable phospholipid specificity with respect to both polar and nonpolar components for the expression of transferase activity (Rothfield and Pearlman 1966). In an extension of these studies, Weiser and Rothfield (1968) showed that LPS would interact in nonstoichiometric ratios with phosphatidylethanolamine. Heating of LPS–phosphatidylethanolamine mixtures to 60 °C greatly enhanced the degree of interaction in the formation of mixed micelles; however, no differences were observed between gradual and rapid cooling. The authors also showed that complex formation could be achieved with phosphatidylcholine even though this phospholipid was completely inactive in the transferase system, although the data suggested the micellar complex to be 'less discrete and more heterogeneous'. Similarly the authors reported micellar complex formation using LPS from smooth organisms, although again the degree of association was much less extensive. These combined experiments suggest that under the appropriate conditions of LPS (rough Rc), phospholipid (phosphatidylethanolamine) and reaction conditions (60 °C, 30 min), mixed micellar complexes can in fact be formed with a maximum LPS content of about 30% of the weight of phospholipid. Electron microscopic examination of such complexes suggested to Rothfield and Horne (1967) that the LPS was being incorporated directly into the phospholipid leaflet structure. While such 'nonspecific' intercalation of LPS into the phosphatidylethanolamine micelle is clearly documented by these studies, the rather rigorous conditions required for extensive intercalation would support the concept of intrinsic LPS macromolecular stability discussed above.

Similar experiments to those of Rothfield and colleagues have also been described by Shands (1973) using LPS from a smooth strain of *S. typhimurium* and phosphatidylcholine vesicles, heated for 10 minutes at either 37 °C or 56 °C. As assessed by electron microscopy, both 'stacking' of LPS particles and vesiculization were proposed as consequences of this interaction, the latter of which could only be demonstrated with 'crude' lecithin preparations. In neither case was there apparent disruption of the laminar patterns of the lecithin micelles. These data are consistent with the earlier conclusions of Rothfield et al (1968) showing limited heterogeneous complex formation of LPS and phosphatidylcholine.

Using a somewhat different approach, Kataoka et al (1971) assessed the relative intrinsic capacity of Re LPS of *Salmonella minnesota* to interact with phospholipids in the formation of liposomes. As assessed by the subsequent capacity of LPS-liposomes to be lysed in the presence of specific LPS antibody and complement, alkali-treated LPS was 10–30-fold more active than native LPS. Similar results were obtained with liposomes prepared in the absence of LPS to which LPS was subsequently added. These results provide support for the concept of increased affinity of LPS for phospholipid vesicles following alkali treatment, results similar to those described above for intact mammalian membranes. It is noteworthy in these experiments that LPS incorporated into liposomes did not activate complement in the absence of antibody, even though the LPS employed in those studies is a good activator of the classical pathway of serum complement (Morrison and Kline 1977). It is possible that incorporation of LPS into the micellar structure of the liposome masks structures required for the effective binding of C1. Alternatively, dispersion of LPS into the liposome effectively reduces the molecular weight of the LPS aggregate to a surface density less than that required to bind and activate C1q (in a manner analogous to the critical spacing requirement for the Fc portion of IgG antibody). Similar alterations in the expression of lipid A activity have been documented following its incorporation into liposomes (Alving and Richardson 1984; Richardson et al 1983). In these latter experiments a critical role for lipid A epitope density on the liposomal surface was suggested. Whatever the reason, the data suggest that LPS in association with phospholipid structures may not express activity identical to that of the isolated phospholipid-free LPS.

Several investigators have probed the consequences of the incorporation of LPS into liposomes on the physical-chemical and functional properties of the liposomal vesicles. In general, these studies support the concept that lipid A or LPS restricts the mobility of the long-chain hydrocarbons and thus decreases fluidity of the lipid bilayers. As reported by Nikaido and Nakae (1973), Rc-type LPS from *Salmonella* increased bilayer surface area and/or interlaminar distance, perhaps by introduction of electrostatic repulsive forces between anionic groups on the inner core of the LPS. ESR studies by Nikaido et al (1977), using either a 5-NO-stearate or 12-NO stearate probe, suggested that the LPS and phospholipid monomers could exist as homogeneous interspersed bilayers which were not separated into large molecularly homogeneous domains. The effective incorporation of LPS into the mixed phospholipid liposomal micelles, however, required the presence of calcium and Na^+. More recent experiments by these investigators (Takeuchi and Nikado 1981), using pure preparations of spin-labeled phospholipids in the absence of unlabeled phospholipids, have provided strong evidence that the formation of mixed micelles by LPS and phospholipids is accomplished via segregated domains of each of the two constituents. Further, these domains were shown to be thermodynamically stable and to persist for long periods of time. Thus, while the formation of homogeneous LPS–phospholipid vesicles can clearly be established under appropriate experimental conditions, the establishment of separate LPS and phospholipid domains, similar to the asymmetric distribution of these two components on the outer membrane of the bacterial surface, may be the preferred molecular orientation under physiological conditions of LPS–mammalian membrane interactions.

A reduction in fluidity of phosphatidylcholine micelles containing isolated lipid A has also been reported by Rottem (1978). (In contrast to the results of Nikaido et al (1977), effective incorporation of lipid A into liposomes did not require the presence of divalent cations). More recent experiments reported by Lui et al (1982) also indicate a decrease in the fluidity of phospholipid micelles with *S. minnesota* Re LPS. In these latter experiments, however, modification of liposomal fluidity was restricted to negatively charged phospholipids (e.g. phosphatidic acid, phosphorylethanolamine) and no effects were noted with phosphatidylcholine. Although this result contrasts with that of Rottem (1978) using lipid A, it is consistent with the earlier observations of Weiser and Rothfield showing a reduced tendency for an Rc-LPS to interact with phosphatidylcholine as compared to phosphatidylethanolamine. The lipid A may, therefore, represent a somewhat unusual situation. In any case, LPS appears to have the general property of reducing the fluidity of the hydrophobic environment of phospholipids with which it interacts with relatively high avidity.

It has been suggested that LPS functions, at least in part, on the bacterial outer membrane, by serving as a permeability barrier (Leive et al 1968; Nikaido and Nakae 1979). The results of experiments designed to examine the effects of LPS on the permeability of model lysosomal membranes have not allowed uniform conclusions to be drawn. As assessed by either the osmotic swelling technique or the release of radiolabeled markers entrapped within mixed phospholipid liposomes, Nikaido and Nakae (1973) concluded that Rc-LPS had no effect on permeability of lysosomes to small hydrophobic molecules, even though by authors' calculations, approximately 25% of the surface area of the liposome was occupied by LPS. Similar conclusions were reached by Onji et al (1981) using an *Escherichia coli* endotoxin incorporated into phosphatidylcholine vesicles. These authors did report that the increased permeability of liposomes which results when liposomes are prepared in the presence of calcium could be abrogated by the simultaneous incorporation of *E. coli* endotoxin. They suggest from these studies that endotoxin increased the molecular packing of phosphatidylcholine vesicles; however, the binding of calcium by LPS and a consequent reduction in the amount of calcium available to participate in liposome formation independent of any interaction of LPS with the permeability function of the liposome cannot be excluded. Davies et al (1978) found that the incorporation of LPS into negatively charged liposomes decreased their permeability as assessed by leakage of ^{51}Cr; however, LPS had no effect on positively charged liposomes. Since Lui et al (1982) have shown a similar selectivity of LPS in modification of liposome fluidity, these combined data suggest a correlation between decreases in permeability and restriction of membrane fluidity. Finally, Rottem (1978), using the osmotic swelling technique, has shown that lipid A decreases the permeability of liposomes to uncharged molecules such as glycerol or erythritol. In some instances, therefore, LPS is without detectable effect on liposomal permeability; where permeability effects have been noted, they have been decreases, which might be anticipated as the result of decreased fluidity of such liposomes.

The studies summarized above provide a firm experimental basis for the concept that, given the appropriate conditions, macromolecular aggregates of LPS can interact with phospholipids and alter the physical-chemical (and perhaps functional) properties of such phospholipid micelles. It is, therefore, relevant to query whether LPS can also interact with phospholipids under conditions in which the latter approximate their configuration in a biologically intact membrane, namely as phospholipid monolayers or bilayers.

The experiments of Rothfield and collaborators to explore the mechanism of action of the galactose transferase in the outer membrane have again provided considerable insight into this question. As convincingly shown by Romeo et al

(1970a), LPS introduced into the aqueous phase will cause significant alterations in the surface pressure of phospholipid monolayers formed at the air-aqueous phase interface. The authors note that the time required to reach equilibrium was temperature-dependent but the net increase in surface pressure was independent of both temperature and, importantly, concentration of LPS (over the range of 20–100 µg/ml). Assuming a monomer Rc-LPS molecular weight of 10 000 daltons, an approximate molar ratio of 5.6:1 for phosphatidylethanolamine:LPS was calculated.

An examination of the LPS and phospholipid parameters which influenced changes in monolayer surface pressure revealed a number of critical variables. LPS derived from rough organisms was more effective than that derived from smooth organisms. Variation in both the phospholipid fatty acid side chains as well as the polar head groups influenced LPS alterations in surface pressure. The authors postulate that the effect of LPS may be the result of absorption onto the polar surface at the aqueous-air interface or an actual penetration of LPS into the phospholipid monolayer. In this respect, the authors noted that phosphatidylethanolamine containing saturated fatty acids resulted in much smaller changes in surface pressure following addition of LPS than did phosphatidylethanolamine with unsaturated fatty acids. Since the polar groups in this case were identical, a role for fatty acids in the interaction with LPS was implicated. These studies indicate the importance of intermolecular phospholipid spacings in the penetration of the monolayer film by LPS. The authors further note, as discussed earlier, that isolated LPS did not form stable monolayers in the absence of phospholipid. Thus, these experiments underscore the important role which phospholipid monolayers may play in facilitating the molecular disassembly of the highly stable LPS macromolecular aggregates.

Experiments similar to those of Romeo et al (1970a) were reported by Benedetto et al (1973), who observed alterations in surface pressure of phospholipid monolayers following the addition of LPS to the aqueous phase. Of importance, treatment of LPS with mild alkali increased its capacity to penetrate monolayers by an order of magnitude as compared to native LPS. These authors confirmed the results of Romeo et al (1970a) indicating an inability of native LPS to form monolayers; treatment of LPS with alkali, however, did result in a preparation of LPS which formed a monolayer film. When examined with a battery of phospholipid monolayers, alkali-treated LPS was shown to penetrate into each, and native LPS into all but cholesterol. It is of interest that the interaction of native LPS with the cholesterol monolayer, which did not result in a change of surface pressure (suggesting a lack of LPS penetration), did nevertheless result in a significant change in surface potential. The authors propose that these data may indicate a role for surface absorption as well as penetration in the interaction of LPS with monolayers.

Of perhaps the most direct relevance to concepts of nonspecific interaction of LPS with mammalian cell membranes are model systems employing phospholipid bilayers. As reported by Schuster et al (1970), at concentrations from 100–700 μg/ml, LPS caused rather dramatic changes in the stability of lecithin cholesterol black lipid membranes. When introduced into the aqueous phase, LPS initiated a rapid change in electrical resistance followed by membrane rupture. If LPS was incorporated into the lipid mixture prior to formation of the bilayer, mean survival times were reduced from 130 minutes to less than 2 minutes. Results similar to these were reported by Benedetto et al (1973), who, in addition, noted changes in the dielectric breakdown potential of the bilayer at concentrations of LPS below the threshold required for detectable decreases in mean survival time. Disruptive effects of high concentrations of LPS and lipid A on cholesterol black lipid membranes have also been reported by Rosenstreich and Blumenthal (1977). These authors also reported that they were unable to detect any changes in the conductance of such membranes following addition of LPS. Since Benedetto et al (1973) have provided data which suggest that LPS does not penetrate cholesterol monolayers, the failure to demonstrate conductance changes might not be unexpected. Although the concentrations of LPS required to achieve these combined effects on lipid bilayers were relatively high, data nevertheless suggest that, given the appropriate experimental conditions, LPS will profoundly disrupt the integrity of phospholipid bilayer membranes.

3. THE INTERACTION OF LIPOPOLYSACCHARIDE WITH ERYTHROCYTE MEMBRANES

The ready availability of mammalian erythrocytes, their relative stability in simple buffers in vitro, and the relative ease of assessing alterations in cellular integrity have made this cell an extremely useful tool to examine the effects of various chemicals and macromolecules on membrane integrity. In this respect considerable information on the interaction of LPS with mammalian cell membranes has been obtained by analyses of LPS–erythrocyte interactions.

The classic experiments of Neter and his colleagues provided the first comprehensive description of the effects of LPS on mammalian erythrocytes. As originally reported by Neter et al (1953), LPS added to sheep erythrocytes will passively sensitize these cells for subsequent agglutination and complement-mediated hemolysis in the presence of homologous anti-LPS antibody. While providing a powerful tool for the serologic analysis of bacterial antigens, these experiments also provided insight into mechanisms by which LPS attaches to the erythrocyte surface. Thus, as shown by Neter et al (1953), membrane constituents such as lecithin and cholesterol were potent inhibitors of the erythrocyte sensitization leading to agglutination. Curiously, however, markedly less inhibition was noted in the effects of these materials on LPS-induced passive hemolysis. Since lipid would not be expected to interfere with binding of antibody to the LPS O-antigen, it is conceivable that complement-mediated lysis was being effected by the 'reactive lysis' mechanism (Thompson and Lachman 1970) in the absence of direct erythrocyte-modifying effects. Nevertheless, these data suggested that erythrocyte mem-

36

brane lipids might serve as the receptor for the 'recognition' of LPS.

Of importance in these studies was the observation that modification of the LPS by heating to 100 °C would markedly enhance its erythrocyte-modifying capacity. In an extension of these studies (Lüderitz et al 1958; Neter et al 1956), it was shown that mild alkali treatment, in addition to heat, would enhance the binding of LPS to erythrocytes, although the former procedure was quantitatively more effective. Since alkali treatment reduced the LPS particle size dramatically without detectable alterations in antigenicity, the authors proposed that 'the groupings opened up by treatment with alkali' might be precisely those which are responsible for attachment to erythrocytes. Since lipid-free polysaccharide from the LPS did not attach to erythrocytes, the authors concluded that the lipid portion of LPS was responsible for attachment.

Additional experiments by these investigators (Neter et al 1956, 1958b) demonstrated two additional classes of molecules, in particular certain antibiotics (streptomycin, humycin, neomycin, polymyxin B) and cationic polypeptides (protamine, histone) which would also inhibit LPS sensitization of erythrocytes (polymyxin B actually may be listed in both classes). In most instances these inhibitors acted by preventing the attachment of LPS to the erythrocyte, and thus were less effective at inhibiting erythrocytes presensitized with LPS. An interesting exception to this mechanism was the inhibition by histone. Treatment of erythrocytes with histone–LPS complexes still resulted in binding of LPS but the modified erythrocytes were insensitive to complement-mediated hemolysis. Treatment of histone–LPS–erythrocyte complexes with trypsin restored hemolytic activity from binding to the LPS. In any case, these experiments provided additional evidence for attachment of LPS to erythrocytes as an initial step in the erythrocyte modification process.

Parenthetically, the demonstration by Neter of the inhibitory effects of polymyxin B on an LPS response provided the first demonstration of what has come to be a more or less standard procedure in the analysis of lipid A-dependent LPS effects. It is now recognized that polymyxin B will bind with high affinity to the lipid A region of LPS (Bader and Teuber 1973; Morrison and Jacobs 1976; Schindler and Osborn 1979) and inhibit many lipid A-dependent LPS immunologic, immunopathologic and toxic effects (Morrison and Ryan 1979; Morrison and Ulevitch 1978). In retrospect, therefore, the experiments of Neter quite clearly documented a role for lipid A in the attachment of LPS to the erythrocyte surface.

More direct evidence for a role of hydrophobic lipid interactions in the attachment of LPS to erythrocytes was provided by the experiments of Hammerling and Westphal (1967). Using stearoyl chloride to derivatize lipid-free polysaccharide derived from LPS (which was shown not to bind erythrocytes), these investigators demonstrated that the stearoyl-modified polysaccharide reacquired its capacity to sensitize sheep erythrocytes for both agglutination and complement hemolysis by homologous anti-polysaccharide antibody. These data provide fairly convincing evidence that nonspecific hydrophobic interaction between LPS and erythrocyte membranes is a sufficient condition for erythrocyte sensitization for antibody-dependent complement hemolysis.

More recent experiments published by Pangburn et al (1980) have shown that LPS will also sensitize erythrocytes for antibody-independent complement hemolysis via the alternative pathway. In these experiments, LPS treated with mild alkali and subsequently incubated with sheep erythrocytes was shown to convert these cells to alternative pathway activators. Evidence was presented to suggest that LPS, localized to the erythrocyte surface, restricted the binding of Factor H and subsequent inactivation of bound C3b by C3bINA. The fact that LPS from the Re mutant of *S. minnesota* was not effective suggested a role for polysaccharide in inhibiting the action of Factor H.

It should be pointed out, nevertheless, that the data summarized above do not address the question of whether such nonspecific interactions are necessary for erythrocyte binding by LPS. In this respect, the demonstration by Springer and colleagues (Springer et al 1970, 1974) of the presence of a glycolipoprotein LPS receptor on erythrocytes would provide an alternative mechanism for the specific attachment of LPS to the erythrocyte membrane. As pointed out earlier, however, a consideration of specific LPS receptors is beyond the scope of this review and is discussed in detail in Chapter 1.

Direct morphological evidence that LPS will bind to erythrocyte membranes was provided by Shands (1973). Using sheep erythrocytes incubated with a relatively high (1.0 mg/ml) concentration of LPS, and subsequent erythrocyte osmotic lysis, Shands demonstrated LPS attachment to the lysed erythrocyte stroma. Of interest, LPS trilaminar particles were shown to attach to erythrocyte stroma in an orientation approximately perpendicular to the membrane bilayer. Shands termed this type of binding an 'edge effect' and speculated that this orientation would favor nonspecific hydrophobic interactions between the LPS and the phospholipid bilayer. More recent experiments by Nygren and Dahlen (1979) and Nygren et al (1979), using horseradish peroxidase-conjugated LPS and cytochemical electron microscopy, have shown uniform distributions of LPS on the erythrocyte membrane surface. It was of interest in the report by Nygren and Dahlen (1979) that a spectrum of binding affinities of LPS to red cells was suggested and that even at high concentrations of LPS, some erythrocytes did not show detectable binding of LPS.

Experiments designed to quantitate the amount of LPS bound to sheep erythrocytes were first reported by Lüderitz et al (1958). These investigators, using ^{32}P-labeled LPS, demonstrated maximal binding of approximately 2 μg of LPS/10^9 erythrocytes. Of importance was that this number was increased about 20-fold following treatment with mild alkali. Based upon an approximate molecular weight of 2 × 10^5 for alkaline-hydrolized LPS, this represents about 100 000 molecules per cell. Similar reports have been obtained by Ciznar and Shands (1971). Using ^{14}C-LPS these authors reported that native LPS bound only with low efficiency to erythrocytes and appeared to be saturable at 0.7 μg/10^9 cells, a figure comparable to that obtained earlier by Lüderitz et al (1958). Mild alkaline hydrolysis of LPS markedly increased both the efficiency of LPS binding and the amount of LPS bound per cell, thus confirming the earlier results of Lüderitz et al (1958). It was also noted in these studies that relatively high concentrations of alkali-treated LPS (> 100 µg/ml) induced a slow time-dependent hemolytic response in the erythrocytes which showed only a minimal dependence upon temperature. Since hemolytic activity was not detected with native LPS, the authors suggested that this activity might well be a laboratory artifact; however, the hemolytic activity clearly suggested that the interaction between alkaline-treated LPS and the erythrocyte membrane was probably more than a simple passive surface absorption phenomenon and may well represent active penetration of the erythrocyte bilayer by the alkaline-treated LPS.

Additional indirect support for an active interaction between alkali-treated LPS and the erythrocyte membrane has been provided by the experiments of Warren and collaborators. These investigators have demonstrated (Warren and Kowalski 1977) a rather pronounced inhibition of concanavalin A (Con A)-induced erythrocyte agglutination by pretreatment with alkali-modified LPS. Of interest was the fact that the inhibition showed unusual specificity in that agglutination induced by two other lectins (soy bean agglutinin and phytohemagglutinin) was not affected. Since LPS treatment had no detectable effect on binding of Con A to the erythrocytes, the authors proposed that LPS was modulating erythrocyte membrane fluidity and/or surface charge. As shown by Warren (1982), the inhibitory effect of LPS on Con A agglutination was abrogated by polymyxin B, suggesting a primary role for lipid A in mediating these responses.

These studies also suggested additional erythrocyte-modifying properties of the alkali-treated LPS. Such cells show a gradual increase in intracellular sodium and decrease in intracellular potassium (Wallas et al 1979; Warren et al 1977). Since no changes were detected in erythrocyte glycolysis and no effects of LPS on the erythrocyte sodium-potassium ATPase activity were detected, the authors concluded that LPS was increasing the nonspecifically erythrocyte permeability to these ions, perhaps by disrupting the integrity of the erythrocyte membrane. Such an interpretation is clearly consistent with the data of Ciznar and Shands (1971) as well as with studies on the effects of LPS on artificial lipid bilayers (see Section 2). These results would, however, contrast with the results of intact LPS incorporated into liposomes where the major effect appears to be one of reduction of permeability to uncharged molecules (Davies et al 1978; Rottem 1978). An additional manifestation of the erythrocyte–LPS interaction studies reported by Warren et al (1983) has been the induction of rather pronounced morphological changes. Depending upon concentration, alkali-treated LPS induces progressive changes of erythrocytes from discocytes to echinocyte to spheroechinocyte. The authors suggest the rather interesting possibility that the alkali-treated LPS may incorporate selectively into the exterior half of the erythrocyte membrane phospholipid bilayer which then results in alterations of membrane curvature. Whatever the mechanism, these combined data provide considerable indirect evidence for an active interaction between LPS and the erythrocyte membrane.

Several recent studies designed to assess the temporal effects of native LPS on erythrocytes have suggested a very dynamic active interaction between LPS and the erythrocyte membrane. In an extensive series of experiments Davies et al (1978) studied the binding characteristics of LPS to a variety of mammalian cells, including erythrocytes. These authors reported a maximal binding of about 13–16 pg of LPS per erythrocyte. Based upon an aggregate LPS molecular weight of 10^6, this is equivalent to 10^7 molecules/cell; a figure orders of magnitude greater than that reported by Lüderitz et al (1958) and Ciznar and Shands (1971). It is of importance that these studies suggested that the binding profile of LPS as a function of erythrocyte number was sigmoidal and that LPS concentration and cell number influence the degree of binding. The authors concluded from these studies that both saturability and steric hindrance of bound LPS might prevent binding of additional LPS molecules.

Perhaps the most striking aspect of the LPS binding studies reported by Davies et al (1978) was the observation that the extent of LPS association with erythrocytes (and other cells as well) was cyclical with a characteristic periodicity. For example, in the case of LPS binding by rabbit erythrocytes, maximal binding was noted at 60, 135 and 215 minutes. A model in which different populations of LPS structures with characteristic structural features and erythrocyte membrane affinities was proposed to explain these results. (Although there is little evidence for different macromolecular populations of LPS, each with a distinct lipid-polysaccharide composition as originally proposed by Davies, there is nevertheless good evidence for such heterogeneous subunit structures within a homogeneous macromolecular LPS aggregate. Such a distinction does not alter the basic conceptual model proposed by Davies et al.) There is, in fact, recent suggestive evidence from several experimental systems (see below and Section 5) which would be entirely consistent with this interpretation. It may be that the novel concept of selective association of lipid-rich LPS subunits with mammalian cell membranes first proposed by Davies may be a generalized mechanism for such interactions.

Recent experiments by Carr and Morrison (1984a, 1984b) have also provided evidence for a dynamic interaction of LPS with erythrocyte membranes. Using LPS from the Re mutant of *S. minnesota*, these investigators made the unusual observation that rabbit erythrocytes, pretreated with LPS at 37 °C, would lyse upon the subsequent addition of polymyxin B. The order of addition was shown to be critical in that, as originally shown by Neter et al (1958b), polymyxin B added to LPS prevented attachment to the erythrocyte. Of particular interest in these studies was the demonstration that the ability of polymyxin B to lyse LPS-treated erythrocytes was both time- and temperature-dependent. At 37 °C, the rate of lysis was reduced to half if the erythrocytes were incubated with LPS for 1 minute prior to polymyxin B addition, and one fourth if the incubation time with LPS was 2 minutes. No lysis was observed if the preincubation time was extended to 10 minutes. The loss of reactivity to polymyxin B was shown to be critically dependent upon temperature; no loss of activity was noted at 4 °C. Since profound time-dependent changes in LPS–erythrocyte lytic responses to polymyxin B were observed in the absence of detectable changes in the binding association of LPS with the erythrocyte, these data were interpreted as time- and temperature-dependent changes in the nature of the association of LPS with the erythrocyte membrane.

The available evidence, therefore, would firmly support the concept that LPS can both actively and passively interact via nonspecific interactions with erythrocyte membranes. Lipid A, and perhaps lipid A-rich subunits contained within the LPS macromolecule, are essential for these interactions. Alkaline hydrolysis of LPS enhances these nonspecific interactions and may result in disruption of the erythrocyte membrane. The existence of LPS receptors on erythrocyte membranes would also argue strongly for specific LPS–erythrocyte membrane interactions. Whether active and/or passive nonspecific interactions influence specific interactions and vice versa is an interesting speculation for future studies.

4. THE INTERACTION OF LIPOPOLYSACCHARIDE WITH PLATELET MEMBRANES

There exists good experimental evidence, from both in vitro as well as in vivo studies, that endotoxins will affect platelet function (reviewed in Morrison and Ulevitch 1978). In the majority of in vitro studies, however, the manifestation of endotoxin effect(s) on platelets required the presence of plasma proteins. The available evidence would indicate, therefore, that in most instances endotoxin serves the function of an activator of plasma constituents (primarily complement). Although the requirement for an actual association of endotoxin with the platelet membrane in this type of indirect endotoxin-mediated platelet response is not readily apparent, experiments have suggested that such passive absorption does facilitate plasma-mediated effects. In addition, several investigators have reported direct effects of LPS on platelets which apparently require an active interaction of LPS with the platelet membrane. The results of these studies will be summarized in the following paragraphs. A detailed discussion of the consequences of LPS interaction with platelets can be found in Chapter 11.

As originally demonstrated by Spielvogel (1967), particulate preparations of endotoxin added to suspensions of rabbit platelets will adhere to the platelet membrane and initiate plasma-dependent morphological alterations in the platelet. These studies, as well as those of Des Prez and Bryant (1966) and Des Prez et al (1961), suggested the complement system as the plasma effector system, a fact later confirmed by Siraganian (1972). These results were extended by MacIntyre et al (1977) and Thorne et al (1977), who proposed that LPS initially bound to the rabbit platelet membrane by a mechanism which did not require calcium, followed by a calcium-requiring interaction of LPS with classical complement components leading to aggregation and secretion.

The more recent studies of Morrison et al (1978) and Morrison and Oades (1979) have provided experimental evidence for binding of LPS to the platelet as an important initial step in the rabbit platelet response. These investigators, however, indicated that complement activation via the alternative pathway leading to platelet lysis was the mechanism responsible for release of platelet constituents. Critical to these experiments was the demonstration that only LPS preparations which activated the alternative pathway were effective at initiating complement-dependent platelet lysis. Of importance, addition of polymyxin B, which binds to the lipid A region of LPS and blocks activation of the classical complement pathway by LPS but has no effect on activation of the alternative complement pathway by LPS (Morrison and Kline 1977), completely inhibited platelet lysis. Additional experiments clearly showed that the effect of polymyxin B was to inhibit the binding of LPS to the platelet membrane in the absence of any effect on alternative pathway activation. These results confirm the general mechanism for LPS–red blood cell interactions reported by Neter et al (1958a). As might be anticipated, mild alkaline hydrolysis of LPS enhanced complement-dependent platelet lysis and addition of lipid A inhibited activity, presumably by competitive binding to sites on the platelet membrane. Thus, these combined data suggest that LPS binds to the platelet via lipid A-dependent interactions. The polysaccharide portion of LPS then serves either as appropriate surface for binding activated C3 or as means of restricting Factor H and C3INA, resulting in complement-dependent platelet lysis.

As suggested above, binding of LPS to the rabbit platelet membrane is essential for the manifestation of complement-dependent responses. Since the primary function of the bound LPS is to passively focus activated complement components on the platelet membrane, it would appear that a passive absorption of LPS via hydrophobic interactions would be a sufficient requirement for the manifestation of activity. In consideration of potential mechanisms for binding, the available evidence does not provide strong support for the existence of LPS receptors on platelet membranes. As reported by Springer and Adye (1975), the most active LPS-binding substances isolated from platelets (and leukocytes) were glycerophosphatides and sphingomyelin, at least as assessed by inhibition of binding of LPS to erythrocytes. These investigators concluded that nonspecific hydrophobic interactions between LPS (lipid A) and platelet membrane lipids would account for most of the binding of LPS to platelets. In support of this conclusion, Morrison and Oades (1979) have suggested that the LPS–platelet association is characterized by a relatively low binding affinity, perhaps due to the limited accessibility of hydrophobic groups within the polysaccharide-containing LPS macromolecular aggregate. In any case, it can be concluded that LPS probably interacts with platelets via passive hydrophobic binding to the platelet membrane.

The results of a series of experiments reported by Hawiger and his colleagues (1975, 1977) have suggested that endotoxins may also actively interact with platelets. Using endotoxin modified by treatment with mild alkali, these investigators demonstrated a dose-dependent increase in secretion of human platelet granule constituents and generation of phospholipid procoagulant platelet factor 3. It was not reported whether a similar effect could be elicited with native endotoxin. Nevertheless, the fact that activity could be blocked with a partly purified membrane fraction suggested that the initial interaction was with the platelet membrane. The authors postulated that the alkaline-hydrolyzed endotoxin induced functional membrane changes in the platelet which they proposed could be accomplished by a molecular rearrangement of platelet membrane phospholipids.

Several recent studies reported by Morrison et al (1980, 1981) have also provided evidence for an active interaction of polysaccharide-deficient LPS preparations with rabbit and murine platelets. In these experiments, the in vitro incubation of washed platelets with *S. minnesota* Re LPS for periods of time up to 90 minutes in divalent cation-free buffer resulted in profound morphologic and functional changes in these cells. As determined by both scanning and transmission electron microscopy, LPS induced the formation of multiple pseudopodia and finger-like projections. These morphological changes appear to be similar to those reported for erythrocytes treated with alkaline-modified LPS described by Warren et al (1983). Importantly, however, the integrity of the platelet membrane was maintained and no loss of intracellular constituents was detected.

Experiments designed to assess functional changes in platelets revealed some rather striking time-dependent alterations following incubation with LPS. After 20 minutes, the addition of calcium to the platelets resulted in the initiation of a noncytotoxic secretory response and the release of granule constituents. These platelet responses were indistinguishable from those initiated by the usual spectrum of immunologic platelet stimuli. Responses were maximal after 90 minutes, and neither magnesium nor barium would substitute for or block the calcium-induced response.

Since the platelet response to calcium was dependent upon time of interaction of the platelet with LPS, and since binding of LPS to the platelet was independent of time, an active 'processing' of the LPS by the platelet membrane subsequent to binding was suggested. Additional evidence for an active interaction of LPS with the platelet membrane was provided by experiments documenting the temperature dependence of the platelet response. Acquisition of calcium-dependent secretion was markedly reduced at 30 °C and was totally abrogated at 23 °C.

It is noteworthy that these platelet responses to Re LPS show a striking similarity to the erythrocyte responses recently described by Carr and Morrison (1984a, 1984b). Thus platelets pretreated with LPS also become subject to lysis by polymyxin B, although the precise temporal relationship between LPS and polymyxin B additions are qualitatively distinct in the erythrocyte and platelet systems. Nevertheless, these combined data would not be inconsistent with the concept of a generalized mechanism for LPS–erythrocyte and LPS–platelet interaction. Thus the platelet, like the erythrocyte, can manifest both passive and active interactions with the lipid A region of LPS. Passive absorption interactions sensitize these cells for complement-mediated hemolysis, and active interactions result in detectable alterations in cellular morphology and function.

5. THE INTERACTION OF LIPOPOLYSACCHARIDE WITH LYMPHOCYTE MEMBRANES

Considerable interest in the mechanism by which LPS interacts with lymphocytes has been generated following the demonstration that LPS will act as a generalized mitogenic signal for the induction of murine B lymphocyte proliferation and differentiation into antibody secreting plasma cells (reviewed in Morrison and Ryan 1979). Stimulatory activity is limited to a subpopulation (approximate one third) of B lymphocytes (Melchers et al 1975), and T lymphocytes, in general, are not induced to proliferate. (An exception to the latter would be the recent report by Vogel et (1983) showing a small subpopulation of T lymphocytes responsive to LPS.) Although B lymphocytes from other species respond in varying degrees to the LPS mitogenic stimulus, it is probably fair to conclude that the mouse B lymphocyte is both one of the most sensitive cells to LPS mitogenic stimulation and one of the best studied with respect to membrane-related events leading to B cell triggering.

Central to such studies has been the question of the specificity of the interaction of LPS with B lymphocytes, particularly with respect to the question of LPS or lipid A receptors. Early studies designed to address this question focused upon potential differences in the interaction of LPS with B and T lymphocytes. However, the recognition and subsequent characterization of the C3H/HeJ (and more recently C57/B110ScN) mouse strain as a 'nonresponder' to the immunostimulatory effects of LPS has provided a powerful tool to explore mechanisms of interaction of LPS with B lymphocytes (as well as other cells). The results of several studies have provided very suggestive evidence for the presence of specific recognition sites on the membranes of LPS reactive cells (e.g. Coutinho et al 1978; Forni and Coutinho 1978; Jacobs et al 1983; Nygren et al 1979). Since these studies are discussed in Chapter 1, the following paragraphs will consider only experimental evidence which would support the concept of nonspecific interactions of LPS with lymphocytes. The actual relevance of such nonspecific interactions will be considered at the end of this section.

It is now well recognized that the lipid A portion of LPS is essential for B-cell responses (reviewed in Morrison and Ryan 1979), and chemical procedures which modify selectively lipid A also alter B-cell activity. As shown by Skidmore et al (1975), treatment of LPS with alkali to modify lipid A abrogates its mitogenic activity. Of particular importance is the observation by Goodman and Sultzer (1977) that alkaline hydrolysis results initially in a marked increase in biological activity of LPS which is followed at longer times by diminution in activity. As discussed earlier, lipid A has an affinity for membrane phospholipids, which is considerably increased by mild alkaline hydrolysis. In addition, as reported by Kabir and Rosenstreich (1977), phosphatidylethanolamine inhibited the binding of LPS to lymphocytes, a result similar to that obtained by Neter et al (1953) on LPS binding to erythrocytes. These observations, therefore, while not excluding other possible interpretations, would not be inconsistent with a nonspecific absorption of LPS to the B cell surface as an initial step perhaps contributing to the subsequent triggering event.

The majority of experiments designed to assess the potential specificity of the interaction of LPS with lymphocytes have until recently yielded only limited information on the nature of the LPS-binding sites. Many investigators have reported equivalent binding of LPS to thymocytes and to B lymphocytes (Bona et al 1976; Moller et al 1973); however, differential binding of purified lipid A to B cells has been noted (Zimmerman et al 1977). Similarly most investigators have noted equivalent binding of LPS to lymphoid cells from responder mice and C3H/HeJ nonresponder mice (Gregory et al 1980; Kabir and Rosenstreich 1977; Sultzer 1976; Watson and Riblet 1975). Thus, it is probably fair to conclude that binding to lymphoid cells can occur under circumstances where such binding does not lead to a cellular triggering response. Since it is unlikely that specific receptors for LPS would exist on all populations of cells which are genetically incapable of responding to LPS, these combined data would strongly suggest that such binding is nonspecific.

Although it is possible to demonstrate nonspecific binding of LPS to lymphocytes, the question of whether such binding is sufficient to initiate a lymphocyte response is more difficult to address. In this respect, the affinity of LPS for membranes in general, which results in nonspecific binding, may mask a more relevant specific interaction with LPS receptors. The recently published morphological analyses of LPS binding to lymphocytes has suggested that, at low concentrations of LPS, a selectivity of binding to responder versus nonresponder B cells (Nygren et al 1979) and B-cell subpopulations (Jacobs et al 1983) can be demonstrated. Whether such selective binding represents the presence of specific LPS receptors and/or alterations as surface charge, hydrophobic areas etc. remains to be determined. Of relevance to this point is the observation of Dumont and Barrios (1976) that B lymphocytes from the nonresponder C3H/HeJ mouse have a significantly lower electrophoretic mobility than B cells from the LPS-responder strains. In view of the contribution of net surface charge of phospholipids to their potential interaction with LPS, it is conceivable that the surface charge of the B cell would have a profound effect on both passive binding and active 'intercalation' of LPS into the membrane. Further, the fact that LPS preparations with reduced subunit heterogeneity can be prepared which do act as potent stimuli for B lymphocytes from C3H/HeJ mice (Vukajlovich and Morrison 1983) would underscore the potential importance of the tertiary structure of the LPS macromolecule in the interaction with the B-cell membrane independent of the presence of putative LPS receptors.

In any case, whether the interaction between LPS and the B cell is specific or nonspecific, it is assumed that the factors which govern an effective triggering signal are defined by membrane-localized components. The recent elegant experiments by Jakobovits et al (1982) have provided convincing evidence in support of this conclusion. Using membrane fusion techniques, the authors transferred LPS reactivity to normally unresponsive T lymphocytes and C3H/HeJ B lymphocytes with membrane fragments from LPS-responder B cells. These results clearly show that LPS reactivity resides in the cytoplasmic membrane compartment of the B cell.

Evidence has also been published to suggest that the interaction between LPS and lymphocytes is not restricted to a passive absorption but rather is an active dynamic event. (Since it is possible to consider the majority of these LPS-related membrane events within the framework of nonspecific interactions, I shall briefly summarize them in the following paragraphs.) In particular, there is evidence to suggest both a 'capping' phenomenon and/or reversible 'shedding' of bound LPS following initial binding, as well as a selective association of LPS subunit structures during the initial binding event.

As reported by Vukajlovich and Morrison (1983), LPS macromolecular aggregates with reduced subunit heterogeneity can be prepared which vary markedly in their capacity to initiate a B lymphocyte proliferative response. LPS macromolecules containing subunits rich in lipid A were shown in these studies to be the most active. Experiments by Goodman and Morrison (1984) and Goodman et al (1984) have recently documented that such LPS subunits contained within heterogeneous LPS aggregates associate selectively with lymphocytes. As shown by sodium dodecylsulfate polyacrylamide gel electrophoresis, lipid-rich subunits are selectively enriched in lymphocyte pellets following interaction with LPS. A similar type of selective association has also been reported by Jacobs et al (1983). Further, as reported by Goodman and Morrison (1983), unbound LPS is significantly altered in its capacity to interact with ionic detergents. These combined data make it highly unlikely that the total interaction between LPS and the lymphocyte membrane can be described as a passive absorption between these two constituents. The data rather suggest, but do not prove, that hydrophobic interactions subsequent to initial binding may take place at the membrane surface.

Support for this concept has been derived from a kinetic analysis of the fate of LPS following its association with the lymphocyte. Several investigators have reported that LPS is bound and subsequently released from the lymphocyte although the time period between binding and release has varied considerably in the hands of different investigators. Thus Symons and Clarson (1979) suggested a very rapid binding to pig lymphocytes (maximal binding at 3–5 minutes) followed by shedding which was virtually complete by 1 hour. Qualitatively similar results were obtained with murine lymphocytes and were equal with responder and nonresponder strains. A somewhat slower binding-and-release profile was reported by Davies et al (1978), who showed cyclical binding to lymphocytes similar to what has been described earlier (Section 3) by these authors for erythrocytes. The selective association with lymphocytes of LPS subunits described above would be entirely consistent with the model proposed by Davies et al (1978) to explain the cyclical binding phenomenon. In both instances, a hydrophobic interaction of LPS with membrane phospholipids and subsequent release of more hydrophilic subunits would provide a simple interpretation for these relatively unusual observations.

A much slower cycle of LPS accumulation and release was reported by Bona et al (1976); however, in these experiments release was preceded by polar accumulation of LPS (capping) and internalization. Of interest, binding and capping was observed on both B and T lymphocytes, but internalization was limited to the former cells. The results of Swartzwelder and Jacobs (1983) have confirmed these results and, in addition, reported an equivalent extent of capping on both responder and nonresponder B lymphocytes. Thus, while such a phenomenon would appear to be a consequence of LPS binding to lymphoid cells, the relevance is certainly not clear. In this respect, the lack of a relationship between surface marker capping and cell triggering events has been clearly documented (Loor 1974).

Finally, there is recent evidence to suggest actual changes in the physical chemical properties of the lymphocyte membrane following interaction with LPS. As originally reported by Kiefer et al (1980), LPS induces a significant decrease in membrane potential which is detectable 3 hours post-stimulation. Such an effect is not noted on T lymphocytes. These results have more recently been confirmed by Monroe and Cambier (1983), who showed that such effects correlate with the percentage of cells undergoing a G_0–G_1 transition. The latter authors postulate that the change in membrane potential may facilitate the opening of calcium channels.

In summary these combined data indicate that LPS can associate with lymphocytes by nonspecific interactions which may be classified as both active and passive. There is good evidence for dynamic LPS–membrane interactions subsequent to binding. At present, there is no compelling evidence to postulate an obligate LPS receptor in the mechanism of triggering of B lymphocytes by LPS.

6. THE INTERACTION OF LIPOPOLYSACCHARIDE WITH CELL MEMBRANES OF OTHER CELLS

It should be readily apparent from the preceding discussion that LPS has an effective capacity to interact with mammalian cell membranes. Although only the interaction of LPS with erythrocytes, platelets and lymphocytes has been considered in detail, there is, in addition, considerable evidence for binding to, and effects on a variety of other cells including, for example, endothelial cells (Harlan et al 1983), macrophages (Esser and Russell 1979; Haeffner-Cavaillon et al 1982), hepatocytes (Ramandori et al 1979), leukocytes (Gimber and Raffer 1969) and both normal and transformed fibroblasts (Bara et al 1973a, 1973b; Brailovsky et al 1977). These interactions will not be discussed in detail since to do so would exceed the space allocated for this review. In general, however, it is probably fair to conclude that the interactions of LPS with cell membranes from these and other cells may be characterized by one or more of the parameters described in detail in the preceding sections for platelets, erythrocytes and lymphocytes. There is evidence for both specific components of binding as well as nonspecific components. Experiments have also suggested alterations in membrane fluidity (Esser and Russell 1979), disruption of membrane integrity (Harlan et al 1983), and alterations of Con A effects (Brailovsky et al 1977). Many of these events would, therefore, in most instances appear not to be singularly unique to a given mammalian cell type but rather reflect the intrinsic affinity of LPS for mammalian cell membranes in general.

7. CONCLUSIONS

Bacterial lipopolysaccharides, isolated from the outer membrane of gram-negative bacteria, have a natural affinity for biological membranes. They exist in the bacterial outer membrane in asymmetric association with phospholipid as a trilaminar structure. When extracted and purified, they naturally form bilayers stabilized by both hydrophobic lipid A interactions and also, most probably, polysaccharide hydrogen bond interactions. When placed in the environment of a mammalian cell membrane, or its hydrophobic constituents, such as monolayers, bilayers, or liposomal vesicles, physical-chemical considerations would dictate detectable nonspecific LPS association with the hydrophobic surface.

The nature of the nonspecific binding interaction may be either passive absorption or active intercalation, depending upon the nature of the LPS (e.g. from rough *vs* smooth bacteria), the composition of its subunits, the physical-chemical properties of the hydrophobic surface (e.g. dispersions *vs* bilayers) and purely physical parameters such as temperature. It would appear reasonable to conclude that active intercalation of LPS (or its subunits) would require that the LPS has a greater affinity for the phospholipid structure than for itself. Such conditions might be achieved naturally when LPS interacts with the appropriate membrane. Alternatively, such interactions might be induced by heat or by treatment of LPS with mild alkali, both of which have been shown to alter interactions with phospholipids and/or membranes.

Although nonspecific absorption and/or intercalation of LPS or its subunit into mammalian cell membranes can clearly be documented with a variety of cell types, the actual relevance of such interactions to the cellular response evoked in LPS-responsive cells is more difficult to evaluate. Many of the phenomena described for LPS–membrane interactions appear to be properties of LPS-unresponsive as well as LPS-responsive cell types. It may be that such interactions are necessary, but not sufficient, for LPS-mediated triggering events. On the other hand, such interactions may be totally irrelevant to cell activation and serve primarily the purpose of distracting the investigator from more important specific interactions at the cell surface. Nevertheless, the potent membranophilic properties of lipopolysaccharides would appear to be more than fortuitous and, as discussed in this chapter, clearly still merit their serious consideration as nonspecific signals in the initiation of mammalian cell responses.

ACKNOWLEDGEMENT

The author acknowledges with appreciation the secretarial assistance of Mrs. Grace Vrell.

REFERENCES

Alving CR, Richardson EC (1984) Mitogenic activities of lipid A and liposome-associated lipid A: effects of epitope density. *Rev. Infect. Dis. 6*, 493-496.

Bader J, Teuber M (1973) Action of polymyxin B on bacterial membranes. I. Binding to O-antigen lipopolysaccharide of S. typhimurium. *Z. Naturforsch. Teil C 28*, 422-430.

Bara J, Lallier R, Brailovsky CA, Nigam VN (1973a) Fixation of a Salmonella minnesota R-form glycolipid on the membrane of normal and transformed rat embryo fibroblasts. *Eur. J. Biochem. 35*, 489-494.

Bara J, Lallier R, Trudel M, Brailovsky C, Nigam VN (1973b) Molecular models on the insertion of a Salmonella minnesota R-form glycolipid into the cell membrane of normal and transformed cells. *Eur. J. Biochem. 35*, 495-498.

Benedetto DA, Shands JW Jr, Shah DO (1973) The interaction of bacterial lipopolysaccharide with phospholipid bilayers and monolayers. *Biochim. Biophys. Acta 298*, 145-157.

Bona C, Juy D, Truffa-Bachi P, Kaplan GJ (1976) Binding, capping and internalization of lipopolysaccharide in thymic and nonthymic lymphocytes of the mouse. Biological and autoradiographic studies. *J. Microbiol. Cell 25*, 47-56.

Brailovsky CA, Nigam VN, Pesant S (1977) Cell surface mobility of a bacterial R-form glycolipid, mR595 after binding to rat cells, and its effect on the mobility of concanavalin A receptors. *Exp. Cell Res. 109*, 389-395.

Burge RE, Draper JC (1967) The structure of the cell wall of the Gram-negative bacterium Proteus vulgaris III. A lipopolysaccharide 'Unit Membrane'. *J. Mol. Biol. 28*, 205-210.

Carr C Jr, Morrison DC (1984a) Lipopolysaccharide interaction with rabbit erythrocyte membranes. *Infect. Immun. 43*, 600-606.

Carr C Jr, Morrison DC (1984b) A two step mechanism for the interaction of Re-lipopolysaccharide with erythrocyte membranes. *Rev. Infect. Dis. 6*, 497-500.

Ciznar I, Shands JW Jr (1971) Effect of alkali treated lipopolysaccharide on erythrocyte membrane stability. *Infect. Immun. 4*, 362-367.

Coutinho A, Forni L, Watanabe T (1978) Genetic and functional characterization of an antiserum to the lipid A-specific triggering receptor on murine B-lymphocytes. *Eur. J. Immunol. 8*, 63-67.

Davies M, Stewart-Tull DES, Jackson DM (1978) The binding of lipopolysaccharide from E. coli to mammalian cell membranes and its effect on liposomes. *Biochim. Biophys. Acta 508*, 260-276.

Des Prez RM, Bryant RE (1966) Effects of bacterial endotoxin on rabbit platelets IV. The divalent ion requirement of endotoxin-induced and immunologically induced platelet injury. *J. Exp. Med. 124*, 971-982.

Des Prez RM, Horowitz HI, Hook EW (1961) Effects of bacterial endotoxin on rabbit platelets I. Platelet aggregation and release of platelet factors in vitro. *J. Exp. Med. 114*, 857-874.

Dumont F, Barrois R (1976) Electrokinetic properties of splenic lymphocytes from the low lipopolysaccharide-responder C3H/HeJ mice. *Folia Biol. (Prague) 12*, 145-150.

Esser AF, Russell SW (1979) Membrane perturbation of macrophages stimulated by bacterial lipopolysaccharides. *Biochem. Biophys. Res. Commun. 87*, 532-540.

Forni L, Coutinho A (1978) An antiserum which recognizes lipopolysaccharide reactive B-cells in the mouse. *Eur. J. Immunol. 8*, 56-62.

Freer JH, Salton MRJ (1971) The anatomy and chemistry of Gram negative cell envelopes.

In: Weinbaum G, Kadis S, Ajl SJ (Eds), *Microbial Toxins Vol 4: Bacterial Endotoxins*, pp 67-126. Academic Press, New York.

Funahara Y, Nikaido H (1980) Asymmetric localization of lipopolysaccharides on the outer membrane of S. typhimurium. *J. Bacteriol. 141*, 1463-1465.

Galanos C (1975) Physical state and biological activity of lipopolysaccharides. Toxicity and immunogenicity of the lipid A component. *Z. Immunitätsforsch. 149S*, 214-229.

Galanos C, Lüderitz O (1975) Electrodialysis of lipopolysaccharides and their conversion to uniform salt forms. *Eur. J. Biochem. 54*, 603-610.

Galanos C, Lüderitz O, Rietschel ETh, Westphal O (1977) Newer aspects of the chemistry and biology of bacterial lipopolysaccharides with special reference to their lipid A component. In: Goodwin TW (Ed), *International Review of Biochemistry, Vol 14: Biochemistry of Lipids II*, pp 239-335. University Park Press, Baltimore, MD.

Gimber PE, Rafter GW (1969) The interaction of E. coli endotoxin with leukocytes. *Arch. Biochem. Biophys. 135*, 14-20.

Goldman RC, Leive L (1980) Heterogeneity of antigenic side chain length in lipopolysaccharides in E. coli O111 and S. typhimurium LT2. *Eur. J. Biochem. 107*, 145-153.

Goodman SA, Morrison DC (1983) Dissociation of lipopolysaccharides by murine spleen cell membranes. *Fed. Proc. 42*, 410.

Goodman SA, Morrison DC (1984) Selective association of lipid-rich R-like lipopolysaccharide (LPS) subunits with murine spleen cells. *Mol. Immunol. 21*, 684-697.

Goodman GW, Sultzer BM (1977) Mild alkaline hydrolysis of lipopolysaccharide endotoxin enhances its mitogenicity for murine B-cells. *Infect. Immun. 17*, 205-214.

Goodman SA, Vukajlovich S, Munkenbeck P, Morrison DC (1984) Selective interaction of lipid A subunits in lipopolysaccharide macromolecular aggregates with lymphocytes. *Rev. Infect. Dis. 6*, 511-518.

Gray GW, Wilkinson SG (1965) The effect of ethylenediaminetetraacetic acid on the cell walls of some Gram negative bacteria. *J. Gen. Microbiol. 39*, 385-399.

Gregory SH, Zimmerman DH, Kern M (1980) The lipid A moiety of LPS is specifically bound to B cell subpopulations of responder and nonresponder animals. *J. Immunol. 125*, 102-107.

Haeffner-Cavaillon N, Chaby R, Cavaillon JM, Szabó L (1982) Lipopolysaccharide receptor on rabbit peritoneal macrophages I. Binding characteristics. *J. Immunol. 128*, 1950-1954.

Hammerling U, Westphal O (1967) Synthesis and use of O-stearoyl polysaccharides in passive hemagglutination and hemolysis. *Eur. J. Biochem. 1*, 46-50.

Harlan JM, Harker LA, Reidy MA, Gajdusek CM, Schwartz SM, Striker GE (1983) Lipopolysaccharide-mediated bovine endothelial cell injury in vitro. *Lab. Invest. 48*, 269-274.

Hawiger J, Hawiger A, Timmons S (1975) Endotoxin-sensitive membrane component of human platelets. *Nature (London) 256*, 125-127.

Hawiger J, Hawiger A, Stekley S, Timmons S, Cheng C (1977) Membrane changes in human platelets induced by lipopolysaccharide endotoxin. *Br. J. Haematol. 35*, 285-299.

Jacobovits A, Sharon N, Zan-Bar I (1982) Acquisition of mitogenic responsiveness by nonresponder lymphocytes upon insertion of appropriate membrane components. *J. Exp. Med. 156*, 1274-1279.

Jacobs DM, Roberts DB, Eldridge JH, Rosenspire AJ (1983) Binding of lipopolysaccharides to murine lymphocytes (Abstract). *Fed. Proc. 42*, 413.

Jann B, Reske K, Jann K (1975) Heterogeneity of lipopolysaccharides: analysis of polysaccharide chain lengths by SDS-polyacrylamide gel electrophoresis. *Eur. J. Biochem. 60*,

239-246.

Kabir S, Rosenstreich DL (1977) Binding of bacterial endotoxin to murine spleen lymphocytes. *Infect. Immun. 15*, 156-164.

Kataoka T, Inoue K, Lüderitz O, Kinsky SL (1971) Antibody and complement dependent damage to liposomes prepared with bacterial lipopolysaccharides. *Eur. J. Biochem. 21*, 80-85.

Kiefer H, Blume AJ, Kaback HR (1980) Membrane potential changes during mitogenic stimulation of mouse spleen cells. *Proc. Natl. Acad. Sci. USA 77*, 2200-2204.

Leive L, Shovlin V, Mergenhagen SE (1968) Physical, chemical and immunological properties of lipopolysaccharides released from E. coli by ethylenediaminetetraacetate. *J. Biol. Chem. 243*, 6384-6391.

Loor F (1974) Binding and redistribution of lectins on lymphocyte membrane. *Eur. J. Immunol. 4*, 210-220.

Lüderitz O, Westphal O, Sievers K, Kroger E, Neter E, Braun OH (1958) Uber die Fixation von ^{32}P-markierten Lipopolysaccharides (Endotoxin) aus E. coli an menschlichen Erythrocyten. *Biochem. Z. 330*, 34-46.

Lui MS, Onji T, Snelgrove NE (1982) Changes in phase transition temperature of phospholipids induced by endotoxin. *Biochim. Biophys. Acta 710*, 248-251.

MacIntyre DE, Allen AP, Thorne KJI, Glavert AM, Gordon JL (1977) Endotoxin-induced platelet aggregation and secretion I. Morphological changes and pharmacological effects. *J. Cell Sci. 28*, 218-225.

Magnusson KE, Stendahl O, Tagesson C, Edebo L, Johansson G (1977) The tendency of smooth and rough Salmonella typhimurium bacteria and LPS to hydrophobic and ionic interaction, as studied in aqueous polymer two phase systems. *Acta Pathol. Microbiol. Scand. Sect. B 85*, 212-218.

Melchers F, von Boehmer H, Phillips RA (1975) B-lymphocyte subpopulations in the mouse: organ distribution and ontogeny of immunoglobulin-synthesizing and of mitogen sensitive cells. *Transplant. Rev. 25*, 26-58.

Moller G, Andersson J, Pohlit H, Sjoberg O (1973) Quantitation of the number of mitogen molecules activating DNA synthesis in T and B lymphocytes. *Clin. Exp. Immunol. 13*, 89-99.

Monroe JG, Cambier JC (1983) B cell activation I. Anti-immunoglobulin-induced receptor cross-linking results in a decrease in the plasma membrane potential of murine B-lymphocytes. *J. Exp. Med. 157*, 2073-2086.

Morrison DC, Jacobs DM (1976) Binding of polymyxin B to the lipid A portion of bacterial lipopolysaccharides. *Immunochemistry 13*, 813-818.

Morrison DC, Kline LF (1977) Activation of the classical and properdin pathways of complement by bacterial lipopolysaccharides. *J. Immunol. 118*, 362-368.

Morrison DC, Oades Z (1979) Mechanism of lipopolysaccharide-initiated rabbit platelet response II. Evidence that lipid A is responsible for binding of lipopolysaccharides to the platelet. *J. Immunol. 122*, 753-758.

Morrison DC, Rudbach JA (1981) Endotoxin-cell membrane interactions leading to transmembrane signaling. In: Inman FP, Mandy WJ (Eds), *Contemporary Topics in Molecular Immunology, Vol 8*, pp 187-218. Plenum Publishing Company, New York.

Morrison DC, Ryan JL (1979) Bacterial endotoxins and host immune responses. *Adv. Immunol. 28*, 293-450.

Morrison DC, Ulevitch RJ (1978) The effects of bacterial endotoxins on host mediation systems. *Am. J. Pathol. 93*, 527-618.

Morrison DC, Kline LF, Oades ZG, Henson PM (1978) Mechanism of lipopolysaccharide-initiated rabbit platelet responses. I. Alternative pathway dependence of the lytic response. *Infect. Immun. 20*, 744-751.

Morrison DC, Oades ZG, Betz SJ (1980) The role of lipid A and lipid A-associated protein in cell degranulation mechanisms. In: Eaker D, Wadstrom T (Eds), *Natural Toxins*, pp 287-294. Pergamon Press, New York.

Morrison DC, Oades ZG, DiPietro D (1981) Endotoxin-initiated membrane changes in rabbit platelets. In: Majde JA, Person RJ (Eds), *Pathophysiological Effects of Endotoxins at the Cellular Level*, pp 47-64. Alan R. Liss, New York.

Mühlradt PF, Golecki JR (1975) Asymmetric distribution and artificial reorientation of LPS in the outer membrane bilayer of S. typhimurium. *Eur. J. Biochem. 51*, 343-352.

Mühlradt PF, Menzel J, Golecki JR, Speth V (1974) Lateral mobility and surface density of lipopolysaccharides in the outer membrane of S. typhimurium. *Eur. J. Biochem. 43*, 533-539.

Nakamura K, Mizushima S (1975) In vitro reassembly of the membranous vesicle from E. coli outer membrane components: role of individual components and magnesium ions in reassembly. *Biochim. Biophys. Acta 413*, 371-393.

Neter E, Zalewski NJ, Zak DA (1953) Inhibition by lecithin and cholesterol of bacterial (Escherichia coli) hemagglutination and hemolysis. *J. Immunol. 71*, 145-151.

Neter E, Westphal O, Lüderitz O, Gorzynski EA, Eichenberger E (1956) Studies of enterobacterial lipopolysaccharides – effects of heat and chemicals on erythrocyte-modifying, antigenic, toxic and pyrogenic properties. *J. Immunol. 76*, 377-385.

Neter E, Gorzynski EA, Westphal O, Lüderitz O (1958a) Effects of antibiotics on enterobacterial lipopolysaccharide (endotoxins) and hemolysis. *J. Immunol. 80*, 66-72.

Neter E, Gorzynski EA, Westphal O, Lüderitz O, Klumpp DJ (1958b) The effects of protamine and histone on enterobacterial lipopolysaccharide and hemolysis. *Can. J. Microbiol. 4*, 371-383.

Nikaido H, Nakae T (1973) Permeability of model membranes containing phospholipids and lipopolysaccharides: some preliminary results. In: Kass EH, Wolff SM (Eds), *Bacterial Lipopolysaccharides*, pp 22-26. University of Chicago Press, Chicago, IL.

Nikaido H, Nakae T (1979) The outer membrane of gram negative bacteria. *Adv. Microb. Physiol. 20*, 163-250.

Nikaido H, Takeuchi Y, Ohnishi SH, Nakae T (1977) Outer membrane of Salmonella typhimurium – electron resonance studies. *Biochim. Biophys. Acta 465*, 152-164.

Niwa M, Milner KC, Ribi E, Rudbach JA (1969) Alteration of physical, chemical and biological properties of endotoxin by treatment with mild alkali. *J. Bacteriol. 97*, 1069-1077.

Nowotny A (1969) Molecular aspects of endotoxic reactions. *Bacteriol. Rev. 33*, 72-98.

Nygren H, Dahlen G (1979) Ultrastructural localization of lipopolysaccharide binding sites with peroxidase conjugated lipopolysaccharide. *J. Immunol. Methods 25*, 355-364.

Nygren H, Dahlen G, Moller G (1979) Bacterial lipopolysaccharide bind selectively to lymphocytes from lipopolysaccharide high responder mouse strains. *Scand. J. Immunology. 10*, 555-561.

Onji T, Liu MS (1981) Effect of E. coli endotoxin on the leakage of ^{14}C-sucrose from phosphatidylcholine liposomes. *Circ. Shock. 8*, 403-410.

Palva ET, Mäkelä PH (1980) Lipopolysaccharide heterogeneity in Salmonella typhimurium analyzed by sodium dodecyl sulfate/polyacrylamide gel electrophoresis. *Eur. J. Biochem. 107*, 137-144.

Pangburn MK, Morrison DC, Schreiber RD, Muller-Eberhard HJ (1980) Activation of the alternative complement pathway: recognition of surface structures on activators by bound C3b. *J. Immunol. 124*, 977-982.

Ramandori G, Hopf U, Meyer zum Buschenfelde KH (1979) Binding sites for endotoxin lipopolysaccharides on the plasma membrane of isolated rabbit hepatocytes. *Acta Hepatogastroenterol. 26*, 368-374.

Richardson EC, Banerji B, Seid RC Jr, Levin J, Alving CR (1983) Interactions of lipid A and liposome-associated lipid A with limulus polypherus amoebacytes. *Infect. Immun. 39*, 1385-1391.

Rietschel ETh, Gottert H, Lüderitz O, Westphal O (1972) Nature and linkages of the fatty acids present in the lipid A component of Salmonella lipopolysaccharides. *Eur. J. Biochem. 28*, 166-173.

Romeo D, Girard A, Rothfield L (1970a) Reconstitution of a functional membrane enzyme system in a monomolecular film. I. Formation of a mixed monolayer of lipopolysaccharide and phospholipid. *J. Mol. Biol. 53*, 475-490.

Romeo D, Hinckley A, Rothfield L (1970b) Reconstitution of a functional membrane enzyme system in a monomolecular film. II. Formation of a functional ternary film of lipopolysaccharide, phospholipid and transferase enzyme. *J. Mol. Biol. 53*, 491-501.

Rosenstreich DL, Blumenthal R (1977) Ionophorous activity and murine B lymphocyte mitogens. *J. Immunol. 118*, 129-136.

Rothfield L, Horne RW (1967) Reassociation of purified lipopolysaccharide and phospholipid of the bacterial cell envelope: electron microscopic and monolayer studies. *J. Bacteriol. 93*, 1075-1721.

Rothfield L, Pearlman M (1966) The role of cell envelope phospholipid in the enzymatic synthesis of bacterial lipopolysaccharide: structural requirements of the phospholipid molecule. *J. Biol. Chem. 241*, 1386-1392.

Rothfield L, Takeshita M (1965) The role of the cell envelope phospholipid in the enzymatic synthesis of bacterial lipopolysaccharide: binding of transferase enzymes to a lipopolysaccharide : lipid complex. *Biochem. Biophys. Res. Commun. 20*, 521-527.

Rottem S (1978) The effect of lipid A on the fluidity and permeability properties of phospholipid dispersions. *FEBS Lett. 95*, 121-124.

Rottem S, Leive L (1977) Effect of variations in lipopolysaccharide on the fluidity of the outer membrane of E. coli. *J. Biol. Chem. 252*, 2077-2081.

Rudbach JA, Anacker RL, Haskins WT, Johnson AG, Milner KC, Ribi E (1966) Physical aspects of reversible inactivation of endotoxin. *Ann. N.Y. Acad. Sci. 133*, 629-643.

Rudbach JA, Milner KC, Ribi E (1967) Hybrid formation between bacterial endotoxins. *J. Exp. Med. 126*, 63-79.

Schindler M, Osborn MJ (1979) Interaction of divalent cations and polymyxin B with lipopolysaccharides. *Biochemistry 18*, 4425-4430.

Schuster BG, Palmer RF, Aronson RS (1970) The effect of endotoxin on thin lipid bilayer membranes. *J. Membr. Biol. 3*, 67-72.

Shands JW Jr (1971a) Evidence for a bilayer structure in gram-negative lipopolysaccharide: relationship to toxicity. *Infect. Immun. 4*, 167-172.

Shands JW (1971b) The physical structure of bacterial lipopolysaccharides. In: Weinbaum G, Kadis S, Ajl SJ (Eds), *Microbial Toxins, Vol 4: Bacterial Endotoxins*, pp 127-144. Academic Press, New York.

Shands JW (1973) Affinity of endotoxin for membranes. In: Kass EH, Wolff SM (Eds), *Bacterial Lipopolysaccharides*, pp 189-193. University of Chicago Press, Chicago, IL.

Shands JW, Chen PW (1980) The dispersion of Gram negative lipopolysaccharides by deoxycholate-subunit molecular weight. *J. Biol. Chem. 255*, 1221-1226.

Shands JW, Graham JA, Nath K (1967) The morphologic structure of isolated bacterial LPS. *J. Mol. Biol. 25*, 15-21.

Siraganian RP (1972) Platelet requirement in the interaction of the complement and clotting systems. *Nature New Biology 239*, 208-210.

Skidmore BJ, Chiller JM, Morrison DC, Weigle WO (1975) Immunologic properties of bacterial lipopolysaccharide (LPS): correlation between the mitogenic, adjuvant and immunologic activities. *J. Immunol. 114*, 770-775.

Spielvogel AR (1967) An ultrastructural study of the mechanisms of platelet-endotoxin interaction. *J. Exp. Med. 126*, 235-250.

Springer GF, Adye JC (1975) Endotoxin binding substances from human leukocytes and platelets. *Infect. Immun. 12*, 978-986.

Springer GF, Huprikar SV, Neter E (1970) Specific inhibition of endotoxin coating of red cells by a human erythrocyte membrane component. *Infect. Immun. 1*, 98-108.

Springer GD, Adye JC, Bezkorovainy A, Jirgensons B (1974) Properties and activity of the lipopolysaccharide receptor from human erythrocytes. *Biochemistry 13*, 1379-1389.

Sultzer BM (1976) Genetic analysis of lymphocyte activation by lipopolysaccharide endotoxin. *Infect. Immun. 13*, 1579-1584.

Swartzwelder F, Jacobs DM (1983) Lipopolysaccharide capping on murine lymphocytes (Abstract). *Fed. Proc. 42*, 413.

Symons DBA, Clarson CA (1979) The binding of lipopolysaccharide to the lymphocyte surface. *Immunology 38*, 503-508.

Takeuchi Y, Nikaido H (1981) Persistence of segregated phospholipid-lipopolysaccharide mixed bilayers: studies with spin labeled phospholipids. *Biochemistry 20*, 523-529.

Thompson RA, Lachmann PJ (1970) Reactive lysis: the complement mediated lysis of unsensitized cells I. The characterization of the indicator factor and its identification as C7. *J. Exp. Med. 131*, 629-642.

Thorne KJ, Oliver RC, MacIntyre DE, Gordon JL (1977) Endotoxin induced platelet aggregation and secretion II. Changes in plasma membrane proteins. *J. Cell. Science 28*, 225-236.

Tripodi D, Nowotny A (1966) Relation of structure to function in bacterial O antigens. V. Nature of the active sites in endotoxic lipopolysaccharides of Serratia marcescens. *Ann. N.Y. Acad. Sci. 133*, 604-621.

Vogel SN, Hilfiker ML, Caulfield MJ (1983) Endotoxin-induced T-lymphocyte proliferation. *J. Immunol. 130*, 1774-1779.

Vukajlovich SW, Morrison DC (1983) Conversion of lipopolysaccharide to molecular aggregates with reduced subunit heterogeneity: demonstration of lipopolysaccharide-responsiveness in 'endotoxin-unresponsive' C3H/HeJ splenocytes, *J. Immunol. 130*, 2804-2808.

Wallas CH, Warren JR, Kowalski MM (1979) Energy metabolism and Na^+, K^+ redistribution in human erythrocytes treated with lipopolysaccharide-endotoxin. *Proc. Soc. Exp. Biol. Med. 161*, 255-259.

Warren JR (1982) Polymyxin B suppresses the endotoxin inhibition of Concanavalin A-mediated erythrocyte agglutination. *Infect. Immun. 35*, 594-599.

Warren JR, Kowalski MM (1977) Inhibition of Con A erythroagglutination by alkali-treated lipopolysaccharide. *Exp. Cell Res. 107*, 462-466.

Warren JR, Kowalski MM, Wallas CH (1977) Effect of alkali-treated lipopolysaccharide on the intracellular cations of human erythrocytes. *Infect. Immun. 17*, 389-394.

Warren JR, Harris AS, Wallas CH (1983) Transformation of human erythrocyte shape by endotoxic lipopolysaccharide. *Infect. Immun. 39*, 431-434.

Watson J, Riblet R (1975) Genetic control of responses to bacterial lipopolysaccharides in mice. II. A gene that influences a membrane component involved in the activation of bone marrow-derived lymphocytes by lipopolysaccharide. *J. Immunol. 114*, 1462-1468.

Weiser M, Rothfield L (1968) The reassociation of lipopolysaccharide, phospholipid and transferase enzymes of the bacterial cell envelope. *J. Biol. Chem. 243*, 1320-1328.

Yokoyama K, Mashimo J, Kasai N, Teraco T, Osawa T (1979) Binding of bacterial lipopolysaccharide to histocompatibility-2-complex proteins of mouse lymphocytes. *Hoppe-Seyler's Z. Physiol. Chem. 360*, 587-595.

Zimmerman DH, Gregory S, Kern M (1977) Differentiation of lymphoid cells: the preferential binding of the lipid A moiety of lipopolysaccharide to B lymphocyte populations. *J. Immunol. 119*, 1018-1023.

Handbook of Endotoxin, Vol. 3: Cellular Biology of Endotoxin
L.J. Berry, editor
© Elsevier Science Publishers B.V., 1985

CHAPTER 3

In vivo distribution and detoxification of endotoxins

ROBERT C. SKARNES

1. INTRODUCTION

During the past 30 years, a great many publications have appeared on the subject of the disposition of bacterial endotoxins in the experimental animal. Two principal aims of these studies have been to relate in vivo distribution patterns of endotoxin to known pathophysiologic effects and to ascertain major sites of detoxification. The earlier literature on endotoxin distribution was reviewed by Braude (1964) and host mechanisms of detoxification by Atkins (1960) and Skarnes and Rosen (1971).

The literature covered in this article deals mainly with studies on endotoxins isolated from smooth-phase microorganisms since these are the toxins most likely encountered by the host in naturally acquired enteric infections. Particular emphasis is placed on the continuing polemic regarding the relative importance of humoral as contrasted to reticuloendothelial mechanisms of detoxification.

2. DISTRIBUTION OF ENDOTOXINS IN THE HOST

There is no doubt that fixed phagocytes of the reticuloendothelial system (RES) exert a major influence on the removal of enteric microorganisms and the toxic components associated with the bacterial cell. Furthermore, it is reasonable to conclude that RES tissues of the normal animal function effectively in the clearance and processing of the small quantities of endotoxins which become detached from the cell during brief encounters with gram-negative bacilli. However, most in vivo distribution studies have concentrated on the fate of partly purified endotoxins following the intravenous administration of moderate to large doses. In these situations, the extent of RES participation is a moot issue. Endotoxins given by the intravenous route are partitioned between the blood compartment and the major reticuloendothelial organs within a short time, although the amounts accumulated in different sites vary considerably. Major determinants influencing results and subsequent interpretations include the source and physical state of the endotoxin, the dose administered and the sensitivity and accuracy of detection methods.

56

2.1. Immunologic and biologic methods

A number of investigators have followed the distribution of endotoxins in experimental animals using specific immunologic methods. These studies have revealed a broad dissemination of antigenic moieties in the walls of peripheral blood vessels and in various phagocytic cells, particularly those of the RES. As expected, the liver was shown to be the paramount organ of entrapment, followed in importance by the spleen and lung.

Using the Coons direct immunofluorescent method, Cremer and Watson (1957) studied the phagocytosis of endotoxin by the RES of rabbits. During a 2-day period after intravenous injection, the somatic antigen was found in Kupffer cells of the liver and in phagocytic cells in both the red pulp of the spleen and the alveoli of the lung. In similar experiments in mice, Golub et al (1968) confirmed these findings and noted after the first hour a gradual decline in the endotoxin levels of RES tissues; they were unable to detect the antigen in spleen macrophages and Kupffer cells beyond 24 hours. By contrast, Tanaka et al (1959) reported the continuing, though diminishing, presence of endotoxins for up to 1 week in the vascular endothelium, 6 weeks in splenic macrophages and up to 12 weeks in Kupffer cells of mice receiving sublethal amounts of endotoxin. Several other tissues showed little or no accumulation of endotoxin. The authors used an indirect immunofluorescence procedure which they claimed was more sensitive than the direct method.

In a recent report, Freudenberg et al (1982) used an immunoperoxidase procedure to study RES distribution of both S (smooth) phase and R (rough) phase lipopolysaccharides (LPS) as a function of time after administration. Within 30 minutes of injection R-LPS was found in the liver, partitioned between Kupffer cells and hepatocytes. On the other hand, S-LPS was slow to accumulate in Kupffer cells (> 2 hr) and was not found in hepatocytes until 3 days post-injection. Antigenic moieties were found in bronchial and alveolar macrophages 7–24 hours later, but little LPS was detected in the kidney during the first 24 hours. Other immunologic methods used to ascertain the presence of endotoxin antigens in tissues or fluids include passive hemagglutination inhibition (Braude 1964; Golub et al 1968) and crossed immunoelectrophoresis (Freudenberg et al 1980).

These studies are in general agreement regarding the major sites of endotoxin deposition but they lack quantitation and do not allow a distinction between toxic and nontoxic antigens. However, an immunodiffusion method has been developed to identify both toxic and nontoxic fractions of endotoxin in the circulating plasma from endotoxin-challenged animals (Chedid et al 1963; Skarnes and Chedid 1964).

Sensitive biologic tests based on toxicity have also been employed to follow blood clearance and/or liver uptake of endotoxins. These include pyrogenicity (Golub et al 1968), the Limulus lysate test (Das et al 1973), and lethality in adrenalectomized (Chedid et al 1963) or actinomycin D-treated (Trejo et al 1972) mice. Other assay methods include lethal toxicity in lead acetate-treated mice (Rippe et al 1974) and rats (Buchanan and Filkins 1976a). In the latter report, blood clearance of the toxic fraction was shown to be delayed, with about 50% present in the circulating plasma 1 hour after intravenous injection. Such methods are of value in determining the presence of biologically active toxins in tissues and fluids and they indicate possible in vivo sites of detoxification. However, data obtained by these methods must be evaluated with caution, unless accompanied by other means of assay or identification of the endotoxin. For example, partial or complete inactivation by plasma and/or tissue components would lead to under-estimates of actual concentrations since nontoxic fragments would not be included. On the other hand, tissue injury induced by endotoxins may lead to the elaboration by the host of toxic metabolites or mediators which could confound interpretations of results.

2.2. Radioisotopes as markers of endotoxin distribution

A major impediment to understanding mechanisms of host detoxification is the difficulty of ascertaining the quantitative distribution of biologically active fractions. The many attempts to do so with the use of isotopically labeled preparations, while providing some useful data, have more often yielded misinformation. Moreover, in only a limited number of studies have attempts been made to discover whether 'endotoxins' detected by radioactivity counts in tissues and fluids are associated with antigenic and biologic activities.

From the published records of the past 25 years, it is apparent that most investigators have chosen to use trichloracetic acid-extracted endotoxins (Boivin preparations) from smooth-phase bacteria for in vivo isotope distribution studies. From earlier studies with crude endotoxins or LPS, it became apparent that the degree of aggregation of the toxin dramatically influenced subsequent distribution in experimental animals. For example, aqueous suspensions of phenol-water-extracted LPS, which appear in the electron microscope as long filamentous particles (Rudbach et al 1966), are cleared from the circulation and appear in the liver in very short order (Gans 1975; Zlydaszyk and Moon 1976). On the other hand, the Boivin preparations form stable colloidal solutions by virtue of their smaller particle size and protein content. When further purified by ultracentrifugation to remove O-polysaccharide fragments and other debris, labeled Boivin preparations exhibit delayed blood clearance and smaller accumulations in RES tissues (Chedid et al 1963).

This review does not include the relatively few tracer studies in which crude phenol-extracted LPS has been used. It is sufficient to note that blood clearance and RES uptake of the latter parallel results obtained with R preparations. However, recent modifications of the original phenol extraction procedure have resulted in purified LPS preparations of smaller and more uniform particle size (Galanos and Lüderitz 1975; Morrison and Leive 1975) which appear to be more suitable for in vivo distribution experiments. For example, blood clearance of finely dispersed preparations of intrinsically labeled LPS is significantly retarded (Munford et al 1981b), particularly if the LPS is administered in a dispersing agent such as triethylamine (Freudenberg et al 1980).

2.2.1. Studies with extrinsic labels

The most widely used methods for determining blood clearance and tissue distribution of S-endotoxins have employed a variety of radioisotope labels. In the earliest published study by Seligman et al (1948), ^{131}I was used to label deproteinized LPS from *Serratia marcescens*. Intravenous injections of sublethal doses in mice resulted in biphasic blood clearance and a broad distribution of the label 1–6 hours later, with most being found in the liver (30%), blood (20%), lung (9%) and kidney (6%). The in vivo stability of the ^{131}I-label is questionable, however, since 24-hour collections of urine contained about 15% of the label. In addition, mouse liver extracts were shown to detach a significant portion of radioactivity during in vitro test incubations.

In another early study with ^{131}I-tagged endotoxin, Barnes et al (1952) reported widespread distribution of the label 3–24 hours after intravenous injection into mice. Large accumulations were found in the liver and lesser amounts in the spleen, lung and other organs. Again, urine was shown to contain large amounts of dialyzable radioactivity. Clearly, such studies, in the absence of other criteria, are of limited value for the quantitative determination of in vivo sites of endotoxin deposition.

Braude et al (1955) introduced the use of ^{51}Cr as a label for *Escherichia coli* endotoxin to study tissue distributions in normal mice and rabbits. Following the intravenous administration of large doses of the ^{51}Cr-labeled endotoxin, a rapid phase of blood clearance was followed by a slow phase which lasted for several hours. During this time, liver uptake of the label ranged from 12 to 17%, and the buffy coat fraction contained 10–12% of the injected dose. Blood clearance and hepatic uptake of radioactivity were quite different in animals receiving small, sublethal doses (10–40 µg). Approximately 90% of the injected dose was cleared from the circulation within 5 minutes, whereas 60% appeared in the liver within 15 minutes (Carey et al 1958).

Subsequent studies by Chedid et al (1963, 1964) and Skarnes and Chedid (1964) with Na$_2$^{51}CrO$_4$-labeled endotoxin from *Salmonella enteritidis* showed that the isotope was attached only to the high molecular weight, toxic antigen. Sedimentation of the toxic fraction by ultracentrifugation permitted its separation from the unlabeled, nontoxic moieties of O polysaccharides which remained in the supernatant. The latter are present in all crude preparations examined in this laboratory, whether extracted by phenol-water or trichloracetic acid procedures. In mice, blood clearance of 10–50 µg doses of the partly purified endotoxin required several hours, and accumulation of radioactivity in the liver did not exceed 25–30% of the dose administered. A small intravenous dose (1 µg) was cleared more rapidly from the circulation concomitant with a greater uptake (~ 50%) by the liver.

The latter results on distribution of radioactive label, although in agreement with those of Braude et al (1955) and Carey et al (1958), brought into question the suitability of the ^{51}Cr-label for in vivo studies. For example, by means of immunodiffusion tests and a sensitive biologic assay, we observed a gradual elimination of the toxic ^{51}Cr-labeled antigen from the circulating plasma. During the first hour following injection, a close correlation was found between toxicity, radioactivity and concentration of the toxic antigen in the circulation. However, significant amounts of unlabeled, nontoxic antigenic fragments were observed in the circulating plasma during the next 24 hours, even though radioactivity had essentially disappeared. Furthermore, urine samples collected during a 24-hour period post-endotoxin contained unlabeled, nontoxic polysaccharide antigen and 15% of the initial radioactivity, most of which was dialyzable. These results indicated that a significant portion of the radioactive tag was removed from the endotoxin by metabolic action of the host. The dialyzable, labeled metabolite could not be fixed to erythrocytes, suggesting that it was not free chromate ion. It was shown subsequently that more than 30% of a dialyzable labeled product could be cleaved from ^{51}Cr-endotoxin during in vitro detoxification in plasma or serum (Skarnes 1966).

DiLuzio and Crafton (1969) and Trejo et al (1972) studied blood clearance and RES tissue uptake of toxic doses of ^{51}Cr-labeled endotoxin in the rat. They confirmed the biphasic nature of the blood clearance of label during the first hour following intravenous injection and reported accumulations of 20–23% in the liver and 3–4% in the spleen and lung. Similar studies in mice showed radioactivity accumulations in the liver (12%), spleen (2%) and lung (1%) 90 minutes after injection of the endotoxin. However, the presence of endotoxin in blood and tissues was not affirmed by other means.

The report of Moreau and Skarnes (1973) provided further evidence that most of the ^{51}Cr-label was cleaved from endotoxin within the vascular compartment of normal mice and rabbits. By radial immunodiffusion assay of plasma samples it was shown that high levels (60–80%) of nontoxic, antigenic moieties remained in the circulation for at least 5 hours post-injection, whereas plasma levels of both radioactivity and the toxic, labeled antigen dropped considerably. In the study of Rippe et al (1974), approximately 38% of the ^{51}Cr-label from a toxic dose of endotoxin was found in the liver of normal mice from 15 minutes to 40 hours after injection. However, toxicity tests on extracted liver homogenates prepared at various time intervals after injection indicated that although toxicity persisted for 18–24 hours, most of the label was separated from the toxin.

The ^{51}Cr-label has also been used to study in vivo distribution in animals rendered tolerant (resistant) to endotoxin (Carey et al 1958; Chedid et al 1964). These studies indicated a rapid blood clearance of the label while liver uptake was higher than that seen in normal animals. In the case of large doses of ^{51}Cr-endotoxins, liver accumulations in tolerant animals ranged from 30–50% within the first hour post-injection but did not increase thereafter. Small doses of 1–10 µg in mice or 30–45 µg in rabbits were cleared from the circulation more quickly, and liver deposition of label ranged from 40–60%, not appreciably different, however, from that of normal animals receiving similar small doses.

In the study of Moreau and Skarnes (1973), 75–80% of an intravenous dose of 50 µg of endotoxin remained in the circulating plasma of both normal and endotoxin-tolerant mice for at least 5 hours, although plasma concentrations of the ^{51}Cr-label dropped precipitously within 10 minutes. Approximately 40% of the original dose (measured by radial immunodiffusion assay) was present in detoxified form in the plasma of both normal and tolerant mice 24 hours later, whereas only 4–5% of the label remained. A similar, though less marked response was observed in tolerant rabbits receiving 500 µg/kg of the ^{51}Cr-endotoxin, that is, the radioactivity content in plasma fell abruptly to 10% within an hour, whereas about 30% of the endotoxin dose was present in the circulation as detoxified antigen 5 hours later.

These results indicate that a significant portion of the rapidly 'cleared' toxic fraction is, in fact, not cleared, but rather unlabeled and detoxified within the vascular compartment. Consequently, assessments of endotoxin concentration made solely on the basis of ^{51}Cr-radioactivity counts underestimate the actual amounts of endotoxin present in circulating plasma. Conversely, radioactivity counts in the liver or other RES organs would overestimate actual toxin concentrations if a detached, labeled metabolite were to accumulate in these tissues. This interpretation would apply to any isotope which can be detached from endotoxin by the action of components in host tissues and fluids. In any case, it is evident that the accuracy of results obtained in distribution studies with labeled endotoxins is directly related to the degree of stability of the labeled preparation in the in vivo environment.

A tritium-labeled endotoxin from *E. coli* was used to study, by radioautography, the intravascular interactions of the toxin with formed elements of the blood (Brunning et al 1964). Following intravenous injections of a toxic dose, heavy platelet labeling became the most prominent feature, but significant numbers of labeled granulocytes were also observed, particularly in the vasculature of the lung, spleen and renal glomeruli. In a subsequent tissue distribution study, Schrader et al (1964) observed a rapid localization of ^{3}H-endotoxin in Kupffer cells of the liver, in sinusoidal cells, in the splenic red pulp and in granulocytes sequestered in pulmonary capillaries. By 24 hours post-injection, radioactivity had decreased in these organs, and little remained in the liver and spleen after 1 week. The quantitative distribution of the tritium label was not investigated in these studies, nor was the endotoxin identified by other means.

Recent distribution studies with a ^{125}I-labeled benzimadate derivative of LPS from *E. coli* confirmed the biphasic nature of blood clearance (Mathison and Ulevitch 1979; Musson et al 1978). An initial rapid clearance was followed by a prolonged phase, the latter associated with the formation of complexes of LPS and high density lipoproteins (HDL). Blood and liver distributions of radioactivity in mice and rabbits were similar to those reported earlier with other labeled endotoxins, although in this report a notable accumulation of LPS in the adrenal glands was also noted. Liver and spleen accumulations of the radioactive preparation between 1–24 hours after injection ranged from 10 to 28%. Autoradiographic studies showed heavy concentrations of ^{125}I in phagocytic vacuoles of Kupffer cells, splenic macrophages and granulocytes. Most, but not all, of the label in the liver

erned by the size of the dose, that is, the larger the dose, the more severe and lasting the episode of vasoconstriction. The early literature on this subject is covered in papers by Atkins (1960), Chien et al (1966) and Hinshaw (1971).

By means of certain contrivances which attenuate vasoconstriction, however, the remarkable phagocytic potential of RES tissues for S-phase endotoxins can be demonstrated. For example, Palmerio et al (1963) and Rutenburg et al (1967) denervated RES organs by surgery or local anesthesia and demonstrated both immediate uptake of substantial amounts of endotoxin and protection of treated animals from an otherwise lethal challenge. Moreau and Skarnes (1973) found that more than 90% of an intravenous dose of endotoxin was sequestered within 10 minutes by the liver and lungs in endotoxin-tolerant rabbits subjected to general anesthesia. It is noteworthy that, in contrast to unanesthetized tolerant controls, neither toxic nor nontoxic antigens could be detected in the circulation from 10 minutes through 6 hours post-injection, an indication that once endotoxin was cleared from the vascular compartment by the RES, it was not subsequently returned. The immediate uptake by the liver of S-endotoxins complexed in vitro with antibody, demonstrated by Chedid et al (1966) and Goodman et al (1969), is most likely due to the increased aggregate size and reduced vasoactivity of the endotoxin–antibody complexes. These special situations result in an increased availability of RES tissues and augment the defensive capability of the host by means which probably have little relevance to in vivo events.

Other attempts to evaluate the contribution of RES cells in defense against endotoxins have not produced a clear understanding of this complex issue. Prolonged interference with phagocytosis of colloids by the liver has been found to occur following injections of large or small doses of endotoxin (Benacerraf and Sebestyen 1957; Biozzi et al 1955; Howard et al 1958). Although depression of RES function by endotoxin is of shorter duration in the tolerant animal (Moses and MacIntyre 1963), it is apparently of sufficient length to allow for an important contribution to detoxification by the activated plasma of tolerant animals (Moreau and Skarnes 1973).

Results of treatments with chemical or biologic agents which either stimulate or depress RES phagocytic function do not relate to the state of resistance to endotoxin challenge (Trejo and DiLuzio 1972), suggesting that phagocytosis is not of major consequence (Benacerraf et al 1959; Stuart and Cooper 1962). Furthermore, blood clearance and tissue distribution of endotoxins do not correlate with changes in resistance to lethal toxicity (DiLuzio and Crafton 1969). It is unfortunate that in these and other similar studies the various agents used to alter RES function have not been examined for their effects on plasma detoxifying mechanisms. For example, it would be of interest to know what effect these chemical agents have on in vivo activation of the plasma detoxifying system.

3.3.1. Detoxification by reticuloendothetial tissues

Although several arguments have been raised against a dominant role in detoxification by the RES in the typical experimental situation, it is clear that the liver and other RES tissues normally assume the burden of defense against endotoxin-producing bacteria by phagocytosis and digestion before a damaging endotoxemia can become established. Moreover, it appears from the experimental studies cited above that RES organs contribute to the detoxification process, albeit in inverse proportion to the dose of endotoxin administered.

Although many reports have appeared which describe the in vitro detoxification of endotoxins by crude extracts of liver and spleen (Atkins 1960; Skarnes and Rosen 1971), there is no consensus regarding the nature of these detoxification reactions. For example, Keene (1962) proposed that the potent detoxifying activity present in water-soluble fractions from liver and spleen extracts was of enzymatic nature. By contrast, Oroszlan et al (1963) described a reversible, nonenzymatic inactivation produced by a partly purified liver fraction containing cationic proteins which combined through electrostatic forces with (anionic) endotoxin. Lamensans et al (1969) also reported a reversible detoxification of ^{51}Cr-labeled endotoxin by crude extracts of mouse spleen. Unlike the plasma detoxification reaction, the chromate label was not detached during the presumed inactivation.

Skarnes et al (1968) isolated a detoxifying fraction from calf spleen by chromatographic procedures. This fraction was rich in carboxylesterase activity, and during the detoxification reaction esterase activity was shown to be associated with the endotoxin following specific precipitation in gel immunodiffusion. This result implicated an esterolytic enzyme, but further purification and endproduct studies are required to establish the enzymatic nature of this detoxification reaction.

Filkins (1971, 1973) described a potent detoxifying activity in liver homogenates and in sonicates of the large granule fraction from macrophages isolated from the liver, lung and peritoneal cavity of the rat. The detoxifying component, probably a lysosomal protein, was most active at acid pH and in this regard, the study of Bona et al (1971) showing that endotoxins are preferentially bound to lysosomes at pH 5 suggests a connection between these two events. The study of Mesrobeanu et al (1969), however, indicated that lysosomes from peritoneal macrophages destroy the erythrocyte-sensitizing property of endotoxins most effectively at neutral pH. Clearly, this subject warrants further study at the biochemical level.

3.3.2. *Granulocytes and detoxification*

The importance of leukocytes in host resistance to endotoxins, though long suspected, has not been established. It is clear that granulocytes have the capacity to ingest endotoxins, but the available evidence on detoxification is both sparse and conflicting. It has been reported that both intact granulocytes and extracts therefrom inactivate the pyrogenic (Collins and Wood 1959) and Shwartzman-provoking (Mesrobeanu et al 1969) effects of endotoxin. Chedid et al (1970) produced a partial inactivation of the lethal action of endotoxin after incubation with extracts of granulocytes. In the latter study, the toxic antigen was still in evidence in 'detoxified' incubates, suggesting a reversible, nonenzymatic reaction. In a comparative study of detoxification by extracts of macrophages and granulocytes, Filkins (1971) concluded that only the macrophage is capable of destroying the lethal property of endotoxin. Having devoted much effort (unpublished) to this subject, we have found no evidence that whole or fractionated extracts from granulocytes can effect an irreversible detoxification of endotoxin.

3.4. Mechanisms of tolerance

The question of host mechanisms of enhanced resistance in the endotoxin-tolerant animal is discussed but briefly in this paper since the subject has received comprehensive treatment in previous reviews by Atkins (1960), Chedid and Parant (1971) and Berry (1977). It is my view, however, that one of two major determinants of enhanced resistance to endotoxins relates to the marked increases in detoxifying potential of circulating plasma (Filkins 1976; Moreau and Skarnes 1973; Skarnes 1970) and RES tissues (Trejo and DiLuzio 1972). The contribution to the tolerant state of the plasma detoxifying system relative to that of the RES tissues appears to be dependent on the dose of endotoxin given and the experimental animal employed. It is presumed from the results of many distribution studies that blood clearance and RES uptake of endotoxins are markedly increased in the tolerant animal. However, endotoxins bearing the unstable ^{51}Cr-tag were employed in virtually all of these studies. When a radial immunodiffusion assay was used to measure blood clearance of moderate doses of endotoxin, differences between normal and tolerant animals proved much less impressive (Moreau and Skarnes 1973). Results from the latter study indicated that humoral detoxification was of paramount importance in the tolerant mouse, whereas RES detoxification was apparently of major importance in the tolerant rabbit.

The second greater determinant in tolerance appears to involve the development of an endotoxin-refractory target cell population (Greisman and DuBuy 1975; Greisman and Woodward 1970) from which the release of various mediators of endotoxicity may be curtailed. For example, Rietschel et al (1980) reported that macrophages from tolerant mice do not release prostaglandins on exposure in vitro to endotoxin, in contrast to results obtained with macrophages from nontolerant control animals. In this context, it is also noteworthy that macrophages from the endotoxin-resistant CH3/HeJ mouse do not release prostaglandins when exposed to endotoxin (Rosenstreich et al 1977).

3.5. Final in vivo disposition of endotoxins

Although the ultimate fate of endotoxins in the experimental host is not fully understood, evidence is available indicating that an unspecified fraction of detoxified antigen is excreted by the kidney, appearing in the urine of mice over a 12- to 24-hour period following intravenous injections (Chedid et al 1963). Although not supported by available evidence from most isotope studies, it is likely that the liver represents the major site for final processing of endotoxins. In addition to dealing with the fraction of endotoxin removed from the circulation during the early granulopectic phase, the liver may also capture the bulk of plasma-detoxified fragments during the prolonged second phase of blood clearance. During this slow phase lasting for 2 days or more (Freudenberg et al 1980), most of the endotoxin is bound to the HDL fraction of plasma. Consequently, clearance of the toxin may be governed primarily by interaction with HDL receptors in the liver, as implied in the report of Munford et al (1981a). Final processing of endotoxins may proceed sequentially from Kupffer cells to hepatocytes and thence through the bile duct to the gut, as suggested by the reports of Mathison and Ulevitch (1979), and Freudenberg and colleagues (1982). In the latter communication, R-phase LPS was shown to concentrate in the hepatocytes initially, inferring that direct uptake and processing by these cells is possible. This eventuality is also supported by the earlier studies of Levy et al (1967), and Zlydaszyk and Moon (1976).

4. SUMMARY

The fate of bacterial endotoxins in vivo has been investigated by means of immunologic and biologic methods and by isotope-labeling techniques. Collectively, these studies have delineated beyond reasonable doubt the major sites of deposition following intravenous administration of S-phase endotoxins. In quantitative terms, however, distribution studies with extrinsically labeled endotoxins have not yielded reliable information. The major difficulty in such experiments, namely the question of label detachment through metabolic action of the host, has not been rigorously controlled. Clearly, in vivo distribution studies with intrinsically labeled endotoxin preparations which include comparisons of specific activity and toxicity of re-isolated fractions with the parent toxin, will provide more accurate information on the stability of the label and also aid in the identification of target sites of endotoxic action and of detoxification.

The exact nature of the predominant detoxifying mechanisms of the host are presently unknown since no specific enzymes have been identified. However, there seems to be little doubt that eventually this issue will be resolved in favor of irreversible enzymic reactions.

It is apparent from the literature on experimental endotoxemia that most detoxification reactions take place in the vascular compartment and in cells of the RES. The relative contributions of intra- and extravascular detoxifying systems are primarily dependent on the dose of S-phase endotoxin administered. With moderate to large doses of endotoxin, the available evidence strongly favors a major contribution by plasma detoxifying mechanisms, whereas small doses appear to be detoxified mainly by RES tissues.

It is not clear to what extent either intravascular or extravascular detoxifying systems participate in naturally acquired enteric bacteremias. For example, at what point and for what reasons does failure of RES function occur and thus lead to the endotoxemic state? Greater understanding of this complex process could conceivably result through an approach designed to mimic more closely the natural situation. Rather than administer bolus injections of S-phase endotoxins to experimental subjects in the usual manner, the use of prolonged intravenous infusions or slow-release (depot) techniques may provide new insights relevant to the prevention and treatment of endotoxin shock.

REFERENCES

Abdelnoor AM, Harvie NR, Johnson AG (1982) Neutralization of bacteria- and endotoxin-induced hypotension by lipoprotein-free human serum. *Infect. Immun. 38*, 157-161.

Atkins E (1960) Pathogenesis of fever. *Physiol. Rev. 40*, 580-646.

Barnes FW, Lupfer H, Henry SS (1952) The biochemical target of Flexner dysentery somatic antigen. *Yale J. Biol. Med. 24*, 384-400.

Bartolf M, Skarnes RC (1985) Purification and characterization of a protein from human plasma which detoxifies bacterial endotoxin. Submitted.

Benacerraf B, Sebestyen MM (1957) Effect of bacterial endotoxin on the reticuloendothelial system. *Fed. Proc. 16*, 860-867.

Benacerraf B, Thorbecke GJ, Jacoby D (1959) Effect of zymosan on endotoxicity in mice. *Proc. Soc. Exp. Biol. Med. 100*, 796-799.

Berry LJ (1977) Bacterial toxins. *CRC Crit. Rev. Toxicol. 5*, 239-318.

Biozzi G, Benecerraf B, Halpern BN (1955) Effect of *Salmonella typhi* and its endotoxin on the reticuloendothelial system. *Br. J. Exp. Pathol. 36*, 226-235.

Bona C, Chedid L, Lamensans A (1971) *In vitro* attachment of radioactive endotoxins to lysosomes. *Infect. Immun. 4*, 532-536.

Braude AI (1964) Absorption, distribution and elimination of endotoxins and their derivatives. In: Landy M, Braun W (Eds), *Bacterial Endotoxins*, pp 98-109. Rutgers University Press, New Brunswick, NJ.

Braude AI, Carey FJ, Zalesky M (1955) Studies with radioactive endotoxin. II. Correlation of physiologic effects with distribution of radioactivity in rabbits injected with lethal doses of *E. coli* endotoxin labelled with radioactive sodium chromate. *J. Clin. Invest. 34*, 858-866.

Brown DL, Lachmann PJ (1972) The behaviour of complement and platelets in lethal endotoxin shock in rabbits. *Int. Arch. Allergy, 45*, 193-205.

Brunning RD, Woolfrey BF, Schrader WH (1964) Studies with tritiated endotoxin. II. Endotoxin localization in the formed elements of the blood. *Am. J. Pathol. 44*, 401-409.

Buchanan BJ, Filkins JP (1976a) Bioassay of endotoxin clearance in vivo and by perfused rat liver. *Am. J. Physiol. 231*, 258-264.

Buchanan BJ, Filkins JP (1976b) Hypoglycemic depression of RES function. *Am. J. Physiol. 231*, 265-269.

Carey FJ, Braude AI, Zalesky M (1958) Studies with radioactive endotoxin. III. The effect of tolerance on the distribution of radioactivity after intravenous injection of *Escherichia coli* endotoxin labelled with ^{51}Cr. *J. Clin. Invest. 37*, 441-457.

Chedid L, Parant M (1971) Role of hypersensitivity and tolerance in reactions to endotoxins. In: Kadis S, Weinbaum G, Ajl SJ (Eds), *Microbial Toxins Vol 5*, pp 415-492. Academic Press, New York.

Chedid L, Skarnes RC, Parant M (1963) Characterization of a ^{51}Cr-labeled endotoxin and its identification in plasma and urine after parenteral administration. *J. Exp. Med. 117*, 561-571.

Chedid L, Parant M, Boyer F, Skarnes RC (1964) Nonspecific host response in tolerance to the lethal effect of endotoxins. In: Landy M, Braun W (Eds), *Bacterial Endotoxins*, pp 500-516. Rutgers University Press, New Brunswick, NJ.

Chedid L, Parant F, Parant M, Boyer F (1966) Localization and fate of ^{51}Cr-labeled somatic antigens of smooth and rough *Salmonellae. Ann. N.Y. Acad. Sci. 133*, 712-726.

Chedid L, Lamensans A, Prixova J (1970) Comparison of the effects of an antibacterial leucocyte extract and the serum endotoxin-detoxifying component of lipopolysaccharides extracted from rough and smooth *Salmonellae. J. Infect. Dis. 121*, 634-639.

Chien S, Chang C, Dellenback RJ, Usumi S, Gregersen MI (1966) Hemodynamic changes in endotoxic shock. *Am. J. Physiol. 210*, 1401-1410.

Collins RD, Wood WB Jr (1959) Studies on the pathogenesis of fever. VI. The interaction of leucocytes and endotoxin *in vitro. J. Exp. Med. 110*, 1005-1016.

Cremer N, Watson DW (1957) Influence of stress on distribution of endotoxin in RES determined by fluorescein antibody technique. *Proc. Soc. Exp. Biol. Med. 95*, 510-513.

Das J, Schwartz AA, Folkman J (1973) Clearance of endotoxin by platelets: role in increasing the accuracy of the *Limulus* gelation test and in combating experimental endotoxemia. *Surgery 74*, 235-240.

DiLuzio NR, Crafton CG (1969) Influence of altered reticuloendothelial function on vascular clearance and tissue distribution of *S. enteritidis* endotoxin. *Proc. Soc. Exp. Biol. Med. 132*, 686-690.

Don MM, Masters CJ, Winzor DJ (1975) Further evidence for the concept of bovine plasma arylesterase as a lipoprotein. *Biochem. J. 151*, 625-630.

Drake WP, Pokorney DR, Kopyta LP, Mardiney MR Jr (1976) In vivo decomplementation of guinea pigs with cobra venom factor and anti-C3 serum: analysis of the requirement of C3 and C5 for the mediation of endotoxin-induced death. *Biomedicine 25*, 91-94.

Filkins JP (1971) Comparison of endotoxin detoxification by leucocytes and macrophages. *Proc. Soc. Exp. Biol. Med. 137*, 1396-1400.

Filkins JP (1973) Acidic pH and the hepatic detoxification of endotoxin. *J. Reticuloendothel. Soc. 13*, 511-517.

Filkins JP (1976) Blood endotoxin inactivation after trauma and endotoxicosis. *Proc. Soc. Exp. Biol. Med. 151*, 89-92.

Fine J (1965) Shock and peripheral circulatory insufficiency. In: *Handbook of Physiology Vol 3: Physiology*, pp 2037-2069. American Physiological Society, Washington DC.

Freudenberg MA, Bog-Hansen TC, Back U, Galanos C (1980) Interaction of lipopolysaccharides with plasma high-density lipoprotein in rats. *Infect. Immun. 28*, 373-380.

Freudenberg MA, Freudenberg N, Galanos C (1982) Time course of cellular distribution of endotoxin in liver, lungs and kidney of rats. *Br. J. Exp. Pathol. 63*, 56-65.

Galanos C, Lüderitz O (1975) Electrodialysis of lipopolysaccharides and their conversion to uniform salt forms. *Eur. J. Biochem.* 603-610.

Gans H (1975) Mechanisms of heparin protection in endotoxin shock. *Surgery 77*, 602-606.

Golub S, Groschel D, Nowotny A (1968) Factors which affect the reticuloendothelial system uptake of bacterial endotoxins. *J. Reticuloendothel. Soc. 5*, 324-339.

Goodman JS, Rogers DE, Koenig MG (1969) The hepatic uptake of bacterial endotoxin I. The influence of humoral and cellular factors. *Proc. Soc. Exp. Biol. Med. 132*, 372-375.

Greisman SE, DuBuy B (1975) Mechanisms of endotoxin tolerance. IX. Effect of exchange transfusion. *Proc. Soc. Exp. Biol. Med. 148*, 675-678.

Greisman SE, Woodward CL (1970) Mechanisms of endotoxin tolerance. VII. The role of the liver. *J. Immunol. 105*, 1468-1476.

Greisman SE, DuBuy JB, Woodward CL (1978) Experimental Gram-negative bacterial sepsis: reevaluation of the ability of rough mutant antisera to protect mice. *Proc. Soc. Exp. Biol. Med. 158*, 482-490.

Gupta JD, Reed CE (1968) The role of serum lipase in the degradation of *Salmonella enteritidis* endotoxin. *J. Immunol. 101*, 308-316.

Hinshaw LB (1971) Release of vasoactive agents and the vascular effects of endotoxin. In: Kadis S, Weinbaum G, Ajl SJ (Eds), *Microbial Toxins Vol 5*, pp 209-260. Academic Press, New York.

Howard JG, Rowley D, Wardlaw AC (1958) Investigations on the mechanism of stimulation of nonspecific immunity by bacterial lipopolysaccharides. *Immunology 1*, 181-203.

Johnson KJ, Ward PA (1972) The requirement for serum complement in the detoxification of bacterial endotoxin. *J. Immunol. 108*, 611-616.

Johnson KJ, Ward PA, Goralnick S, Osborn MJ (1977) Isolation from human serum of an inactivator of bacterial lipopolysaccharide. *Am. J. Pathol. 88*, 559-574.

Keene WR (1962) Detoxification of bacterial endotoxin by soluble tissue extracts. *J. Lab. Clin. Med. 60*, 433-438.

Lamensans A, Chedid L, Laurent M, Deslandres A (1969) Étude comparative de la déto-xification d'une endotoxine radioactive (^{51}Cr) par du sérum citraté et de l'extrait de rate. *Ann. Inst. Pasteur (Paris) 117*, 756-767.

Levy E, Path FC, Ruebner BH (1967) Hepatic changes produced by a single dose of endotoxin in the mouse. *Am. J. Pathol. 51*, 269-278.

Mathison JC, Ulevitch RJ (1979) The clearance, tissue distribution and cellular localization of intravenously injected lipopolysaccharide in rabbits. *J. Immunol. 123*, 2133-2143.

Mathison JC, Ulevitch RJ (1981) In vivo interaction of bacterial lipopolysaccharide (LPS) with rabbit platelets: modulation by C3 and high density lipoproteins. *J. Immunol. 126*, 1575-1580.

May JE, Kane MA, Frank MM (1972) Host defense against bacterial endotoxemia-contribution of the early and late components of complement to detoxification. *J. Immunol. 109*, 893-895.

McCracken JA, Glew ME, Scaramuzzi RJ (1970) Corpus luteum regression induced by prostaglandin F_2alpha. *J. Clin. Endocrinol. Metab. 30*, 544-546.

Mesrobeanu L, Mesrobeanu I, Bonna C, Vranialici D (1969) Le destin des endotoxines thermostables pinocytées par les leucocytes. In: *La Structure et les Effets Biologiques des Produits Bactériens Provenant de Germes Gram-négatifs*, pp 429-438. Editions C.N.R.S., Paris.

Moreau SC, Skarnes RC (1973) Host resistance to bacterial endotoxemia: mechanisms in endotoxin-tolerant animals. *J. Infect. Dis. 128*, S122-133.

Morrison DC, Leive L (1975) Fractions of lipopolysaccharide from *Escherichia coli* O111:B4 prepared by two extraction procedures. *J. Biol. Chem. 250*, 2911-2915.

Moses JM, MacIntyre WJ (1963) Effect of endotoxin on simultaneously determined cardiac output and hepatic blood flow in rabbits. *J. Lab. Clin. Med. 61*, 484-493.

Munford RS, Andersen JM, Dietschy JM (1981a) Sites of tissue binding and uptake in vivo of bacterial lipopolysaccharide-high density lipoprotein complexes. *J. Clin. Invest. 68*, 1503-1513.

Munford RS, Hall CL, Dietschy JM (1981b) Binding of *Salmonella typhimurium* lipopolysaccharides to rat high density lipoproteins. *Infect. Immun. 34*, 835-843.

Munford RS, Hall CL, Lipton JM, Dietschy JM (1982) Biological activity, lipoprotein-binding behavior and in vivo disposition of extracted and native forms of *Salmonella typhimurium* lipopolysaccharides. *J. Clin. Invest. 70*, 877-888.

Musson RA, Morrison DC, Ulevitch RJ (1978) Distribution of endotoxin (lipopolysaccharide) in the tissues of lipopolysaccharide-responsive and unresponsive mice. *Infect. Immun. 21*, 448-457.

Nolan JP (1975) The role of endotoxins in liver injury. *Gastroenterology 69*, 1346-1356.

Oroszlan SI, Mora PT, Shear MJ (1963) Reversible inactivation of an endotoxin by intracellular protein. *Biochem. Pharmacol. 12*, 1131-1146.

Oroszlan S, McFarland VW, Mora PT, Shear MJ (1966) Reversible inactivation of an endotoxin by plasma proteins. *Ann. N.Y. Acad. Sci. 133*, 622-628.

Palmerio C, Zetterstrom B, Shammish J, Euchbaum E, Frank E, Fine J (1963) Denervation of the abdominal viscera for the treatment of traumatic shock. *N. Engl. J. Med. 269*, 709-712.

Rietschel ET, Schade U, Lüderitz O, Fischer H, Peskar BA (1980) Prostaglandins and endotoxicosis. In: Schlessinger D (Ed), *Microbiology – 1980*, pp 66-76. American Society for Microbiology, Washington DC.

Rippe DF, Soltas JG, Berry LJ (1974) In vivo detoxification of endotoxin by mouse liver. *J. Reticuloendothel. Soc. 16*, 175-183.

Rosen FS, Skarnes RC, Landy M, Shear MJ (1958) Inactivation of endotoxin by a humoral component. III. Role of divalent cation and a dialyzable component. *J. Exp. Med. 108*, 701-711.

Rosenstreich DL, Glode M, Wahl LM, Sandberg AL, Mergenhagen SE (1977) Analysis of the cellular defects of endotoxin-unresponsive C3H/HeJ mice. In: Schlessinger D (Ed), *Microbiology – 1977*, pp 314-320. American Society for Microbiology, Washington DC.

Rousselot C, Parant M, Chedid L (1972) Rôle des anticorps naturels dans l'épuration sanguine d'endotoxines hybridées, extraites de souches smooth and rough. *Ann. Inst. Pasteur (Paris) 122*, 179-191.

Rowley D, Ali W, Jenkin CR (1958) A reaction between fresh serum and lipopolysaccharides of Gram-negative bacteria. *Br. J. Exp. Pathol. 39*, 90-101.

Rudbach JA, Johnson AG (1964) Restoration of endotoxin activity following alteration by plasma. *Nature (London) 202*, 811-812.

Rudbach JA, Anacker RL, Haskins WT, Johnson AG, Milner KC, Ribi E (1966) Physical aspects of reversible inactivation of endotoxin. *Ann. N.Y. Acad. Sci. 133*, 629-642.

Rutenburg S, Skarnes RC, Fine J (1967) Detoxification of endotoxin by perfusion of liver and spleen. *Proc. Soc. Exp. Biol. Med. 125*, 455-459.

Schrader WH, Woolfrey BF, Brunning RD (1964) Studies with tritiated endotoxin. III. The local Shwartzman reaction. *Am. J. Pathol. 44*, 597-605.

Schultz DR, Becker EL (1967) The alteration of endotoxin by postheparin plasma and its purified fraction. II. Relationship of the endotoxin detoxifying activity of euglobulin from postheparin plasma to lipoprotein lipase. *J. Immunol. 98*, 482-489.

Seligman AM, Shear MJ, Leiter J, Sweet B (1948) Chemical alteration of polysaccharide from *Serratia marcescens*. I. Tumor-necrotizing polysaccharide tagged with radioactive iodine. *J. Natl Cancer Inst. 9*, 13-18.

Skarnes RC (1966) The inactivation of endotoxin after interaction with certain proteins of normal serum. *Ann. N.Y. Acad. Sci. 133*, 644-662.

Skarnes RC (1968) *In vivo* interaction of endotoxin with a plasma lipoprotein having esterase activity. *J. Bacteriol. 95*, 2031-2034.

Skarnes RC (1970) Host defense against bacterial endotoxemia: mechanism in normal ani-

Handbook of Endotoxin, Vol. 3: Cellular Biology of Endotoxin
L.J. Berry, editor
© Elsevier Science Publishers B.V., 1985

CHAPTER 4

Genetic control of endotoxin response: C3H/HeJ mice*

DAVID L. ROSENSTREICH

1. INTRODUCTION

Endotoxin produces an extremely large number of seemingly different biological effects. These effects are secondary to the activation, triggering or stimulation of different cell types (i.e. fibroblasts, macrophages or lymphocytes) or enzymatic reactions (i.e. complement or kinins). Some effects probably involve several of these systems acting in concert.

Despite the apparent complexity of these processes, analysis of the mechanisms underlying the biological effects of endotoxin has been greatly aided by the existence of genetically responsive and nonresponsive inbred mouse strains. The availability of such strains has made a critical difference in understanding several of these complex biological phenomena, including adjuvanticity, toxicity and resistance to infection.

The most commonly used, endotoxin-nonresponsive inbred mouse strain is the C3H/HeJ. These mice are nonresponsive or hyporesponsive to most of the biological effects of endotoxin that are due to the lipid A moiety, and nonresponsiveness has been found to be due to a mutation in a single chromosome 4 gene. Several other mouse strains that are descended from the C3H/HeJ are also endotoxin-nonresponsive (i.e. C3H/He ORL, C3H/BtS and several of BXH recombinant inbred strains).

Other mouse strains, such as the C57BL/10ScCr and C57BL10/ScN, are also endotoxin-nonresponsive and have been used in a number of experimental studies. Nonresponsiveness in these strains is also thought to be due to a mutation in the same chromosome 4 gene that is abnormal in the C3H/HeJ mouse, although historical and genetic evidence suggests that the two mutations occurred independently of each other.

This chapter will discuss the genetic basis of endotoxin sensitivity and review all those biological effects of endotoxin that have been shown to be under genetic control.

---

* This work was supported by National Institutes of Health Grant No. AI17934.

2. GENETIC CONTROL OF ENDOTOXIN RESPONSIVENESS

2.1. Early studies

Differences between mouse strains in susceptibility to the toxic effects of endotoxin were first demonstrated by Hill et al (1940) over 40 years ago. These workers noted that A strain mice were very sensitive to LPS toxicity ($LD_{50} = 200$ μg), while a second strain, RIII, was resistant ($LD_{50} = 700$–1000 μg) (Hill and Weiss 1964). Low responsiveness in the RIII strain was found to be dominant since (A × RIII)F_1 hybrids were also nonresponsive (Heppner and Weiss 1965). The endotoxin-response defect of RIII mice has not received much study since this latter report. The dominant inheritance of nonresponsiveness as well as the presence of an intact B cell mitogenic response to endotoxin (personal observation) suggest that the RIII defect is probably distinct from that of the C3H/HeJ mice. However, a recent report demonstrating a poor antibody response to T-independent antigens in RIII mice suggests some phenotypic similarity between these two genetic defects.

2.2. C3H/HeJ mice: initial studies

The decreased sensitivity to the lethal effects of endotoxin of C3H/HeJ mice was first reported by Heppner and Weiss (1965). Sultzer pursued this observation in a series of studies. He confirmed that C3H/HeJ mice were very resistant to endotoxin-induced lethality and demonstrated that this was associated with a marked alteration in the intraperitoneal accumulation of leukocytes in response to endotoxin (Sultzer 1968, 1969). He found that in comparison to normal mice, C3H/HeJ mice exhibited an elevated number of mononuclear cells and a decreased number of polymorphonuclear leukocytes 24 hours after the intraperitoneal inoculation of endotoxin.

Sultzer also performed a Mendelian genetic analysis of the inheritance of responsiveness to the lethal and extravascular leukocyte-inducing effects of endotoxin (Sultzer 1972). He found that (A × C3H/HeJ)F_1 hybrids exhibited intermediate responses in both assays and concluded on the basis of this finding as well as on the statistical analysis of responsiveness of F_2 and backcross mice that the responses were under polygenic control.

2.3. Genetic defect in C3H/HeJ mice: the *Lps* gene locus

Progress in this area accelerated after the introduction of genetic analyses based on in vitro assays. Endotoxin is a potent inducer of murine B cell mitogenesis in vitro (Gery et al 1972), and this response is mediated by the lipid A component of the molecule (Rosenstreich et al 1973). In 1974, Watson and Riblet reported that C3H/HeJ mice were unresponsive to the B cell mitogenic effects of endotoxin, and that this was due to a defect in a single, autosomal dominant gene. Subsequent studies by other groups, using other breeding partners, confirmed that B cell mitogenesis in this strain was due to a defect in a single autosomal gene (Coutinho et al 1975b; Glode and Rosenstreich 1976; Sultzer 1976).

Watson and his coworkers subsequently mapped the endotoxin-response gene to chromosome 4. They first analyzed a set of recombinant inbred mice that were derived from the C3H/HeJ strain (BXH), and found a concordance between en-dotoxin responsiveness and the major urinary protein locus, *Mup*-1, which suggested that the endotoxin gene was on chromosome 4 (Watson et al 1977). This location was subsequently confirmed by a backcross linkage analysis which located the gene close to the brown coat color locus (*b*) (Fig. 1) (Watson et al 1978a). This gene has been assigned the designation *Lps*, with n and d representing the normal and defective alleles respectively. Thus, C3H/HeJ mice carry the Lps^d allele.

By analyzing the endotoxin responsiveness of 11 C3H strains, Glode and Rosenstreich (1976) estimated that the Lps^d mutation in the C3H/HeJ mice occur-red some time between 1960 and 1968 (Fig. 2). It is noteworthy in this regard that Oldstone and Dixon (1973) reported a dramatic change in the susceptibility of C3H/HeJ mice to lymphocytic choriomeningitis (LCM) virus that occurred some time between 1967–1972. However, no connection between endotoxin responsive-ness and LCM susceptibility has ever been established.

The inheritance of the *Lps* locus is somewhat controversial. Some studies indi-cate that the Lps^n allele is expressed in a fully dominant fashion. Other studies have demonstrated codominant expression of these alleles, especially when careful dose–response analyses were performed (Glode and Rosenstreich 1976) (Fig. 3). The inheritance of the *Lps* gene has been nicely reviewed by Scibienski (1981).

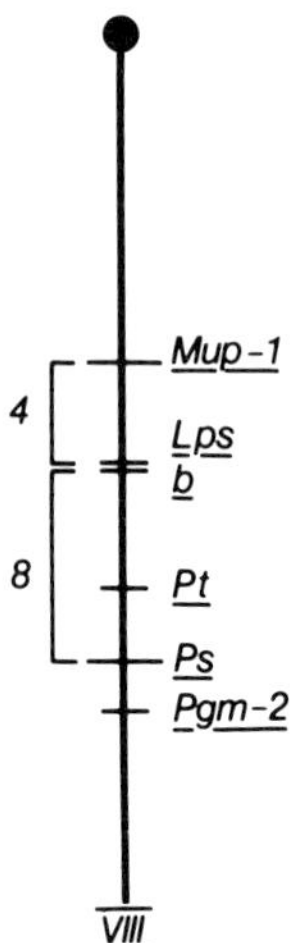

Fig. 1 *Schematic diagram of mouse chromosome 4. The Lps locus is located close to the brown coat color locus, b; and about 4 centimorgans from the major urinary protein locus Mup-1.*

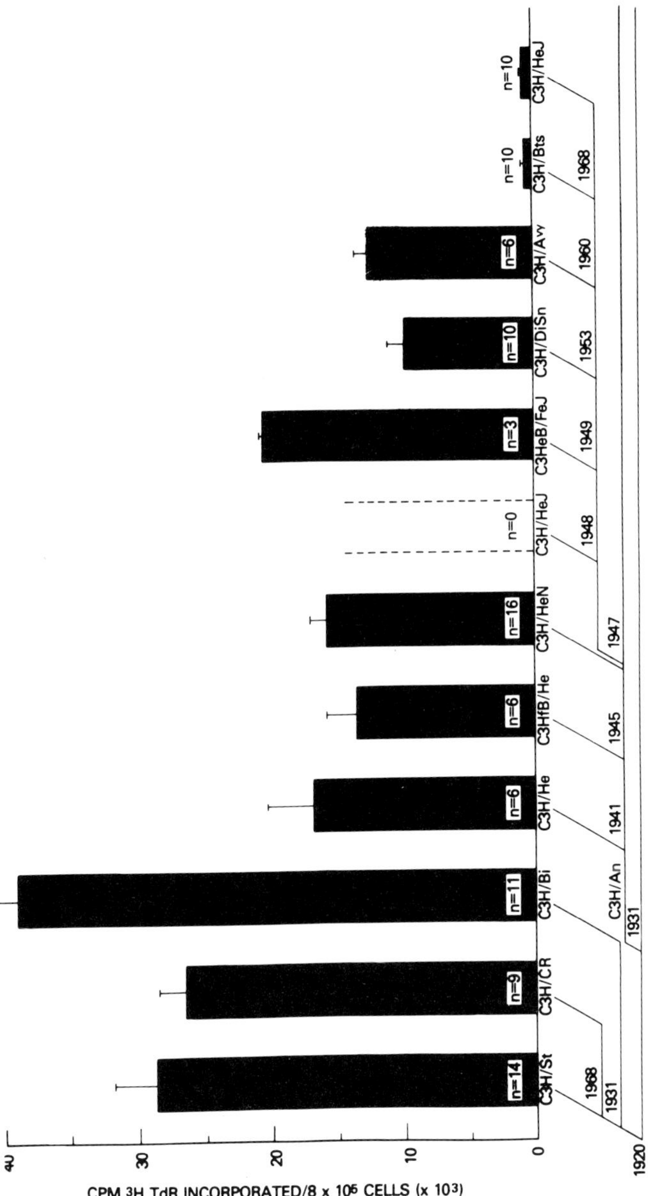

Fig. 2 *Response of C3H substrains to the in vivo mitogenic effects of endotoxin. Results illustrate the response of spleen cells to E. coli K235 (Ph) endotoxin (reproduced from Glode and Rosenstreich 1976, with permission).*

85

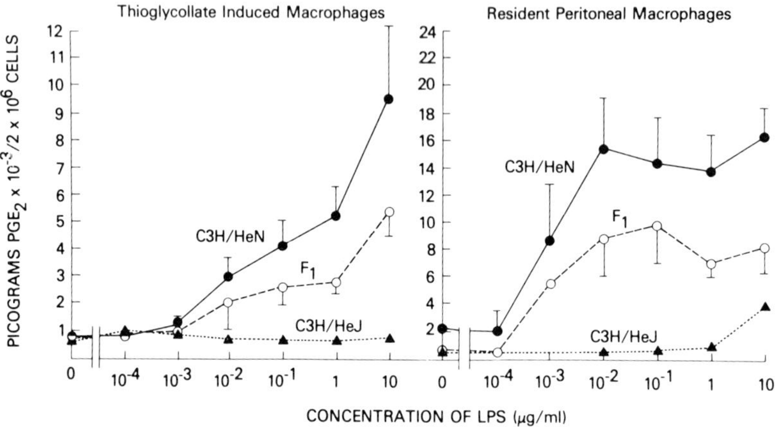

Fig. 3 *Response of parental and F₁ mouse macrophages to endotoxin. Parental and F₁ macrophages were cultured in vitro with E. coli K235 (Ph) endotoxin for 24 hours. Endotoxin sensitivity was determined by measurement of PGE₂ secretion into the culture supernatants after 24 hours (reproduced from Rosenstreich et al 1978, with permission).*

2.4. Genetic defect in C57BL/10ScCr mice

The second commonly used endotoxin-nonresponsive mouse strain, the C57BL/10ScCr, was first described by Coutinho et al (1977). These workers had produced a rabbit antibody against a putative 'LPS receptor', and in screening the B cells of a number of inbred mouse strains noted that cells from two strains, the C3H/HeJ and the C57BL/10ScCr, were negative. Like the C3H/HeJ, the C57BL/10ScCr was found to be unresponsive to the B cell mitogenic effects of endotoxin.

The progenitor strain of these mice, the C57BL/10ScN, has also been found to be endotoxin-nonresponsive (Vogel et al 1979a). Phenotypically, the C57BL/10ScCr mice behave like the C3H/HeJ, since they do not respond to lipid A, but do respond normally to other B cell mitogens such as poly I:C and 8-bromo-cGMP (McAdam and Ryan 1978).

Evidence also indicates that the C57BL/10ScCr mouse possesses an abnormality in the *Lps* gene locus since the F₁ hybrids derived from a cross of C57BL/10ScCr and C3H/HeJ mice remain endotoxin-nonresponsive. The defect in these mice has also been mapped to chromosome 4 (Coutinho and Meo 1978).

Two pieces of evidence suggest that the C57BL/10ScCr mutation occurred independently of the C3H/HeJ. First, the C57 and C3H lines have been separate and distinct for over 50 years. Secondly, F₁ hybrids of crosses between C57BL/10ScCr and C57BL/10Sn (responder) mice are always fully endotoxin-responsive (Coutinho and Meo 1978), while F₁ hybrids of crosses between C3H/HeJ and responder mice are often intermediate in endotoxin responsiveness (Glode and Rosenstreich 1976). However, this latter argument is indirect and has not yet been proven in studies where the same responder parent has been used for both crosses.

3.　BIOLOGICAL ACTIVITIES REGULATED BY THE *Lps* GENE LOCUS

Over the past 10 years, Lps^d mice (primarily C3H/HeJ) have been tested for reactivity to endotoxin in a large variety of biological assays and for the most part have been found to be abnormally responsive in all of these. For some of these assays, evidence that observed differences are due to the Lps^d allele have been provided by genetic linkage studies using B cell mitogenicity or linkage to chromosome 4 markers. In other assays, single gene control has been reported, but no formal linkage to the *Lps* locus has been shown. For many biological assays, however, no genetic studies have been performed nor has there been formal proof that the observed abnormality is due to the Lps^d allele. Nevertheless, it has been assumed that when C3H/HeJ mice differ from a closely related, but Lps^n strain (i.e. C3H/HeN or C3Heb/FeJ), the difference is due to the Lps^d gene. To date, this hypothesis has not yet been disproven, and it has proved to be a useful starting point for many studies. The sections that follow will briefly review the known defects in C3H/HeJ mice.

3.1.　Actions on individual cell types

3.1.1.　B lymphocytes

As discussed previously, one of the earliest in vitro observations on C3H/HeJ mice was that their B lymphocytes did not proliferate in response to endotoxin (Rosenstreich and Glode 1975; Watson and Riblet 1974). This was shown to be due to a defect in the B lymphocyte population and not to a defective accessory cell (i.e. T cell or macrophage) (Glode et al 1976a; Rosenstreich and Mergenhagen 1975; Watson and Riblet 1975).

B lymphocytes of C3H/HeJ mice are also unresponsive to other effects of endotoxin, including polyclonal activation (Coutinho et al 1975a, 1975b; Watson and Riblet 1974) and the induction of surface Ia antigen (Watson 1977). These direct B cell effects have been shown to be regulated by the *Lps* gene locus.

3.1.2.　Macrophages

Cells of the monocyte/macrophage lineage are also endotoxin-sensitive, and many studies have shown that macrophages from Lps^d mice are abnormal in this regard.

Macrophages are killed in vitro by relatively high concentrations of endotoxin (Weiner and Levanon 1968). Glode et al (1977) were the first to demonstrate that C3H/HeJ macrophages were resistant to the cytotoxic effects of endotoxin and that this was a direct effect of endotoxin on macrophages. As illustrated in Figure 4, macrophages from Lps^n mice (C3H/HeN) are killed by two endotoxin preparations, while C3H/HeJ macrophages are not (Fig. 4). A butanol-extracted endotoxin was not very cytotoxic in either strain. Normal, but not C3H/HeJ macrophages also secrete the monokine interleukin-1 (IL-1) after endotoxin stimulation

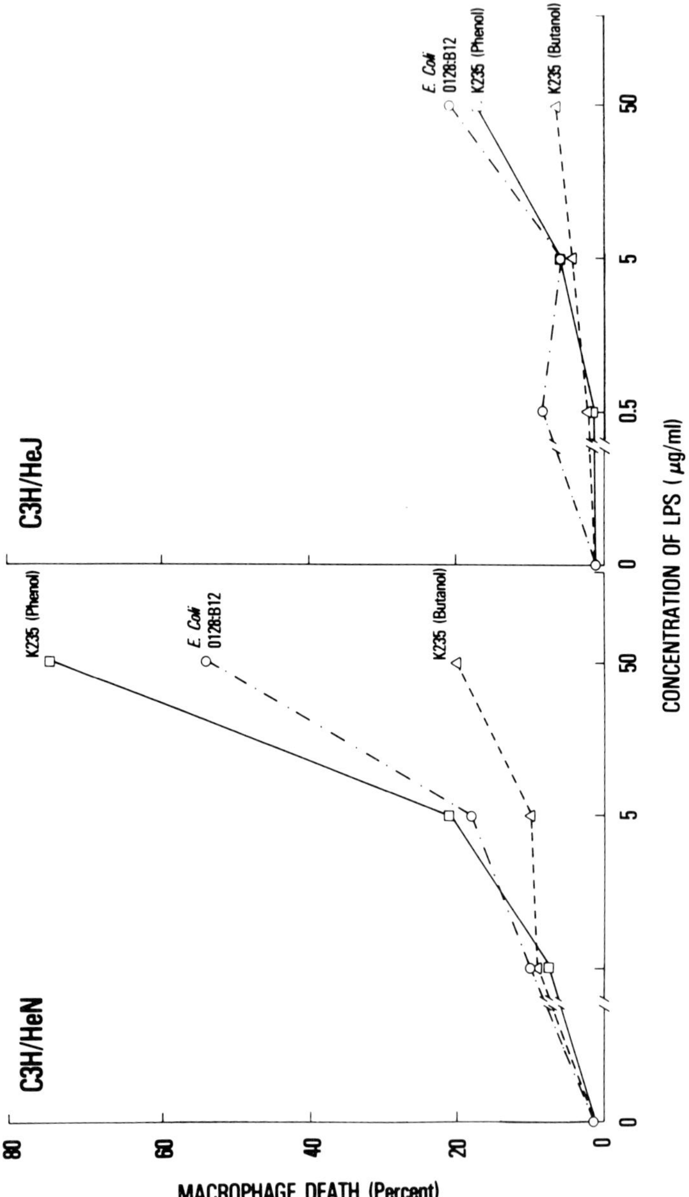

Fig. 4 *Cytotoxic effect of endotoxin on macrophages in vitro. Results represent the mean percent dead cells after 24 hours of culture in the presence of endotoxin in comparison to control cultures (reproduced from Glode et al 1977, with permission).*

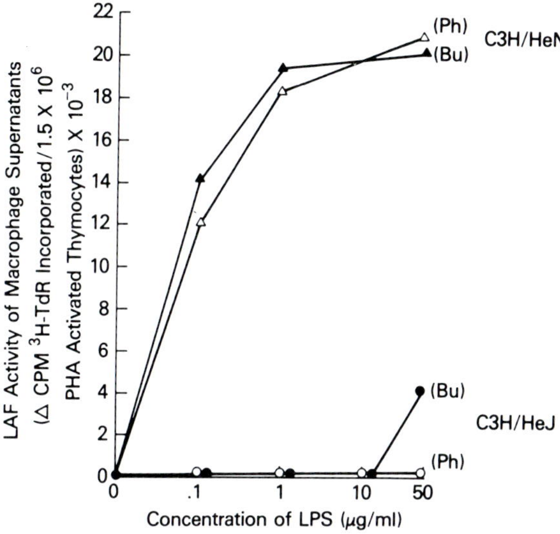

Fig. 5 *Induction of macrophage interleukin-1 (IL-1) production by endotoxin. Results represent the IL-1 activity of 72-hour culture supernatants of thioglycolate-induced macrophages stimulated with phenol-extracted (Ph) or butanol-extracted (Bu) E. coli K235 endotoxin (reproduced from Rosenstreich et al 1978, with permission).*

(Rosenstreich et al 1978) (Fig. 5). Both endotoxin-induced macrophage cytotoxicity and IL-1 production have been shown to be regulated by the *Lps* gene locus (Rosenstreich et al 1978).

C3H/HeJ macrophages also exhibit defective responses in endotoxin-induced prostaglandin synthesis (Wahl et al 1979), glucose metabolism (Ryan et al 1979), inhibition of phagocytosis (Vogel et al 1979a) and tumor cytotoxicity (Doe and Henson 1979).

3.1.3.　T lymphocytes

In contrast to its effects on B cells and macrophages, the effect of endotoxin on T lymphocytes is not as striking or as well studied. Koenig et al (1977) were the first to demonstrate that endotoxin acted directly on a T cell precursor to induce the expression of new surface antigens and that this effect was decreased in C3H/HeJ mice. That T cells respond to endotoxin was subsequently confirmed by McGhee et al (1979), who found that LPS-induced adjuvanticity in vitro required T lymphocytes and that C3H/HeJ T cells were ineffective. Normal, but not C3H/HeJ T cells have also been found to enhance endotoxin-induced macrophage IL-1 production in vitro (Vogel et al 1982b). More recently, it has been shown that endotoxin is actually mitogenic for T lymphocytes. In an in vivo study, Mita et al (1982) showed that in mice immunized with sheep red blood cells, endotoxin induced a twofold increase in splenic T lymphocyte DNA synthesis. This was confirmed by Vogel et al (1983b), who demonstrated in an in vitro study that endotoxin was mitogenic for a small subpopulation of splenic T cells. This effect was absent in C3H/HeJ mice (Fig. 6).

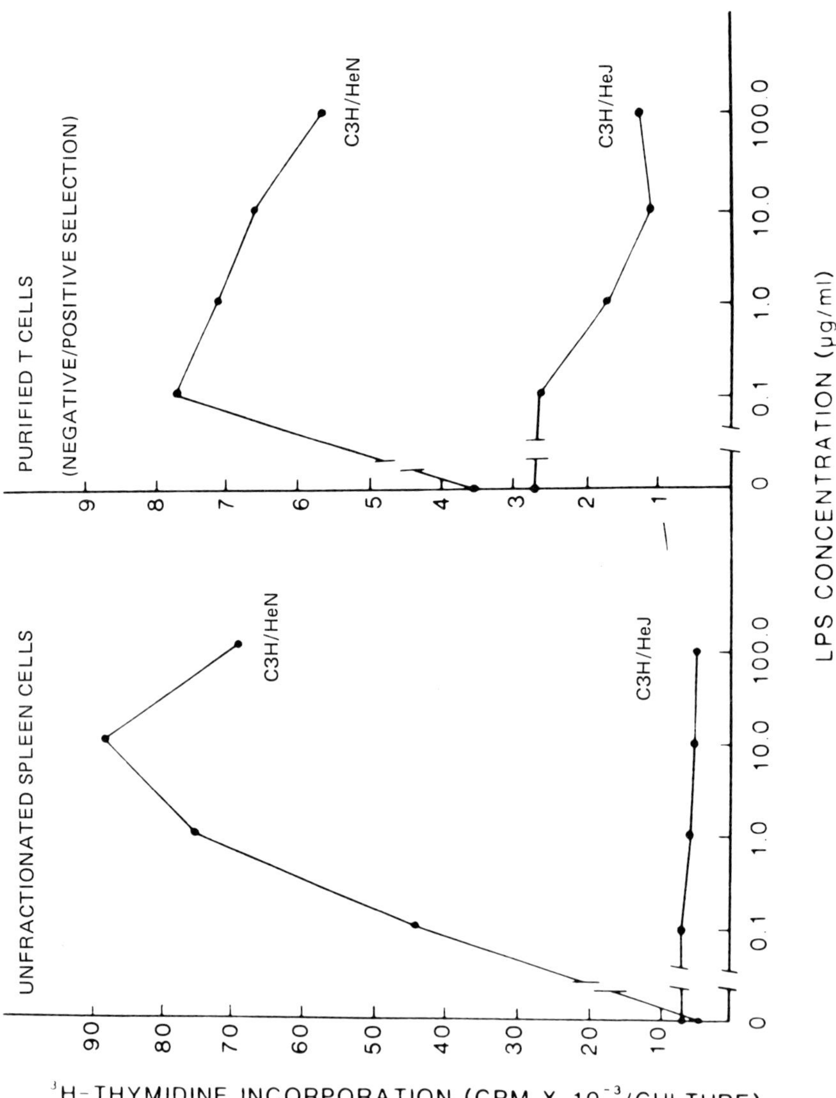

Fig. 6 *Endotoxin-induced stimulation of murine T lymphocytes. Results represent the incorporation of ^{3}H-TdR into new DNA (reproduced from Vogel et al 1983b, with permission).*

3.1.4. Fibroblasts

Endotoxin exerts a number of effects directly on fibroblasts, including the acceleration of proliferation in vivo (Vaheri et al 1973). Ryan and McAdam (1977) were the first to demonstrate that endotoxin-induced enhancement of fibroblast glucose metabolism was also deficient in C3H/HeJ mice.

3.1.5. Complement system

All the biological properties cited thus far involve the interaction of endotoxin with cells of some type. Endotoxin is also capable of activating the complement cascade (Morrison and Kline 1977). However, complement levels and the activation of complement appear to be normal in C3H/HeJ mice (Curry and Morrison 1979). These findings support the hypothesis that the defect specified by the Lps^d allele involves the interaction of endotoxin with cell membranes.

Hoffmann (1978) made the interesting observation that C3H/HeJ mouse serum would not allow the activation and binding of C3b onto antibody-coated erythrocytes. The mice were found to have functional levels of C3, but were missing some heat-stable serum factor. This factor has not been further identified, nor has this interesting observation been pursued.

3.2. Biological activities of endotoxin that are deficient in C3H/HeJ mice: in vivo studies

Lps^d mice have proved to be abnormally responsive to almost all of the in vivo biological effects of endotoxin for which the lipid A moiety plays a central role. This is not surprising but rather predictable given the widespread distribution of the endotoxin-response defect among different cell types. Nevertheless, the use of C3H/HeJ mice to study these in vivo phenomena has proved to be very useful since it has allowed a partial delineation of the relevant cellular and molecular mechanisms. As with the majority of in vitro studies, for the in vivo effects a general working hypothesis has been that those phenomena in which C3H/HeJ mice differ from closely related strains are due to the effects of the *Lps* gene locus. Only for a minority of effects has an actual genetic linkage to the *Lps* gene been established.

3.2.1. Endotoxin-induced lethality

Endotoxin is lethal for most mouse strains within 72 hours. Resistance to the lethal effects of endotoxin was one of the first phenomena demonstrated in C3H/HeJ mice (Heppner and Weiss 1965; Sultzer 1968, 1969). This resistance is very dramatic. For example, while 100 µg of *Escherichia coli* K235 endotoxin will kill 50% of C3H/HeN mice, 100% of C3H/HeJ mice survive the administration of 8 mg of this preparation (Glode et al 1976b). Endotoxin will also cause fetal death and abortion in pregnant mice. As illustrated in Figure 7, C3H/HeJ mice are resistant to this effect as well. One manifestation of endotoxin toxicity in mice is a rapid development of hypothermia. This parameter has been linked genetically to the *Mup*-1 locus on chromosome 4, suggesting its regulation by the *Lps* gene (Watson et al 1978b).

The cellular basis of toxicity has also been studied with C3H/HeJ mice. C3H/HeJ and C3H/HeN mice are histocompatible and do not reject mutual skin grafts (Rosenstreich and Glode 1975), which allows adoptive transfer studies to be performed with this strain pair (Fig. 8). Glode et al (1976b) showed that irradiated

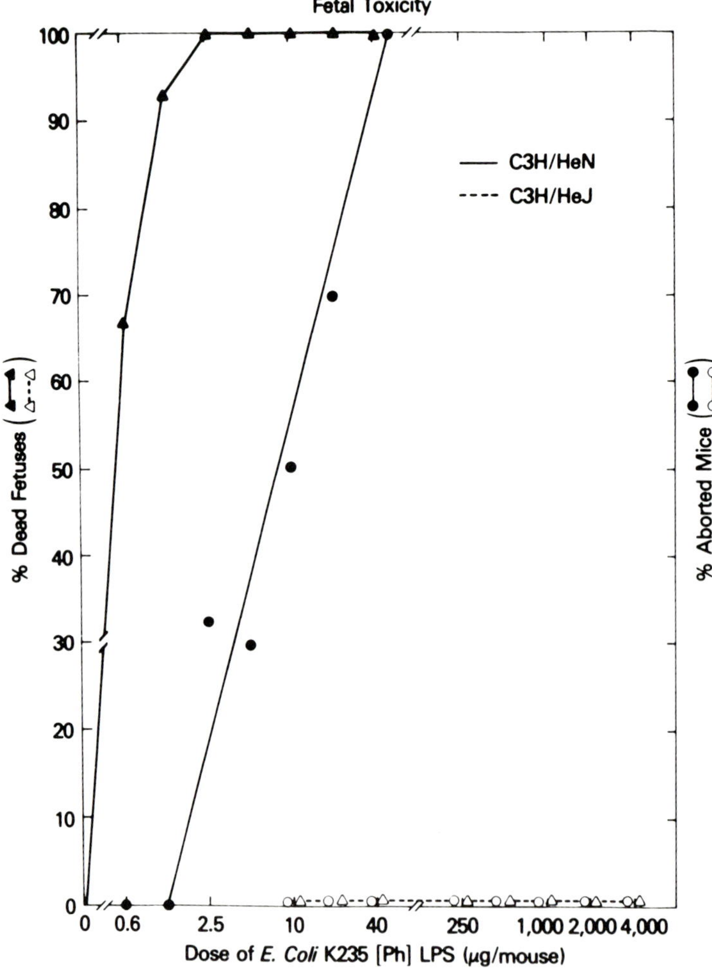

Fig. 7 *Induction of abortion in mice by endotoxin. Mice were examined for signs of abortion 24 hours after the intraperitoneal administration of endotoxin.*

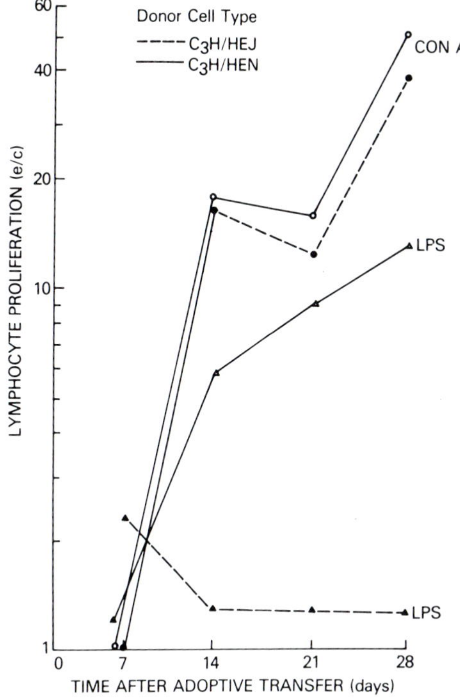

Fig. 8   *Induction of mitogenic responsiveness in C3H/HeJ mice by the adoptive transfer of C3H/HeN spleen cells. C3H/HeJ mice were sublethally irradiated and transfused with 10 × 10⁶ C3H/HeJ or C3H/HeN spleen cells. The mitogenic response of spleen cells in vitro to either concanavalin A (Con A) or endotoxin (LPS) was determined at various times after transfusion (reproduced from Rosenstreich and Glode 1975, with permission).*

C3H/HeJ mice could be rendered sensitive to the lethal effects of endotoxin by the adoptive transfer of Lps^n (C3H/HeN) spleen cells and conversely, that irradiated C3H/HeN mice were rendered nonresponsive by the adoptive transfer of C3H/HeJ spleen cells (Fig. 9).

It was subsequently shown that bone marrow cells were more efficient in transferring endotoxin sensitivity then were spleen cells (Michalek et al 1980). These findings are extremely important since they demonstrate that a complex phenomenon such as lethality was mediated by the effect of endotoxin on a bone marrow-derived cell or cells, and not directly on the structural cells of the target host.

3.2.2. *Endotoxin-induced tumor necrosis*

One of the most striking effects of endotoxin is the induction of hemorrhagic necrosis in subcutaneous sarcomas that occurs 18–72 hours after intravenous or

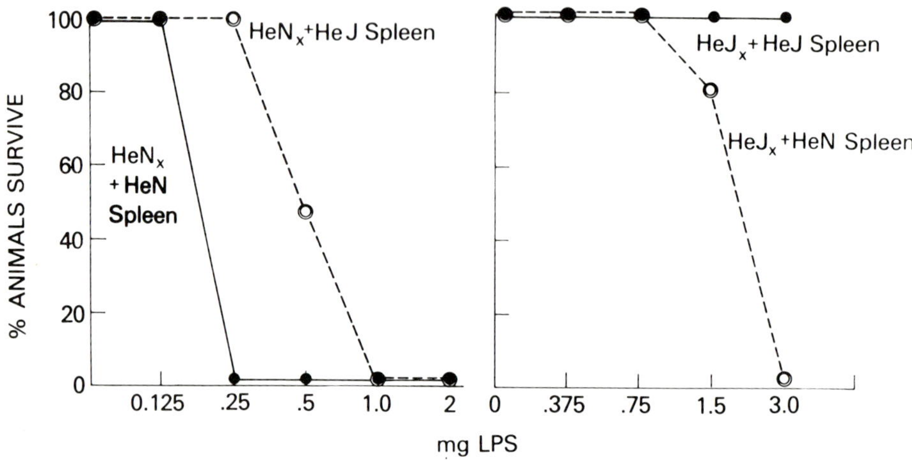

Fig. 9   *Responsiveness of chimeric mice to endotoxin-induced lethality. Mice were adoptively transfused with 50 × 10⁶ spleen cells. Eight weeks later they were tested for sensitivity to the lethal effects of E. coli K235 endotoxin (reproduced from Glode et al 1976b, with permission).*

intraperitoneal administration of endotoxin. C3H/HeJ mice, however, do not exhibit this response (Mannel et al 1979, 1980). In an interesting set of experiments, Mannel et al showed that this phenomenon was mediated by an effect on a bone marrow-derived cell(s) of the host and not by a direct effect on the tumor itself. Fibrosarcomas were induced with methylcholanthrene in C3H/HeJ or C3H/HeN mice. When the C3H/HeN tumor was grafted onto C3H/HeJ mice, no necrosis was induced by endotoxin. In contrast, when the C3H/HeJ tumor was grafted on C3H/HeN mice, endotoxin-induced tumor necrosis was observed. In addition, as with toxicity, C3H/HeJ mice could be rendered capable of manifesting endotoxin-induced tumor necrosis by the adoptive transfer of C3H/HeN bone marrow cells, and vice versa.

3.2.3.   Extravascular accumulation of leukocytes

The extravascular accumulation of leukocytes induced by endotoxin was one of the first phenomena found to be abnormal in C3H/HeJ mice (Sultzer 1968). More recently, Snyderman and his coworkers have reinvestigated this phenomenon. In response to intraperitoneal endotoxin, C3H/HeJ mice exhibit a rapid (12 hrs) influx of polymorphonuclear leukocytes, followed by an influx of macrophages that is maximal at 48 hours (Moeller et al 1978). These responses are earlier and considerably greater in magnitude in the C3H/HeJ mice than in other *Lps*ⁿ strains. This effect was shown to be controlled by a dominant gene but has not been formally linked to the *Lps* locus (Verghese et al 1980). One reason for the increased magnitude of the response in C3H/HeJ mice may be that endotoxin is antiinflammatory in most mouse strains and blocks the intraperitoneal accumulation of cells in response to other cell-attracting agents (Verghese and Snyderman 1981). In contrast, in C3H/HeJ (and C57BL/10ScCr) mice, endotoxin does not inhibit the effects of these other agents. This differential migration has been confirmed using an in vitro model system. Verghese and Snyderman (1982) showed that endotoxin increased the random migration of Lps^d macrophages but either decreased or had no effect on movement of cells from Lps^n mice. This was due to a direct effect of endotoxin on the cells and not secondary to the generation of C5a, since it occurred in serum-free medium. The migration of both types of macrophages in zymosan- or endotoxin-activated serum was the same. In addition, endotoxin-induced migration of polymorphonuclear leukocytes from both types of mice was similar, suggesting that the Lps^d defect is not expressed in this class of cell.

These findings are interesting in a number of regards. One of the most important, however, is that the presence of the Lps^d allele does not always indicate that a given endotoxin-induced response will be absent or diminished. As in this case, the response of Lps^d mice or cells may actually be greater than that of Lps^n mice. Presumably, in vivo endotoxin acts directly on C3H/HeJ macrophages to induce their migration into the peritoneal cavity. The molecular mechanisms underlying this effect are thus far unexplained.

3.2.4. Immunological phenomena

Endotoxin exhibits a number of immunoenhancing properties including: adjuvanticity, antibody production, interferon production, and enhancement of nonspecific resistance to infection. C3H/HeJ mice are hyporesponsive to all these actions of endotoxin, and all can be restored by the adoptive transfer of C3H/HeN bone marrow cells (Michalek et al 1980).

Talcott et al (1975) demonstrated that C3H/HeJ mice did not exhibit endotoxin-induced adjuvanticity, and that this was due to a dominant gene. Linkage to the Lps locus was suggested by the subsequent studies of Skidmore et al (1976). Hoffman et al (1977) showed that the defective adjuvant response in C3H/HeJ mice could be restored by macrophages from Lps^n mice, but that the B cell mitogenic response remained low. This finding is consistent with the hypothesis that endotoxin-induced adjuvanticity is a macrophage- and T cell-mediated phenomenon (McGhee et al 1979) while B cell mitogenicity results from a direct effect on B cells. C3H/HeJ mice, in contrast to other mice, also do not make a specific secondary response to endotoxin (Rudbach and Reed 1977), and do not produce IgM rheumatoid factor in response to endotoxin (Izui et al 1979).

Other endotoxin-related immunological phenomena in vivo that are deficient in C3H/HeJ mice are: interferon production (Apte et al 1977; Michalek et al 1980) and colony stimulating factor production (Michalek et al 1980).

3.2.5. *Enhancement of nonspecific resistance to infection*

Endotoxin will protect mice against infections with a variety of unrelated organisms (Parant et al 1976). Chedid et al (1976) showed that C3H/HeJ mice were not protected from infection with *Klebsiella pneumoniae* by phenol-extracted endotoxin. Interestingly, they also were not protected by trichloroacetic acid(TCA)-extracted endotoxins that were mitogenic for C3H/HeJ B cells. These findings are consistent with the hypothesis that endotoxin-induced enhanced resistance to infection is mediated by macrophage and not by B cell stimulation. In a subsequent genetic study, Parant et al (1977) demonstrated that both enhancement of resistance and sensitivity to lethality were controlled by a single, autosomal dominant gene.

As with most Lps^d gene mediated defects, the failure to develop enhanced nonspecific resistance to infection is relatively specific for endotoxin. This was shown by Eisenstein and Angerman (1978), who demonstrated that C3H/HeJ mice could be protected against a lethal infection with *Salmonella typhimurium* with a vaccine that contained both endotoxin as well as ribosomal elements. C3H/HeJ mice are also protected as well as an Lps^n strain by the immunostimulator *Propionobacterium acnes* from lethal infection with *Babesia microti* (Wood and Clark 1982).

However, other studies indicate that C3H/HeJ mice exhibit a submaximal response to immunostimulants other than endotoxin. Gangemi et al (1980) showed that unlike C3H/HeN mice, C3H/HeJ mice did not develop tumoricidal macrophages after infection with *P. acnes*. In this same study, natural killer (NK) cell activation after *P. acnes* was normal in the C3H/HeJ. Subsequently, these workers found that *P. acnes* did not protect C3H/HeJ mice equally well as C3H/HeN from influenza virus infection (Gangemi et al 1983).

The exact interpretation of these results is not clear. It is possible that C3H/HeJ mice do not respond as well as normal mice to some immunostimulatory component of *P. acnes* which may or may not be lipid A-like. On the other hand, it is much more likely that this abnormality is related to the other C3H/HeJ macrophage defects that are described in later sections, and which can be linked conceptually to a lipid A-recognition defect.

3.2.6. *Induction of inflammation*

Endotoxin is a potent inducer of acute-phase serum proteins in mice, but C3H/HeJ mice do not exhibit a rise in one of these, the serum amyloid-associated (SAA) protein, after administration of normal amounts of endotoxin (McAdam and Sipe 1976). As mentioned previously, the SAA response to endotoxin has been linked genetically to a chromosome 4 gene (Watson et al 1978b).

The cellular basis of this response has been studied using C3H/HeJ mice as a model. It was first shown that C3H/HeJ mice would respond to endotoxin with a rise in SAA protein after the adoptive transfer of Lps^n (C3H/HeN) bone marrow cells (Rosenstreich and McAdam 1979), indicating that the SAA response was not due to a direct effect of endotoxin on the liver (Fig. 10). Sipe et al (1979) next demonstrated that serum from endotoxin-treated C3H/HeN mice, or culture supernatants of endotoxin-stimulated C3H/HeN macrophages, would induce an SAA elevation when injected into C3H/HeJ mice. These macrophage supernatants were subsequently shown to act directly on hepatocytes to induce SAA production in vitro (Selinger et al 1980). Finally Sztein et al (1981) demonstrated in an in vivo study that the SAA-inducing factor was identical or closely related to IL-1. On the basis of these studies, it is now felt that endotoxin stimulates macrophages to release IL-1, which in turn acts on hepatocytes and induces their production of SAA.

3.3. Biological differences in C3H/HeJ mice that are unrelated to the exogenous or experimental administration of endotoxin

The discussion so far has focused on the defective response of C3H/HeJ mice to endotoxin in a variety of in vitro and in vivo assays. The message of all these studies seems clear: that because of a genetic defect in responsiveness to the lipid

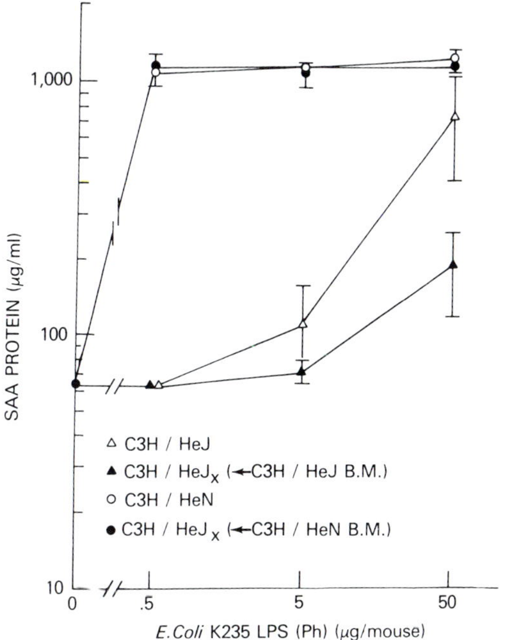

Fig. 10 *Endotoxin-induced serum amyloid A (SAA) production in chimeric mice. Mice were lethally irradiated and adoptively transfused with 10×10^6 bone marrow cells. The rise in serum SAA levels after endotoxin administration was measured 8 weeks after transfusion (reproduced from Rosenstreich and McAdam 1979, with permission).*

A moiety of endotoxin, C3H/HeJ mice exhibit abnormal responses to virtually all lipid A-mediated biological events.

Over the past 5 years, however, it has become increasingly clear that C3H/HeJ mice respond differently than their Lps^n counterpart strains in a variety of in vivo and in vitro assays in which no endotoxin is experimentally administered. These phenomena include; decreased macrophage tumoricidal capacity, differentiation, and response to lymphokine; decreased suppressor cell activity; decreased T cell helper activity; and altered susceptibility to a variety of infectious organisms. The mechanisms underlying these abnormalities have not yet been elucidated. For some assays, the abnormal function has been formally shown by genetic analysis to be regulated by the *Lps* gene locus, but for most the role of this locus is presumed. The most likely explanation for all these abnormalities is that small amounts of naturally occurring endotoxin or endotoxin-like substances are actually required for these biological phenomena to be expressed, and there is some evidence that supports this hypothesis. On the other hand it remains a possibility that the *Lps* gene regulates other cell responses in addition to recognition of endotoxin.

3.3.1. *Abnormalities in macrophage function*

One of the first of these abnormalities to be studied was the failure of C3H/HeJ macrophages to become tumoricidal after infection with Bacillus Calmette-Guérin (BCG). When mice are infected with *Mycobacterium bovis*-BCG, their macrophages will kill tumor target cells in vitro. Ruco and Meltzer (1978) were the first to report that C3H/HeJ macrophages did not become tumoricidal after BCG infection in vivo or after in vitro treatment with the lymphokine macrophage activating factor (MAF) (Fig. 11). The defect was formally shown to be linked to the *Lps* gene locus (Ruco et al 1978).

The mechanism underlying this defect has not yet been established. Ruco and Meltzer (1978) did not feel that it was related to the presence of endotoxin. However, in a very careful study, Weinberg et al (1978) demonstrated that macrophage tumoricidal activity in vitro was dependent on the very small amounts of endotoxin that are normally found in reagents such as calf serum. They noted that BCG-activated C3H/HeJ macrophages could be rendered tumoricidal, but that they required 250 times more endotoxin than did the C3H/HeN cells. These findings suggest that the tumoricidal defect in C3H/HeJ mice may be another manifestation of the abnormal recognition of lipid A by the Lps^d mutant allele.

A related macrophage abnormality is the decreased response of C3H/HeJ and C57BL/10ScCr macrophages to the lymphokine migration inhibitory factor (MIF) (Tagliabue et al 1978, 1979), and to interferon (Boraschi and Meltzer 1980; Boraschi and Tagliabue 1980). A third in vitro macrophage-deficiency is the failure of C3H/HeJ macrophages to differentiate normally in vitro. Macrophages from normal mice spontaneously increase their surface Fc receptors over a 48-hour period in culture (Bianco et al 1975), while macrophages from C3H/HeJ mice either remain the same or decrease in number (Fig. 12) (Vogel et al 1979c). This abnor-

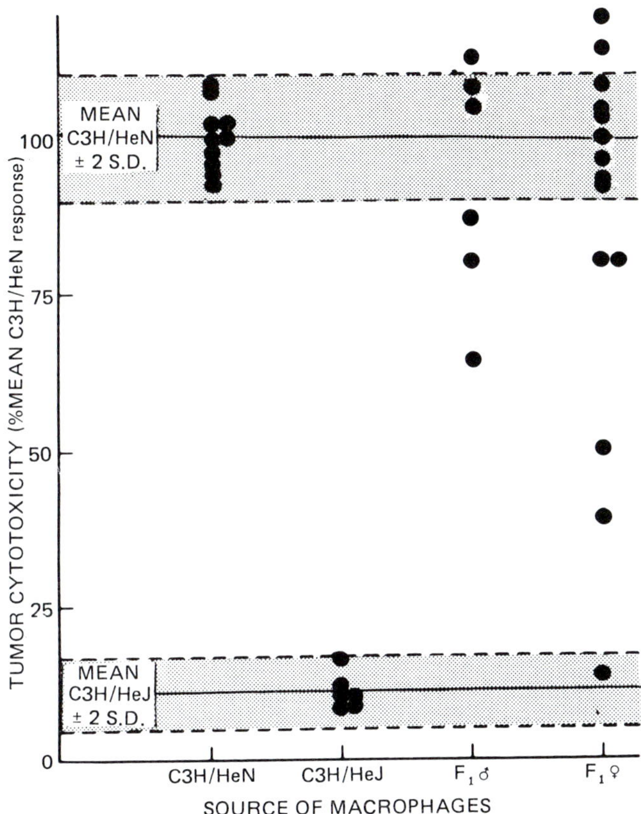

Fig. 11   *BCG-induced macrophage tumoricidal capacity in C3H/HeN and C3H/HeJ and F_1 hybrid mice. The in vitro tumoricidal capacity of phosphate-buffered saline-elicited macrophages was tested 12 days after the intraperitoneal inoculation of 10^7 live BCG organisms (reproduced from Ruco et al 1978, with permission).*

mality can be reversed by activating the C3H/HeJ macrophages with a lymphokine (Vogel et al 1979a), or by raising intracellular levels of cyclic AMP within the macrophages (Vogel et al 1981) (Fig. 13). It has subsequently been shown that the lymphokine that reverses the C3H/HeJ macrophage Fc receptor abnormality is γ-interferon (Vogel et al 1982a, 1983a).

An interesting, and probably related phenomenon is the B cell mitogenic activity of Fc fragments in vitro. Morgan and Weigle (1979) demonstrated that Fc fragments were mitogenic for murine B cells and that this response was abnormal in C3H/HeJ mice. However, C3H/HeJ spleen cells would respond to Fc fragments in the presence of adherent cells from Lps^n mice (C3H/HeN or C3H/ST). In a subsequent study they found that macrophages cultured with Fc fragments produced a 14 000 dalton B cell mitogen (Morgan and Weigle 1980). Thus, this apparent

C3H/HeJ B cell abnormality appears to be a macrophage defect that is probably related to the macrophage Fc receptor deficiency in C3H/HeJ mice described above.

The most logical explanation for these findings is derived from the study of Weinberg et al (1978): macrophage differentiation in vitro is driven by small amounts of naturally occurring endotoxin or endotoxin-like substances. *Lps*^{n} macrophages are capable of being stimulated by this low level contamination while *Lps*^{d} macrophages are not. However, after activation with lymphokine or cAMP, *Lps*^{d} macrophages are rendered more sensitive to endotoxin and therefore stimulated to differentiate 'spontaneously'. This hypothesis has not yet been either proven or disproven.

There is also a single but interesting report on the abnormal response of C3H/HeJ mice to the intraperitoneal administration of prostaglandins (Gurtler and Rank 1976). These workers noted that intraperitoneal injection of prostaglandins E_1, E_2 and $F_{2\alpha}$ caused an influx of neutrophils in *Lps*^{n} (C3H/HaN) mice, but that this response was greatly diminished in C3H/HeJ mice. The relationship of this abnormality to the defect in endotoxin response is not known. It is noteworthy

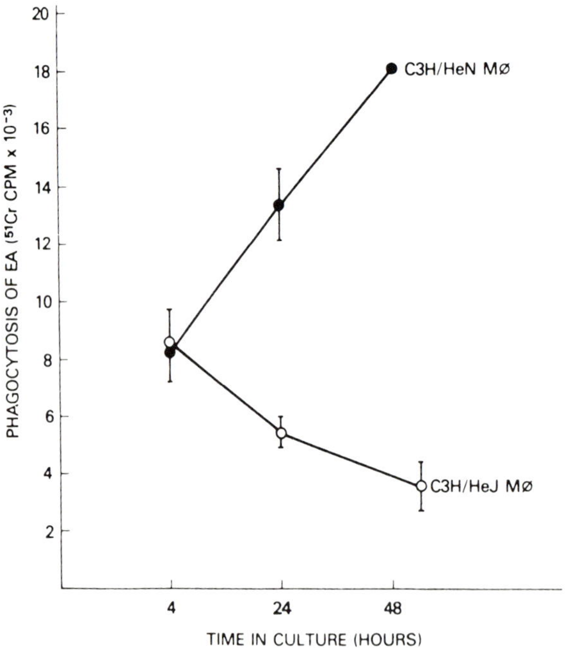

Fig. 12 *Spontaneous alteration of macrophage Fc receptors with time in culture. Thioglycollate-elicited macrophages were cultured in the presence of fetal calf serum, but in the absence of exogenously administered endotoxin. Fc receptors were estimated by the ability of cultured cells to phagocytose antibody coated sheep erythrocytes (EA) (reproduced from Vogel and Rosenstreich 1979, with permission).*

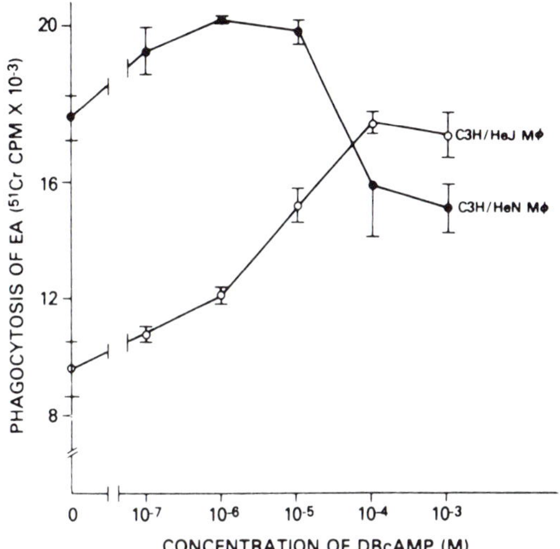

Fig. 13 *Effect of dibutyrylcyclic AMP (DB-cAMP) on macrophage Fc receptor activity. Thioglycollate-elicited macrophages were cultured in the presence of different concentrations of DB-cAMP for 48 hours. Fc receptor activity was then estimated by measuring the phagocytosis of antibody-coated sheep erythrocytes (EA) (reproduced from Vogel et al 1981, with permission).*

that C3H/HeJ macrophages are hyporesponsive to cAMP-elevating agents in vitro (Vogel et al 1981). It is also possible that Lps^d polymorphonuclear leukocytes are also hyporesponsive, and that this abnormality is present in vivo as well as in vitro.

3.3.2. *T and B lymphocyte abnormalities*

A number of aberrant responses of T and B lymphocytes have been observed in C3H/HeJ mice. Again, these occur under natural or experimental situations where the role of endotoxin is not immediately apparent. One of the earliest observations were those of Kincade (1977), who noted that C3H/HeJ mice had a subnormal incidence of colony forming B cells in vitro.

Reports of selective T lymphocyte defects in C3H/HeJ mice are more extensive. Bick et al (1977) observed that C3H/HeJ spleen cells responded poorly in vitro to low concentrations of the T cell mitogen concanavalin A. It was not clear, however, whether this was due to an intrinsic T cell abnormality or to a defect in an accessory cell such as a macrophage.

Goodman and Weigle (1979) noted that T cells enhanced the secretion of immunoglobulin by B cells in response to endotoxin, but that C3H/HeJ T cells were deficient in this regard. This abnormality is probably easier to relate to the defective response of T cells to endotoxin that was described in an earlier section.

Among the most interesting reports in this area was the finding of Ruco et al

(1981) that the in vivo antibody response of C3H/HeJ mice to a number of T-dependent immunogens (sheep red blood cells, trinitrophenylated horse red blood cells, dinitrophenylated keyhole-limpet hemocyanin) was less than that of C3H/HeN mice. This observation was pursued by D'Agostaro et al (1982) who found in an in vitro study that this abnormality was due to a deficiency in T helper cells. C3H/HeJ macrophage antigen presentation was found to be normal.

The work of McGhee and his colleagues provides a potential link between the *Lps*^{d} gene and these T cell alterations. They observed that oral administration of antigen resulted in higher IgA responses in C3H/HeJ mice than in the C3H/HeN strain (Kiyono et al 1980a). This was shown to be due to an increase in T helper cell activity in the gut-associated lymphoid tissues (GALT). This group then showed that C3H/HeJ mice were resistant to the induction of tolerance by oral administration of antigen, and this was due to a decreased generation of T suppressor cells (Kiyono et al 1982).

Evidence from experiments using germfree mice suggests that the response to endotoxin derived from gut flora may explain these results. Germfree mice have been found to behave like C3H/HeJ mice in many ways. They exhibit a 5-fold decrease in response to endotoxin-induced lethality, but show a marked increase in antibody response (splenic plaque forming cells) to orally administered trinitrophenylated LPS (Kiyono et al 1980b). One possible explanation is that endogenous gut endotoxin enhances T suppressor cell activation in the GALT, and that these T suppressor cells are responsible for minimizing antibody responses or generating tolerance after oral administration of antigen. C3H/HeJ T suppressor cell activation would be less than in normal mice because of their inability to be stimulated by ambient levels of endotoxin in the gut. Germfree T suppressor cells would not be stimulated because of the absence of gut endotoxin. This was demonstrated directly by Wannemuehler et al (1982), who showed that the oral administration of endotoxin to germfree mice would render them susceptible to the oral induction of tolerance.

On the basis of these observations, it is reasonable to assume that gut flora-derived endotoxin acts as a natural 'adjuvant'. The decreased in vivo antibody response in C3H/HeJ mice that was described by Ruco et al (1981), and the decreased in vitro response described by D'Agostaro et al (1982) could therefore be ascribed to the failure of their T cells to be stimulated by ambient concentrations of endotoxin.

3.3.3. Natural resistance to infection

One of the most interesting and useful results of research on the C3H/HeJ mice over the past few years has been the observation that the *Lps* gene locus exerts a profound influence on the natural resistance of mice to a number of different pathogenic organisms. Most of the work in this area has involved the study of *S. typhimurium*. In mice, this organism produces a systemic and often lethal infection that is the murine equivalent of typhoid fever. Robson and Vas (1972) were the first to note the increased susceptibility of C3H/HeJ mice to *S. typhimurium* but at the time they did not relate this to an endotoxin response defect. This observation was pursued by VonJeney et al (1977), who noted a correlation between natural resistance of inbred mouse strains to intraperitoneal infection with this organism and the ability of their spleen cells to respond to the mitogenic effect of endotoxin in vitro.

The first formal linkage between the *Lps* gene locus and susceptibility to *S. typhimurium* was demonstrated by O'Brien et al (1980). They noted that C3H/HeJ mice were significantly more susceptible to this organism than all other C3H sub-strains and demonstrated that this was due to a gene that was identical or very closely linked to the *Lps* locus. After intraperitoneal or subcutaneous inoculation, this organism grows rapidly in the spleens of C3H/HeJ mice, and death supervenes 10–14 days later. In contrast, growth in the spleens of C3H/St mice is slower, and the organism begins to decrease in number after day 19 (Fig. 14). In subsequent studies, these workers showed that susceptibility was partly due to a failure of macrophages to become activated in the C3H/HeJ strain (O'Brien et al 1982), as well as to a defect in the production of anti-salmonella antibodies (A. O'Brien, personal communication).

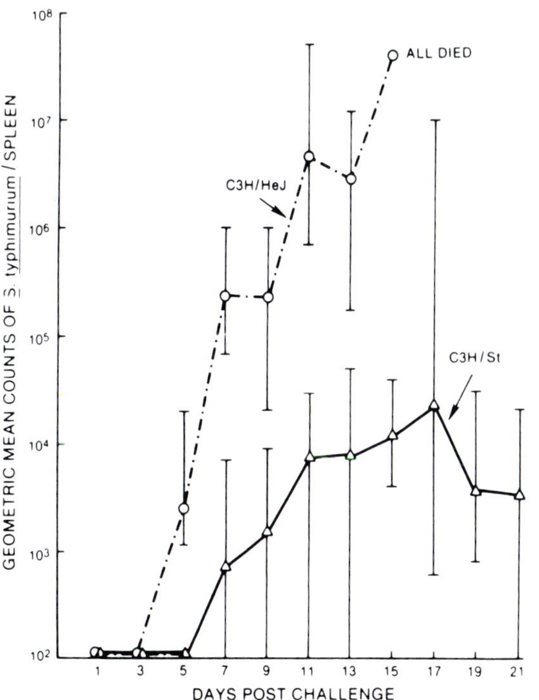

Fig. 14   *Growth of S. typhimurium in the spleens of C3H mice. Mice were inoculated with* 10^2 *organisms intraperitoneally, and spleen counts were determined at various times (reproduced from O'Brien et al 1980, with permission).*

It has now been shown that the *Lps* gene locus regulates susceptibility to a number of other gram-negative bacteria. Thus, Parant et al (1977) noted that 5 hours after inoculation, intravascular levels of *Klebsiella pneumoniae* were higher in C3H/HeJ mice than in an *Lps*^{n} strain although subsequent mortality was the same in both strains.

One of the most interesting observations in this area was made by Streeter and Corbeil (1981). These workers noted that C3H/HeJ mice were *more* resistant to both local and disseminated infection by *Neisseria gonorrheae* than were *Lps*^{n} mice. C3H/HeJ mice clear gonococci from the blood more rapidly and exhibit a greater peritoneal leukocyte response to gonococci. This enhanced inflammatory response to gonococci is thought to be the basis of the enhanced resistance of C3H/HeJ mice. Presumably, this biological effect is similar to the enhanced extravascular accumulation of leukocytes in response to endotoxin that has been described previously (Moeller 1978; Sultzer 1968). The resistance to local colonization was felt by Streeter and Corbeil to be due to both the enhanced inflammatory response as well as to the predominance of inhibitory gram-negative genital flora in the C3H/HeJ strain.

C3H/HeJ mice are also more susceptible to other types of bacterial organisms. Anderson and Osterman (1980) were the first to note that C3H/HeJ mice were the most susceptible among a group of 25 inbred strains to *Rickettsia akari*. This susceptibility was confirmed by Meltzer and Nacy (1980), who found in a separate study that susceptibility was associated with a failure of C3H/HeJ macrophages to become rickettsiacidal after natural infection, or after treatment with other macrophage-activating agents (BCG or lymphokines) (Nacy and Meltzer 1982). A related observation has been made by Ivins and Wyrich (1978), who found that C3H/HeJ mice were more susceptible to *Chlamydia psittaci*, and that this was associated with a decreased reactivity of their peritoneal macrophages to these organisms in vitro.

C3H/HeJ mice also exhibit differential susceptibility to other pathogens. Kirchner et al (1978) noted that C3H/HeJ mice were 100 times more resistant to herpes simplex virus-1 (HSV-1) than were the C3H/HeN mice. This was associated with a failure of HSV-1 to grow within the C3H/HeJ peritoneal leukocytes. C3H/HeJ mice are also more susceptible than C3H/Sn mice to infection with *Babesia microti* (Wood and Clark 1982).

With the possible exception of the HSV-1 phenomenon, it is possible to relate the role of the *Lps* gene locus in the regulation of susceptibility to this heterogeneous group of pathogens, to the recognition of lipid A. All the bacteria discussed contain cell surface-associated endotoxin or endotoxin-like molecules. The increased susceptibility of C3H/HeJ mice to *S. typhimurium*, *Rickettsia akari* and *C. psittaci* is probably secondary to a defect in recognition of lipid A and a subsequent failure to develop the 'activated' macrophages that are required to eliminate these organisms. Similarly, the experimental finding that C3H/HeJ mice exhibit a marked increased cellular influx in response to lipid A seem to account for their increased resistance to *N. gonnorheae*.

The findings with HSV-1 are somewhat harder to explain since these organisms clearly do not contain endotoxin. However, Kirchner et al (1978) showed that the HSV-1 growth within spleen cells in vitro did not occur unless the cells were simultaneously activated with a B cell mitogen such as endotoxin. It is therefore possible that a failure of C3H/HeJ cells to be stimulated by the natural levels of B cell mitogens within the body would result in a poorer growth environment for the virus, the demonstrated decreased recovery of virus from peritoneal cells, and the subsequent resistance to infection.

4. MECHANISM OF ACTION OF THE *Lps* GENE

4.1. Specificity

The nuclear and cellular events that are controlled by the *Lps* gene are only minimally understood at the present time, but all the available evidence suggests that its primary effect is to regulate the recognition and response to the lipid A moiety of enterobacterial endotoxin. This hypothesis is based on the decreased response of Lps^d cells to lipid A derived from endotoxin or to glycolipid preparations made from O polysaccharide-deficient mutants (McAdam and Ryan 1978; Wahl et al 1979). In contrast, the response of C3H/HeJ mice or cells to most other similar biological agents has been shown to be normal or near normal.

Most studies have analyzed B cell mitogenicity and have found that C3H/HeJ cells respond normally to a large variety of stimulants. These include the purified protein derivative of tuberculin (PPD), dextran sulfate, polynucleotides, the membrane lipoprotein from enterobacteriaciae, and mitogens derived from *Listeria monocytogenes*, *Nocardia*, *Actinomyces viscosus* and *Actinomyces laidlawii* (Bona et al 1974; Cohen et al 1975; Coutinho et al 1975c; Engel et al 1977; Kelly and Watson 1979; Kirchner et al 1977; Melchers et al 1975; Rosenstreich and Blumenthal 1977; Sultzer and Nillson 1972). C3H/HeJ mice also exhibit normal antibody responses to pneumococcal polysaccharide (Braley and Freeman 1971).

More importantly, C3H/HeJ mice are responsive to some types of endotoxin as well. Thus, their B cells respond mitogenically to endotoxins from *Brucella abortus* (Moreno and Berman 1979), *Pseudomonas aeruginosa* (Pier et al 1981), and *Bacteroides fragilis* (Joiner et al 1982). C3H/HeJ mice also respond to TCA- and butanol-water-extracted enterobacterial endotoxins but not to phenol-water-extracted preparations (Goodman et al 1978; Skidmore et al 1975) (Fig. 15). The response to the butanol- and TCA-extracted enterobacterial endotoxins is partly due to contamination with a mitogenic lipid A-bound protein (Betz and Morrison 1977; Morrison et al 1976). However, C3H/HeJ mice also exhibit B cell mitogenic responses to ostensibly protein-free preparations of enterobacterial endotoxin. Thus, endotoxin from *E. coli* O111:B4 retains mitogenic activity even after a series of hot phenol-water extractions that renders the preparation protein- and nucleic acid-free (McIntyre et al 1967) (Fig. 16).

There have been several studies comparing the chemical structure of C3H/HeJ in stimulatory and nonstimulatory preparations. When Vukajlovich and Morrison (1983) disaggregated non-C3H/HeJ stimulatory enterobacterial endotoxin with deoxycholate and fractionated the material by molecular sieve chromatography, they were able to isolate a lipid-rich, low molecular weight fraction that stimulated C3H/HeJ spleen cells. Endotoxin from *P. aeruginosa* has also been found to be mitogenic for spleen cells of C3H/HeJ mice (Pier et al 1981). Chemical analysis and comparison of this lipid A to that of a lipid A derived from *E. coli* that was nonmitogenic for C3H/HeJ mice, revealed differences in the carbon chain length of the fatty acid that is directly linked to the glucosamine backbone. Vogel et al (1983c) have studied the properties of a lipid A precursor molecule derived from a conditional mutant of *S. typhimurium*. The complete endotoxin or lipid A from this organism did not stimulate C3H/HeJ spleen cells or macrophages. In contrast, the lipid A precursor molecule deficient in KDO (3-deoxy-D-*manno*-2-octulosonic acid) stimulated both these cell types. Comparative chemical analyses of the C3H/HeJ stimulatory lipid A precursor indicated that it lacked β-hydroxymyristic acid-associated linked lauryl and myristoyl residues, as well as 4-amino-4-deoxy-L-arabinose and phosphorylethanolamine residues that are present in normal (and C3H/HeJ nonstimulatory) lipid A preparations.

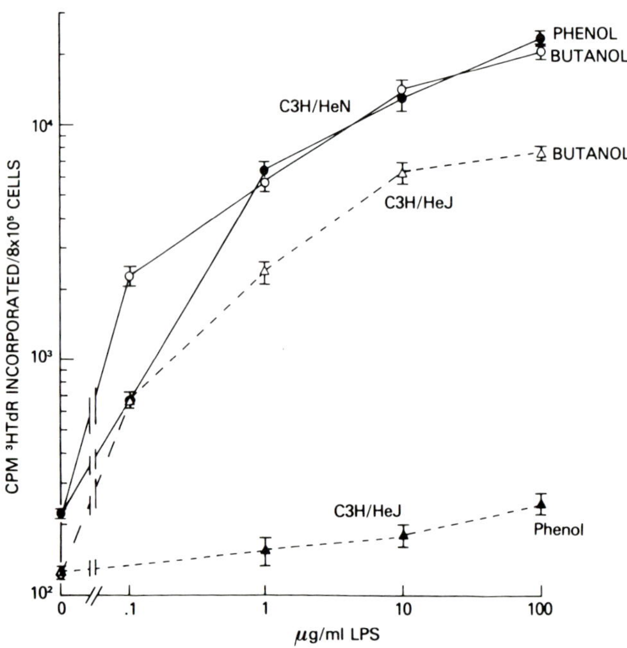

Fig. 15 *Mitogenic response of C3H/HeN and C3H/HeJ spleen cells to phenol- and butanol-extracted E. coli K235 endotoxin. Spleen cells were cultured serum-free for 48 hours, and their mitogenic response determined by the incorporation of ³H-TdR into new DNA.*

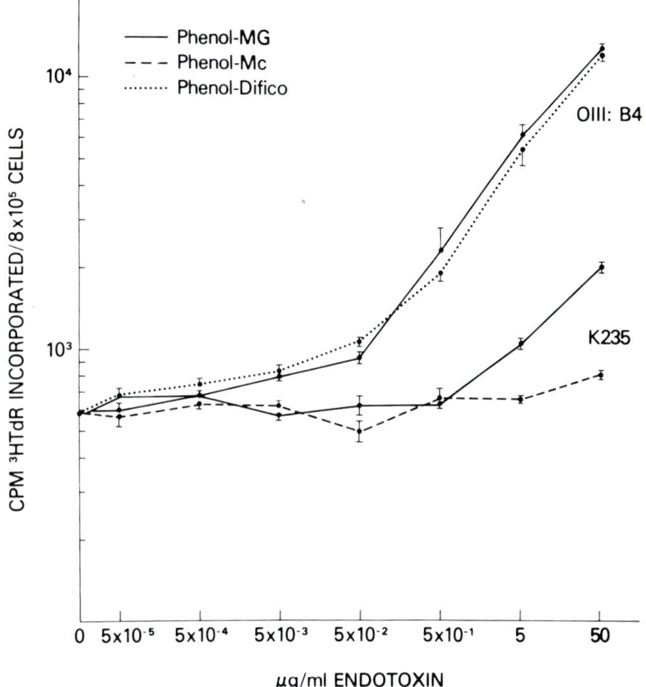

Fig. 16   *Mitogenic response of C3H/HeJ spleen cells to different enterobacterial endotoxins. C3H/HeJ spleen cells were cultured serum-free for 48 hours in the presence of endotoxin from either E. coli O111:B4 or E. coli K235 that had been prepared by extraction with hot phenol-water (MG & MC) (McIntyre et al 1967) or by a commercial Westphal preparation (DIFCO). The mitogenic response was determined by the measurement of the incorporation of ^{3}H-TdR into new DNA.*

On the basis of these interesting and provocative findings, Vogel et al (1983c) have hypothesized that the Lps^d defect is related to an inability to convert lipid A into a stimulatory form. Presumably, this conversion involves the removal of various residues that produce a molecule structurally similar to the lipid A precursor. This hypothesis is supported by earlier work of Truffa-Bachi et al (1977) who found that Lps^n spleen cells would convert endotoxin in vitro into a low molecular weight entity ($< 40\,000$ daltons) that was mitogenic for C3H/HeJ spleen cells.

Although the 'processing' hypothesis has not yet been proven, the results of all these studies taken together strongly suggest that the Lps^d defect is specific for certain chemical configurations of lipid A.

4.2.   Cell membrane components that recognize lipid A

It is logical to assume that the *Lps* gene regulates the expression of some membrane component that controls recognition of lipid A, and there is some evidence in support of this hypothesis. The most convincing evidence is from the studies of Jakobovits et al (1982), who found that they could render C3H/HeJ B cells responsive to endotoxin by fusing them with membranes from the cells of endotoxin-responsive mice (C3H/HeJ). The work of Watanabe and Ohara (1981) also localized the Lps^d defect to the membrane and not the nucleus. These authors inserted C3H/HeJ nuclei into C3H/HeN cells and found that the fused cells retained endotoxin responsiveness.

This hypothesis is also consistent with earlier work of Forni and Coutinho (1978), who produced a rabbit antiserum that bound to the B cells of endotoxin-responsive strains, but not to the cells of nonresponder mouse strains. In a subsequent study, Coutinho et al (1978) found that lipid A would block the binding of the antiserum, and that the antiserum itself was mitogenic for B cells. On the basis of these findings, these workers concluded that the antiserum was directed against a lipid A-specific triggering receptor on B cells, and that this receptor was missing from Lps^d cells.

Unfortunately, other laboratories have been unable to reproduce these findings (Watson et al 1980). One possible explanation for this discrepancy is that Coutinho's antiserum is directed against a membrane component that is poorly immunogenic (i.e. phospholipid). Alternatively, the antiserum may be present in the membranes of all mice but may be decreased in quantity in the membranes of Lps^d mouse cells.

However, if such a membrane component exists, available evidence suggests that it is not merely responsible for the binding of lipid A to cells. Kabir and Rosenstreich (1977) studied the binding of intrinsically labeled endotoxin

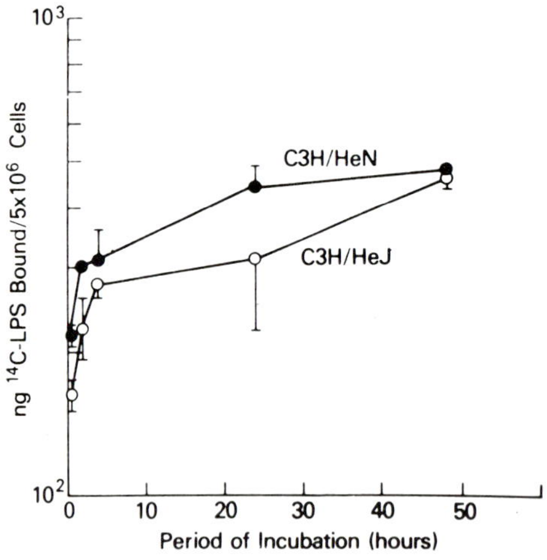

Fig. 17 *Binding of endotoxin to C3H/HeJ and C3H/HeN spleen cells. Results represent the binding of ^{14}C-labeled endotoxin by erythrocyte- and macrophage-depleted spleen cells (reproduced from Kabir and Rosenstreich 1977, with permission).*

glycolipid to murine lymphocytes. They noted that B lymphocytes bound more of this glycolipid than did T lymphocytes, but C3H/HeJ cells bound equally well as did normal cells (Fig. 17). Other workers, using I^{125}-labeled lipid A, have confirmed these findings (Gregory et al 1980; Jacobs et al 1983; Zimmerman et al 1977).

4.3. Alteration of endotoxin sensitivity of animals and cells

The phenotypic expression of the *Lps* gene locus can be altered experimentally, so that Lps^d mice can be rendered endotoxin-sensitive. It has been known for many years that infection of mice with BCG will increase their sensitivity to endotoxin (Suter et al 1958). Vogel et al (1980) showed that BCG infection would increase some, but not all, responses of C3H/HeJ mice to endotoxin. Thus, BCG-infected C3H/HeJ mice exhibited increased endotoxin-induced lethality, hypoglycemia and serum levels of interferon and the acute-phase reactant SAA protein (Fig. 18). However, anti-endotoxin antibody production was not changed. These findings suggested that BCG infection was increasing sensitivity of macrophages and/or T cells but not that of B cells. BCG infection will also increase the resistance of C3H/HeJ mice to *S. typhimurium* (O'Brien et al 1982) (Fig. 19).

In a subsequent in vitro study, Vogel et al (1982b) demonstrated that BCG infection produced a population of activated T cells which in turn activated macrophages and rendered them endotoxin-sensitive. It is noteworthy that isolated macrophages from BCG-infected C3H/HeJ mice remained endotoxin-unrespon-

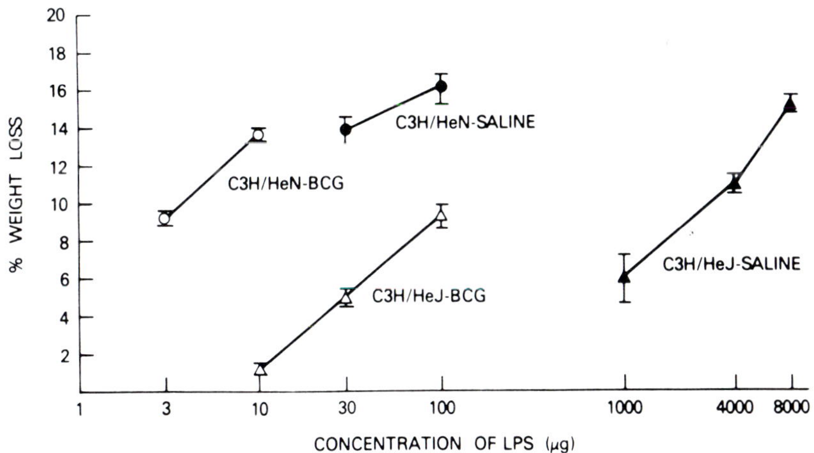

Fig. 18 *Increase in C3H/HeJ endotoxin sensitivity induced by infection with M. bovis (BCG) infection. Mice were infected with 10^7 live BCG organisms. 14 days later they were tested for sensitivity to endotoxin-induced weight loss. Mice were injected intraperitoneally with different amounts of E. coli K235 (Ph) endotoxin and were then weighed daily for 72 hours (reproduced from Vogel et al 1980, with permission).*

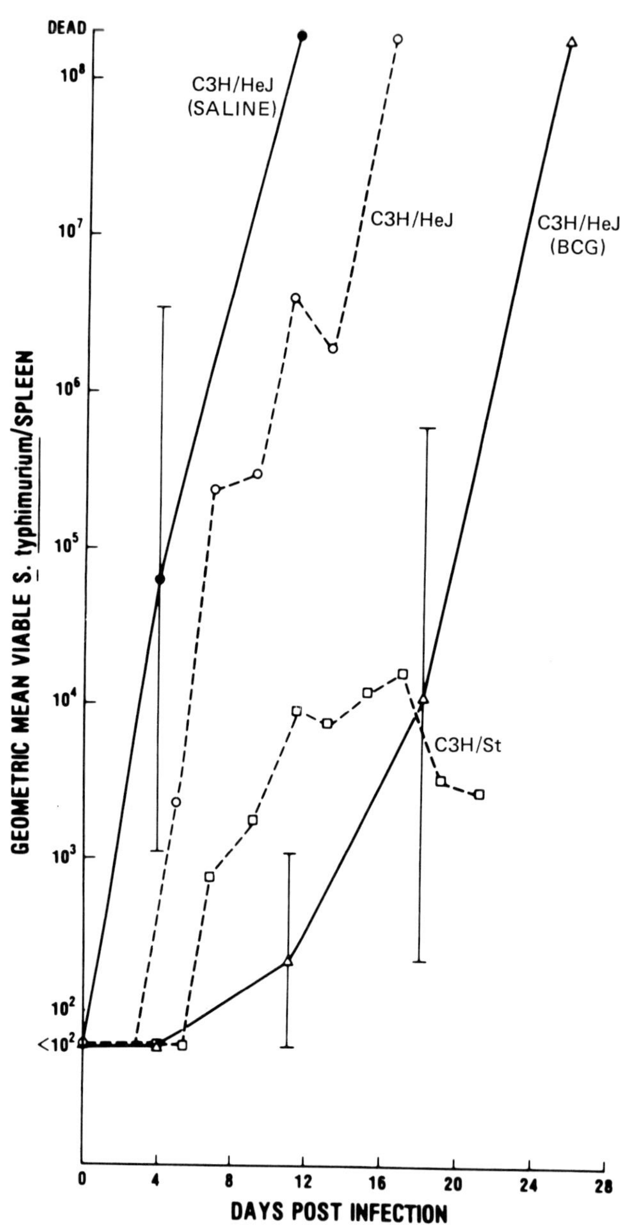

Fig. 19 *Increased resistance of C3H/HeJ mice to S. typhimurium induced by M. bovis (BCG) infection. Mice were infected with 10^7 live BCG organisms. 11 days later they were infected with 10^2 S. typhimurium and daily spleen counts initiated (reproduced from O'Brien et al 1982, with permission).*

sive unless they were cultured together with T cells from the same animals or with T cells from Lps^n mice. Peavy et al (1979) also noted that macrophages from BCG-infected C3H/HeJ mice remained endotoxin-unresponsive in vitro. However, in their studies, BCG infection did not render the Lps^d mice more endotoxin-sensitive. The reasons for this discrepancy have not yet been determined.

A reasonable working hypothesis to explain the increase in endotoxin sensitivity of C3H/HeJ mice demonstrated in the previous experiments is that 'activation' of macrophages or T cells produces a general increase in membrane receptors, including the component that is responsible for recognition of lipid A. The data indicating an increase in macrophage membrane Fc receptors after 'activation' are consistent with this hypothesis (Vogel et al 1981, 1983a). There is also evidence that T cell activation will increase their endotoxin sensitivity, so that T cells from sheep red blood cell-immunized mice will respond to the mitogenic effects of endotoxin (Mita et al 1982), as well as the T cells from mice that are in the process of rejecting an allogeneic graft (V. Tuohy, personal communication). Presumably, this is the result of an activation-induced increase in the expression of a 'receptor' on the T cell surface.

4.4. Recognition of lipid A by Lps^d cells

The hyporesponsiveness of Lps^d cells is most easily explained by a failure to be stimulated by the concentrations of lipid A that are stimulatory for Lps^n cells. The finding that Lps^d mice will exhibit a typical response when exposed to high concentrations of endotoxin is certainly consistent with this hypothesis. This pattern is seen with several in vivo assays such as lethality. In vitro, however, this is much more difficult to demonstrate, since as endotoxin concentrations are raised experimentally to the levels that would be expected to stimulate Lps^d cells, nonspecific toxic effects supervene and no cellular response can be measured. This may be the case with most B cell and macrophage assays.

However, it is important to be aware that there is evidence suggesting that Lps^d cells do respond to or recognize normal amounts of lipid A, but that their response to this material is abnormal. Glode et al (1976a) and Rosenstreich et al (1977) noted that endotoxin, at concentrations that were stimulatory for normal B cells, would inhibit the response of C3H/HeJ B cells to other B cell mitogens such as poly I:C (Fig. 20). This finding was confirmed and extended by Haas et al (1978).

The findings of Verghese and Snyderman (1982) also demonstrate that Lps^d cells do 'recognize' endotoxin, but that their response is different from that of Lps^n cells.

5. OTHER ENDOTOXIN RESPONSE GENES

The bulk of this review has dealt with the Lps gene locus on mouse chromosome 4, since this has been studied most extensively. However, as was mentioned earlier, there is evidence for the existence of genes distinct from Lps that regulate aspects

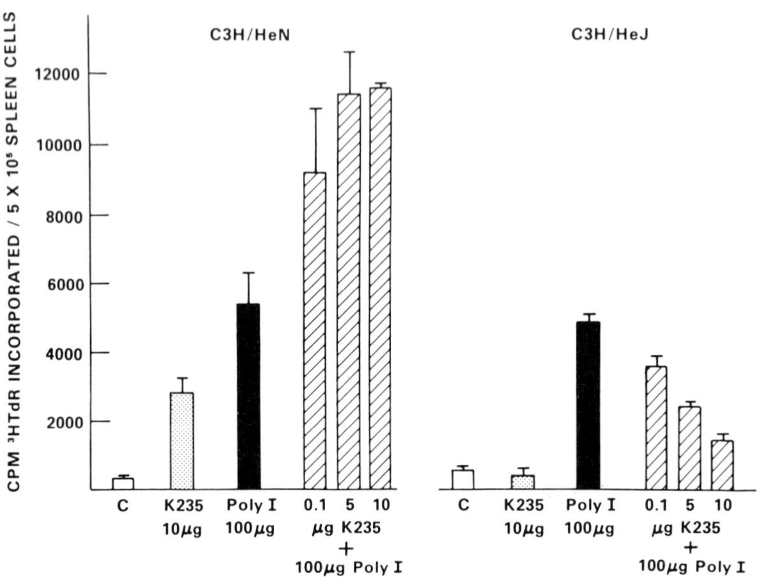

Fig. 20   *Inhibitory effect of endotoxin on the response of C3H/HeJ B cells to another B cell mitogen, polyinosinic acid (Poly I). C3H/HeN or C3H/HeJ spleen cells were cultured for 48 hours in vitro in the presence of the indicated mitogens. Responses were determined by the measurement of ³H-TdR incorporation into new DNA (reproduced from Glode et al 1976a, with permission).*

of endotoxin responsiveness in mice.

One such gene was described by Hill and Weiss (1964); it renders RIII mice hyporesponsive to the lethal effects of endotoxin. As was discussed earlier, it is likely that this gene is distinct from *Lps*, but no formal genetic analysis has yet been done to establish this difference.

Breeding studies on C3H mouse strains also suggested the existence of genes or alleles different from *Lps* that regulated in vitro responses to endotoxin in mice (Glode and Rosenstreich 1976) (Fig. 2). In this study, C3H strains could be categorized as high responders to endotoxin (C3H/St and C3H/Bi), intermediate responders (C3H/HeN and C3Heb/FeJ) and low responders (C3H/HeJ). This pattern of responsiveness as well as a genetic analysis indicated that a mutation in a distinct gene or in the *Lps* locus was responsible for the difference in endotoxin responsiveness between C3H/HeN and C3H/Bi mice.

These findings have not yet been pursued in order to determine the location of this gene and its relationship to *Lps*. Moreover, since it was based solely on in vitro mitogenic responsiveness, it is possible that the observed differences had nothing at all to do with endotoxin, but were merely a reflection of other variables such as splenic B cell number or survival of spleen cells in culture.

112

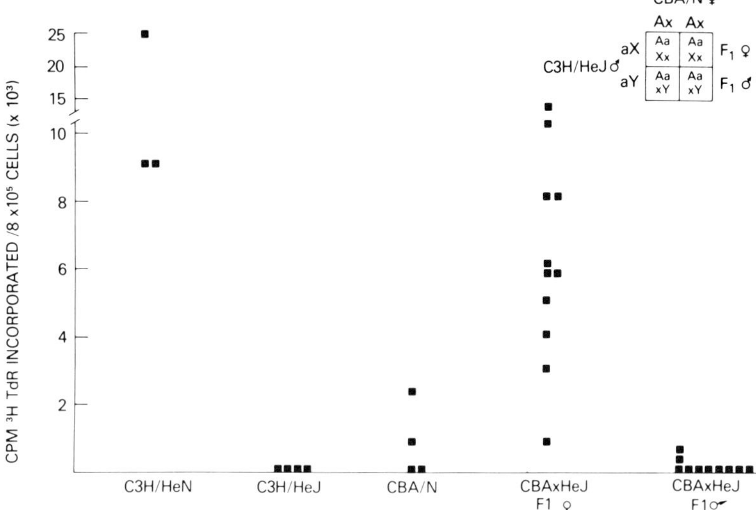

Fig. 21　*In vitro endotoxin responsiveness of C3H/HeJ and CBA/N mice and their F₁ progeny. The in vitro mitogenic response of spleen cells to E. coli K235 (Ph) endotoxin was determined for each strain (reproduced from Glode and Rosenstreich 1976, with permission).*

The work of O'Brien et al (1980) and Eisenstein et al (1982) may shed some light on these other genes. These workers noted that C3H strains varied in their susceptibility to *S. typhimurium*. In both these studies, there was a rough direct correlation between the resistance of a mouse strain to *Salmonella* and its previously described endotoxin response, suggesting that genes other than *Lps* are responsible for both these effects. As yet, however, little else is known about this gene(s).

There is a recessive X-linked gene in mice that regulates B cell maturation. The abnormal allele of this gene, termed *xid* (X-linked immunodeficiency) is present in CBA/N mice. Because of the *xid* gene-regulated absence of a B cell subpopulation, CBA/N mice respond poorly to endotoxin in vitro (Rosenstreich et al 1978a). The distinction between *xid* and *Lps* is shown by the presence of gene complementation in the F₁ hybrid progeny of CBA/N and C3H/HeJ parents (Fig. 21).

6. SUMMARY AND CONCLUSIONS

In mice, several gene loci have been described that regulate in vivo or in vitro responsiveness to endotoxin. By far the best studied of these is *Lps*, since a mutation in this locus has resulted in a profound abnormality in endotoxin responsiveness.

The *Lps* gene locus, which is located on mouse chromosome 4, seems to regulate the response of cells to the lipid A moiety of endotoxin. Mice that carry the defective mutant allele of *Lps* (*Lps*^{d}) respond abnormally to most endotoxic lipid A preparations, but respond most abnormally to lipid A from certain enterobacteriaciae, or to related glycolipids.

Endotoxin responsiveness controlled by the *Lps* gene is inherited in either a dominant or a codominant manner and is expressed in several cell types including B cells, T cells, macrophages and fibroblasts.

Mice that are homozygous for *Lps*^{d} respond abnormally to all the in vivo and in vitro effects of lipid A. In addition, such mice exhibit biological abnormalities that are unrelated to the exogenous administration of endotoxin including a variety of T cell, B cell and macrophage functions in vitro and in vivo as well as natural resistance to several different pathogens.

The mechanism of action of the *Lps* gene locus is not yet understood. There is evidence suggesting that its effects are expressed in the cell membrane, but direct measurements of lipid A-binding indicate that the *Lps*^{d} gene does not alter the quantitative binding of lipid A to cells. It has been hypothesized that certain lipid A configurations are either nonstimulatory or inhibitory to cells, and that *Lps*^{d} mice lack the ability to 'process' lipid A to a stimulatory form. If this hypothesis is true, it raises the possibility that the *Lps* gene regulates levels of some membrane-bound esterase or lipase. Such a mechanism would fit most of the observations about this gene.

The existence of other genes or gene loci that control in vivo or in vitro endotoxin responsiveness in mice is supported by scattered pieces of evidence. These include gene(s) that may be abnormal in RIII mice and other endotoxin response genes or alleles in C3H mice. The *xid* gene also results in diminished endotoxin responsiveness which appears to be secondary to an abnormal B cell subpopulation.

Endotoxin is a ubiquitous molecule in mammals, and a great deal of evidence supports its importance in the homeostasis of the immune system as well as its pathogenic potential. Genetic studies of this molecule have led to some interesting and potentially exciting advances in our understanding of the immune system in health and disease. The value of this analytical and conceptual approach has been proved several times over, and should continue to prove useful in the future.

ACKNOWLEDGEMENT

I would like to thank Mrs. Betty Donovan for her expert secretarial assistance.

REFERENCES

Anderson GW Jr, Osterman JV (1980) Host defenses in experimental rickettsialpox: genetics of natural resistance to infection. *Infect. Immun. 28*, 132-136.

Apte RN, Ascher O, Pluznik DH (1977) Genetic analysis of generation of serum interferon by bacterial lipopolysaccharide. *J. Immunol. 119*, 1898.

Betz SJ, Morrison DC (1977) Chemical and biologic properties of a protein-rich fraction of bacterial lipopolysaccharides. I. The *in vitro* murine lymphocyte response. *J. Immunol. 119*, 1475-1481.

Bianco C, Griffin FM, Silverstein SC (1975) Studies of the complement receptor. Alteration of receptor function upon macrophage activation. *J. Exp. Med. 141*, 1278.

Bick PH, Persson U, Smith E, Moller E, Hammarstrom L (1977) Genetic control of lymphocyte activation: lack of response to low doses of concanavalin A in lipopolysaccharide-nonresponder mice. *J. Exp. Med. 146*, 1146-1151.

Bona C, Damais C, Chedid L (1974) Blastic transformation of mouse spleen lymphocytes by a water-soluble mitogen extracted from Nocardia. *Proc. Natl Acad. Sci. USA 71*, 1602-1606.

Boraschi D, Meltzer MS (1980) Defective tumoricidal capacity of macrophages from mice: tumoricidal defect involves abnormalities in lymphokine-derived activation stimuli and in mononuclear phagocyte responsiveness. *J. Immunol. 125*, 777-782.

Boraschi D, Tagliabue A (1980) Interferon-induced enhancement of macrophage-mediated tumor cytolysis and its difference from activation by lymphokines. *Eur. J. Immunol. 11*, 110.

Braley HC, Freeman MJ (1971) Strain differences in the antibody plaque-forming cell responses for inbred mice to pneumococcal polysaccharide. *Cell Immunol. 2*, 73.

Chedid L, Parant M, Damais C, Parant F, Juy D, Galelli A (1976) Failure of endotoxin to increase nonspecific resistance to infection of lipopolysaccharide low-responder mice. *Infect. Immun. 13*, 722-727.

Cohen JJ, Rodriguez GE, Kind PD, Campbell PA (1975) Listeria cell wall fraction: a B cell mitogen. *J. Immunol. 114*, 1132-1134.

Coutinho A, Meo T (1978) Genetic basis for unresponsiveness to lipopolysaccharide in C57B1/10Cr mice. *Immunogenetics 7*, 17.

Coutinho A, Gronowicz E, Sultzer BM (1975a) Genetical control of B cell responses I. Selective unresponsiveness to lipopolysaccharides. *Scand. J. Immunol. 4*, 139.

Coutinho A, Moller G, Gronowicz E (1975b) Genetical control of B cell responses IV. Inheritance of the unresponsiveness to lipopolysaccharides. *J. Exp. Med. 142*, 253.

Coutinho A, Forni L, Melchers F, Watanabe T (1977) Genetic defect in responsiveness to the B cell mitogen lipopolysaccharide. *Eur. J. Immunol. 7*, 325-328.

Coutinho A, Forni L, Watanabe T (1978) Genetic and functional characterization of an antiserum to the lipid A-specific triggering receptor on murine B lymphocytes. *Eur. J. Immunol. 8*, 63-67.

Curry BJ, Morrison DC (1979) Role of complement in endotoxin initiated lethality in mice. *Immunopharmacology 1*, 125-135.

D'Agostaro G, Ruco LP, Uccini S, Garavini M, Baroni CD, Doria G (1982) Defective helper T-cell activity in LPS-unresponsive C3H/HeJ mice. *Cell. Immunol. 70*, 231-240.

Doe WF, Henson PM (1979) Macrophage stimulation by bacterial lipopolysaccharides. III. Selective unresponsiveness of C3H/HeJ macrophages to the lipid A differentiation signal. *J. Immunol. 123*, 2304-2310.

Engel D, Plagett J, Page R, Williams B (1977) Mitogenic activity of Actinomyces viscosus. I. Effects on murine B and T lymphocytes, and partial characterization. *J. Immunol. 118*, 1466-1471.

Eisenstein TK, Angerman CR (1978) Immunity to experimental salmonella infection: studies on the protective capacity and immunogenicity of lipopolysaccharide, acetone-killed cells, and ribosome-rich extracts of *Salmonella typhimurium* in C3H/HeJ and CD-1 mice. *J. Immunol. 121,* 1010-1014.

Eisenstein TJ, Deakins LW, Killar L, Saluk PH, Sultzer BM (1981) Dissociation of innate susceptibility to salmonella infection and endotoxin responsiveness in C3Heb/FeJ mice and other strains in the C3H lineage. *Infect. Immun. 36,* 696.

Forni L, Coutinho A (1978) An antiserum which recognizes lipopolysaccharide-reactive B cells in the mouse. *Eur. J. Immunol. 8,* 56-62.

Gangemi JD, Ghaffar A, Trauger RJ, Sigel MM (1980) Natural killer activation in lipopolysaccharide responsive and non-responsive mice by viral and bacterial agents. *J. Reticuloendothel. Soc. 27,* 525-533.

Gangemi JD, Hightower JA, Jackson RA, Maher MH, Welsh MG, Sigel MM (1983) Enhancement of natural resistance to influenza virus in lipopolysaccharide-responsive and non-responsive mice by propioni. *Infect. Immun. 39,* 726-735.

Gery I, Kruger GJ, Speisel SZ (1972) Reactions of thymus-deprived mice and karyotypic analysis of dividing cells in mice bearing T_6T_6 thymus grafts. *J. Immunol. 108,* 1088-1091.

Glode LM, Rosenstreich DL (1976) Genetic control of B cell activation by bacterial lipopolysaccharide is mediated by multiple distinct genes or alleles. *J. Immunol. 117,* 2061-2066.

Glode LM, Scher I, Osborne B, Rosenstreich DL (1976a) Cellular mechanism of endotoxin unresponsiveness in C3H/HeJ mice. *J. Immunol. 116,* 454-461.

Glode LM, Mergenhagen SE, Rosenstreich DL (1976b) Significant contribution of spleen cells in mediating the lethal effects of endotoxin *in vivo. Infect. Immun. 14,* 626-630.

Glode LM, Jacques A, Mergenhagen SE, Rosenstreich DL (1977) Resistance of macrophages from C3H/HeJ mice to the *in vitro* cytotoxic effects of endotoxin. *J. Immunol. 119,* 162-166.

Goodman MG, Weigle WO (1979) T cell regulation of polyclonal B cell responsiveness. I. Helper effects of T cells. *J. Immunol. 122,* 2548.

Goodman MG, Parks DE, Weigle WO (1978) Immunologic responsiveness of the C3H/HeJ mouse: differential ability of butanol-extracted lipopolysaccharide (LPS) to evoke LPS-mediated effects. *J. Exp. Med. 147,* 800-813.

Gregory SH, Zimmerman DH, Kern M (1980) The lipid A moiety of lipopolysaccharide is specifically bound to B cell subpopulations of responder and non-responder animals. *J. Immunol. 125,* 102.

Gurtler LG, Rank WR (1976) Genetic differences between endotoxin-sensitive and resistant C3H mice. Intraperitoneal cell response to lipopolysaccharide and prostaglandins. *Z. Immunitaetsforsch. 15,* 420-429.

Haas GP, Johnson AG, Nowotny A (1978) Suppression of the immune response in C3H/HeJ mice by protein-free lipopolysaccharide. *J. Exp. Med. 148,* 1081.

Heppner G, Weiss DW (1965) High susceptibility of strain A mice to endotoxin and endotoxin-red blood cell mixtures. *J. Bacteriol. 90,* 696-703.

Hill GJ, Weiss DW (1964) Relationships between susceptibility of mice to heat-killed salmonella and endotoxin and the affinity of their red blood cells for killed organisms. In: Landy M, Braun W (Eds), *Bacterial Endotoxins,* pp 422-427. Institute of Microbiology, Rutgers State University, New Brunswick, New Jersey.

Hill AB, Hatswell JM, Topley WWC (1940) The inheritance of resistance, demonstrated by the development of a strain of mice resistant to experimental inoculation with a bacterial endotoxin. *J. Hyg. 40,* 538.

Hoffmann MK (1978) Serum from LPS nonresponder C3H/HeJ mice does not support the formation of functional EAC reagents. *J. Immunol. 121,* 619-621.

Hoffmann MK, Galanos C, Koenig S, Oettgen HF (1977) B-cell activation by lipopolysaccharide. Distinct pathways for induction of mitosis and antibody production. *J. Exp. Med. 146*, 1640-1647.

Ivins BE, Wyrick PB (1978) Response of C3H/HeJ and C3H/HeN mice and their peritoneal macrophages to the toxicity of *Chlamydia psittaci* elementary bodies. *Infect. Immun. 22*, 620.

Izui S, Eisenberg RA, Dixon FJ (1979) IgM rheumatoid factors in mice injected with bacterial lipopolysaccharides. *J. Immunol. 122*, 2096.

Jacobs DM, Roberts DB, Eldridge JH, Rosenspire AJ (1983) Binding of bacterial lipopolysaccharide to murine lymphocytes. *Ann. N.Y. Acad. Sci. 409*, 72-81.

Jakobovits A, Sharon N, Zan-Bar I (1982) Acquisition of mitogenic responsiveness by nonresponding lymphocytes upon insertion of appropriate membrane components. *J. Exp. Med. 156*, 1274.

Joiner FA, McAdam K, Kasper DL (1982) Lipopolysaccharides from *Bacteroides fragilis* are mitogenic for spleen cells from endotoxin responder and non-responder mice. *Infect. Immun. 36*, 1139.

Kabir S, Rosenstreich DL (1977) Binding of bacterial endotoxin to murine spleen lymphocytes. *Infect. Immun. 15*, 156-164.

Kelly K, Watson J (1979) Genetic and biochemical evidence for the involvement of a bacterial component in the mitogenic properties of polyribonucleotides on murine B lymphocytes. *J. Immunol. 122*, 2304-2308.

Kincade PW (1977) Defective colony formation by B lymphocytes from CBA/N and C3H/HeJ mice. *J. Exp. Med. 145*, 249-263.

Kirchner H, Brunner H, Ruhl H (1977) Effect of *A. laidlawii* on murine and human lymphocyte cultures. *Clin. Exp. Immunol. 29*, 176-180.

Kirchner H, Hirt HM, Rosenstreich DL, Mergenhagen SE (1978) Resistance of C3H/HeJ mice to lethal challenge with herpes simplex virus (39983). *Proc. Soc. Exp. Biol. Med. 157*, 29-32.

Kiyono H, Babb JL, Michalek SM, McGhee JR (1980a) Cellular basis for elevated IgA responses in C3H/HeJ mice. *J. Immunol. 125*, 732-737.

Kiyono H, McGhee JR, Michalek SM (1980b) Lipopolysaccharide regulation of the immune response: comparison of responses to Lps in germfree, *Escherichia coli*-monoassociated and conventional mice. *J. Immunol. 124*, 1603-1611.

Kiyono H, McGhee JR, Wannemuehler MJ, Michalek SM (1982) Lack of oral tolerance in C3H/HeJ mice. *J. Exp. Med. 155*, 605-610.

Koenig S, Hoffmann MK, Thomas L (1977) Induction of phenotypic lymphocyte differentiation in LPS unresponsive mice by an LPS-induced serum factor and by lipid A-associated protein. *J. Immunol. 118*, 1910-1911.

Mannel DN, Rosenstreich DL, Mergenhagen SE (1979) Mechanism of lipopolysaccharide-induced tumor necrosis: requirement for lipopolysaccharide-sensitive lymphoreticular cells. *Infect. Immun. 24*, 573-576.

Mannel DN, Moore RN, Mergenhagen SE (1980) Endotoxin-induced tumor cytotoxic factor. In: Schlessinger DS (Ed), *Microbiology – 1980*, p 141. American Society for Microbiology, Washington D.C.

McAdam KPWJ, Ryan JL (1978) C57BL/10/CR mice: nonresponders to activation by the lipid A moiety of bacterial lipopolysaccharide. *J. Immunol. 120*, 249-253.

McAdam KPWJ, Sipe JD (1976) Murine model for human secondary amyloidosis: genetic variability of the acute-phase serum protein SAA response to endotoxins and casein. *J.*

Exp. Med. 144, 1121-1127.

McGhee JR, Farrar JJ, Michalek SM, Mergenhagen SE, Rosenstreich DL (1979) Cellular requirements for lipopolysaccharide adjuvanticity: a role for both T lymphocytes and macrophages in *in vitro* responses to particulate antigens. *J. Exp. Med. 149*, 793.

McIntyre FC, Sievert HW, Barlow GH, Finley RA, Lee AY (1967) Chemical, physical and biological properties of a lipopolysaccharide from *Escherichia coli* K235. *Biochemistry 6*, 2363-2372.

Melchers F, Braun V, Galanos C (1975) The lipoprotein of the outer membrane of *Escherichia coli*: a B-lymphocyte mitogen. *J. Exp. Med. 142*, 473-482.

Meltzer MS, Nacy CA (1980) Macrophages in resistance to rickettsial infection: susceptibility to lethal effects of *Rickettsia akari* infection in mouse strains with defective macrophage function. *Cell Immunol. 54*, 487.

Michalek SM, Moore RN, McGhee JR, Rosenstreich DL, Mergenhagen SE (1980) The primary role of lymphoreticular cells in the mediation of host responses to bacterial endotoxin. *J. Infect. Dis. 141*, 55-63.

Mita A, Ohta H, Mita T (1982) Induction of splenic T cell proliferation by lipid A in mice immunized with sheep red blood cells. *J. Immunol. 128*, 1709.

Moeller GR, Terry L, Snyderman R (1978) The inflammatory response and resistance to endotoxin in mice. *J. Immunol. 120*, 116-123.

Moreno E, Berman DT (1979) *Brucella abortus* lipopolysaccharide is mitogenic for spleen cells of endotoxin-resistant C3H/HeJ mice. *J. Immunol. 123*, 2915.

Morgan EL, Weigle WO (1979) The requirement for adherent cells in the Fc fragment induced proliferative response of murine spleen cells. *J. Exp. Med. 150*, 256-266.

Morgan EL, Weigle WO (1980) Regulation of Fc fragment-induced murine spleen cell proliferation. *J. Exp. Med. 151*, 1-11.

Morrison DC, Kline LF (1977) Activation of the classical and properdin pathways of complement by bacterial lipopolysaccharides (LPS). *J. Immunol. 118*, 362-368.

Morrison DC, Betz SJ, Jacobs DM (1976) Isolation of a lipid A bound polypeptide responsible for 'LPS initiated' mitogenesis of C3H/HeJ spleen cells. *J. Exp. Med. 144*, 840.

Nacy CA, Meltzer MS (1982) Macrophages in resistance to Rickettsial infection: strains of mice susceptible to the lethal effects of *Rickettsia akari* show defective macrophage rickettsicidal activity *in vitro*. *Infect. Immun. 36*, 1096-1101.

O'Brien AD, Rosenstreich DL, Scher I, Campbell GH, MacDermott RP, Formal SB (1980) Genetic control of susceptibility to *Salmonella typhimurium* in mice. Role of the *Lps* gene. *J. Immunol. 124*, 20-24.

O'Brien AD, Metcalf ES, Rosenstreich DL (1982) Defect in macrophage function confers *Salmonella typhimurium* susceptibility on C3H/HeJ mice. *Cell. Immunol. 67*, 325-333.

Oldstone MBA, Dixon FJ (1973) Change in susceptibility of C3H/HeJ mice to LCM virus infection. *J. Immunol. 111*, 1613-1615.

Parant M, Galelli A, Parant F, Chedid L (1976) Role of B-lymphocytes in nonspecific resistance to *Klebsiella pneumoniae* infection of endotoxin-treated mice. *J. Infect. Dis. 134*, 531-539.

Parant M, Parant F, Chedid L (1977) Inheritance of lipopolysaccharide-enhanced nonspecific resistance to infection and of susceptibility to endotoxic shock in lipopolysaccharide low-responder mice. *Infect. Immun. 16*, 432-438.

Peavy DL, Baughn RE, Musher DM (1979) Effects of BCG infection on the susceptibility of mouse macrophages to endotoxin. *Infect. Immun. 24*, 59-64.

Pier GB, Markham RB, Eardley D (1981) Correlation of the biologic responses of C3H/HeJ mice to endotoxin with the chemical and structural properties of the lipopolysaccharides from *Pseudomonas aeruginosa* and *Escherichia coli. J. Immunol. 127*, 184.

Rank WR, Flugge U, DiPauli R (1969) Inheritance of the lipoid-A-induced LPS plaque forming cell response in mice: evidence for three antigen recognition mechanisms. *Behringwerk-Mitt. 49 Suppl*, 222.

Robson GG, Vas SJ (1972) Resistance of inbred mice to *Salmonella typhimurium. J. Infect. Dis. 126*, 378.

Rosenstreich DL, Blumenthal R (1977) Ionophorous activity and murine B cell mitogens. *J. Immunol. 118*, 129.

Rosenstreich DL, Glode LM (1975) Difference in B cell mitogen responsiveness between closely related strains of mice. *J. Immunol. 115*, 777-780.

Rosenstreich DL, McAdam KPWJ (1979) The role of lymphoid cells in endotoxin-induced production of the amyloid-related serum protein, SAA. *Infect. Immun. 23*, 181.

Rosenstreich DL, Mergenhagen SE (1975) Interaction of endotoxin with cells of the lymphoreticular system: cellular basis of adjuvanticity. In: Schlessinger D (Ed), *Microbiology – 1975*, pp 320-326. American Society of Microbiology, Washington D.C.

Rosenstreich DL, Nowotny A, Chused T, Mergenhagen SE (1973) *In vitro* transformation of mouse bone-marrow-derived (B) lymphocytes induced by the lipid component of endotoxin. *Infect. Immun. 8*, 406-411.

Rosenstreich DL, Glode LM, Wahl LM, Sandberg AL, Mergenhagen SE (1977) Analysis of the cellular defects of the endotoxin unresponsive C3H/HeJ mouse. In: Schlessinger D (Ed), *Microbiology – 1977*, pp 314-320. American Society of Microbiology, Washington D.C.

Rosenstreich DL, Vogel SN, Jacques AR, Wahl LM, Oppenheim JJ (1978a) Macrophage sensitivity to endotoxin: genetic control by a single codominant gene. *J. Immunol. 121*, 1664-1670.

Rosenstreich DL, Vogel SN, Jacques A, Wahl LM, Scher I, Mergenhagen SE (1978b) Differential endotoxin sensitivity of lymphocytes and macrophages from mice with an x-linked defect in B cell maturation. *J. Immunol. 121*, 685-690.

Ruco LP, Meltzer MS (1978) Defective tumoricidal capacity of macrophages from C3H/HeJ mice. *J. Immunol. 120*, 329-334.

Ruco LP, Meltzer MS, Rosenstreich DL (1978) Macrophage activation for tumor cytotoxicity: control of macrophage tumoricidal capacity by the *Lps* gene. *J. Immunol. 121*, 543.

Ruco LP, Uccini S, D'Agostoro G, DiMichele A, Doria G, Baroni CD (1981) Low antibody responsiveness to T-dependent antigens in C3H/HeJ mice. *Cell. Immunol. 59*, 1.

Rudbach JA, Reed ND (1977) Immunological responses of mice to lipopolysaccharide: lack of secondary responsiveness by C3H/HeJ mice. *Infect. Immun. 16*, 513-517.

Ryan JL, McAdam KPWJ (1977) Genetic non-responsiveness of murine fibroblasts to bacterial endotoxin. *Nature (London) 269*, 153.

Ryan JL, Glode LM, Rosenstreich DL (1979) Lack of responsiveness of C3H/HeJ macrophages to lipopolysaccharide: the cellular basis of LPS stimulated metabolism. *J. Immunol. 122*, 932-935.

Scibienski RJ (1981) Defects in murine responsiveness to bacterial lipopolysaccharide. The C3H/HeJ and C57BL/10ScCr strains. In: Gershwin ME, Merchant B (Eds), *Immunologic Defects in Laboratory Animals 2*, p 241. Plenum Press, New York.

Selinger MJ, McAdam KPWJ, Kaplan MM, Sipe JD, Vogel SN, Rosenstreich DL (1980) Monokine-induced synthesis of serum amyloid A protein by hepatocytes. *Nature (Lon-*

don) 285, 498.

Sipe JD, Vogel SN, Ryan JL, McAdam KPWJ, Rosenstreich DL (1979) Detection of a mediator derived from endotoxin in stimulated macrophages that induces the acute phase SAA response in mice. *J. Exp. Med. 150*, 597.

Skidmore BJ, Morrison DC, Chiller JM, Weigle WO (1975) Immunologic properties of bacterial lipopolysaccharide (LPS). II. The unresponsiveness of C3H/HeJ mouse spleen cells to LPS-induced mitogenesis is dependent on the method used to extract LPS. *J. Exp. Med. 142*, 1488-1508.

Skidmore BJ, Chiller JM, Weigle WO, Riblet R, Watson J (1976) Immunologic properties of bacterial lipopolysaccharide (LPS). III. Genetic linkage between the *in vitro* mitogenic and *in vivo* adjuvant properties of LPS. *J. Exp. Med. 143*, 143-150.

Streeter PR, Corbeil LB (1981) Gonococcal infection in endotoxin-resistant and endotoxin-susceptible mice. *Infect. Immun. 32*, 105-110.

Sultzer BM (1968) Genetic control of leucocyte responses to endotoxin. *Nature (London) 219*, 1253-1254.

Sultzer BM (1969) Genetic factors in leucocyte responses to endotoxin: further studies in mice. *J. Immunol. 103*, 32-38.

Sultzer BM (1972) Genetic control of host responses to endotoxin. *Infect. Immun. 5*, 107-113.

Sultzer BM (1976) Genetic analysis of lymphocyte activation by lipopolysaccharide endotoxin. *Infect. Immun. 13*, 1579-1584.

Sultzer BM, Nillson BS (1972) PPD tuberculin a B cell mitogen. *Nature New Biol. 240*, 199.

Suter E, Ullman GE, Hoffman RG (1958) Sensitivity of mice to endotoxin after vaccination with BCG (Bacillus Calmette-Guérin) *Proc. Soc. Exp. Biol. Med. 99*, 167.

Sztein MB, Vogel SN, Sipe JD, Murphy PA, Mizel SB, Oppenheim JJ, Rosenstreich DL (1981) Role of macrophages in the acute-phase response: SAA inducer is closely related to lymphocyte activating factor and endogenous pyrogen. *Cell. Immunol. 63*, 164.

Tagliabue A, McCoy JL, Herberman RB (1978) Refractoriness to migration inhibitory factor of macrophages of LPS non-responder mouse strains. *J. Immunol. 121*, 1223.

Tagliabue A, Herberman RB, McCoy JL (1979) Variation among mouse strains in responsiveness to migration inhibitory factor. *Cell. Immunol. 45*, 464-468.

Talcott JA, Ness DB, Grumet FC (1975) Adjuvant effect of LPS in *in vivo* antibody response: genetic defect in C3H/HeJ mice. *Immunogenetics 2*, 507.

Truffa-Bachi P, Kaplan JG, Bona C (1977) The mitogenic effect of lipopolysaccharide metabolic processing of lipopolysaccharide by mouse lymphocytes. *Cell. Immunol. 30*, 1.

Vaheri A, Ruoslahti E, Sarvas M, Nurminen M (1973) Mitogenic effect by lipopolysaccharide and pokeweed lectin on density-inhibited chick embryo fibroblasts. *J. Exp. Med. 138*, 1356.

Verghese MW, Snyderman R (1981) Differential anti-inflammatory effects of LPS in susceptible and resistant mouse strains. *J. Immunol. 127*, 288.

Verghese MW, Snyderman R (1982) Endotoxin (Lps) stimulates *in vitro* migration of macrophages from Lps-resistant but not from Lps-sensitive mice. *J. Immunol. 128*, 608.

Verghese MW, Prince M, Snyderman R (1980) Genetic control of peripheral leukocyte response to endotoxin in mice. *J. Immunol. 124*, 2468-2473.

Vogel SN, Rosenstreich DL (1979a) Defective Fc receptor-mediated phagocytosis in C3H/HeJ macrophages. I. Correction by lymphokine induced stimulation. *J. Immunol. 123*, 2842-2850.

Vogel SN, Hansen CP, Rosenstreich DL (1979b) Characterization of a congenitally LPS resistant, athymic mouse strain. *J. Immunol. 122*, 619.

Vogel SN, Marshall ST, Rosenstreich DL (1979c) Analysis of the effects of lipopolysaccharide in macrophages: differential phagocytic responses of C3H/HeN and C3H/HeJ macrophages *in vitro. Infect. Immun. 25*, 328-336.

Vogel SN, Moore RN, Sipe JD, Rosenstreich DL (1980) BCG-induced enhancement of endotoxin sensitivity in C3H/HeJ mice. I. *In vivo* studies. *J. Immunol. 124*, 2004-2009.

Vogel SN, Weedon LL, Oppenheim JJ, Rosenstreich DL (1981) Defective Fc-mediated phagocytosis in C3H/HeJ macrophages. II. Correction by cAMP agonists. *J. Immunol. 126*, 441.

Vogel SN, Weedon LL, Moore RN, Rosenstreich DL (1982a) Correction of defective macrophage differentiation in C3H/HeJ mice by an interferon-like molecule. *J. Immunol. 128*, 380.

Vogel SN, Weedon LL, Wahl LM, Rosenstreich DL (1982b) BCG-induced enhancement of endotoxin sensitivity in C3H/HeJ mice. II. T cell modulation of macrophage sensitivity to LPS *in vitro. Immunobiology, 160*, 479.

Vogel SN, Finbloom DS, English KE, Rosenstreich DL, Langreth SG (1983a) Interferon-induced enhancement of macrophage Fc receptor expression: α-interferon treatment of C3H/HeJ macrophages results in increased numbers and density of Fc receptors. *J. Immunol. 130*, 1210.

Vogel SN, Hilfiker ML, Caulfield M (1983b) Endotoxin-induced T lymphocyte proliferation. *J. Immunol. 130*, 1774-1779.

Vogel SN, Madonna GS, Wahl LM, Rick PD (1984) *In vitro* stimulation of C3H/HeJ B cells and macrophages by A lipid A precursor molecule derived from *Salmonella typhimurium. J. Immunol. 132*, 347-353.

VonJeney N, Gunther E, Jann K (1977) Mitogenic stimulation of murine spleen cells: relation to susceptibility to Salmonella infection. *Infect. Immun. 15*, 26-33.

Vukajlovich SW, Morrison DC (1983) Conversion of lipopolysaccharides to molecular aggregates with reduced subunit heterogeneity demonstration of Lps-responsiveness in 'endotoxin in unresponsive' C3H/HeJ golenocytes. *J. Immunol. 130*, 2804.

Wahl LM, Rosenstreich DL, Glode LM, Sandberg AL, Mergenhagen SE (1979) Defective prostaglandin synthesis by C3H/HeJ mouse macrophages stimulated with endotoxin preparations. *Infect. Immun. 23*, 8-13.

Wannemuehler MJ, Kiyono H, Babb JL, Michalek SM, McGhee JR (1982) Lipopolysaccharide (Lps) regulation of the immune response: Lps converts germ free mice to sensitivity to oral tolerance induction. *J. Immunol. 129*, 959-965.

Watanabe T, Ohara J (1981) Functional nuclei of LPS-nonresponder C3H/HeJ mice after transfer into LPS-responder C3H/HeN cells by cell fusion. *Nature (London) 290*, 58.

Watson J (1977) Differentiation of B lymphocytes in C3H/HeJ mice: the induction of Ia antigens by lipopolysaccharide. *J. Immunol. 118*, 1103-1108.

Watson J, Riblet R (1974) Genetic control of responses to bacterial lipopolysaccharides in mice. I. Evidence for a single gene that influences mitogenic and immunogenic responses to lipopolysaccharides. *J. Exp. Med. 140*, 1147.

Watson J, Riblet R (1975) Genetic control of responses to bacterial lipopolysaccharides in mice. II. A gene that influences a membrane component involved in the activation of bone marrow derived lymphocytes by lipopolysaccharides. *J. Immunol. 114*, 1462-1468.

Watson J, Riblet R, Taylor BA (1977) The response of recombinant inbred strains of mice to bacterial lipopolysaccharides. *J. Immunol. 118*, 2088-2093.

Watson J, Kelly K, Largen M, Taylor BA (1978a) The genetic mapping of a defective LPS response gene in C3H/HeJ mice. *J. Immunol. 120*, 422-424.

Watson J, Largen M, McAdam KPWJ (1978b) Genetic control of endotoxic responses in mice. *J. Exp. Med. 147*, 39-49.

Watson J, Kelly K, Whitlock C (1980) Genetic control of endotoxin sensitivity. In: Schlessinger DS (Ed), *Microbiology – 1980*, p 4. American Society for Microbiology, Washington, D.C.

Weinberg JB, Chapman HA Jr, Hibbs JB Jr (1978) Characterization of the effects of endotoxin on macrophage tumor cell killing. *J. Immunol. 121*, 72-80.

Weiner E, Levanon D (1968) The *in vitro* interaction between bacterial lipopolysaccharide and differentiating monocytes. *Lab. Invest. 19*, 584-590.

Wood PR, Clark IA (1982) Genetic control of propionobacterium acnes-induced protection of mice against *Babesia microti. Infect. Immun. 35*, 52-57.

Zimmerman DH, Gregory S, Kern M (1977) Differentiation of lymphoid cells: the preferential binding of the lipid A moiety of lipopolysaccharide to B lymphocyte populations. *J. Immunol. 119*, 1018-1023.

Handbook of Endotoxin, Vol. 3: Cellular Biology of Endotoxin
L.J. Berry, editor
© Elsevier Science Publishers B.V., 1985

CHAPTER 5

Glucocorticoid-antagonizing factor*

ROBERT N. MOORE, GREGORY M. SHACKLEFORD AND L. JOE
BERRY

1. INTRODUCTION

The study of bacterial endotoxins and their toxic activities has been revitalized
over the past few years by several observations. These include the discovery and
use of mouse strains inherently defective in their capacity to respond to the
lipopolysaccharide (LPS) component of endotoxin (Coutinho et al 1977; Skidmore
et al 1975; Sultzer and Nilsson 1972), and the identification of host-derived
molecules that mediate certain responses generated by injection of LPS. By the
use of this information, it has been demonstrated that the toxic nature of LPS is
determined by lymphoreticular cells of the host rather than being an innate property of LPS (Michalek et al 1980) and that a key determinant of toxicity is the
interaction of LPS with cells of the mononuclear phagocyte system, particularly
macrophages. LPS-stimulated macrophages produce a variety of active molecules
that may be responsible for much of the pathology associated with the toxin. These
include hydrolytic enzymes (Bradley 1979; Wahl et al 1974), toxic oxygen metabolites (Pabst and Johnston 1980), procoagulant activity (Niemetz and Morrison
1977) and other molecules that can influence both the physiological and metabolic
state of the host (Filkins 1982; Kawakami et al 1982; Moore et al 1978a). The role
of mononuclear phagocytes in the response to LPS has been further emphasized
by recent observations that minute amounts of LPS (approaching 10^{-15} M)
can influence macrophage activities in vitro (Moore et al 1980; Pabst and Johnston
1980). There remain, however, many unanswered questions regarding toxicity associated with both endotoxemia and gram-negative bacterial sepsis. First, the precise nature of responses involved in in vivo toxicity remains to be elucidated.
Regardless of all the intricate in vitro experimentation, there is in fact a paucity
of direct evidence to establish whether macrophages and their products influence
the pathology associated with LPS injection or gram-negative bacterial infection.
Second, there remains a disturbing problem in that transfer of LPS-
induced mediators fails to generate the typical toxic response. Either factors are
missing which remain to be discovered or cannot be transferred effectively, or

---

* The work reported in this chapter was supported in part by Grant AI-10087 from the
National Institute of Allergy and Infectious Diseases.

there is a critical combination of events involved in toxicity which remains to be elucidated and cannot be identified by this approach.

The present chapter will deal with these problems through discussion of a mediator which appears to be directly involved in inducing the lethal state of endotoxin poisoning without being toxic itself. This mediator, glucocorticoid antagonizing factor (GAF), will be discussed not only in the context of its role in endotoxin poisoning but also as it relates to the pathogenesis of gram-negative infection. In addition, evidence accumulated recently regarding the mode of action of this mediator will be presented.

2. GLUCOCORTICOID-ANTAGONIZING FACTOR AND ENDOTOXEMIA

It is amusing to discuss historical aspects of a mediator that was discovered only 9 years ago. Such, however, is the nature of science that GAF, which was a breakthrough discovery only a few short years ago, is now just a member of a growing family of LPS-induced mediators. The discovery of GAF and delineation of its cellular origin, the macrophage (Moore et al 1976), was important for two reasons. First, it was only the second metabolically active molecule found that could be directly linked to a pathological condition associated with endotoxin poisoning. Leukocytic endogenous mediator (LEM), also known as endogenous pyrogen (EP), was the first such mediator described (Atkins 1960; Kampschmidt 1980). Second, it was the first clear finding of macrophage involvement in a pathological response to LPS. This early work is reviewed in detail elsewhere (Moore et al 1978a).

GAF is apparently responsible for the steroid-unresponsive state associated with endotoxin injection (Berry 1971; Moore et al 1980). This protein mediator can be detected in serum with an M_r of approximately 150 000 (Moore et al 1978b). It is produced within 2 hours of an LPS injection (Moore et al 1978b) and exerts at least some of its effects on hepatocytes during these first few hours (Goodrum and Berry 1979). It mimics LPS in that it inhibits corticosteroid-induced synthesis of the gluconeogenic enzyme phosphoenolpyruvate carboxykinase (PEPCK). This effect is demonstrable as suppression of enzyme induction elicited by pharmacologic doses of steroid, as suppression of normal circadian rhythm of the enzyme or as suppression of induction by imposed fasting (Moore et al 1978a; Rippe and Berry 1972). This effect of LPS and GAF helps to explain the derangement in carbohydrate metabolism associated with endotoxemia. Injection of LPS is associated with a progressive hypoglycemia whose major cause has been established as an impairment in gluconeogenesis (Berry 1975). Associated with the suppression of PEPCK induction, there is an enhancement of activity of the glycolytic enzyme pyruvate kinase (Snyder et al 1971) resulting in a ratio of enzymatic activities unfavorable for gluconeogenesis.

The pathophysiology associated with altered gluconeogenesis is further complicated by hyperinsulinemia (Yelich and Filkins 1980), which may further act to suppress blood glucose levels. Interestingly, Filkins (1982) has reported that LPS-stimulated macrophages secrete both an insulin-like activity and a molecule that stimulates release of insulin from the B cells within the pancreas. In addition, there is a dramatic effect of LPS on lipid metabolism in the intact host. Within 20 hours of an LPS injection, there is a rise in serum triglyceride levels and a decrease in adipose tissue levels of lipoprotein lipase activity, the enzyme responsible for this activity (Kawakami and Cerami 1981). Again, macrophages were identified as the source of a mediator responsible for this activity (Kawakami et al 1982). The relationship of GAF to these other mediators has not been investigated. However, it is apparent that at least a major portion of the metabolic disorders induced by LPS is the result of the products of stimulated macrophages.

3. GLUCOCORTICOID-ANTAGONIZING FACTOR IN GRAM-NEGATIVE BACTERIAL INFECTION

Pathological effects observed in endotoxemic animals are often assumed to accurately predict the pathophysiological state of animals infected with gram-negative bacteria. On the basis of our current knowledge, however, such generalizations must be regarded as unfounded. As Shands (1975) has recently stated, the relationship between artificially induced endotoxemia and gram-negative sepsis is unclear. It is therefore advisable to investigate effects produced by endotoxemia and by infection under carefully controlled experimental conditions to determine whether a correlation does exist. This is particularly important in the light of recent findings which demonstrate that responses of the host rather than direct actions of the toxin are responsible for several endotoxin effects. With regard to GAF and infection, two early reports suggested the presence of the mediator during experimental murine typhoid elicited by injection of *Salmonella typhimurium* (Berry et al 1975; Moore et al 1977a). Subsequent investigation not only confirmed that GAF-like activity was present in body fluids of infected mice but also demonstrated the connection between GAF production and the response of the mononuclear phagocyte system to the infection.

Evidence of a GAF-like effect elicited in mice infected with 50 LD_{50} of *S. typhimurium* is given in Figure 1A. At this infective dose, Ha/ICR mice begin to die within 96 hours, and there is approximately 80% mortality within 7 days. Activity levels of hepatic PEPCK remain relatively constant during the first 5 days of the infection, an observation that could lead to the erroneous conclusion that regulation of this enzyme's synthesis is not influenced during the infection. In actuality, there is a dramatic effect. First, as shown in Figure 1A, there is no induction of the enzyme in response to at least one stressful aspect of the infection, namely fasting. There is a severe diminution of food consumption during this 5-day interval (Moore et al 1977a). When control animals are allowed to eat only that amount consumed by infected animals during respective 24-hour intervals, there is a significant enhancement of PEPCK activity. This enhancement occurs between

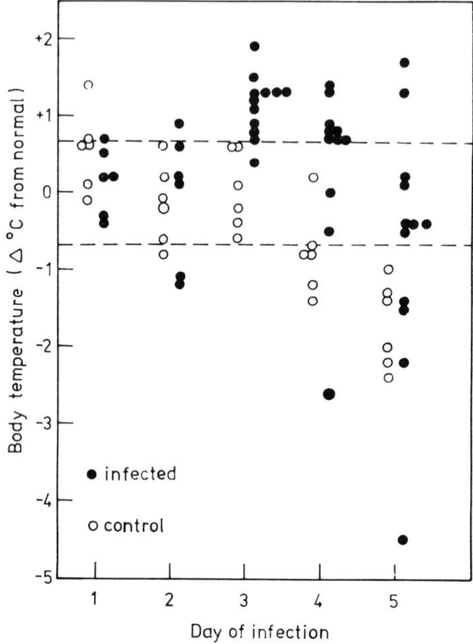

Fig. 3   *Body temperatures of infected and semifasted control mice following injection of S. typhimurium. Conditions are the same as those described in Figure 1. Dashed lines enclose the mean ± 1 standard deviation (35.8 ± 0.6 °C) for 25 normal mice measured at 1400 hr for 5 consecutive days.*

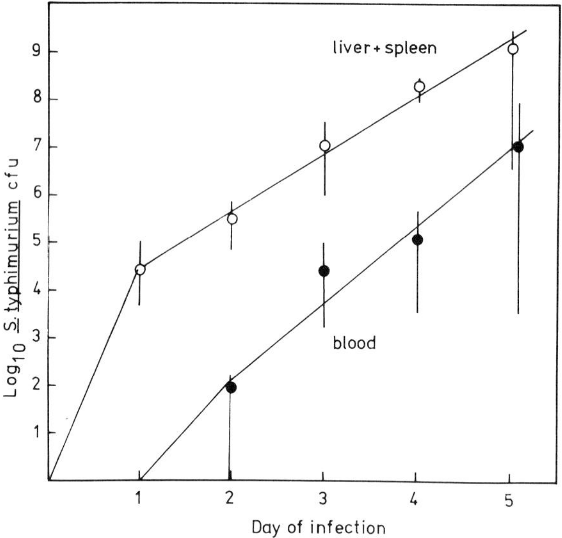

Fig. 4   *Bacterial counts in combined liver and spleen and in the blood of S. typhimurium-infected mice. Conditions are the same as described in Figure 1. Data points represent the mean and range of at least 6 mice per group.*

128

stage of the infection immediately preceding shock and death (Fig. 5), and there is an association between the severity of septicemia and the onset of hypoglycemia (Fig. 6). There is, of course, a reverse side to this argument, which is that blood-borne cfu result from the escape of bacilli from their macrophage reservoirs and represent the degree of bacterial multiplication within macrophage-rich organs. Correlations between blood-borne cfu and alterations in the physiological or metabolic state of the host would, therefore, be merely coincidental. In the remainder of this section, the argument will be advanced that the critical feature of this acute infection is septicemia and that the toxic effects result from the reassociation of blood-borne bacilli or their products with the mononuclear phagocyte system.

Treatment of animals with immunostimulatory agents significantly enhances their response to the toxic effects of LPS (Benacerraf et al 1959; Suter et al 1958; Vogel et al 1980). It is, therefore, important to note that one of the earliest detectable responses of mice to infection with *S. typhimurium* is a significant enhancement of phagocytic activity (Bohme et al 1959) (Fig. 7). This enhancement, measured as increased intravascular clearance of colloidal carbon (Bohme et al 1959), is dramatic and can be detected within 8 hours after injection of viable bacilli. The development of this enhancement is so rapid that it tends to exclude the involvement of a specific cellular immune response, and indeed, O'Brien and Metcalf (1982) have reported recently that early phagocyte function in murine typhoid does not require the participation of T lymphocytes or their products. Associated with the enhanced phagocytic activity is concomitant sensitization to LPS. The LD_{50} for Boivin-type *S. typhimurium* endotoxin given by the intravenous route decreased from 250 µg in control mice to 12.5 µg in mice infected for 2 days. This enhanced sensitivity is even more impressive in that the LD_{50} for control animals was determined over a 72-hour period while that for infected animals represents a 12-hour observation time. Sensitization to endotoxin during murine typhoid has been reported previously (Suter 1962), but an association with enhanced phagocytic activity has not been noted.

The argument, therefore, proceeds in the following manner. First, as an innate response to the presence of viable typhoid bacilli, the murine mononuclear phagocyte system is activated. This activation provides the host with an elevated level of nonspecific resistance. However, the activation process has a potentially dangerous consequence as well as a beneficial function. If the initially infected phagocytes are unable to kill or significantly delay bacterial multiplication, a progressive septicemia ensues. Blood-borne bacteria or their products (endotoxin) encounter previously uninfected but newly activated macrophages. The response of these secondarily exposed macrophages is greater than that of the initial phagocytes. Thus a small dosage of endogenously produced endotoxin can effectively stimulate the GAF and LEM effects associated with the onset of septicemia. In such an extreme acute infection as the one described, activation of the phagocyte system may actually contribute to the demise of the host rather than protect it.

Positive support for the macrophage activation argument would be a demonstration of enhanced secretory response to LPS in *S. typhimurium*-infected animals.

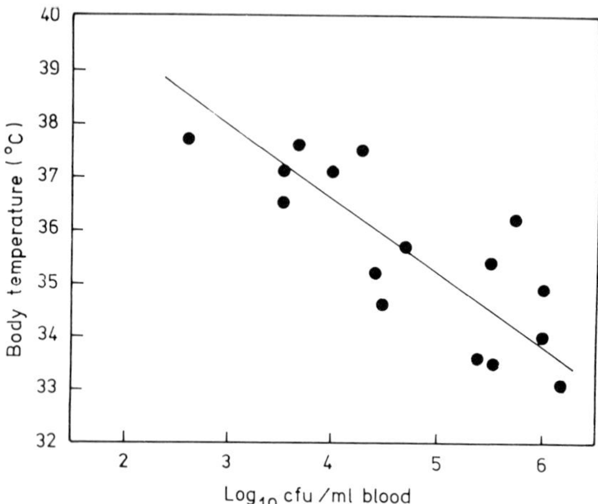

Fig. 5 *Relationship between blood-borne S. typhimurium and body temperature of 3- to 5-day infected mice. A significant correlation (p ≤ 0.001) between cfu and body temperature is indicated by linear regression analysis.*

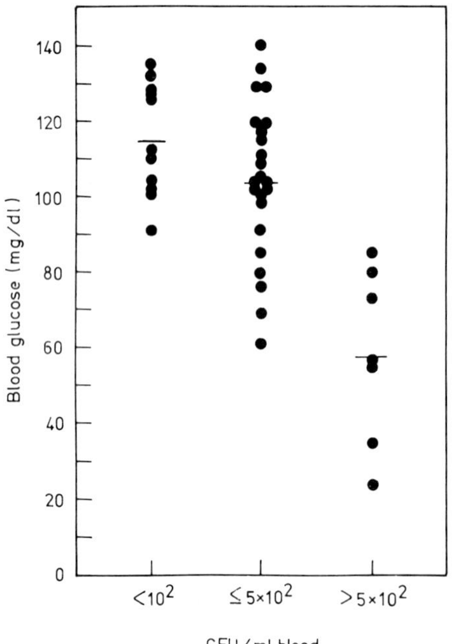

Fig. 6 *Relationship between blood-borne S. typhimurium and blood glucose levels in infected mice measured between 48 and 72 hours after injection of bacteria. Glucose levels were not significantly different between mice with $< 10^2$ and $\leq 5 \times 10^2$ cfu/ml. However, significant difference (p ≤ 0.001) was noted between mice having $\leq 5 \times 10^2$ and $> 5 \times 10^2$ cfu/ml.*

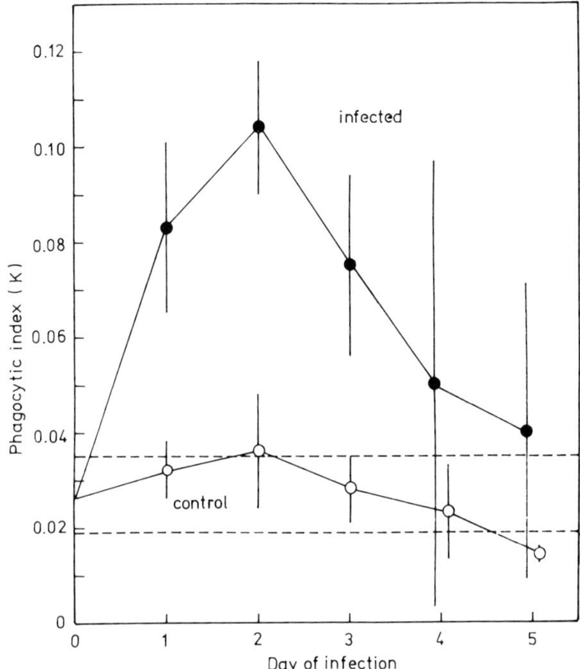

Fig. 7   *Phagocytic activity (K index), measured by intravascular clearance of colloidal carbon, of S. typhimurium and semifasted control mice. Conditions are those described in Figure 1. Data points represent the mean ± 1 standard deviation of a minimum of 6 mice per group. Dashed lines enclose the mean ± 1 standard deviation of 25 control normal mice measured 5 times a day over 5 consecutive days.*

Such data are given in Table 1. This experiment is based on the observation that athymic (nu/nu) mice do not exhibit a GAF response upon LPS treatment, that is, PEPCK remains steroid-inducible (Moore et al 1977b). Mice infected for 2 days and their semifasted controls were given 25 µg of LPS intravenously and bled 2 hours later. Previous findings had shown that 2 hours post-LPS serum contained maximum detectable levels of GAF (Moore et al 1978b). Upon injection into steroid-treated nu/nu mice, GAF activity was detectable in the serum of both LPS-treated infected and control mice. The serum of the LPS-treated, infected mice, however, was significantly more inhibitory for PEPCK induction indicating enhanced GAF production. Thus mice exhibiting elevated phagocytic activity in the early portion of an infection (Fig. 7), prior to exhibiting their own GAF response (Fig. 1), are primed to produce enhanced levels of the mediator upon exposure to LPS. Priming for production of LPS-stimulated macrophage products by infection with BCG has been reported for other mediators including interferon (Younger and Stinebring 1965), tumor necrosis factor (Carswell et al 1975) and interleukin (Mannel et al 1980). This is, however, the first report of a similar

TABLE 1   *GAF production in 48-hour S. typhimurium-infected mice injected 2 hours previously with 25 µg of S. typhimurium endotoxin*

Treatment**	Hepatic PEPCK activity ± SEM* in athymic (nu/nu) mice
None	115±7
0.5 ml control serum*** + 1 mg hydrocortisone	220±10
	NS
0.5 ml endotoxin-treated control serum + 1 mg hydrocortisone	189±22
0.5 ml infected mouse serum + 1 mg hydrocortisone	228±11
	$p \leq 0.001$
0.5 ml endotoxin-treated infected mouse serum + 1 mg hydrocortisone	155±13

*   6 mice per assay group.
**   pooled serum from 5 treated or untreated mice.
***   semifasted control mice.
NS = not significant.

priming effect in a gram-negative bacterial infection.

A last observation regarding GAF in murine typhoid is related to its production during the infection. Results have been presented indicating that a GAF-like response is detectable (Fig. 1) and that enhanced levels of GAF are produced following LPS injection (Table 1). Serum transfer experiments were, however, too insensitive to detect small amounts of GAF. Therefore, serum from infected mice was assayed for GAF activity by measuring its inhibition of steroid-induced PEPCK synthesis by the Reuber hepatoma cell line (Goodrum and Berry 1979). As shown in Figure 8, serum GAF-like activity was detectable during one sampling period of an extremely acute infection. In this experiment, there was 100% mortality during the third and fourth day. The presence of GAF in the serum, however, did correspond with the appearance of significant bacilli in the blood of infected animals. Thus an activity functionally defined as GAF can be detected in the serum of mice infected with *S. typhimurium*.

On the basis of the preceding data, it can be concluded that GAF is produced during murine typhoid and that its production correlates, at least temporally, with the onset of hypoglycemia. In addition, it appears that GAF production (as well as LEM production) is triggered by the presence of bacilli or their products in the blood and that production of at least GAF is augmented by prior activation of the mononuclear phagocyte system during the initial response to the infection. Production of this mediator and subsequent induction of a pathological state appears, therefore, to occur as a result of failure of mononuclear phagocytes to prevent the bacilli or their toxic products from entering the blood. It is, therefore, important

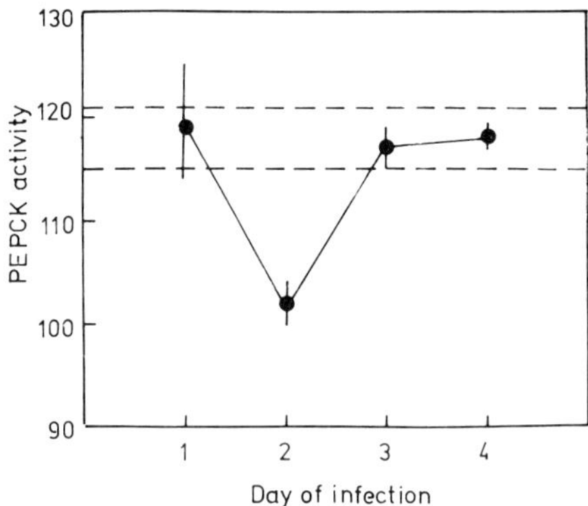

Fig. 8   *Effect of serum from S. typhimurium-infected mice on steroid induction of PEPCK in Reuber hepatoma cells. Mice were bled for serum (10 mice pooled per group) and 2% serum from each day added to the hepatoma cells at the time of enzyme induction by 10^{-6} M hydrocortisone-21-sodium succinate. Enzyme activity (nmol $NaH^{14}CO_3$ fixed/min at 37°C) was assayed 8 hours after addition of serum and steroid. Enzyme activity of nontreated cells was 73±6 units.*

not only to continue to study the mediators of endotoxic effects but also to investigate the mechanism responsible for macrophage activation, a process which can both protect the host or contribute to its demise if inadequate for the challenge.

4.   OBSERVATIONS ON THE MODE OF ACTION OF GLUCOCORTICOID-ANTAGONIZING FACTOR

All of the preceding in vivo studies of GAF had the uncertainties of whole animal research and were therefore deemed unsuitable for investigation of the mode of action of this mediator. To circumvent this difficulty, an assay system for GAF activity was devised with the use of Reuber H35 rat hepatoma cells in culture. These cells have retained glucocorticoid-inducible PEPCK and tyrosine aminotransferase (TAT) and were ideal for our purposes. Table 2 shows that the addition of endotoxin to hepatoma cells had no effect on the level of PEPCK nor on its inducibility (Goodrum and Berry 1978). However, the addition of GAF-rich serum to hepatoma cells completely blocked induction. The assumption that inhibition of PEPCK induction by endotoxin in the livers of mice was not a direct action of the poison but rather a mediated effect was confirmed with these findings using hepatoma cells. The results also show that the mediator in mouse serum retains activity with cells derived from a heterologous species. Moreover, it has been shown that rabbits also produce an endotoxin-induced mediator that is active in the rat hepatoma cell assay (Berry et al 1980). Thus, there does not appear to be a species specificity of the type that has been observed with interferon.

Also of interest is the fact that the induction of PEPCK in hepatoma cells by cyclic AMP (Wicks et al 1974) is not inhibited by the concomitant addition of GAF-rich serum (Table 2). This corroborates the evidence from in vivo studies of endotoxin-treated mice discussed above and suggests that the process of induction by hydrocortisone differs from that by cAMP. There is good experimental evidence to indicate that PEPCK induction by glucocorticoids results from enhanced production of specific enzymic mRNA (Iynedjian and Hanson 1977b; Yoo-Warren et al 1981). Increased levels of PEPCK mRNA have also been detected in animals treated with dibutyryl cAMP measured in both cell-free protein synthesizing systems (Beale et al 1981; Iynedjian and Hanson 1977a) and by DNA–RNA hybridization (Beale et al 1982; Yoo-Warren et al 1981). Thus, both glucocorticoids and cAMP can apparently induce PEPCK at a pretranslational step, presumably at transcription. On this basis, GAF would appear to interfere specifically with the process of transcriptional regulation by glucocorticoids which could occur at one of several points. This will be analyzed further later in this chapter.

Not only does GAF fail to inhibit cAMP induction of PEPCK, it also fails to inhibit the glucocorticoid induction of another hepatic enzyme retained by hepatoma cells, TAT. This can be seen in Table 3 taken from Goodrum and Berry (1978). The nearly fourfold increase in TAT activity that results from the addition of GAF-rich serum alone is presumed to be due to the presence of endogenous glycocorticoid in the serum. This finding with hepatoma cells confirms an earlier observation in mice where hepatic TAT induction by glucocorticoid was unaltered by endotoxin (Berry et al 1966). These two findings lead one to believe that the glucocorticoid induction of TAT mRNA (Nickol et al 1978; Roewekamp et al 1976) is, for reasons that are not clear, not inhibited by treatments with endotoxin

TABLE 2 *Induction of PEPCK in hepatoma cells exposed to endotoxin and GAF-rich serum*

Treatment	PEPCK activity following addition inducer to medium*		
	None	1 μM hydrocortisone	0.5 mM dibutyryl cAMP − 1.0 mM theophylline
None	40±2 (6)	96±2 (6)	90±2 (6)
Endotoxin 10 μg/ml	41±2 (6)	94±2 (6)	88±2 (6)
GAF-rich serum 2% v/v	39±2 (6)	41±1 (6)	88±2 (6)

* Expressed as nmoles $NaH^{14}CO_3$ fixed/min/mg protein ± SEM for 8-hour induction period. ( ) number of replicate experiments.
Modified from Goodrum and Berry (1978).

TABLE 3   *Effect of GAF-rich serum on the induction of TAT in hepatoma cells*

Treatment	TAT activity	
	Control cells	GAF-treated cells
None	36±2 (6)	140±9 (6)
Hydrocortisone (1μM)	276±10 (6)	249±9 (6)

() number of experiments.
Modified from Goodrum and Berry (1978).

or GAF. This point will be considered again later.

The technique used to maximize the amount of GAF present in serum has been described (Goodrum and Berry 1979; Moore et al 1978a). The priming of mice with zymosan is believed to stimulate proliferation of the reticuloendothelial system, including macrophages. The administration of endotoxin to these mice results in the appearance of a peak level of GAF in blood 2 hours later (Moore et al 1978b). It may be relevant to point out that mice so primed are greatly sensitized to the lethal effect of endotoxin, and it was of interest to establish whether the amount of circulating GAF varied under different experimental conditions. This was determined by Goodrum and Berry (1978) through the use of an assay developed for the purpose. The GAF-rich serum was added to cultures of hepatoma cells at concentrations of 0.25% (v/v), 0.5%, 0.75%, 1.08% and 2%, then the percent induction of PEPCK was measured and graphed. A linear decline in induction was obtained with increasing amounts of serum up to about 1% of the total culture volume. This linear graph permitted the calculation of the 50% inhibitory dose (ID_{50}) of GAF. The total number of ID_{50} units per ml would then be estimated with the results shown in Table 4 (taken from Goodrum and Berry 1979). Several interesting facts emerge. GAF is detectable in normal serum even though it is not present in an amount sufficient to inhibit PEPCK induction when 0.2 ml is injected into endotoxin-tolerant mice (Moore et al 1976) or C3H/HeJ mice (Moore et al 1978b). The administration of endotoxin to normal mice raises the GAF titer to more than double the control amount. This titer in circulating blood is sufficient to block the glucocorticoid induction of PEPCK in mice. Priming mice with zymosan prior to injecting them with endotoxin results after 2 hours in the appearance of the largest titer of GAF measured (35 ID_{50}/ml of serum). These mice, as already mentioned, are highly sensitive to endotoxin and behave very much as if they were adrenalectomized. An injection of the high-titered serum into control mice does not result, however, in manifestations of endotoxin poisoning, but mice so treated cannot withstand the stress of exposure to 5 °C cold as has been demonstrated by Couch et al (1979). GAF, therefore, is not a toxic substance but its antagonism of glucocorticoids seems to deny the animal its ability to cope with alterations in its external or internal environments. The high degree of sensitivity of zymosan-primed mice to the lethal effect of endotoxin cannot be duplicated by an injection of GAF-rich serum into normal recipients. This leads one to believe that some lethal factor is responsible for the prompt death of these mice. If such a factor exists and is distributed systematically via the circulation, it must either leave the blood stream very rapidly or else be unusually labile since it does not appear to be demonstrable in serum. The dilution factor involved in the injection of small volumes of serum into mice could also explain its nondetection by these methods. The concept of a lethal factor, which can only be inferred, is nevertheless intriguing and helps to explain otherwise confusing observations.

Two antiinflammatory agents are known to protect animals against at least some of the biological effects of endotoxin. These are indometacin and hydrocortisone. As the data of Table 4 make evident, both compounds suppressed the titer of GAF that appears in the serum of zymosan-primed mice injected with endotoxin. In both situations the titer is less than that measured in serum of normal mice given endotoxin. The extent to which prostaglandins are involved in the production of GAF is not known but the possibility that they are involved cannot be overlooked in light of these results. Finally, the extremely low titers of GAF that were found in endotoxin-tolerant mice (Table 4) may help to explain why these animals are so refractory to endotoxin. Until more is known about the sequence of events that leads to the production and release of GAF from macrophages, the induced titer

TABLE 4   *Titration of GAF with hepatoma cell cultures expressed as ID_{50}/ml of serum*

Source of serum	Titer (ID_{50}/ml)
Normal mice	
untreated	6 ± 2 (5)
endotoxin-injected (50 µg i.v.)	15 ± 1 (4)
Zymosan-treated mice	
control	6 ± 1 (3)
endotoxin-injected (50 µg i.v.)	35 ± 4 (4)
Indometacin-pretreated mice (24 hr)	
control	4 (1)
endotoxin-injected (50 µg i.v.)	8 (1)
zymosan-primed, endotoxin-injected (50 µg i.v.)	14 (1)
Hydrocortisone-pretreated mice (24 hr)	
control	4 (1)
endotoxin-injected (50 µg i.v.)	12 (1)
zymosan-primed, endotoxin-injected (50 µg i.v.)	13 (1)
Endotoxin-tolerant mice	
control	5 (1)
endotoxin-injected (50 µg i.v.)	4 ± 1 (3)

() number of separate determinations.

observed after an injection of endotoxin into tolerant mice will remain purely phenomenological.

The data contained in Table 5, taken from a paper by Goodrum and Berry (1978), provide convincing experimental evidence that adherent peritoneal exudate cells in culture produce highly significant amounts of GAF. These cells were derived from normal animals (not zymosan-primed ones) and there were 1.27×10^7 cells per culture dish. The supernatant fluid was collected, concentrated 10-fold and desalted before it was added to the hepatoma cells in culture. Assays for PEPCK were carried out 8 hours after the addition of hydrocortisone. The results indicate that the addition of 10 μg/ml of endotoxin to the macrophages in culture for a period of 4 hours elicited as much GAF as was found after 24 hours of contact. Unless the assay system used here is saturated with GAF activity in both cases, these data suggest that the synthesis of GAF by macrophages may not be a continuing process but rather one of limited duration. If so, then it is possible that the macrophages in endotoxin-tolerant mice are depleted of GAF and other mediators for the period of tolerance. Although this has not been determined experimentally, such investigations may help provide insight into the biochemical (metabolic?) basis of tolerance.

Experiments by Moore et al (1978b) suggest that GAF, when precipitated with ammonium sulfate between 33% and 67% saturation and then chromatographed on Sephadex G200, has a molecular weight in the region between 100 000 and 200 000 daltons (see above). GAF was inactivated by heating to 75 °C for 1 hour or by digesting with trypsin at 37 °C for 2 hours. These findings were obtained in studies with C3H/HeJ mice which respond to GAF but not to endotoxin itself. Goodrum and Berry (1979) confirmed these characteristics of GAF in experiments with hepatoma cells. On the basis of these observations, GAF must either be a large protein molecule, a polymerized smaller molecule, or else a small molecule bound to a larger protein carrier.

TABLE 5 *Effect of macrophage culture supernatants on the induction of PEPCK in hepatoma cells*

Additions to medium	PEPCK activity*
None	39±3 (6)
1 μM hydrocortisone	80±4 (6)
1 μM hydrocortisone + 10% v/v macrophage supernate from	
4-hr untreated cells	58±1 (6)
4-hr endotoxin-treated cells	51±3 (6)
24-hr untreated cells	64±3 (6)
24-hr endotoxin-treated cells	50±1 (6)

* Expressed as nmoles $NaH^{14}CO_3$ fixed/min per mg protein ± SEM for 8-hour induction period for the number of determinations shown in parentheses.

It is relevant to pose the question of whether some or all of the properties of GAF are dependent upon either the concurrent presence of insulin in the GAF-rich samples of serum or the association of insulin with the molecule we refer to as GAF. In an attempt to elucidate the matter, the experiments summarized in Table 6 were carried out by Goodrum and Berry (1979). The important facts that emerge from these data are (a) both insulin and GAF-rich serum inhibit PEPCK induction by hydrocortisone while (b) only insulin inhibits induction of the enzyme by cAMP. The latter observation discriminates between the action of the two substances. Other differences have also been found. Heating GAF-rich serum at 70 °C for 30 minutes inactivates it, but similar treatment of insulin is without effect on its ability to inhibit PEPCK induction (Goodrum and Berry 1979). Moreover, the hormonal induction of renal PEPCK in mice is inhibited by GAF-rich serum but not by insulin (unpublished observations). The lack of uniform action by insulin and GAF, in the experiments described, leads us to believe that the two substances are distinct entities. In addition, the clear evidence that macrophages produce GAF, or at least some substance with an action that is identical to some effects of endotoxin, virtually eliminates insulin as the primary mediator even though Filkins (1982) has shown that macrophages synthesize a factor that has insulin-like activity. If insulin were a normal product of these cells, diabetes would be a disease with an etiology quite different from B cell dysfunction in the pancreas. Moreover, GAF and insulin would behave differently when filtered on Sephadex G200 unless the hormone were complexed with a protein carrier, which seems unlikely in view of the ease with which insulin in serum can be quantitated by the radio-immunoassay technique.

The thought of an endogenous protein (GAF) blocking the induction of certain hepatic and renal enzymes rather than endotoxin itself makes the concept more acceptable since the mediator, even though a large molecule, is still smaller than endotoxin by nearly an order of magnitude. However, it must be remembered that a number of experiments have indicated that the site of attack of GAF is not the enzyme directly but rather its synthesis (Rippe and Berry 1972). In other words,

TABLE 6   *Effect of insulin on PEPCK induction in hepatoma cells*

Inducer added to culture medium	PEPCK activity*		
	Control cells	Insulin-treated cells	GAF-rich serum-treated cells
Hydrocortisone (1 µM)	96±4 (5)	50±2 (5)	39±1 (4)
Dibutyryl cAMP (0.5 mM) + theophylline (1 mM)	70±3 (5)	41±4 (5)	72±5 (5)

*   Expressed as nmoles $NaH^{14}CO_3$ fixed/min per mg protein ± SEM for 8-hour induction period for number of separate determinations shown in parentheses.

GAF is not an enzyme poison. It intervenes in some way to prevent the glucocorticoid from augmenting the total amount of enzyme produced.

In order to understand where GAF might interfere, it is necessary to list the sequence of events known to occur (for a review, see Higgins and Gehring 1978) during the glucocorticoid induction of proteins (Fig. 9). First, the steroid hormone diffuses through the cell membrane and binds with high affinity to specific cytoplasmic receptor proteins. Second, this steroid–receptor complex then undergoes a conformational change, termed activation, which enables it to enter the nucleus and bind to acceptor sites on the chromatin. These sites have yet to be determined, but certain DNA sequences are known to play a role (Tata 1982) which may help to explain the high degree of specificity involved in the response. Third, the consequence of the binding of the hormone complex to chromatin is an augmentation in the transcription of specific protein RNAs. Only a very small subset (less than 1%) of active genes are affected by glucocorticoids (Ivarie and O'Farrell 1978). The molecular mechanism of this transcriptional regulation is unknown. Fourth, an increased level of specific protein results from the translation of the induced mRNA for that protein.

To begin answering questions concerning the mechanism of the blockage by GAF of the glucocorticoid induction of PEPCK, each step of the induction sequence described above was analyzed for endotoxin-induced changes. As Table 7 shows, the amount of tritium-labeled dexamethasone that binds to cytoplasmic receptors from adrenalectomized mice is not reduced by the injection of an amount of endotoxin that blocks PEPCK induction. From this it appears that GAF alters neither the number nor the binding characteristics of glucocorticoid receptors. Thus, the first step, above, does not appear to be the point of attack of GAF.

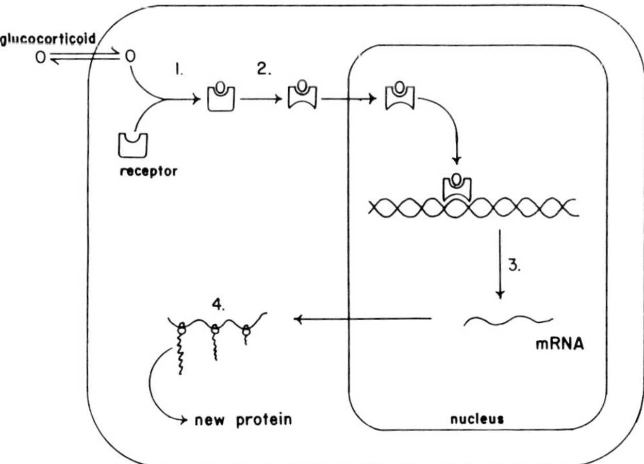

Fig. 9 *Events in the glucocorticoid induction of proteins. (1) Glucocorticoid–receptor complex formation; (2) activation of the complex; (3) induced transcription of affected genes; (4) translation of the induced mRNA.*

TABLE 7   *Effect of endotoxin treatment on hepatic glucocorticoid receptor complex formation*

Treatment	Specific binding of ^{3}H-dexamethasone (cpm/ml cytosol)*
None	61 450
Endotoxin 50 µg, 4 hr	62 890

* 33% cytosol prepared from pooled livers of at least 3 adrenalectomized ICR mice. Binding was measured in void volume of P-6 desalting columns and corrected for nonspecific binding in the presence of 1000-fold excess unlabeled hormone.

Contradictory results have recently been reported by McCallum and Stith (1982), but despite their claims of a significant difference between the control level of binding and slightly lower binding in the endotoxin group, our calculations of their published data show no significance (at $p = 0.95$) using the one-tailed or the two-tailed Student's t test. In a later publication by the same investigators (Stith and McCallum 1983), the conclusion that down-regulation of glucocorticoid receptors (i.e., lowered measured binding) is causally related to the inhibition of PEPCK induction is accompanied by an important contradiction in their data. At a dose (100 µg endotoxin) and time (3 hours post-treatment) when PEPCK induction is significantly inhibited, no significant reduction in the binding of labeled glucocorticoid to receptors could be measured.

The effect of endotoxin on the second step, activation, has been evaluated with the use of DEAE ion-exchange chromatography which can separate activated from unactivated complexes (Munck and Foley 1979; Schmidt et al 1980). Figure 10 illustrates the results from a typical experiment where no difference in the activation capabilities of receptors isolated from untreated or endotoxin-treated mouse livers can be detected.

In order to evaluate the effect of GAF on step three, it was necessary to isolate mRNA from livers of mice 4 hours after an injection of either hydrocortisone alone or of hydrocortisone and endotoxin given concurrently. The RNA was added to a rabbit reticulocyte lysate preparation containing ^{35}S-methionine and the other constituents necessary (Pelham and Jackson 1976) for in vitro protein synthesis. An aliquot of the supernatant fluid from each type of reaction mixture was subjected to sodium dodecylsulfate polyacrylamide gel electrophoresis (SDS-PAGE) followed by autoradiography. As shown in the autoradiograph of small aliquots of total synthesized proteins (Fig. 11), each RNA preparation was able to synthesize proteins to a similar extent except for minor differences due to the induction of mRNAs in the endotoxin-treated groups (acute-phase proteins) or in the hydrocortisone-treated groups. To assay functional PEPCK-mRNA, rabbit anti-PEPCK antibodies were used to precipitate PEPCK out of the in vitro translation mixtures with the aid of protein A-bearing *Staphylococcus aureus*. The immunoprecipitated

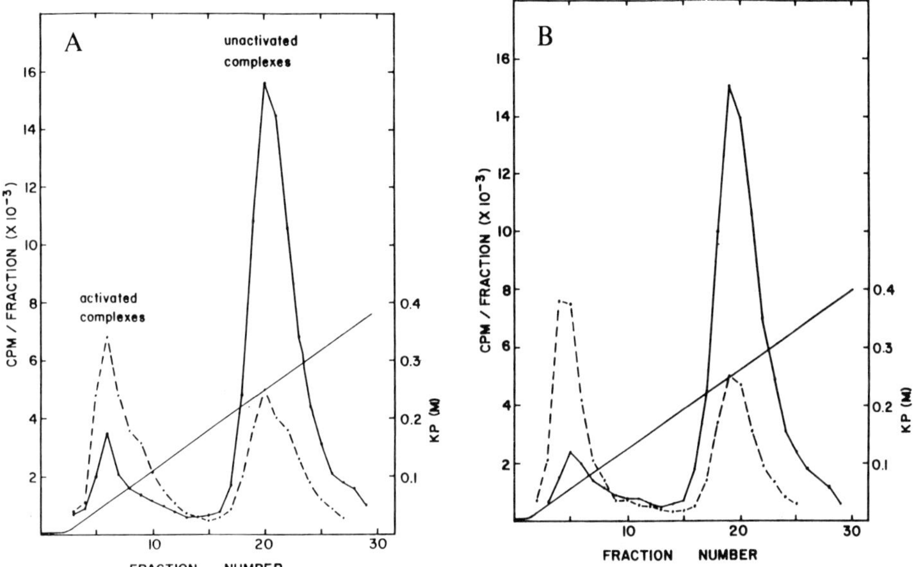

Fig. 10 *Activation of hepatic glucocorticoid receptor complexes from untreated and endo-toxin-treated adrenalectomized mice. The cytosols from pooled livers of untreated (A) or endotoxin-treated (50 μg, 4 hr) (B) mice were allowed to bind excess ^{3}H-dexamethasone and split into two equal aliquots one of which was activated (23 °C, 30 min) while the other was not (0 °C, 30 min). These were desalted to remove unbound hormone and applied to a DEAE-Sephadex column. After washing, activated and unactivated complexes elute sequentially with a 5mM-400mM potassium phosphate gradient. Activated cytosol (·—·—·—·); unactivated cytosol (——).*

enzyme was eluted from the washed *Staphylococci* and subjected to SDS-PAGE. The bands corresponding to PEPCK in the hydrocortisone-treated groups are, as expected, more intense than those in the control groups. RNA preparations from livers of poisoned mice (4 hours after endotoxin and hydrocortisone were injected) produced either no corresponding bands or barely perceptible ones. This indicates that endotoxin blocks at some point and in some way the ability of the glucocorticoids to stimulate functional PEPCK-mRNA synthesis. Of course, these experiments cannot rule out the possibility that endotoxin treatment has affected PEPCK-mRNA stability or its ability to be translated (step four). These possibilities are being investigated with the use of recombinant DNA techniques to measure PEPCK-mRNA directly by nucleic acid hybridization. Unfortunately, as mentioned above, little is known of the mechanism of transcriptional induction by glucocorticoid; however, this newly discovered effect of endotoxin, or actually GAF, may serve as an important probe for further analysis of how mRNA synthesis is increased by adrenocortical hormones. This information must be in hand

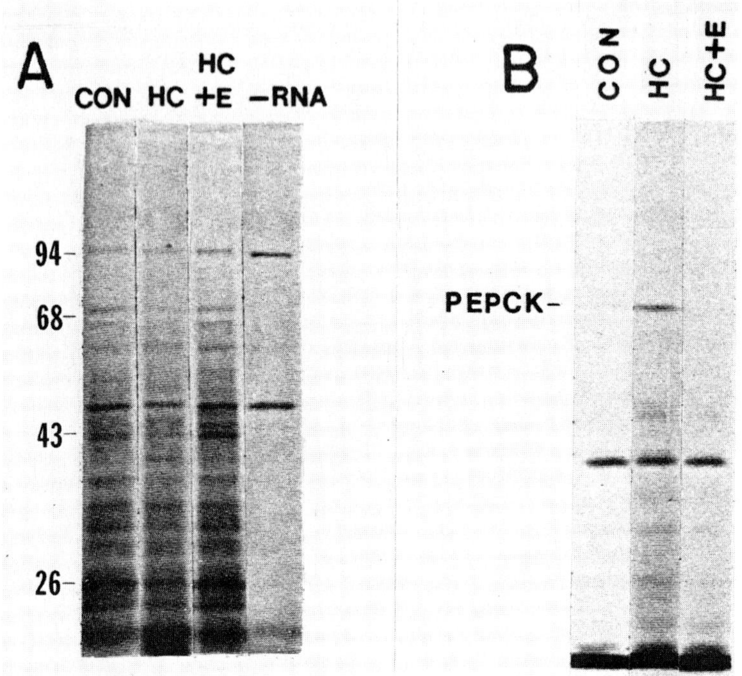

Fig. 11   *Effect of endotoxin on the glucocorticoid induction of functional PEPCK mRNA. (A) Autoradiogram of ^{35}S-methionine–labeled translation products of total mouse liver RNA separated by SDS-PAGE. Molecular weights ($\times$ 10^{-3}) are indicated. (B) Autoradiogram of labeled PEPCK immunoprecipitated from the translation assays and separated from cross-reacting or nonspecifically precipitated products by SDS-PAGE. The position of a purified mouse liver PEPCK standard is indicated. Con = control; HC = hydrocortisone; E = endotoxin.*

before the precise nature of the inhibitory specificity can be understood, that is, why GAF blocks PEPCK induction and not that of TAT.

Up to this point, GAF has been considered only as an inhibitor of glucocorticoid induction of certain hepatic and renal enzymes. If the name of the mediator, glucocorticoid-antagonizing factor, is to be appropriate, it must also act to antagonize other effects of the hormone(s). The earliest observed interaction between cortical hormones and endotoxin was the antilethal property of the hormone. Mice given an LD_{90} dose of endotoxin survived when injected immediately thereafter with hormone (Berry and Smythe 1964). A large dose is required for protection and the interval between the two injections cannot be more than about 1 hour. Why endotoxemic shock becomes irreversible within this brief time and before most of the overt symptoms of poisoning appear is not fully understood. It has been known for years (Berry and Smythe 1964), however, that the induction of enzymes is no longer possible at the same time that glucocorticoids no longer protect. Since this happens with endotoxin, it should happen when GAF is substituted. Data to show these relationships are presented in Table 8. It is evident that 150 μg of endotoxin killed two-thirds of the mice. Concurrent injections of 1 mg hydrocortisone protected them but 2 hours after the endotoxin was given, hydrocortisone was nonprotective. When GAF-rich serum was given 2 hours before 150 μg of endotoxin, all mice died and at this time they could not be protected when hydrocortisone was given concurrently. Obviously, GAF had interacted with host cells in such a way that the effects of the hormone were neutralized. One of these effects is the ability of the hormone to induce specific enzymes (see Table 2).

Some years ago Previte and Berry (1963) found that mice exposed to 5 °C (singly housed without bedding) were sensitized 100–1000-fold to the lethal effect of endotoxin compared to mice kept at 23 °C. More recently, Couch et al (1979) confirmed these results and noted that 50 ng of endotoxin killed 11 of 12 mice within 24 hours. In fact, almost half were dead after 10 hours. When GAF-rich serum alone was injected into endotoxin-tolerant mice exposed to the cold, all of 25 mice were dead after 24 hours. One-third of the uninjected controls were dead at this time. Aminoglutethimide, a drug that blocks the secretion of adrenal cortical hormones, was given at a dosage of 3 mg to mice at the time of exposure to cold and all of 12 mice died. The net result of GAF and aminoglutethimide is the same even though it is known (Berry and Smythe 1961) that endotoxin (and GAF), rather than blocking cortical hormone secretion, stimulates its release, probably through an increase in ACTH from the pituitary.

To show that endotoxin stimulates the release of endogenous glucocorticoids, endotoxin-tolerant mice were used to obtain the results in Table 9. Notice first that the stress of cold alone increased PEPCK activity by 70% above the value found at room temperature. A small amount of endotoxin injected into the tolerant mice at 25 °C nearly doubled PEPCK activity while a smaller dose at 5 °C did not significantly alter the inductive effect of the cold. GAF-rich serum, however, inhi-

TABLE 8 *Effect of delay on the protective ability of hydrocortisone in endotoxemic or GAF-injected mice*

Treatment	No. survivors/total
150 μg endotoxin	3/9
150 μg endotoxin + 1 mg hydrocortisone	5/5
150 μg endotoxin + 1 mg hydrocortisone 2 hr later	3/10
0.4 ml GAF-rich serum + 150 μg endotoxin 2 hr later	0/17
0.4 ml GAF-rich serum + 150 μg endotoxin + 1 mg hydrocortisone 2 hr later	0/18

TABLE 9 *Induction of PEPCK in tolerant mice given a small dose of endotoxin or an injection of GAF-rich serum and exposed to cold*

Experimental treatment	PEPCK activity in mice housed for 4 hr at	
	25 °C	5 °C
Control	127±22 (6)	212±3 (6)
Endotoxin, 25 µg at 25 °C and 10 µg at 5 °C	243±72 (11)	192±29 (12)
0.4 ml GAF-rich serum	123±28 (12)	130±33 (11)

bited the induction of PEPCK at both 5 °C and 25 °C. The very large induction that is seen at 25 °C after the injection of endotoxin alone most likely reflects the release of endogenous glucocorticoids and the formation by tolerant mice of as little GAF as is indicated by the data in Table 4. Thus the magnitude of induction of PEPCK in mice placed in the cold and in tolerant mice given a small dose of endotoxin proves that endogenous glucocorticoid stimulates as much enzyme synthesis as an injection of 1 mg hydrocortisone. The exogenous hormone has low solubility and while the dosage used approximates a significant percentage of the total weight of the adrenal cortical tissue, the amount that reaches the circulation must be little more than that secreted by the glandular tissue.

Another way of demonstrating that endogenous glucocorticoid is present in sufficient amount to saturate glucocorticoid receptors in liver of endotoxin-treated animals and hence give full inductive impetus to PEPCK and other enzymes is shown in Figure 12. It can be seen that liver preparations of intact mice to which radiolabeled dexamethasone was added failed to bind any of the hormone when they were derived from cold-stressed mice or from endotoxin-treated mice (due to presaturation by endogenous released hormone) while those from control mice showed the expected binding. However, as stated earlier, liver preparations from endotoxin-treated animals which have been adrenalectomized showed normal binding of the labeled steroid (Table 7).

5. CONCLUSIONS AND CONJECTURES

Several things now known about GAF will be reiterated. It has the characteristics of a protein and while its molecular weight appears to be in the 150 000 dalton range, it could be in polymerized form or in some type of association with another molecule. Attempts made to further purify it have not met with success. We do

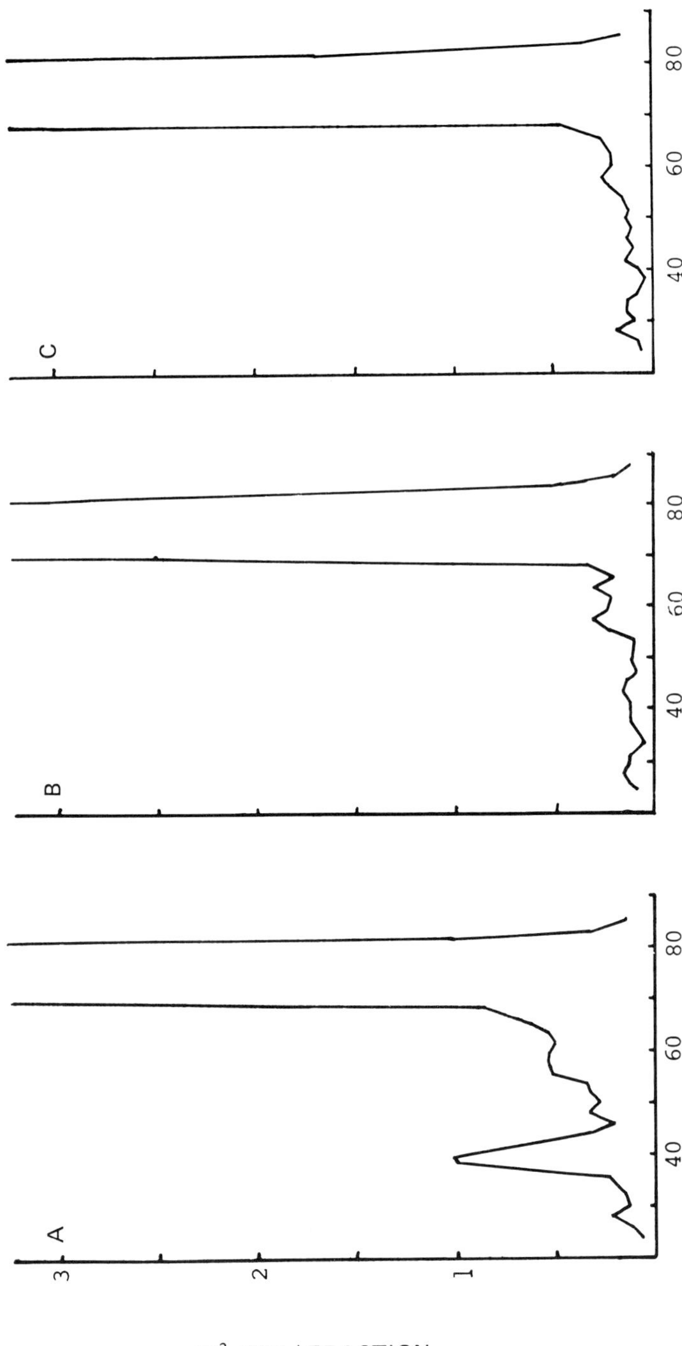

Fig. 12   *Effect of endotoxin treatment and cold stress on saturation of hepatic glucocorticoid receptors by endogenous hormone. Liver cytosols from untreated intact mice (A), endotoxin-treated (50 μg, 4 hr) mice (B), and cold stressed (4°C, 2 hr) mice (C) were incubated with excess ^{3}H-dexamethasone to allow binding of the label to glucocorticoid receptors not prebound by endogenous hormone. Cytosols were applied to an Ultrogel AcA 34 gel filtration column. Labeled complexes elute at fraction 40; excess label elutes between fractions 70 and 80.*

know, however, that it is of macrophage origin and appears to be synthesized in response to endotoxin rather than being preformed, stored and released. This idea is based on the 2-hour delay between injecting endotoxin into mice or adding it to macrophages in culture and the appearance of the peak titer of GAF in blood or supernatant fluid.

GAF is not a lethal agent. Animals injected with GAF-rich serum show no distress nor any of the signs of endotoxemia. They are, in fact, similar to adrenalectomized mice that are maintained under nonstressful conditions. Both GAF-treated mice and adrenalectomized mice die rapidly in the cold and are greatly sensitized to the lethal effect of endotoxin. GAF does not appear to block adrenocortical secretion but it does seem to interfere with those body responses that an animal must mount if it is to cope with the stresses of infection, cold, or noxious agents like endotoxin. Unfortunately, the role played by glucocorticoids in an animal subjected to stress has never been described in any detail. Is enzyme induction a part of the necessary response to stress? The only answer than can be given is that it does not occur after an injection of GAF and it does not occur after adrenalectomy, both of which sensitize animals to stress. Moreover, the rapid loss of carbohydrate reserves that accompanies an injection of endotoxin may be due in part to the lack of PEPCK induction, the enzyme that is rate-limiting in gluconeogenesis (Rognstad 1979). Insulin undoubtedly plays a role in carbohydrate metabolism in stressed animals.

Other characteristics of glucocorticoid action include antiinflammatory and immunosuppressive effects. These have not been measured with and without injections of GAF-rich serum. Whether they are involved in diminishing the undesirable aspects of stress is not known but each could be assessed experimentally. The fact that indometacin can protect lead acetate-sensitized mice against the lethal action of endotoxin (at least over a period of about 12 hours) (Goodrum et al 1978) suggests that the suppression of the inflammatory response is beneficial under these conditions. Since acetylsalicylic acid also protects against lethality (Fletcher et al 1976), it is consistent to assume that prostaglandins play a part in inflammation. (See Volume 2, Chapters 9 and 10 for discussion of arachidonic acid products in endotoxemia.) How immunosuppression contributes to survival in endotoxin-poisoning can only be conjectured. There are no studies addressing this question and, therefore, it must remain moot.

It is not immediately apparent why GAF is produced. It appears to be a normal constituent of blood so it must have survival value to have appeared through evolutionary time. Hormones tend to have antagonists and at the trivial level it could be said to have arisen as an antiglucocorticoid. But as indicated in the paragraph above, the extent of its antiglucocorticoid effect is undetermined. Since endotoxin has been known for nearly 25 years to block gluconeogenesis (at least in part because it inhibits the induction of PEPCK and other enzymes), GAF would be viewed as a protein-sparing compound. If it is also able to diminish the immunosuppressive property of glucocorticoid, it might have survival value as a result. A number of challenging questions remain unanswered; any scientist who accepts the challenge of further elucidating the role of GAF in normal and in pathological states will find ample rewards.

REFERENCES

Atkins E (1960) Pathogenesis of fever. *Physiol. Rev. 40*, 580-646.

Beale EG, Katzen CS, Granner DK (1981) Regulation of rat liver phosphoenolpyruvate carboxykinase (GTP) messenger ribonucleic acid activity by N^6, $O^{2'}$-dibutyryladenosine 3', 5'-phosphate. *Biochemistry 20*, 4878-4883.

Beale EG, Hartley JL, Granner DK (1982) N^6, $O^{2'}$-dibutyryl cyclic AMP and glucose regulate the amount of messenger RNA coding for hepatic phosphoenolpyruvate carboxykinase (GTP). *J. Biol. Chem. 257*, 2022-2028.

Benacerraf B, Thorbecke GJ, Jacoby D (1959) Effect of zymosan on endotoxin toxicity in mice. *Proc. Soc. Exp. Biol. Med. 100*, 796-799.

Berry LJ (1971) Metabolic effects of bacterial endotoxin. In: Kadis S, Weinbaum G, Ajl SJ (Eds), *Microbial Toxins Vol 5*, pp 165-208. Academic Press, New York.

Berry LJ (1975) Metabolic effects of endotoxin. In: Schlessinger D (Ed), *Microbiology – 1975*, pp 315-319. American Society for Microbiology, Washington DC.

Berry LJ, Smythe DS (1961) Effects of bacterial endotoxins on metabolism IV. Renal function and adrenocortical activity as factors in the nitrogen excretion assay for endotoxin. *J. Exp. Med. 114*, 761-778.

Berry LJ, Smythe DS (1964) Effect of bacterial endotoxins on metabolism. VII. Enzyme induction and cortisone protection. *J. Exp. Med. 120*, 721-732.

Berry LJ, Smythe DS, Colwell LS (1966) Inhibition of inducible enzymes by endotoxin and actinomycin D. *J. Bacteriol. 92*, 107-115.

Berry LJ, Laney MW, Moore RN (1975) Effect of endotoxin on metabolism and the metabolic changes in bacterial shock. In: Urbaschek B, Urbaschek R, Neter E (Eds), *Gram-negative Bacterial Infections and Mode of Endotoxin Actions*, pp 271-277. Springer-Verlag, Austria.

Berry LJ, Goodrum KJ, Ford CW, Resnick IG, Shackleford GM (1980) Partial characterization of glucocorticoid antagonizing factor in hepatoma cells. In: Schlessinger D (Ed), *Microbiology – 1980*, pp 77-81. American Society for Microbiology, Washington DC.

Bohme DH, Schneider HA, Lee JM (1959) Some physiopathological parameters of natural resistance to infection in murine salmonellosis. *J. Exp. Med. 110*, 9-25.

Bradley SG (1979) Cellular and molecular mechanisms of action of bacterial endotoxins. *Annu. Rev. Microbiol. 31*, 67-94.

Carswell EA, Old LJ, Kassel RL, Green S, Fiore N, Williamson B (1975) An endotoxin-induced serum factor that causes necrosis of tumors. *Proc. Natl Acad. Sci. USA 72*, 3666-3670.

Couch RE, Moore RN, Berry LJ (1979) Sensitization of tolerant mice to cold with a serum factor induced by endotoxin. *J. Appl. Physiol. 46*, 14-18.

Coutinho A, Forni L, Melchers F, Watanabe T (1977) Genetic defect in responsiveness to the β cell mitogen lipopolysaccharide. *Eur. J. Immunol. 7*, 325-329.

Filkins JP (1982) Role of the RES in the pathogenesis of endotoxic hypoglycemia. *Circ. Shock 9*, 269-280.

Fletcher JR, Herman CM, Ramwell PW (1976) Improved survival in endotoxemia with aspirin and indomethacin in pretreatment. *Surg. Forum 27*, 11-12.

mechanisms of altered triglyceride kinetics in endotoxemia – involving rates of appearance and removal by peripheral tissues and resulting in hypertriglyceridemia – remains to be fully elucidated.

7. METABOLIC STUDIES WITH ISOLATED ADIPOCYTES

The isolated adipocyte preparation lends itself well to metabolic investigations of the mediated as well as direct effects of endotoxin (J.A. Spitzer 1980). Several cellular and molecular mechanisms have been examined in an attempt to elucidate their respective potential role in initiating and sustaining the pathophysiologic events of endotoxemia. A summary of such investigations is presented in Table 1.

Using both 2-deoxyglucose and 3-O-methylglucose, Leach and J.A. Spitzer (1981) assessed endotoxin-induced alterations in *glucose transport* in isolated adipocytes. The changes observed included significantly increased basal 3-O-methylglucose uptake and cytochalasin B-insensitive uptake of both 2-deoxyglucose and 3-O-methylglucose. Endotoxin reduced insulin-stimulated (V decreased, K_M unchanged) and spermine-stimulated 2-deoxyglucose entry to a similar extent, but phosphorylation of 2-deoxyglucose was seemingly unimpaired. The results were consistent with the hypothesis that (a) a site of endotoxin-induced insulin resistance is at the cell membrane level and may reflect a decrease in number or activity of effective carrier units, rather than alterations in affinity; (b) endotoxin does not compromise the hexokinase system; (c) the cell membrane-localized effect of endotoxin on hexose transport is not necessarily mediated by the insulin receptor, and (d) two separate transport systems may operate for 3-O-methylglucose and 2-deoxyglucose entry, and these are modulated by endotoxin, in different ways.

Some of the studies on *lipolysis* and *cAMP accumulation* by isolated adipocytes of endotoxic animals are discussed under the heading of 'lipid mobilization'.

Alterations in *insulin responsiveness* and *sensitivity* also have been explored using fat cell preparations (Holley and J.A. Spitzer 1980; J.A. Spitzer and Holley 1979). Two experimental modes have been employed for delivery of the endotoxin insult: in vivo – when cells were isolated from epididymal fat pads several hours after the injection of endotoxin – and in vitro – when control, untreated adipocytes were incubated with endotoxin and some of the ensuing metabolic and hormonal alterations were monitored. Insulin action was evaluated on the basis of stimulation of glucose oxidation and suppression of catecholamine-stimulated lipolysis (antilipolytic effect). The glucose oxidation studies revealed an insulin-like effect on basal glucose oxidation, but a diminished response to insulin stimulation, thus lending support to earlier findings in other systems, referring to the insulin-like action of endotoxin (Berry 1977; Cline et al 1968; Woods et al 1961). A similar insulinomimetic effect of endotoxin on basal glucose oxidation in epididymal fat pads was also reported by Witek-Janusek and Filkins in 1979.

Although endotoxin induced an apparent insulin-resistant state in terms of stimu-

TABLE 1 *Endotoxin-induced changes in cellular function in isolated adipocytes**

Function	Change	Experiment	
		in vivo	in vitro
Hormone-stimulated lipolysis	increase	J.A. Spitzer et al 1973	J.A. Spitzer 1974
cAMP accumulation	increase		J.A. Spitzer 1974
3-O-Methylglucose transport	increase	Leach and J.A. Spitzer 1981	
Glucose oxidation	increase (in absence insulin stimulation)	Holley and J.A. Spitzer 1980	J.A. Spitzer and Holley 1979
Insulin sensitivity	decrease (for glucose oxidation)	Holley and J.A. Spitzer 1980; J.A. Spitzer 1979	Holley and J.A. Spitzer 1980; J.A. Spitzer and Holley 1979
	increase (for antilipolysis)	J.A. Spitzer 1979	J.A. Spitzer and Holley 1979
Insulin binding	increased affinity at low insulin concentrations	Holley and J.A. Spitzer 1980	
Ca^{2+} fluxes	increased binding to cell membrane; some similarities to action of ionophore A23187	Nelson and J.A. Spitzer 1981, 1985	Nelson and J.A. Spitzer 1985; Hikawyj-Yevich and J.A. Spitzer 1977b

* Endotoxin used was *Escherichia coli* O127:B8 lipopolysaccharide, isolated by the Boivin procedure (Difco Laboratories, Detroit).

lation of glucose oxidation, at the same time adipocytes were rendered more sensitive to the *antilipolytic effect of insulin* (J.A. Spitzer and Holley 1979; J.A. Spitzer 1979). These results provided evidence for nonuniform alterations by endotoxin of the sensitivities to various physiologic actions of insulin in the same target cell. Modulation of adipocyte insulin sensitivity in endotoxemia may also be accomplished in part through changes in insulin receptor affinity, since Holley and J.A. Spitzer (1980) have demonstrated a tendency for higher insulin-binding affinity at low insulin concentrations.

7.1   α-Adrenergic receptors and calcium in the mediation of endotoxin action

The alterations in lipolytic patterns brought about by endotoxin in isolated adipocytes are species-dependent (Hikawyj-Yevich and J.A. Spitzer 1977a). In attempting to elucidate the underlying reason(s) for species variability, the possible role of adrenergic receptors in mediating endotoxin action on adipocytes was explored by Hikawyj-Yevich and J.A. Spitzer (1977b). With the use of human adipocytes that have both β- and α-adrenergic receptors, it was demonstrated that endotoxin depressed the enhancement, induced by the alpha blocker phentolamine, of norepinephrine-stimulated glycerol release and cAMP concentration, while the cellular response to the β-blocker propranolol remained unaltered. These results lend credence to the hypothesis that in human adipocytes – or possibly in other cells that possess this type of adrenergic receptors – endotoxin treatment leads to increased α-adrenergic activity.

Hikawyj-Yevich and J.A. Spitzer (1977b) have extended their studies on human fat cells to include the potential contribution of calcium fluxes to the cellular expression of endotoxin action. Calcium fluxes are important regulators of both basal and catecholamine-stimulated lipolytic processes (Akgün and Rudman 1969; Efendic et al 1970). Furthermore, evidence obtained by different groups of investigators indicated changes in Ca^{2+} fluxes across plasma (Conner et al 1973) and mitochondrial membranes (Nicholas et al 1972) being associated with endotoxin-evoked functional and metabolic changes. A comparison of the actions of endotoxin and the divalent cation ionophore A23187 on lipolysis and cAMP concentrations in human adipocytes suggests that endotoxin may exert its influence at the cellular level by affecting calcium movements (Hikawyj-Yevich and J.A. Spitzer 1977b).

More recently the question whether endotoxin was capable of perturbing adipocyte calcium homeostasis in a fashion which is consistent with concurrent metabolic alterations was addressed by Nelson and J.A. Spitzer (1981). Their studies have shown that adipocytes isolated from epididymal fat pads of rats injected with endotoxin 6 hours earlier had accumulated significantly more [45]Ca than cells of saline-injected animals. In addition, endotoxin also altered the desaturation kinetics of the fat cells. In an attempt to differentiate between surface-bound and intracellular [45]Ca, the cells were exposed to the trivalent cation lanthanum, which is assumed to displace [45]Ca from its extracellular binding sites and to block inward calcium movements. By definition, the amount of ^{45}Ca remaining associated with lanthanum-treated adipocytes was taken to be located intracellularly and was increased in adipocytes from endotoxin-treated rats (Nelson and J.A. Spitzer 1984). The amount of ^{45}Ca displaced by lanthanum was also increased. These findings suggested that the endotoxin-induced increase in ^{45}Ca accumulation included both the lanthanum-sensitive portion of the cell, that is, the cell surface, and intracellular binding sites.

In order to assign a role to a disruption of cellular calcium homeostasis in the metabolic sequelae of endotoxin action, it is necessary that (a) endotoxin perturb cellular calcium homeostasis, (b) calcium be implicated in the regulation of the metabolic pathway being perturbed by endotoxin, and (c) the changes in cellular calcium homeostasis and metabolic activity are in accord. The results of the ^{45}Ca studies with adipocytes seem to satisfy these criteria. First, perturbation of calcium homeostasis by endotoxin was demonstrated. Second, endotoxin enhances the oxidation of glucose to carbon dioxide and the rate of lipolysis in adipocytes (J.A. Spitzer 1980; Witek-Janusek and Filkins 1981). Calcium is thought to be involved in the regulation of these metabolic pathways. Dehaye et al (1979) reported that lipolytic agents stimulated the net uptake of ^{45}Ca and the requirements for extracellular calcium varied among the different agents. Endotoxin appeared to interfere only with calcium bound to the cell surface without measurably perturbing La^{3+}-resistant intracellular calcium pools. This is consistent with the concept that exogenous calcium may be more significant in mediating some actions of endotoxin than a perturbation of intracellular calcium stores. Indeed, the stimulatory action of endotoxin on glucose oxidation can be eliminated with the inclusion of EGTA in the medium (Witek-Janusek and Filkins 1981). Verapamil, an inhibitor of calcium entry, also inhibits endotoxin-stimulated lipolysis in perfused rat hearts (Williamson et al 1981). Third, the increase in calcium associated with fat cells treated with endotoxin either in vivo or in vitro, is in agreement with the enhanced rate of glucose oxidation, and lipolysis present in these cells.

Extrapolating results obtained with isolated cell systems in vitro to a much more complex in vivo situation is always difficult and fraught with hazards. Yet, there are some aspects of the studies on endotoxin-treated adipocytes that merit special attention and may contribute to a better understanding of the metabolic plight of the septic patient or experimental animal model.

The greater sensitivity of adipocytes of endotoxemic rats to the antilipolytic action of insulin may be instrumental in bringing about the inadequate lipolytic/ketonemic responses observed, thus obliging the infected host to cannibalize its muscle protein stores. The ability of endotoxin to increase glucose uptake and glucose oxidation is likely to be a contributing factor to the scenario of mismatched glucose supply and utilization, which ultimately precipitates lethal hypoglycemia. Finally, the implications of α-adrenergic receptors and calcium fluxes (especially membrane-bound calcium) in the mediation of the cellular effects of endotoxin conform quite well to current concepts of some hormonal actions (Exton 1981). Perhaps it is not too far-fetched to propose that the impact of endotoxin on some components of the endocrine system may be achieved through involvement of α-adrenergic or similar receptors and calcium movement.

REFERENCES

Agarwal MK, Lazar G (1977) Metabolic basis of endotoxicosis. *Microbios 20*, 183-214.

Akgün S, Rudman D (1969) Relationships between the mobilization of free fatty acids from adipose tissue and the concentrations of calcium in the extracellular fluid and in the tissue. *Endocrinology 84*, 926-930.

Armstrong DT, Steele R, Altszuler N, Dunn A, Bishop JS, DeBods RC (1961) Regulation of plasma free fatty acid turnover. *Am. J. Physiol. 201*, 9-15.

Bagby G, Spitzer JA (1980) Lipoprotein lipase activity in rat heart and adipose tissue during endotoxic shock. *Am. J. Physiol. 238*, H325-H330.

Bagby GJ, Spitzer JA (1981) Decreased myocardial extracellular and muscle lipoprotein lipase activities in endotoxin-treated rats. *Proc. Soc. Exp. Biol. Med. 168*, 395-398.

Ballard K, Alliapoulos JC (1975) The response of blood flow to altered perfusion pressure in canine adipose tissue. *Proc. Soc. Exp. Biol. Med. 150*, 65-70.

Beisel WR, Bruton J, Anderson VD, Sawyer WD (1967) Adrenocortical responses during tularemia in human subjects. *J. Clin. Endocrinol. Metab. 27*, 61-69.

Berry LJ (1977) Bacterial toxins. *CRC Crit. Rev. Toxicol. 5*, 239-318.

Blackard WG, Anderson JH, Spitzer JJ (1976) Hyperinsulinemia in endotoxic shock. *Metabolism 25*, 675-684.

Bulow J, Madsen J (1981) Influence of blood flow on fatty acid mobilization from lipolytically active adipose tissue. *Pfluegers Arch. 390*, 169-174.

Cahill GF Jr, Owen OE, Morgan AP (1967) The consumption of fuels during prolonged starvation. *Adv. Enzyme Regul. 6*, 143-150.

Cline MJ, Melmon KL, Davis WC, Williams HE (1968) Mechanism of endotoxin interaction with human leukocytes. *Br. J. Haematol. 15*, 539-547.

Clowes GHA, O'Donnell TF, Ryan NT, Blackburn GL (1974) Energy metabolism in sepsis: treatment based on different patterns of shock and high output stage. *Ann. Surg. 179*, 684-696.

Conner J, Fine J, Kusano K, McCrea MJ, Parnas I, Poosser CL (1973) Potentiation by endotoxin of responses associated with increases in calcium conductance. *Proc. Natl Acad. Sci. USA 70*, 3301-3304.

Daniel AM, Pierce CH, Shizgal HM, MacLean LD (1978) Protein and fat utilization in shock. *Surgery 84*, 588-594.

Dehaye JP, Winand J, Poloczek P, Christophe J (1979) Relationship between lipolysis and calcium in epididymal adipose tissue of obese-hyperglycaemic mice. *Diabetologia 16*, 399-408.

Efendic S, Alin B, Low H (1970) Effects of Ca^{++} on lipolysis in human omental adipose tissue in vitro. *Horm. Metab. Res. 2*, 287-291.

Exton JH (1981) Molecular mechanisms involved in α-adrenergic responses. *Mol. Cell. Endocrinol. 23*, 233-264.

Ferguson JL, Spitzer JJ, Miller HI (1978) Effects of endotoxin on regional blood flow in the unanesthetized guinea pig. *J. Surg. Res. 25*, 236-243.

Fiser RH, Denniston JC, Beisel WR (1972a) Infection with *Diplococcus pneumoniae* and

Salmonella typhimurium in monkeys: changes in plasma lipids and lipoproteins. *J. Infect. Dis. 125*, 54-60.

Fiser RH, Denniston JC, Beisel WR (1972b) Host fuel interrelationships during infection. *Pediatr. Res. 6*, 398.

Fredholm BB (1971) The effect of lactate in canine subcutaneous adipose tissue in situ. *Acta Physiol. Scand. 81*, 110-113.

Gallin JI, Kaye D, O'Leary WM (1969) Serum lipids during infection. *N. Engl. J. Med. 281*, 1081-1086.

Griffiths J, Groves AC, Leung FYT (1972) The relationship of plasma catecholamines to serum triglycerides in canine gram-negative bacteremia. *Surg. Gynecol. Obstet. 134*, 795-798.

Hikawyj-Yevich I, Spitzer JA (1977a) Endotoxin influence on lipolysis in isolated human and primate adipocytes. *J. Surg. Res. 23*, 106-113.

Hikawyj-Yevich I, Spitzer JA (1977b) The role of adrenergic receptors and Ca^{2+} in the action of endotoxin on human fat cells. *J. Surg. Res. 23*, 233-238.

Hirsch RL, McKay DG, Travers RI, Skraly RK (1964) Hyperlipidemia, fatty liver and bromsulfophthalein retention in rabbits injected with bacterial endotoxins. *J. Lipid Res. 5*, 563-568.

Hochachka PW, Neely JR, Driedzic NR (1977) Integration of lipid utilization with Krebs cycle activity in muscle. *Fed. Proc. 36*, 2009-2014.

Holley DC, Spitzer JA (1980) Insulin action and binding in adipocytes exposed to endotoxin in vitro and in vivo. *Circ. Shock 7*, 3-12.

Hutcherson JD, Lequire VS, Hamilton R (1958) Inhibition of postheparin clearing factor by the sera of rabbits made lipemic by hemorrhage or by the injection of Shear's antigen. *Anat. Rec. 130*, 317.

Issekutz B Jr, Miller HI, Paul P, Rodahl K (1965) Effect of lactic acid on free fatty acids and glucose oxidation in dogs. *Am. J. Physiol. 209*, 1137-1144.

Jansen H, Kalkman C, Birkenhager J, Hulsmann P (1980) Demonstration of a heparin-releasable liver-lipase-like activity in rat adrenals. *FEBS Lett. 1120*, 30-34.

Kaufmann RL, Matson CE, Beisel WR (1976) Hypertriglyceridemia produced by endotoxin: role of impaired triglyceride disposal mechanisms. *J. Infect. Dis. 133*, 548-555.

Kawakami M, Cerami A (1981) Studies of endotoxin-induced decrease in lipoprotein lipase activity. *J. Exp. Med. 154*, 631-639.

Kovach AGB, Kovach E, Sandor P, Spitzer JA, Spitzer JJ (1976) Metabolic responses to localized ischemia in adipose tissue. *J. Surg. Res. 20*, 37-44.

Krauss RM, Windmueller M, Levy RI, Fredrickson DS (1973) Selective measurement of two different triglyceride lipase activities in rat postheparin plasma. *J. Lipid Res. 14*, 286-295.

Leach GJ, Spitzer JA (1981) Endotoxin-induced alterations in glucose transport in isolated adipocytes. *Biochim. Biophys. Acta 648*, 71-79.

Linder C, Cheruick SS, Fleck TR, Scow RO (1976) Lipoprotein lipase and uptake of chylomicron triglyceride by skeletal muscle of rats. *Am. J. Physiol. 231*, 860-864.

Liu MS, Spitzer JJ (1977) Myocardial fatty acid and lactate metabolism after *E. coli* endotoxin administration. *Circ. Shock 4*, 191-200.

Liu MS, Long WM, Spitzer JJ (1981) Influence of *Escherichia coli* endotoxin on palmitate, glucose and lactate utilization by isolated dog heart myocytes. In: Majde JA, Person RJ (Eds), *Pathophysiological Effects of Endotoxins at the Cellular Level*, pp 115-121. Progress in Clinical and Biological Research, Vol 62. Alan R. Liss, New York.

Neely JR, Rovetto MJ, Oram JF (1972) Myocardial utilization of carbohydrates and lipids. *Prog. Cardiovas. Dis. 15*, 289-329.

Nelson KM, Spitzer JA (1981) Calcium uptake and efflux in adipocytes after endotoxin administration. *Circ. Shock 8*, 224.

Nelson KM, Spitzer JA (1985) Alteration of adipocyte calcium homestasis by *E. coli* endotoxin. *Am. J. Physiol. 248*, R331-R338.

Neufeld HA, Pace JA, White FE (1976) The effect of bacterial infections on ketone concentrations in rat liver and blood and on free fatty acid concentrations in rat blood. *Metabolism 25*, 877-884.

Neufeld HA, Pace JG, Kaminski MV, George DT, Jahrling PB, Wannemacher RW Jr, Beisel WR (1980) A probable endocrine basis for the depression of ketone bodies during infections or inflammatory state in rats. *Endocrinology 107*, 596-601.

Nicholas GG, Mela LM, Miller LD (1972) Shock induced alterations of mitochondrial membrane transport. *Ann. Surg. 176*, 579-584.

Nolan JP (1975) The role of endotoxin in liver injury. *Gastroenterology 69*, 1346-1356.

Oehler G, Hassinger R, Schmahl FW, Huth K, Roka L (1975) Changes of lipoprotein lipase (LPL) after intravenous injection of endotoxin. In: Urbaschek B, Urbaschek R, Nefer E (Eds), *Gram-Negative Bacterial Infections*, pp 301-305. Springer Verlag, Berlin-Heidelberg-New York.

Robinson DS (1970) The function of the plasma triglycerides in fatty acid transport. In: Florkin C, Stotz M (Eds), *Comprehensive Biochemistry Vol 18*, pp 51-116. Elsevier, New York.

Romanosky AJ, Bagby GJ, Bockman EL, Spitzer JJ (1980a) Increased skeletal muscle glucose uptake and lactate release following E. coli endotoxin administration. *Am. J. Physiol. 239*, E311-E316.

Romanosky AJ, Bagby GJ, Bockman EL, Spitzer JJ (1980b) Free fatty acid utilization by skeletal muscle following endotoxin administration. *Am. J. Physiol. 239*, E311-E316.

Romanosky AJ, McGuinness O, Spitzer JJ (1982) Ketone body (KB) clearance following endotoxin (ET) administration in dogs. *Fed. Proc. 41*, 1133.

Sakaguchi O, Sakaguchi S (1979) Alterations of lipid metabolism in mice injected with endotoxin. *Microbiol. Immunol. 23*, 71-85.

Shudman GI, Williams PE, Liljenquist JE, Lacy WW, Keller V, Cherrington AD (1980) Effect of hyperglycemia independent of changes in insulin or glucagon on lipolysis in the conscious dog. *Metabolism 29*, 317-321.

Spitzer JA (1974) Endotoxin-induced alterations in isolated fat cells: effect on norepinephrine-stimulated lipolysis and cyclic 3',5'-adenosine monophosphate accumulation. *Proc. Soc. Exp. Biol. Med. 145*, 186-191.

Spitzer JA (1979) Altered insulin sensitivity in endotoxemia. *Fed. Proc. 38*, 1193.

Spitzer JA (1980) Endotoxin induced metabolic alterations in isolated adipocytes. In: Agarwal MK (Ed), *Bacterial Endotoxins and Host Response*, pp 341-360. Elsevier/North-Holland, Amsterdam-New York-Oxford.

Spitzer JA, Holley DC (1979) Alterations in insulin action by endotoxin in vitro. In: Schumer W, Spitzer JJ, Marshall BE (Eds), *Advances in Shock Research Vol 2*, pp 129-136. Alan R. Liss, New York.

Spitzer JA, Archer L, Greenfield LJ, Hinshaw LB, Spitzer JJ (1972) Metabolism of the nonhepatic splanchnic area in baboons and the effects of endotoxin. *Proc. Soc. Exp. Biol. Med. 141*, 21-25.

Spitzer JA, Kovach AGB, Rosell S, Sandor P, Spitzer JJ, Storck R (1973) Influence of endotoxin on adipose tissue metabolism. In: Kovach AG, Stoner HB, Spitzer JJ (Eds), *Neurohumoral and Metabolic Aspects of Injury*, pp 337-344. Plenum Press, New York.

Spitzer JA, Bagby GJ, Elahi D (1976) Lipoprotein lipase in endotoxic shock. *Fed. Proc.* *35*, 415.

Spitzer JJ, Bechtel AA, Archer LT, Black MR, Hinshaw LB (1974) Myocardial substrate utilization in dogs following endotoxin administration. *Am. J. Physiol. 227*, 132-136.

Spitzer JJ, Ferguson JL, Hirsch HJ, Loo S, Gabbay KW (1980) Effects of E. coli endotoxin on pancreatic hormones and blood flow. *Circ. Shock 7*, 353-360.

Tan MH, Santa T, Havel RJ (1977) The significance of lipoprotein lipase in rat skeletal muscles. *J. Lipid Res. 18*, 363-370.

Van Tol A, Van Gent R, Jansen H (1980) Degradation of high density lipoprotein by heparin-releasable liver lipase. *Biochem. Biophys. Res. Commun. 94*, 101-108.

Williamson JR, Cooper RH, Hoek JB (1981) Role of calcium in the hormonal regulation of liver metabolism. *Biochim. Biophys. Acta 639*, 243-295.

Witek-Janusek L, Filkins JP (1979) Glucocorticoid antagonism of in vitro stimulation of adipose tissue glucose oxidation by endotoxin. *Physiologist 22*, 134.

Witek-Janusek L, Filkins JP (1981) Insulin-like action of endotoxin: antagonism by steroidal and nonsteroidal anti-inflammatory agents. *Circ. Shock 8*, 573-583.

Woods MW, Burk D, Howard T, Landy M (1961) Insulin-like action of endotoxins in normal and leukemic leukocytes and other tissues. *Proc. Am. Assoc. Cancer Res. 3*, 279.

Yelich MR, Filkins JP (1980) Mechanism of hyperinsulinemia in endotoxicosis. *Am. J. Physiol. 239*, E156-E161.

Handbook of Endotoxin, Vol. 3: Cellular Biology of Endotoxin
L.J. Berry, editor
© Elsevier Science Publishers B.V., 1985

CHAPTER 7

Effect of endotoxin on mitochondrial function*

LEENA MELA-RIKER AND HASSAN TAVAKOLI

1. INTRODUCTION

Animals administered bacterial endotoxins develop metabolic abnormalities which are related to mitochondrial dysfunction in several organs. These abnormalities develop slowly, often several hours after the administration of the endotoxin. They occur in most organs of the body, including those that accumulate only minor quantities of endotoxin following its systemic administration. These findings suggest that the in vivo effects of bacterial endotoxins on mitochondrial function are indirect, secondary to effects of endotoxins on some other primary events. These primary events can occur in the cell whose mitochondrial function will secondarily deteriorate. They can also occur in other organ systems distant from the final targets of mitochondrial functions. In vitro administration of bacterial endotoxin on purified fractions of isolated mitochondria have been shown to damage mitochondrial structure and function. The in vitro endotoxin effects, however, are not identical with those induced by in vivo administration of endotoxin. These findings provide further evidence in support of the suggestion that the in vivo metabolic effects of endotoxin are not due to direct effects of endotoxin on mitochondrial membrane functions, but are mediated by other primary factors.

In this article we attempt to critically review the available information on in vivo and in vitro effects of endotoxin on mitochondrial structure and function. The mechanisms of endotoxin actions will be emphasized. We will try to correlate the in vivo effects of endotoxin on mitochondrial function with other biochemical or hemodynamic effects.

Mitochondria have two important functions in the cell. In mammalian cells the generation of high energy phosphate compounds, and thus cellular energy, is primarily a mitochondrial function. A net production of only two ATP molecules from a molecule of glucose is possible through anaerobic glycolysis. Mitochondrial electron transfer reaction, however, generates 18 times as many ATP molecules from one molecule of glucose. This synthesis of ATP requires the function of the

---

* The work performed in the authors' laboratory was supported in part by grants from the National Institutes of Health, GM 19867, GM 28889 and NS 10939.

166

mitochondrial electron transfer chain. The electron transfer reaction is initiated by the entry of Krebs cycle intermediates into the chain in the form of various mitochondrial respiratory substrates, pyruvate, glutamate, β-hydroxybutyrate, and succinate being some of the important ones. The electron transfer chain consists primarily of dehydrogenases, NADH, iron sulfa proteins, coenzyme Q, and the cytochromes b, c_1, c, a, and a_3. The final reaction of the electron transfer chain is that of cytochrome oxidase with molecular oxygen. At three points along the electron transfer chain, from the dehydrogenases to oxygen, a sufficient amount of chemical energy is generated to form a high energy phosphate bond. These three phosphate bonds are used to form the high energy-rich compound ATP in the presence of ADP and inorganic phosphate.

The respiratory chain activity is low in the absence of ADP when ATP synthesis does not occur. This state of respiration was named by Chance and Williams (1955; also Chance 1959) State 4 respiration, the resting respiratory state. In the presence of ADP, the respiratory activity is increased several times due to the rapid utilization of high energy bonds to form ATP. This state of respiration is called the acting state, or State 3 respiratory state. The activity of mitochondrial respiration in State 3 provides important information about the capability of the respiratory chain to respire in a coupled manner and to generate ATP during the respiratory activity. The ratio of the active State 3 respiration to the resting State 4 respiration has been termed the respiratory control ratio. This ratio provides some information about how well respiration is coupled to oxidative phosphorylation. The higher the ratio, the better the coupling. Another important mitochondrial function is accumulation of calcium. This accumulation is energy-dependent and is enhanced by the presence of permeant anions such as phosphate or acetate. Sufficient experimental evidence now exists to support the suggestion that mitochondrial calcium accumulation is important in controlling the level of intracellular free calcium in some tissues. Under normal physiologic conditions, the steady-state level of free intracellular calcium is controlled primarily by mitochondrial uptake and release of calcium in certain cells, particularly in the liver, the heart, and the brain.

Because of the importance of the mitochondrial ATP synthesis, the State 3 respiration and respiratory control ratios, and the calcium transport activities for supporting normal cellular metabolism, particular attention has been paid to the effects of in vivo or in vitro administrations of endotoxin on these mitochondrial parameters.

2. DIRECT EFFECTS OF ENDOTOXIN ON MITOCHONDRIAL FUNCTION

2.1. In vitro studies

Mager and Theodor (1957) first demonstrated that fractions of bacterial endotoxins added to normal liver mitochondria in vitro depress oxygen consumption and oxidative phosphorylation. They used the nontoxic polysaccharidic hapten isolated from *Shigella paradysenteriae* type 3 bacteria. The oxygen consumption by pyridine nucleotide-linked substrates measured under State 4 conditions (in the absence of oxidative phosphorylation) was inhibited by more than 80%, while only a slight inhibition of succinate-induced respiration was found. These data are difficult to interpret since State 4 conditions were used and rotenone was not present during succinate oxidation. A clear-cut inhibition of oxidative phosphorylation by hapten preparations was demonstrated by the inhibition of oxygen utilization during State 3 respiration (percent inhibition by native hapten calculated to be 65% with pyruvate as substrate), and by the decline in phosphate/oxygen (P/O) ratios to almost zero during both pyruvate and succinate oxidation. Berry et al (1959) demonstrated an increased hydrolysis of ATP in mouse liver mitochondria after an addition of *Salmonella typhimurium* endotoxin. This effect appears to be similar to the uncoupler-induced activation of the mitochondrial ATPase.

In most of the recent studies, lipopolysaccharides extracted from the cell wall of the gram-negative organism *Escherichia coli* have been used. Preparations used differed in degree of purity, most of them being the crude preparations commercially available from Difco Laboratories (Detroit, MI).

2.2. Substrate oxidation and electron transfer activity

In intact normal mitochondria isolated from rat liver, substrate oxidation and electron transfer activity in the absence and presence of *E. coli* endotoxin O127:B8 (Difco) were studied. At lower concentrations of added endotoxin it was found that succinate oxidation was specifically inhibited (Mela et al 1970a). This inhibition was seen when oxygen utilization rates were measured in liver mitochondria which were supplied with succinate as the respiratory substrate. Spectrophotometric measurements of cytochrome *b* redox changes indicated an inhibition of succinate-induced electron transfer in the presence of rotenone which blocks NADH-linked electron transfer reactions (Fig. 1). Half-maximal inhibition of cytochrome *b* reduction by succinate was obtained at 35 μg of endotoxin per mg of mitochondrial protein. Electron microscopic studies of the succinate-supported electron transfer-induced conformational changes of rat liver mitochondria were also shown to be blocked by *E. coli* endotoxin (White et al 1971). Higher concentrations of endotoxin were required to block maximal rates of the oxidation of pyridine nucleotide-linked substrates (Mela et al 1970a). A 50% inhibition was achieved at 150 μg of endotoxin per mg of mitochondrial protein. The inhibition induced at higher endotoxin concentrations appears to be a nonspecific increase in mitochondrial membrane permeability, presumably resulting in cycling of cations in and out of the mitochondrial matrix space. The evidence to support this conclusion is based on the findings that EDTA and EGTA partly block this inhibition (Mela et al 1970a). It is also supported by the finding that increased State 4 respiratory activity, induced at the same endotoxin concentrations, is partly blocked by EDTA and EGTA.

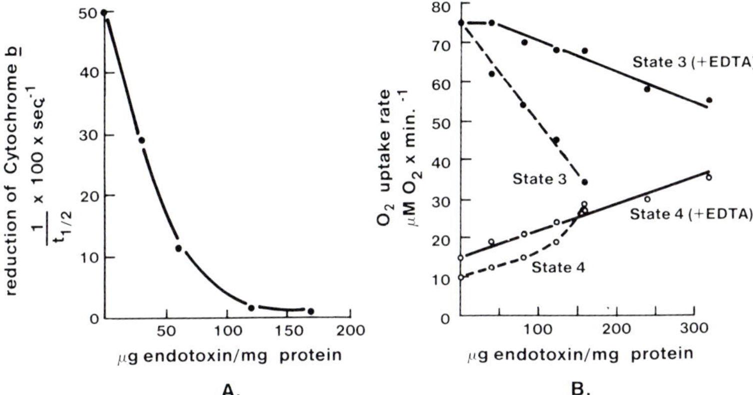

Fig. 1.   *Influence of E. coli endotoxin in vitro on mitochondrial succinate-induced reduction of cytochrome b (Fig. 1A) and State 3 and 4 respiratory activities with the use of glutamate and malate as substrates at 10 mM concentration (Fig. 1B). Rat liver mitochondria were suspended at 1.8–2.4 mg protein per ml in mannitol-sucrose Tris-phosphate buffer pH 7.4. In A, 6 mM succinate was used as substrate in the presence of 3 μM rotenone, and the halftime of reduction of cytochrome b upon addition of succinate was recorded in a dual wavelength spectrophotometer set at 430–410 nm. In B, mitochondrial respiratory activity was measured by an oxygen electrode in the absence (State 4) and presence (State 3) of 530 μM ADP in the absence and presence of 5 mM EDTA and E. coli endotoxin as indicated in the figure. Data from Mela et al (1970a).*

Using mitochondria isolated from rat and human liver, Schumer et al (1970, 1971a) studied the effects of various endotoxin preparations on mitochondrial respiratory activity in State 4 (without added ADP) with succinate, glutamate, pyruvate + malate, citrate and α-ketoglutarate as substrates. The authors found that citrate and α-ketoglutarate oxidation were particularly sensitive to endotoxin and exhibited an 80–90% inhibition in the presence of *E. coli* endotoxin. Glutamate and pyruvate-malate oxidation were inhibited much less (30–50%). Succinate oxidation under these conditions was not affected at all. It should be noted, however, that these studies were not performed under conditions of maximal respiratory activities, since ADP was not added. Thus, the interpretation of these findings becomes difficult. Schumer et al (1970) also found large variations in the inhibitory efficacy of the various endotoxin preparations used, which indicates that the physical and chemical differences in the toxins due to different extraction procedures become important factors in their ability to cause inhibition of mitochondrial functions. These findings, of course, question the relevance of in vitro studies of endotoxin effects on mitochondrial function.

2.3. Kinetics of calcium accumulation

The kinetics of the accumulation of calcium by mitochondria has been studied by monitoring the rate of disappearance of extramitochondrial calcium into the mitochondrial compartment by the use of calcium-sensitive indicators (Mela 1981; Nicholas et al 1972). The in vitro effects of endotoxin on the mitochondrial calcium transport system are complex. Nicholas et al (1972) showed that both the energy-dependent accumulation and the energy-independent 'binding' are affected. At low endotoxin concentrations (up to 50 μg of *E. coli* O127:B8 endotoxin per mg of mitochondrial protein), increased calcium uptake rates were observed both with and without an energy supply available. This increase was blocked by La^{3+} at a concentration sufficient to inhibit the Ca^{2+} carrier function by 80%. This observation implies that the increased Ca^{2+} uptake rate is perhaps due to increased binding of Ca^{2+} to the mitochondrial membrane or due to increased 'leakage' of Ca^{2+} into the mitochondria via some carrier-independent mechanism. At concentrations of endotoxin above 50 μg per mg of protein, a progressive decrease in mitochondrial energy-linked Ca^{2+} uptake occurs, reaching total inhibition of Ca^{2+} accumulation at about 125 μg *E. coli* endotoxin per mg of mitochondrial protein in the liver preparations (Nicholas et al 1972). Thus, mitochondrial Ca^{2+} transport is affected by endotoxin at concentrations similar to those required to inhibit substrate oxidation. In intact mitochondria isolated from the kidney, however, a differential inhibitory pattern was found (Mela 1981). A 50% inhibition of energy-dependent Ca^{2+} transport by kidney mitochondria was found at about 100 μg of *E. coli* endotoxin per mg of mitochondrial protein while a 50% inhibition of State 3 respiratory activity was reached only at 500 μg of endotoxin per mg of protein. Similar results were obtained in studies of isolated intact brain mitochondria. Thus, it appears that in most organs the Ca^{2+} uptake function is more sensitive to endotoxin than the respiratory function of the mitochondria is. It is possible that the inhibition of calcium accumulation is at least partly due to an increased rate of calcium efflux from mitochondria resulting in a lowered net flux into the mitochondria.

2.4. Mitochondrial membrane permeability

There are several membrane toxins which affect the permeability of the mitochondrial membrane. The data on Ca^{2+} transport of the mitochondria indicate that bacterial endotoxins can cause unspecific leakiness of the mitochondrial membrane. Bradley and collaborators (Bradley 1981; McGivney and Bradley 1979a, 1979b) determined changes in the permeability of liver mitochondrial membrane directly by measuring the leakage of intramitochondrial enzymes as a function of time during endotoxin incubation. They found that a 45-minute incubation of mouse liver mitochondria with *E. coli* endotoxin caused a significant leakage of malate dehydrogenase (26%), succinate dehydrogenase (16%) and isocitrate dehydrogenase (14%) from the mitochondria at 10–25 μg of endotoxin per ml. A similar leakage of mitochondrial enzymes was also achieved by incubating intact

170

cultured mouse liver cells with 10 µg *E. coli* endotoxin per ml of cell suspension. Significant amounts of both malate and succinate dehydrogenase left the cells, while no significant leakage of cytoplasmic or other subcellular enzymes, such as hexokinase, β-glucuronidase or acid phosphatase, occurred. These data suggest that endotoxins induce increased permeability of mitochondrial membranes at lower concentrations than required for damaging other subcellular or cell membrane permeabilities.

2.5. Oxidative phosphorylation

Several mitochondrial reactions involved in ATP synthesis have been studied to determine the effects of endotoxin on this important mitochondrial function. Mela and collaborators characterized the endotoxin effects on State 3 respiration, ADP stimulation of the electron transfer, and ATPase activities (Mela 1981; Mela et al 1970a; Nicholas et al 1972). Even at endotoxin concentrations that have no effect on State 4 respiration, State 3 respiration becomes inhibited (see Section 2.2.). In liver mitochondria at concentrations as low as 80 µg per mg of mitochondrial protein, additions of ADP caused no stimulation of the respiratory rate over the State 4 level, thus resulting in a loss of respiratory control (Mela et al 1970a). The ATPase activity is nearly totally inhibited at 50 µg of endotoxin per mg of mitochondrial protein (Mela et al 1970a). Similar inhibition of ATPase by *E. coli* endotoxin was also reported by Lavine et al (1971) in rat liver mitochondria, while DePalma et al (1977) observed an endotoxin-induced inhibition of State 3 respiration and a decreased respiratory control in liver mitochondria.

In addition to the inhibition of ATPase activity, adenine nucleotide translocase activity is an important supporting function in oxidative phosphorylation. Takeda and Liu (1980) examined the translocase activity directly in heart mitochondria after incubation with *E. coli* endotoxin. At 50 µg of endotoxin per mg of mitochondrial protein, they found a 21% inhibition of adenine nucleotide exchange across the mitochondrial membrane. At higher concentrations of endotoxin (100 µg/mg), leakage of labeled adenine nucleotides from mitochondria increased by 100%. Thus the inhibition of ATP synthesis by endotoxin is at least partly attributable to its direct inhibitory effects on the mitochondrial ATPase-ATP synthetase enzyme and adenine nucleotide translocase activities.

2.6. Amplification of endotoxin effects by improper ionic environment

Alterations of cellular ion homeostasis are important contributing factors in endotoxin-induced cell injury. Experiments have been performed on isolated mitochondria to identify any cation-induced modulations of the endotoxin effects. Increased Na^+ concentrations above normal intracellular levels parallel with decreased K^+ and acid pH significantly amplify the effects of endotoxin on mitochondrial function in vitro (Fig. 2) (Mela et al 1972a). Additions of Ca^{2+} at a level slightly above normal intracellular free Ca^{2+} concentrations (10^{-5} M) have a

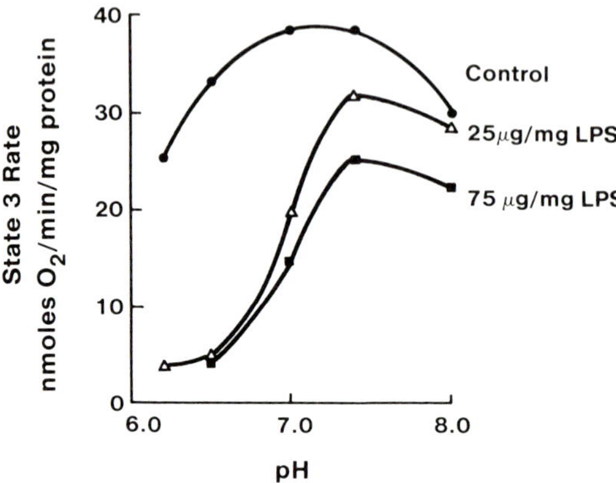

Fig. 2.   *Amplification of endotoxin-induced inhibition of mitochondrial respiratory activity in State 3 by low pH. Rat liver mitochondria were suspended at 1.3 mg protein per ml in mannitol-sucrose Tris-phosphate buffer in the presence of 8 mM glutamate and malate as substrates and 520 μM ADP as phosphate acceptor. Their oxygen utilization rates were determined at varying pH in the absence and presence of added E. coli endotoxin (LPS) as indicated in the figure. Data from Mela et al (1972).*

similar modulating effect (Mela et al 1972a). Thus, although at first it appears that very high concentrations of endotoxin are required in vitro to significantly alter mitochondrial capacity to synthesize ATP, utilize ADP, fuels and oxygen in the altered ionic environment simulating intracellular changes induced in endotoxin shock, these changes occur in fact at significantly lower concentrations of endotoxin.

3.   EFFECTS OF ENDOTOXIN ON ENERGY METABOLISM AS STUDIED IN ISOLATED ORGANS AND CELLS

The early experiments of Rush and Hsieh (1968) on dog liver slices indicated that endotoxin did not deleteriously affect the cellular energy metabolism in vitro. Rush and Hsieh used *E. coli* endotoxin (Difco) and the dose was adjusted to 12 mg per kilogram, a large dose which, when given in vivo to dogs, induced a significant depression of oxygen consumption by the liver. Harken et al (1975), however, using rabbit liver slices and homogenates, found that in vitro effects of endotoxin on liver oxygen consumption could be demonstrated. An addition of *E. coli* endotoxin at 0.5 mg/g of tissue decreased the liver oxygen consumption in slices by 45% and in homogenates by 50%. Both of these effects were significant ($p < 0.001$).

Using isolated intact myocardial cells, Liu et al (1981) found that in vitro incubations of the cells with endotoxin resulted in increased glucose and lactate utilization by the cells at endotoxin concentrations of 80–400 µg/ml. These data are in agreement with those of Holley and Spitzer (1980) on isolated rat adipocytes, in which a 30-minute incubation with 500 µg/ml of *E. coli* endotoxin induced a 48% increase in glucose oxidation to carbon dioxide. Liu et al (1981), however, found that palmitate oxidation by isolated myocytes was inhibited 58% by 200 µg/ml of *E. coli* endotoxin. This amount corresponds to about 40 µg of endotoxin per mg of myocyte protein. Studies of the cellular uptake of the metabolically inactive fatty acid analog hexadecanol and glucose analog 3-O-methyl-D-glucose indicated that the reason for the altered metabolism was not altered transport across the plasma membrane. On the basis of this finding and the independent findings of Lavine et al (1971) that palmitate oxidation by liver mitochondria is specifically inhibited by *E. coli* endotoxin, Liu et al (1981) proposed a direct mitochondrial effect of endotoxin on fatty acid oxidation.

Investigations of Silver (1981) with various cell types in culture, including fibroblasts, macrophages and neuroblasts, demonstrated that in vitro additions of *E. coli* endotoxin ranging from 1–1000 µg/ml induced parallel mitochondrial structural changes, leakage of K^+ from the cells and decreased pH measured adjacent to the cell surface. These alterations were absent when cells were protected from endotoxin effects by prior additions of glucocorticoids to the cell suspension.

Incubation of mouse neuroblastoma cells in culture with *E. coli* O111:B4 lipopolysaccharide was shown to cause early alterations of mitochondrial energy metabolism (Kilpatrick-Smith et al 1981). As seen in Table 1, at 0.6–0.7 mg of endotoxin per mg of dry weight of cells, a 30-minute incubation induced a significant increase in cellular ADP concentration parallel with a decrease in cellular ATP. Thus, the cellular phosphate potential, $[ATP]/[ADP][P_i]$, declined from a control of 1491 to 913 during 30 minutes of incubation with endotoxin (a 40% decrease). This decline progressed further when cells were incubated 60 or 120 minutes with endotoxin, while control incubation without endotoxin caused an increase in the $[ATP]/[ADP][P_i]$ ratio. These changes in the cellular phosphate potential occurred parallel with large increases in the intramitochondrial $[NAD^+]/$

TABLE 1   *Effect of E. coli endotoxin on the energy metabolic parameters of neuroblastoma cells in vitro*

	[ATP] (mM)	[ADP] (mM)	$\dfrac{[ATP]}{[ADP]\cdot[P_i]}$	$\dfrac{Lactate}{Pyruvate}$	$\dfrac{[NAD+]}{[NADH]}$
Control	4.98	0.53	1491	33.0	9.05
Endotoxin	4.16	0.79	860	36.2	19.18

The cells were incubated in the absence (control) and presence of 0.7 mg of endotoxin per mg of dry weight for 60 min. Cellular adenine nucleotides, lactate/pyruvate and NAD+/NADH were determined at the end of the incubation period. Data from Kilpatrick-Smith et al (1981).

[NADH] ratio. During a 30-minute incubation with endotoxin, the [NAD$^+$]/ [NADH] increased by 20% and reached more than twice the control level after a 60-minute incubation. During these incubations no significant alterations were found in the [lactate]/[pyruvate] ratio. On the basis of these findings, the authors concluded that the early effects of endotoxin on cells in vitro are mitochondrial in nature, resulting in the decline of cellular energy stores (Kilpatrick-Smith et al 1981). The authors postulate that the primary mitochondrial response to endotoxin is an inhibition of substrate influx. The mechanism of this inhibition, however, is not clear at the present time.

Recently, Manny et al (1980) used a cross-perfusion of a kidney from a donor dog and infused *E. coli* endotoxin at 0.5 mg/kg directly into the arterial inflow to the kidney under constant perfusion pressure of 100 mmHg. They examined the kidney morphologically 3 hours later. Mitochondrial destruction and swelling parallel with increased numbers of lysosomes with some membrane disruption was evident in the distal and proximal tubules. No significant structural alterations were evident in the glomeruli. Histochemical examination of the proximal and distal convoluted tubules also revealed markedly decreased mitochondrial succinic dehydrogenase activity and increased lysosomal acid phosphatase and glucuronidase activities. Thus, direct introduction of endotoxin into the kidney deleteriously affects the integrity and activity of mitochondria in the proximal and distal convoluted tubules, even in the presence of adequate tissue perfusion.

Our unpublished negative results on the effects of endotoxin on liver energy metabolism in vitro add to the apparently controversial findings on in vitro effects of endotoxin on cell and organ metabolism. In our studies (Mela and Scholtz, unpublished observations), perfusion of rat liver with *E. coli* endotoxin in Krebs-Ringer solution failed to alter liver pyruvate utilization, lactate output or oxygen consumption during a perfusion period of 1 hour. Several variables need be considered when judging the relevance of the in vitro endotoxin effects on the mitochondrial function in organs and cells. Although accumulation of endotoxin by several cell types has been demonstrated (Maier et al 1981; Silver 1981; Zlydaszyk and Moon 1976), differences in the amount and rate of uptake are evident. Thus, the amount of endotoxin added to isolated cells or organs should be adjusted accordingly. In many studies these conditions have not been properly controlled. As is evident from this section and Section 2 of this chapter dealing with the in vitro effects of endotoxins on mitochondrial function, the amounts of endotoxin required for deleterious effects are quite high. Thus, the relevance of these in vitro effects to the endotoxin-induced mitochondrial alterations in vivo is in question.

4. IN VIVO ENDOTOXIN EFFECTS ON MITOCHONDRIAL FUNCTION

On the basis of experiments on dogs given a lethal dose of *E. coli* endotoxin intravenously, which caused an increasing oxygen deficit without direct correlation with increasing lactate levels, Rosenberg and Rush (1966) postulated an endotoxin-induced 'histotoxic anoxia', which was a result of a direct endotoxin interference with cellular metabolic processes. Staples et al (1969) gave rats a lethal dose of *E. coli* endotoxin intraperitoneally. At preterminal stages, they determined tissue ATP content in several organs, including liver, brain, kidney, heart and skeletal muscle. They found a significantly lowered tissue ATP content in all these organs except the heart. In skeletal muscle samples from patients in sepsis, significantly lowered ATP and phosphocreatine levels and total adenine nucleotides were also found (Liaw et al 1980). These findings provide indirect evidence of a cellular metabolic block by in vivo endotoxemia or sepsis. To induce such a dramatic lowering of cellular energy stores, this block is most likely located in the pathways of oxidative metabolism.

4.1. Liver mitochondrial capacity to synthesize ATP after in vivo endotoxemia

To characterize the capacity of mitochondria to synthesize ATP after an in vivo endotoxic insult, several investigators have studied the energy-linked functions of mitochondria isolated from endotoxin-infected animals. Early studies were performed by Fonnesu and Severin (1956a, 1956b) with *Salmonella typhimurium* and diphtheria toxins injected into rats and guinea pigs. They demonstrated that after 24 hours of endotoxemia, the mitochondria isolated from morphologically affected livers of these animals exhibited lowered P/O ratios with both succinate and α-ketoglutarate as substrates. Yet, oxygen consumption in State 3 was practically unchanged by endotoxemia. Later studies have demonstrated that the characteristic functional alterations found in liver mitochondria are (a) significantly lowered State 3 respiratory activities with pyruvate-malate, glutamate-malate, α-ketoglutarate, succinate or palmitylcarnitine as substrates, (b) normal or increased State 4 respiratory activities with all substrates, leading to consequently lowered respiratory control ratios, and (c) an inhibited ATPase activity (DePalma et al 1970; Mela et al 1970b, 1971; Mela 1975; Sayeed et al 1970; Schumer et al 1971a). A profound inhibition of liver mitochondrial State 3 respiration and, thus, ATP synthesis rates, was reported by all these investigators. Respiration supported by α-ketoglutarate was inhibited particularly severely (up to 90%). This inhibition, however, was partly corrected by an addition of Mg^{2+} ions, thus suggesting a loss of mitochondrial bound Mg^{2+} during endotoxemia (Mela et al 1970b, 1971). Mitochondrial ATPase is also strictly Mg^{2+}-dependent. Liver mitochondrial ATPase activity is reduced to 30% of control in late endotoxemia (Mela et al 1971). This severe reduction of ATPase activity, however, cannot be corrected by externally added Mg^{2+} ions. Although loss of liver mitochondrial bound Mg^{2+} occurs during in vivo endotoxemia (Mela et al 1972b), other mitochondrial alterations responsible for inhibition of mitochondrial ATP synthesis reactions have been identified. In addition to the inhibition of ATPase function, mentioned above, liver mitochondrial adenine nucleotide translocase activity declines in late endotoxemia. This defect was characterized by determining the redox responses of liver mitochondria to additions of ADP (Mela et al 1971, 1972b) and by determining the transport of adenine nucleotides into the mitochondria (Jones and Kramer 1978; Nicholas et al 1974). Both studies indicate that adenine nucleotide transport in mitochondria becomes defective only in late endotoxemia, corresponding with the time of onset of the inhibition of State 3 respiration and ATPase activity and the loss of mitochondrial structural integrity seen by electron microscopy (White et al 1973).

4.2. Early alterations of liver mitochondrial membrane transport in endotoxemia

In vivo endotoxemia causes a severe inhibition of liver mitochondrial Ca^{2+} uptake function (Mela et al 1972b). This inhibition occurs early in endotoxemia. A 50% inhibition of liver mitochondrial Ca^{2+} uptake capacity was demonstrated at the time when State 3 respiration and ATP synthetic capacity was still fully functional (Nicholas et al 1974). As the Ca^{2+} uptake capacity declines, mitochondrial membranes become increasingly permeable to K^+ ions (Mela 1975). This permeability change could be responsible for the increased swelling of liver mitochondria seen to occur in situ during endotoxemia (Mela et al 1972b; White et al 1973).

4.3. Kidney and brain mitochondria in endotoxemia

Mitochondria isolated from the kidneys and the brain of endotoxin-infected animals exhibit inhibitory patterns very similar to those seen in the liver. State 3 respiratory and Ca^{2+} transport activities of kidney and brain mitochondria were studied carefully after an LD_{100} dose (at 3 days) of *E. coli* endotoxin in rats and guinea pigs (Fig. 3) (Mela et al 1974b; Mela and Miller 1983). This dose of endoto-

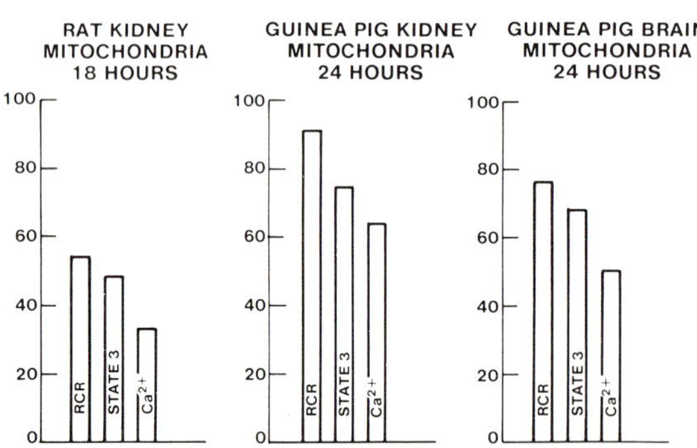

Fig. 3.   *Effect of in vivo endotoxemia on kidney and brain mitochondrial function. An LD_{70} (at 24 hours) dose of E. coli endotoxin was injected intraperitoneally in rats and guinea pigs. Mitochondria were isolated at 18–24 hours after the injection and analyzed for their State 3 respiratory activities (10 mM glutamate + malate, 500 μM ADP in 10 mM Tris Cl, 10 mM Tris-phosphate buffer pH 7.4 osmotically adjusted with 0.3 M mannitol-sucrose), respiratory control ratios (RCR) and Ca^{2+} accumulation rates. The data are presented as percentage of the same parameters determined in control animals.*

xin causes 60% mortality in 24 hours. Kidney and brain mitochondria were isolated after 17 and 24 hours of endotoxemia. State 3 respiratory activities were only slightly inhibited in brain mitochondria at 17 hours and not at all in kidney mitochondria. Both brain and kidney mitochondria showed severe inhibition of State 3 respiration and ATP synthesis rates at 24 hours in surviving animals (about 40% of those injected). Ca^{2+} transport activities in both brain and kidney mitochondria, at both time points studied, showed a more pronounced decline than did State 3 respiration. When glucocorticoid administration was used to protect against mortality and mitochondrial deterioration, the capacity of brain mitochondria to function 24 hours after endotoxemia correlated closely with the severity of endotoxemia (mortality) (Mela and Miller 1983). These data indicate that brain mitochondrial function is a particularly sensitive indicator of the deleterious effects of systemic endotoxemia.

Mitochondrial deterioration, induced by in vivo administrations of endotoxin, are prominent only at later stages of endotoxemia. Schmahl et al (1980) studied high energy metabolites in cat brain after an intravenous and intracisternal injection of endotoxin under conditions of unaltered brain blood flow at mean arterial pressures above 70 mmHg and at lowered mean arterial blood pressure (40–55 mmHg) resulting in lowered brain blood flow. They found that intracisternal injections of endotoxin caused a slow, gradual fall in the mean arterial blood pressure and cerebral blood flow. When the arterial pressure was still above 70 mmHg and cerebral blood flow and oxygen pressure were not significantly altered, a significant decline of brain tissue phosphocreatine concentration, [phosphocreatine]/[creatine] ratio and glucose concentration occurred, but ATP concentration and [ATP]/[ADP] ratio remained within normal range. After an intravenous injection at mean arterial blood pressure, no significant changes were found in any of the above metabolites. When the measurements were made at a time when mean arterial pressure had fallen to 40–55 mmHg and the brain blood flow was significantly reduced, brain phosphocreatine concentration, [phosphocreatine]/[creatine] ratio, [ATP] and [ATP]/[ADP] ratio were all significantly below normal. Simultaneously, brain tissue lactate concentration increased more than fivefold and [lactate]/[pyruvate] ratio doubled. On the basis of these findings, Schmahl et al (1980) concluded that endotoxin affects the energy metabolism of the brain both directly and indirectly. The direct endotoxin effects occur without any change in tissue perfusion. These direct metabolic effects of endotoxin after intracisternal injections correlated with morphologic evidence of swelling of mitochondria and increased numbers of lysosomes in the cortical nerve cells.

4.4. Heart and skeletal muscle mitochondria in endotoxemia

The available data on in vivo endotoxin effects on heart and skeletal muscle mitochondrial function are controversial (Mela 1982). Studies by Schumer and collaborators (Reed et al 1970; Schumer and Erve 1971; Schumer et al 1971b) indicated that intraperitoneal injections of an LD_{50} dose of *E. coli* endotoxin

(O26:B6) in rats resulted in cardiac and skeletal muscle mitochondrial deterioration in 18 hours. When these mitochondria were isolated and analyzed for their function, the respiratory capacities in State 3, and thus their ATP synthetic capacities (rates), were significantly below normal when α-ketoglutarate was used as substrate. Mela et al (1974a), however, found no significant deterioration in cardiac mitochondrial function 4.5–6.5 hours after an intravenous injection of an LD_{60-80} dose of *E. coli* endotoxin in dogs. These authors monitored myocardial function throughout the experiment and studied the ultrastructure of the myocardium and the function of the isolated mitochondria at the end of the experiments. In spite of the onset of myocardial performance failure, the myocardial blood flow and oxygen consumption remained normal. The ultrastructural examination revealed normal mitochondrial morphology without swelling. The T-system, however, appeared dilated. The mitochondrial State 3 respiratory capacity, respiratory control ratios, efficiency of oxidative phosphorylation and ATPase activities were all normal. Thus, this study suggests that the late failure in myocardial performance induced by endotoxin is not a result of a mitochondrial failure to provide sufficient cellular energy stores. Mela (1975) reported similar mitochondrial findings in rat hearts after 6 hours of *E. coli* endotoxemia. Mela (1975) also was unable to demonstrate any decrease in mitochondrial function of skeletal muscle after 6 hours of lethal endotoxemia (LD_{90}). On the contrary, in the experiments on rats, an increased rate of mitochondrial State 3 respiratory capacity of heart and skeletal muscle was found.

The data on various organs in endotoxemia suggest that the lethal in vivo effects of endotoxin on mitochondrial function may not be direct but are most likely a result of changes in tissue perfusion and the resulting tissue ischemia. The studies by Schmahl et al (1980) support this conclusion.

5. MITOCHONDRIAL FUNCTION IN BACTEREMIA AND SEPSIS

5.1. Acute bacteremia

Some studies have employed intravenous or intraperitoneal injections of lethal doses of live gram-negative bacteria. Mitochondrial function was analyzed within the first 24 hours following bacterial injections. Tanaka et al (1982) injected 1.25–1.5×10^9 *E. coli* organisms/100 g of body weight intravenously in rats. This dose of bacteria resulted in a 100% mortality in 24 hours. At time points from 3 to 14 hours after the onset of bacteremia, liver mitochondria were isolated and analyzed for their function. The authors found no decline in the mitochondrial capacity to utilize oxygen or synthesize ATP with either glutamate, glutamate + succinate, or β-hydroxybutyrate as substrates. On the contrary, at all time points studied, a slight increase in these parameters was found. Similarly, an increase in calcium-stimulated respiration was found. Mela (1979), in collaborative studies with Professor Lerner Hinshaw, studied the effects of *E. coli* bacteremia in dogs, induced by an intravenous injection of $8.6 \times 10^9 - 1.2 \times 10^{10}$ live organisms per kg of body weight. This dose of *E. coli* results in 100% mortality within 24 hours. Both kidney and brain mitochondria were isolated preterminally. Kidney mitochondrial State 3 respiratory capacity was not different from control, although the respiratory control ratios were reduced. Brain mitochondria, however, exhibited significantly reduced respiratory capacities in State 3 and thus, reduced ATP synthetic activity. These changes were completely eliminated by early administration (during the first hour) of gentamicin and glucocorticoids which, as Hinshaw has shown, completely prevent mortality in dogs and baboons suffering from acute live *E. coli* bacteremia (Hinshaw et al 1979, 1981).

5.2. Acute peritonitis

Acute peritonitis was induced in rats by Fry et al (1979) by cecal ligation and puncture. This procedure induces 100% mortality in 8 hours. At 2, 4 and 6 hours, liver mitochondria were isolated and the respiratory function studied with glutamate, pyruvate or succinate as substrates. A slight increase was found in State 3 respiratory activities with no significant change in State 4 respiration and no change in ADP/oxygen ratios. Decker et al (1971) after an intraperitoneal injection of *Klebsiella pneumoniae* (LD_{70} at 24 hours) similarly found no alterations of rat liver mitochondrial function studied 18 hours after the onset of peritonitis. Thus, it appears that acute bacteremia or peritonitis does not have significant deleterious effects on liver or kidney mitochondrial function. Brain mitochondria, however, appear to be altered in their function during acute bacteremia.

5.3. Subacute septicemia

Tavakoli and Mela (1982) studied mitochondrial function in liver and skeletal muscle in rats that were made septic for 6 days by cecal ligation. All animals studied developed bacteremia within 18–24 hours. The blood cultures were positive for *E. coli*, *Proteus* and *Klebsiella*. The septic animals developed a 'protein wasting' syndrome, losing a considerable proportion of their muscle mass during the 6-day period.

The liver mitochondria isolated after 6 days of sepsis were normal in all other respects except for increased State 4 respiratory rates resulting in decreased respiratory control ratios (Tavakoli and Mela 1982). The skeletal muscle mitochondria, however, exhibited several abnormalities. State 3 respiration and the calculated ATP synthesis rates with pyruvate and malate as substrates were significantly below those of sham-operated controls. No significant changes in these parameters were found with other substrates studied (glutamate + malate, α-ketoglutarate, β-glycerophosphate), although State 4 rates were above normal in all cases. In addition, skeletal muscle mitochondrial cytochrome concentrations per mg of total mitochondrial protein or per gram of muscle tissue declined in septic animals, indicating that specific muscle proteins were catabolized at rates too fast for protein synthesis to compensate for.

It thus seems that sepsis lasting several days can cause mitochondrial alterations reflecting the protein catabolic changes so characteristic of human sepsis. The inhibition of pyruvate-supported respiration and ATP synthesis suggests a specific alteration of pyruvate dehydrogenase activity in skeletal muscle mitochondria after subacute sepsis.

6. EFFECT OF ENDOTOXIN ON MITOCHONDRIAL FUNCTION – DIRECT OR INDIRECT

At the present time it is not clear how endotoxin, administered in vivo, causes the mitochondrial alterations described in the previous sections (4 and 5). Several possibilities have been proposed. Correlation of the in vitro and in vivo endotoxin effects on mitochondria clearly suggests that the in vivo effects are not due to the sole direct action of endotoxin on mitochondrial membranes (Mela 1981). This conclusion is based on the large amounts of endotoxin required in vitro to affect the mitochondrial membranes. The amounts required in vitro are 200–1000 times greater than could be expected to reach any cell, including the liver cells, in endotoxemia (Mela 1981; Zlydaszyk and Moon 1976). The finding that endotoxin in vitro specifically inhibits succinate oxidation, which effect has not been demonstrated after in vivo endotoxemia, also speaks against the role of direct actions of endotoxin on mitochondrial membranes in endotoxemia (Mela et al 1970a, 1972a).

The hemodynamic disturbances caused by endotoxemia affect the perfusion of several organs, causing tissue ischemia. Comparisons of liver, kidney and brain mitochondrial alterations caused by endotoxemia or endotoxin shock with those seen after hemorrhagic shock or tissue ischemia attracted attention to the close similarity of these alterations (Mela et al 1971, 1972b; Mela, 1975, 1982). It is quite possible that ischemic injury plays an important role in the mitochondrial deterioration caused by in vivo administrations of endotoxin. It should be noted, however, that controlled hypoxia, without lowered tissue perfusion, does not cause mitochondrial deterioration as seen in endotoxemia or tissue ischemia (Mela 1982; Mela et al 1972a, 1973; White et al 1973).

In addition to mitochondrial deterioration caused by tissue ischemia, primary or secondary toxic effects on mitochondrial membranes could play a major role in endotoxemia. The direct effect of released lysosomal enzymes on mitochondrial function has been considered a possible etiologic factor (Bradley 1981; McGivney and Bradley 1979b; Mela et al 1972a, 1973; Nicholas et al 1972). Protection of mitochondrial deterioration in endotoxemia by glucocorticoids occurs parallel with protection against lysosomal alterations (Mela and Miller 1983). This finding indicates the close correlation between lysosomal and mitochondrial changes. In spite of several attempts, however, it has not been possible to determine the cause and effect relationship of these two alterations. Decreased cell pH and altered intracellular cation concentrations in combination with endotoxin or released lysosomal enzymes can cause changes similar to those seen after in vivo endotoxemia (Mela

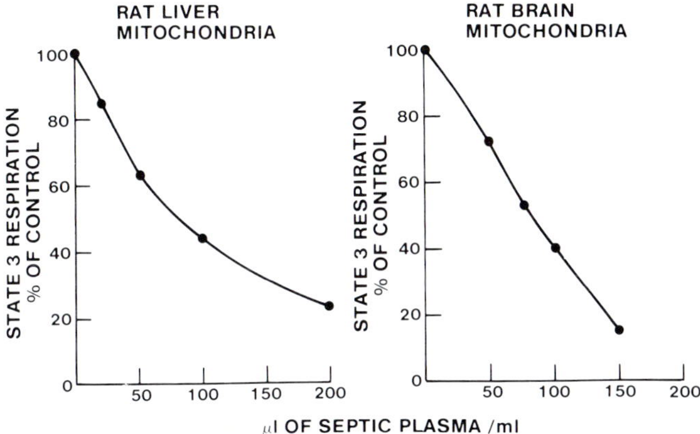

Fig. 4. *Effect of plasma peptide from septic patients on rat liver and brain mitochondrial respiratory activity in State 3. Normal, intact isolated rat liver and brain mitochondria were suspended in mannitol-sucrose Tris-phosphate buffer at pH 7.4 in the presence of 10 mM glutamate and malate and 500 μM ADP. The rates of State 3 (+ ADP) respiration were monitored and plotted as a function of the concentration of septic plasma added to the cuvette.*

1982; Mela et al 1972a, 1972b, 1973, 1974b). Although endotoxin alone in the normal intracellular environment might not cause the kinds of changes seen in endotoxemia in vivo, the altered cellular environment brought about by the hemodynamic effects of endotoxin can 'sensitize' the mitochondrial membranes sufficiently to cause serious functional deterioration. Thus, in the altered intracellular environment of circulatory shock, direct endotoxin effects on mitochondrial function are possible.

Mediated mitochondrial effects of endotoxin have also been suggested (Mela 1981). Direct evidence now exists to support this hypothesis. We studied the effects of two types of plasma mediators from septic patients and rats. Figure 4 illustrates the inhibitory effects of a plasma peptide fraction from patients in sepsis (courtesy of Dr. George Clowes, Harvard University) on rat liver and brain mitochondrial responses to ADP. Similar changes were obtained in mitochondria isolated from guinea pig brain. Plasma from rats made septic by cecal ligation also caused a similar inhibition of mitochondrial function (Mela and Tavakoli, unpublished observations). It is thus feasible that endotoxemia and sepsis cause their deleterious mitochondrial alterations indirectly via the action of primary circulating mediators, which could be the result of the body's early immune responses to bacterial toxins.

ACKNOWLEDGEMENT

The authors thank Barbara Lantzy for her invaluable help in the preparation of the manuscript.

REFERENCES

Berry LJ, Smythe DS, Young LG (1959) Effects of bacterial endotoxin on metabolism. I. Carbohydrate depletion and the protective role of cortisone. *J. Exp. Med. 110*, 389-405.

Bradley SG (1981) Direct action of bacterial endotoxin on cells, mitochondria and lysosomes. *Prog. Clin. Biol. Res. 62*, 3-14.

Chance B (1959) Quantitative aspects of the control of oxygen utilization. In: Wolstenholme GEW, O'Connor CM (Eds), *Ciba Foundation Symposium on Regulation of Cell Metabolism*, pp 91-121. J & A Churchill, London.

Chance B, Williams GR (1955) Respiratory enzymes in oxidative phosphorylation. III. The Steady State. *J. Biol. Chem. 217*, 409-427.

Decker GAG, Blevings S, Maclean LD (1971) Effect of peritonitis on mitochondrial respiration. *J. Surg. Res. 11*, 528-532.

DePalma RG, Harano Y, Robinson AV, Holden WD (1970) Structure and function of hepatic mitochondria in hemorrhage and endotoxemia. *Surg. Forum 21*, 3-6.

DePalma RG, Glickman MH, Hartman P, Robinson AV (1977) Prevention of endotoxin-induced changes in oxidative phosphorylation in hepatic mitochondria. *Surgery 82*, 68-73.

Fonnesu A, Severi C (1956a) Oxidative phosphorylation in mitochondria from livers showing cloudy swelling. *J. Biophys. Biochem. Cytol. 2*, 293-299.

Fonnesu A, Severi C (1956b) Lowered P/O ratios with mitochondria isolated from livers showing cloudy swelling. *Science 123*, 324.

Fry DE, Silver BB, Rink RD, VanArsdall LR, Flint LM (1979) Hepatic cellular hypoxia in murine peritonitis. *Surgery 85*, 652-661.

Harken AH, Lillo RS, Hufnagel HV (1975) Direct influence of endotoxin on cellular respiration. *Surg. Gynec. Obstet. 140*, 858-860.

Hinshaw LB, Beller BK, Archer LT, Fluornoy DJ, White GL, Phillips RW (1979) Recovery from lethal Escherichia coli shock in dogs. *Surg. Gynec. Obstet. 149*, 545-553.

Hinshaw LB, Archer LT, Beller-Todd BK, Benjamin B, Fluornoy DJ, Passey R (1981) Survival of primates in lethal septic shock following delayed treatment with steroid. *Circ. Shock 8*, 291-300.

Holley DC, Spitzer JA (1980) Insulin action and binding kinetics in adipocytes exposed to endotoxin in vitro and in vivo. *Circ. Shock 7*, 3-12.

Jones GRN, Kramer R (1978) Translocase inhibition in endotoxic shock: a late step in the sequence of lethal events. *Biochem. Pharmacol. 27*, 2931-2937.

Kilpatrick-Smith L, Erecinska M, Silver IA (1981) Early cellular responses in vitro to endotoxin administration. *Circ. Shock 8*, 585-600.

Lavine L, Harano Y, DePalma RG (1971) ATPase activity and palmitate oxidation of hepatic mitochondria in energy production in endotoxemia. *Surg. Forum 22*, 5-7.

Liaw KY, Askanazi J, Michelson CB, Kantrowitz LR, Furst P, Kinney JM (1980) Effect of injury and sepsis on high-energy phosphates in muscle and red cells. *J. Trauma 20*, 755-759.

Liu M-S, Long WM, Spitzer JJ (1981) Influence of *Escherichia coli* endotoxin on palmitate, glucose, and lactate utilization by isolated dog heart myocytes. In: Majde J, Person R (Eds), *Pathophysiological Effects of Endotoxins at the Cellular Level*, pp 115-121. Alan R. Liss, New York.

Mager J, Theodor E (1957) Inhibition of mitochondrial respiration and uncoupling of oxidative phosphorylation by fractions of the *Shigella paradysenteriae* Type III somatic antigen. *Arch. Biochem. Biophys. 67*, 169-177.

Maier RV, Mathison JC, Ulevitch RJ (1981) Interactions of bacterial lipopolysaccharides with tissue macrophages and plasma lipoproteins. *Prog. Clin. Biol. Res. 62*, 133-155.

Manny J, Livni N, Schiller M, Guttman A, Boss J, Rabinovici N (1980) Structural changes in the perfused canine kidney exposed to the direct action of endotoxin. *Isr. J. Med. Sci. 16*, 153-161.

McGivney A, Bradley SG (1979a) Action of bacterial endotoxin and lipid A on mitochondrial enzyme activities of cells in culture and subcellular fractions. *Infect. Immun. 25*, 664-671.

McGivney A, Bradley SG (1979b) Effects of bacterial endotoxin on lysosomal and mitochondrial enzyme activities of established cell cultures. *J. Reticuloendothel. Soc. 26*, 307-316.

Mela L (1975) Mitochondrial metabolic alterations in experimental circulatory shock. In: Urbaschek B, Urbaschek R, Neter E (Eds), *Gram-negative Bacterial Infections and Mode of Endotoxin Actions*, pp 288-295. Springer-Verlag, Wien.

Mela L (1979) Reversibility of mitochondrial metabolic response to circulatory shock and tissue ischemia. *Circ. Shock Suppl. 1*, 61-67.

Mela L (1981) Direct and indirect effects of endotoxin on mitochondrial function. *Prog. Clin. Biol. Res. 62*, 15-21.

Mela L (1982) Mitochondrial function in shock, ischemia and hypoxia. In: Trump B, Cowley A (Eds), *Pathophysiology of Shock, Anoxia and Ischemia*, pp 84-95. Williams and Wilkins, Baltimore, MD.

Mela L, Miller LD (1983) Efficacy of glucocorticoids in preventing mitochondrial metabolic failure in endotoxemia. *Circ. Shock 10*, 371-381.

Mela L, Miller LD, Diaco JF, Sugerman HJ (1970a) Effect of E. coli endotoxin on mitochondrial energy-linked functions. *Surgery 68*, 541-549.

Mela L, Bacalzo LV, White RR, Miller LD (1970b) Shock induced alterations of mitochondrial energy-linked functions. *Surg. Forum 21*, 6-8.

Mela L, Bacalzo LV, Miller LD (1971) Defective oxidative metabolism of rat liver mitochondria in hemorrhagic and endotoxin shock. *Am. J. Physiol. 220*, 571-577.

Mela L, Miller LD, Nicholas GG (1972a) Influence of cellular acidosis and altered cation concentrations on shock-induced mitochondrial damage. *Surgery 72*, 102-110.

Mela L, Miller LD, Bacalzo LV, Olofsson K, White RR (1972b) Alterations of mitochondrial structure and energy-linked functions in hemorrhagic shock and endotoxemia. *Adv. Exp. Biol. Med. 33*, 231-242.

Mela L, Miller LD, Bacalzo LV, Olofsson K, White RR (1973) Role of intracellular variations of lysosomal enzyme activity and oxygen tension in mitochondrial impairment in endotoxemia and hemorrhage in the rat. *Ann. Surg. 178*, 727-735.

Mela L, Hinshaw LB, Coalson JJ (1974a) Correlation of cardiac performance, ultrastructural morphology, and mitochondrial function in endotoxemia in the dog. *Circ. Shock 1*, 265-272.

Mela L, Nicholas GG, Laskowski R, Miller LD (1974b) Glucocorticoid protection against endotoxin-induced cellular shock. *Surg. Forum 25*, 77-79.

Nicholas GG, Mela L, Miller LD (1972) Shock-induced alterations of mitochondrial membrane transport: effects of endotoxin and lysosomal enzymes on calcium transport. *Ann. Surgery 176*, 579-584.

Nicholas GG, Mela L, Miller LD (1974) Early alterations in mitochondrial membrane transport during endotoxemia. *J. Surg. Res. 16*, 375-383.

Reed PC, Erve PR, DasGupta TK, Schumer W (1970) Endotoxemic effect of Escherichia coli on cardiac and skeletal muscle mitochondria. *Surg. Forum 21*, 13-14.

Rosenberg JC, Rush BF (1966) Lethal endotoxin shock. Oxygen deficit, lactic acid levels, and other metabolic changes, *J. Am. Med. Assoc. 196*, 87-89.

Rush BF, Hsieh J (1968) In vivo and in vitro effects of endotoxin on tissue metabolism. *Surgery 63*, 298-300.

Sayeed MM, Wurth M, Planner J, Baue AE (1970) Alpha-ketoglutarate respiration in liver during shock. *Fed. Proc. 29*, 712.

Schmahl FW, Betz E, Heckers H, Reinhard U, Schlote W, Urbaschek B (1980) Metabolic and morphologic brain reactions in shock induced by intracisternal injection of endotoxin. *Adv. Shock Res. 4*, 113-117.

Schumer W, Erve PR (1971) Bovine serum albumin effect on endotoxin-challenged mitochondria. *Surgery 69*, 699-701.

Schumer W, Erve P, Kapica SK, Moss GS (1970) Endotoxin effect on respiration of rat liver mitochondria. *J. Surg. Res. 10*, 609-612.

Schumer W, DasGupta TK, Moss GS, Nyhus LM, Erve PR (1971a) Effect of endotoxemia on liver cell mitochondria in man. *Z. Exp. Chir. 4*, 329-337.

Schumer W, Erve PR, Obernolte RP (1971b) Endotoxemic effect on cardiac and skeletal muscle mitochondria. *Surg. Gynecol. Obstet. 133*, 433-436.

Silver IA (1981) Some effects of Escherichia coli endotoxin on cells in culture. *Prog. Clin. Biol. Res. 62*, 81-95.

Staples D, Topuzlu C, Blair E (1969) A comparison of adenosine triphosphate levels in hemorrhagic and endotoxic shock in the rat. *Surgery 66*, 883-885.

Takeda H, Liu M-S (1980) Effect of Escherichia coli endotoxin on adenine nucleotide translocation in canine heart mitochondria. *Arch. Biochem. Biophys. 204*, 153-160.

Tanaka J, Kono Y, Shimahara Y, Sato T, Jones RT, Cowley RA, Trump BF (1982) A study of oxidative phosphorylative activity and calcium-induced respiration of rat liver mitochondria following living Escherichia coli injection. *Adv. Shock Res. 7*, 77-90.

Tavakoli H, Mela L (1982) Alterations of mitochondrial metabolism and protein concentrations in subacute septicemia. *Infect. Immun. 38*, 536-541.

White RR, Mela L, Miller LD, Berwick L (1971) Effect of E. coli endotoxin on mitochondrial form and function: inability to complete succinate-induced condensed-to-orthodox conformational change. *Ann. Surg. 174*, 983-990.

White RR, Mela L, Bacalzo LV, Olofsson K, Miller LD (1973) Hepatic ultrastructure in endotoxemia, hemorrhage and hypoxia: emphasis on mitochondrial changes. *Surgery 73*, 525-534.

Zlydaszyk JC, Moon RJ (1976) Fate of ^{51}Cr-labeled lipopolysaccharide in tissue culture cells and livers of normal mice. *Infect. Immun. 14*, 100-105.

Handbook of Endotoxin, Vol. 3: Cellular Biology of Endotoxin
L.J. Berry, editor
© Elsevier Science Publishers B.V., 1985

CHAPTER 8

Effect of endotoxin on cytochrome P-450 activity

NOBUHIKO KASAI AND KIYOSHI EGAWA

1. INTRODUCTION

Since Cooper et al (1965) and Omura et al (1965) established that cytochrome P-450 is the terminal oxidase of the hepatic microsomal drug-metabolizing enzyme system, the biochemical background in so-called cytochrome P-450-dependent mixed function oxidase (MFO) systems or monooxygenases has been studied by many investigators. The existence of a simple or complex electron flow system utilizing NADPH and/or NADH as the electron donor and the occurrence of the multiple forms of cytochrome P-450 have been shown (Conney 1967; Ichikawa 1981; Imai and Sato 1982; Lu and West 1980; Nebert 1979; Sato and Omura 1978). In this chapter, studies concerning the effect of endotoxin on the cytochrome P-450-dependent MFO systems will be described.

2. HISTORICAL BACKGROUND

Searching for a metabolic explanation of endotoxin toxicity, Berry (1964) found that the lethality of endotoxin in mice was significantly enhanced when actinomycin D was injected concurrently with endotoxin. A similar enhanced toxicity was observed by Bradley and coworkers (Karp and Bradley 1968; Rose and Bradley 1968; Rose et al 1972) when an antitumor drug, such as vincristine, pactamycin or polyinosinic-polycytidylic acid (poly I:C), was injected at the same time as endotoxin. To assess the effects that endotoxin and these chemotherapeutic drugs may have on the metabolism of each other, Rose et al (1972) measured the duration of sleeping time induced by pentobarbital in mice receiving endotoxin and found that the barbiturate-induced sleeping time in mice receiving endotoxin prior to, but not simultaneously with, pentobarbital was significantly increased. They speculated that prolongation of barbiturate-induced sleeping time resulted from decreased hepatic microsomal MFO activity, either as a subtle effect on enzyme activity or synthesis, or as a generalized cytotoxic effect. They also suggested that the inhibition of microsomal enzymes by endotoxin, which was indicated by an increase in sleeping time, could account for the enhanced lethality of combinations of endotoxin and drug if hepatic microsomal MFO activity was related to the ability to degrade drug.

On the other hand, Gemsa et al (1974) reported that heme oxygenase in rat liver, which catalyzes the degradation of heme to biliverdin and carbon monoxide, was markedly increased after treatment with endotoxin. They suggested that the endotoxin-related stimulation of heme oxygenase activity may account for the rise in bilirubin formation in endotoxin-treated animals which was found by Eddington and Kampfschmidt (1968).

In connection with these studies, Bissell and Hammaker (1976a) investigated the degradation of cytochrome P-450 heme in the rat liver by a pulse labeling experiment. Hepatic heme was labeled by administration of a trace pulse of [5-^{14}C]δ-aminolevulinic acid (ALA), and its degradation was analyzed in terms of the excretion of labeled carbon monoxide (^{14}CO), which is a specific degradation product of the labeled heme. Within minutes after administration of [5-^{14}C]ALA, ^{14}CO was detectable and increased after 2 hours to an 'early peak', reflecting the elimination of labeled heme from a rapidly changing pool in the liver. Beyond the early peak, the rate of ^{14}CO production decreased in a log-linear manner, consistent with the degradation of heme in stable hepatic hemoproteins. From the rate at which ^{14}CO production declined during this phase, from the predominant labeling of cytochrome P-450 by the administered [5-^{14}C]ALA, and from the known turnover characteristics of this hemoprotein in the liver, it could be inferred that production of ^{14}CO – between 16 and 30 hours after administration of labeled ALA – largely reflected degradation of cytochrome P-450 heme. By this approach, the authors indicated that endotoxin caused marked acceleration of the degradation of cytochrome P-450 heme, the effect occurring over the same dose range as that for stimulation of hepatic heme oxygenase. Since ALA synthetase, which catalyzes the condensation of glycine and succinyl-CoA to ALA, was a rate-limiting enzyme for heme synthesis in the liver, Bissell and Hammaker (1976b) also studied the effect of endotoxin on the induction of ALA synthetase by allylisopropylacetamide and showed that endotoxin is highly effective in blocking the induction of ALA synthetase. These studies suggest that the regulation of ALA synthetase and of heme oxygenase in the liver are closely interrelated, and that endotoxin perturbs the regulation mechanism.

In order to clarify the effect of endotoxin (LPS) on the hepatic drug-metabolizing enzyme activity, several investigators have independently explored the ability of livers removed from LPS-treated animals to metabolize drugs.

Vainio (1973), who studied drug metabolism in microsomes of rat liver after a single intraperitoneal injection of a lethal dose (100 mg/kg) of *Escherichia coli* (O127:B4) endotoxin, showed that, about 1.5 hours after the administration of endotoxin, the oxidation of aniline, a type II substrate of the MFO, was slightly enhanced, but this was followed by a decline in activity. On the other hand, the aryl hydrocarbon hydroxylase, catalyzing the oxidation of a type I substrate, 3,4-benzpyrene, was only inhibited after endotoxin. The author suggests that a defective drug metabolism occurred in livers of rats during endotoxin shock.

Gorodischer et al (1976) assumed that the assessment of the capability of liver to metabolize drugs during the course of the endotoxemia that accompanies gram-negative bacterial infections, was essential for the rational formulation of drug dosage. They accordingly measured microsomal enzyme activities in rat liver 24 hours after the administration of 1 mg of a purified LPS from *E. coli* O26:B6, and found that this treatment resulted in a significant decrease in the activities of aniline hydroxylase, benzpyrene hydroxylase, bilirubin UDPGA transferase, and cytochrome P-450 content, but not in nitroreductase and azoreductase.

Kasai (1976) and Egawa and Kasai (1979) found that a single intraperitoneal injection of 2–20 µg per mouse of the glycolipid or free lipid A from *Salmonella minnesota* Re mutant caused a significant depression of hepatic aminopyrine N-demethylase activity. The enzyme activity in the liver decreased to about 50% of the saline control 14–24 hours after the endotoxin injection and then recovered to the normal level after 48 hours. A similar pattern of changes was observed in the aniline hydroxylase activity. The dosage of the endotoxin (glycolipid) was only one-hundredth of the LD_{50}, but a considerable loss of appetite and decrease in body weight in the test animals were observed. Since starvation has a prominent effect on drug-metabolizing enzyme activities (Dixon et al 1960; Kato 1977; Kato and Gillette 1965), the question of whether or not the depression of enzyme activities resulted from starvation remained. However, the results of additional experiments, in which food was removed for the duration of the experiment, indicated that depression of the enzyme activities was not due to starvation, but to the effect of the glycolipid.

Renton and Mannering (1976) also described a significant depression of hepatic ethylmorphine N-demethylase, aniline hydroxylase, cytochrome P-450 and cytochrome b_5 levels in rats given interferon-inducing agents. Endotoxin (*E. coli*) was the most potent of the agents tested. The authors suggested that the depression of enzyme activities may be a general property of interferon-inducing agents.

On the basis of the investigations just described, it is clear that endotoxin is not an inducer of hepatic drug-metabolizing enzymes, but a potent depressor.

3. MODE OF ACTION

In view of the inhibitory activity of endotoxin on hepatic drug-metabolizing function, it is reasonable to expect that the rate of drug metabolism will be altered in gram-negative bacterial infections and in endotoxemia. It is also likely that the toxicity and/or pharmacological properties of a drug will be enhanced by the action of endotoxin. Accordingly, it is important to learn in practice about the mechanism by which endotoxin causes depression of hepatic MFO systems.

As mentioned above, Bissell and Hammaker (1976a, 1976b) presented evidence that the heme moiety of cytochrome P-450 dissociates reversibly from its apoprotein and, prior to its degradation, mixes with endogenously synthesized heme to form a pool that regulates ALA synthetase activity. They also reported (Bissell and Hammaker 1977) that endotoxin causes a relatively specific enlargement of the hepatic 'free' heme pool (rather than a nonspecific toxic effect), and that endotoxin is probably only one of a large group of compounds that, at pharmacological doses, cause dissociation of heme from cytochrome P-450. These observations suggest that the depression of hepatic drug metabolizing function after endotoxin administration results from metabolic alterations in microsomal cytochrome P-450.

In connection with the metabolism of microsomal cytochrome P-450, Williams et al (1979) studied the effect of administering phenobarbital (80 mg/kg) and endotoxin (2 mg/kg), singly and in combination, to rats. The activity of tryptophan oxygenase, heme oxygenase and cytochrome P-450 content in liver microsomes was then evaluated. Total tryptophan oxygenase activity, but not holoenzyme activity, was decreased 4 hours after the administration of either endotoxin or of the endotoxin–phenobarbital combination. Thus, an increased percentage of heme saturation of tryptophan oxygenase was seen in these animals. At the early time point, heme oxygenase and cytochrome P-450 were not affected by endotoxin or phenobarbital, but cytochrome P-450 was slightly decreased by the combination. At 12 hours after the injection, tryptophan oxygenase activity for all treated groups was not different from control values. Heme oxygenase was increased and cytochrome P-450 decreased at 12 and 24 hours after injection of rats with either endotoxin or endotoxin and phenobarbital in combination. Phenobarbital-pretreated animals showed a 20 and 100% increase in cytochrome P-450 at these time points. Microsomal preparations from endotoxin- or endotoxin–phenobarbital-treated animals did not contain cytochrome P-420. Although the microsomal preparations from phenobarbital-treated animals contained about 20% apo-P-450, no apoprotein could be detected in preparations from endotoxin- or endotoxin–phenobarbital treated animals. On the basis of these data the authors concluded that endotoxin may block the synthesis of the apoproteins of both tryptophan oxygenase and cytochrome P-450.

Yoshida et al (1982) also showed time-related changes in hepatic ALA synthetase, heme oxygenase and cytochrome P-450 levels following intraperitoneal injection of *S. minnesota* Re type endotoxin glycolipid at a dose of 20 μg per mouse. The levels of cytochrome P-450 and ALA synthetase activity decreased progressively with a maximum reduction appearing at about 24 hours. On the other hand, heme oxygenase activity markedly increased with maximum levels occurring about 16 hours after the injection. The effects of endotoxin on the ALA synthetase and heme oxygenase activities were dose-dependent for the range from about 1 to 20 μg. These observations suggest that the degradation of hepatic cytochrome P-450 heme is closely related to the stimulation of heme oxygenase activity and the regulation of ALA synthetase activity.

From in vitro studies (Schacter et al 1973), it has been suggested that the cytochrome P-450 heme may be degraded by a peroxidative process. However, Bissell and Hammaker (1976a) believed that this mechanism may not be operative in vivo since pretreatment of animals with an antioxidant in doses sufficient to prevent lipid peroxidation failed to prevent the effect of endotoxin on hepatic heme oxygenase activity. Moreover, the decrease in cytochrome P-450 and increase in heme oxygenase that occur in hepatocyte cultures are unaffected by the addition of a variety of antioxidants to the incubation medium.

Yoshida et al (1982) also showed that the depression of hepatic aminopyrine N-demethylase activity and cytochrome P-450 levels by endotoxin administration in mice was not abrogated by treatment with α-tocopherol, reduced glutathione or indometacin, but a significant reversal of enzyme levels was observed after treatment with cortisone. In contrast, Sasaki et al (1979) reported that LPS-induced changes of hepatic microsomal drug-metabolizing enzyme activities and cytochrome P-450 content in mice were prevented by pretreatment with glutathione (200 mg/kg/day for 3 days). It is unclear whether or not the discrepancy in these studies is due to the dose of the drug or the timing of treatment.

Since previous studies (Gemsa et al 1974; Renton and Mannering 1976) showed that administration of endotoxin tends to decrease the microsomal cytochrome c reductase activity as well as levels of cytochrome P-450 and cytochrome b_5, it appears likely that the depression of these heme enzymes by endotoxin administration may be related to the alteration of the cytochrome P-450-dependent electron flow system. It has been reported that there are three distinct pathways in the microsomal cytochrome P-450-dependent electron transport system (Ichikawa 1981):

(1) NADPH $\rightarrow$ NADPH cytochrome c reductase $\rightarrow$ cytochrome P-450 $\rightarrow$ O$_2$
(2) NADH $\rightarrow$ NADH   cytochrome c reductase $\rightarrow$ cytochrome P-450 $\rightarrow$ O$_2$
(3) NADPH $\rightarrow$ NADPH cytochrome c reductase $\rightarrow$ cytochrome P-450

$$\qquad\qquad\qquad\qquad\qquad\qquad\qquad \Big|\!\underline{\qquad\nearrow}\text{substrate}$$

$$\text{cytochrome P-450 substrate complex} \rightarrow \text{O}_2$$

NADH $\rightarrow$ NADH   cytochrome b_5 reductase $\rightarrow$ cytochrome b_5

Sasaki et al (1981) showed that the administration of *E. coli* LPS (5 mg/kg, intravenously) caused a decrease in both hepatic microsomal NADPH- and NADH-dependent cytochrome c reductase activities in addition to the reduction in the cytochrome P-450 and cytochrome b_5 levels in mice. Egawa et al, however, (unpublished observations) failed to detect any significant decrease in hepatic microsomal cytochrome b_5 by the administration of endotoxin in doses sufficient to cause the depression of cytochrome P-450 levels.

It is of interest now to determine whether the effect of endotoxin on the hepatic MFO systems is produced by a direct action or by an indirect mechanism via mediator(s).

Yaffe and Sonawane (1978) suggested that inhibition of the MFO system might be mediated via a metabolite of endotoxin. However, the in vitro incubation of endotoxin (100 μg/mg microsomal protein) did not reduce aminopyrine N-demethylase or aniline hydroxylase activity, while preincubation with a NADPH-generating system supplemented with EDTA (1 mM) plus microsomes for 30 minutes resulted in a significant reduction in aminopyrine N-demethylase activity and cytochrome P-450 levels. The authors assumed that this in vitro effect resulted from lipid peroxidation in the liver microsomal membrane that was enhanced by endotoxin.

Williams et al (1980) studied the ability of endotoxin to depress the hepatic MFO activity in the C3H/HeJ mouse strain which is generally considered to be unresponsive to the biological effects of endotoxin (Berry 1977). When these mice were injected with 0.5 mg/kg of *E. coli* LPS (O26:B6), significant decreases in the activity of ethylmorphine N-demethylase and in the levels of cytochrome P-450 and cytochrome b_5 comparable to that observed in the endotoxin-sensitive C3H/ HeN mice were observed. These data suggested that cellular constituents other than B lymphocytes or macrophages are probably involved in eliciting the response, since considerable evidence indicates that these cells of the C3H/HeJ mouse are unresponsive to endotoxin. Moreover, preliminary studies failed to demonstrate an effect on the activity of aminopyrine N-demethylation or aniline hydroxylation in isolated hepatic parenchymal cells that had been incubated with endotoxin for 8 hours.

In a recent study, Egawa et al (1981) found that serum from mice administered with Re glycolipid contained an unidentified serum factor that depressed hepatic ALA synthetase activity and cytochrome P-450 levels. This serum factor appeared in mouse blood soon after the glycolipid was injected, but the greatest activity was found after 9 hours. Hepatic ALA synthetase activity and cytochrome P-450 levels, as mentioned previously, showed a maximum decrease about 24 hours after the endotoxin was administered. The effect of this serum factor was dose-dependent. Since the serum factor was stable after being heated at 56 °C for 30 minutes, the presence of residual endotoxin in the serum cannot be excluded. Nevertheless, the authors assumed that the serum factor was of host origin since the hepatic ALA synthetase and cytochrome P-450 levels in mice whose reticuloendothelial system (RES) had been partly blocked by pretreatment with carrageenan, could be decreased by an injection of the serum factor but not by endotoxin. In addition, hepatic aminopyrine N-demethylase activity in untreated mice could be diminished by endotoxin but not by the serum factor. Since the serum factor is relatively heat-stable, it seems to be different from endotoxin-induced heat-labile interferon. Furthermore, the times at which the serum factor appears after an injection of endotoxin differ from those of other known endotoxin mediators, including glucocorticoid antagonizing factor (Moore et al 1978), colony-stimulating factor (Quesenberry et al 1972), tumor necrosis factor (Carswell et al 1975), and serum amyloid A mediator (Sipe et al 1979). The origin of the serum factor has not yet been determined.

Azhary and Mannering (1979) have assumed that the depression in activity of ALA synthetase and in the level of cytochrome P-450 might be due to interferon, since inducers such as poly I:C or tilorone produce similar changes in mice. Singh and Renton (1981) also presented evidence that suggests a direct involvement of interferon through the use of inbred strains of mice that carry four distinct genetic loci that regulate the levels of circulating interferon produced by specific viruses.

For Newcastle disease virus (NDV), one autosomal locus (IF-1) determines a 10-fold difference in serum interferon levels. In C57BL/6J mice that contain the high production allele at IF-1^h, cytochrome P-450 and aminopyrine N-demethylase activity are decreased 24 hours after NDV injection. In C3H/HeJ mice with the low production allele at IF-1^l, no significant change was observed in these enzyme levels after NDV administration. By contrast poly I:C, which induces interferon via loci other than IF-1, stimulates high levels of interferon production and depresses cytochrome P-450 and aminopyrine N-demethylase activity in both strains of mice.

It remains to be elucidated whether or not the depression of MFO activity by endotoxin is due to interferon or to some other factor.

Williams and Szentivanyi (1982) have tried to determine whether the endogenous mediators (catecholamines, prostaglandins, etc.) that are released in vivo following LPS administration are responsible for blocking the dexamethasone induction of hepatic tryptophan oxygenase and for the LPS-induced decrease in hepatic MFO activity. Pretreatment of animals with phenoxybenzamine, but not with propranolol or indometacin, prevented the LPS inhibition of tryptophan oxygenase induction. Administration of phenoxybenzamine, propranolol, indometacin, dexamethasone or polymyxin B, prior to LPS, did not alter the LPS effect on the MFO activity. Indometacin and polymyxin B significantly reduced MFO activity in rats. The activity of aniline hydroxylase and ethylmorphine N-demethylase, the levels of cytochrome P-450 and the heme oxygenase activity in LPS-tolerant rats were not altered by an injection of LPS. From these results, Williams and Szentivanyi (1982) suggested that α-adrenergic mechanisms may be involved in the inhibition of tryptophan oxygenase induction by LPS but that cellular mechanisms associated with the RES may be involved in the decrease in MFO activity by endotoxin.

In addition, apart from endotoxin, administration of immunomodulators such as cord factor (Toida 1974), *Corynebacterium parvum* (Soyka et al 1976), BCG (Farquhar et al 1976), zymosan (Hojo et al 1976), glucan (Hojo et al 1976), statolon (Renton and Mannering 1976), *Bordetella pertussis* vaccine (Egawa and Kasai 1978; Williams and Szentivanyi 1975, 1977), and streptococcal preparation (OK-432), BCG cell wall component, or a protein-bound polysaccharide (Hojo and Hashimoto 1977) to mice and rats has also been shown to produce various degrees and somewhat different modes of depression of MFO activity. In these investigations, an important role of RES (macrophages or other lymphoreticular cells such as T and/or B cells) (Soyka et al 1979) in the depression of MFO activity has been emphasized.

Whether the mechanism by which such immunostimulants and endotoxin cause a decrease in the hepatic MFO activity is common or different, remains to be determined. Identification of cellular and humoral mediators which may be involved in the effect of endotoxin will be of interest. In addition, endotoxin extensively affects hepatic mitochondrial functions (see Chapter 7), but there is as yet no direct evidence to indicate that mitochondrial alterations are involved in the decrease in hepatic MFO activity.

Unfortunately, very little is known about how the ability of the liver to metabolize drugs is altered upon administration of endotoxin. Higuchi et al (1978) have shown that endotoxin may alter the metabolism of 6-mercaptopurine, an anti-leukemic agent, in mice. 6-Mercaptopurine is converted to a biologically active thioinosinic acid by hypoxanthine-guanine phosphoribosyltransferase (HGPRTase) and then to an inactive thiouric acid by xanthine oxidase. They found a signi-ficant increase in thioinosinic acid in the liver of mice given [8-^{14}C]6-mercapto-purine 6 hours after the administration of *E. coli* LPS (4 mg/kg), whereas the levels of HGPRTase were similar in endotoxin-treated and control mice. The time-course of the increases in thioinosinic acid was consistent with the prolonga-tion of pentobarbital sleeping time after endotoxin. Significant increases in xanth-ine oxidase activity were also observed 24 and 32 hours after endotoxin administ-ration. Higuchi et al (1978) concluded that the metabolism of 6-mercaptopurine might be modified as a result of changes in the activity of drug-metabolizing en-zymes in the liver, but the mechanism of enhanced toxicity of combined endotoxin and 6-mercaptopurine (Marecki and Bradley 1974) has not been determined.

Sasaki et al (1981) also studied the effect of *E. coli* LPS on the concentration of drugs in blood and found that aminopyrine and pentobarbital given to mice 24 hours after the administration of endotoxin were each significantly higher than in untreated mice. After the administration of cyclophosphamide, a significantly low blood level of normustard, an active metabolite of the drug, was found in LPS-treated mice. Alterations in blood concentration of these drugs were believed to result from the depression of hepatic drug-metabolizing enzymes, a decrease in heme proteins, and an inhibition of the cytochrome P-450-dependent electron transport system in liver microsomes.

It is of interest to note that the depression of hepatic MFO activity by endotoxin, as seen in adult animals, has also been observed in young and pregnant animals, even though their response to endotoxin varies during pregnancy and postnatal development (Sonawane and Yaffe 1980). Whether or not the results obtained in these studies can be extended to human beings is unknown, even though im-munotherapy with *Corynebacterium parvum* or BCG have been reported to de-press drug metabolism in man (Rios et al 1977).

4. ACTIVE PRINCIPLE

As mentioned above, several microbial preparations of diverse chemical structure inhibit hepatic MFO activities in experimental animals. Most of the preparations also have immunomodulating or interferon-inducing properties. It is important, therefore, to identify the structural units responsible for host responses, not only because of chemical and biological interests, but for practical reasons as well.

Our earlier study (Kasai 1976) found that the Re type endotoxic glycolipid, consisting mainly of 3-deoxy-D-*manno*-2-octulosonic acid (KDO) and lipid A, exhibited activity as a potent depressor of hepatic drug-metabolizing enzymes such as aminopyrine N-demethylase and aniline hydroxylase. Egawa et al (Egawa and Kasai 1978, 1979; Egawa et al 1980) observed similar effects on cytochrome P-450 levels in mice treated with heat-killed bacterial cells (approximately 0.1 mg per mouse), with LPS preparations (approximately 20 μg per mouse) derived from various gram-negative bacteria, including *E. coli*, *Salmonella* (S and Ra-Re mutants), *Pseudomonas*, *Proteus*, *Bordetella pertussis* and *Caulobacter*. Inhibitory activity as great as that seen with endotoxin was not observed with the gram-positive bacteria that have been tested, even though some heat-killed cells such as *Propionibacterium avidum* O575 showed a significant activity when administered at a dose of 1.2 mg per mouse.

Among the chemical degradation products of LPS that have been tested, free lipid A derived from a mild acetic acid hydrolysate was the most active. Its effect was weaker, however, than the Re glycolipid (Egawa and Kasai 1979). When the effect of Re glycolipid, and its degradation products, on the activities of hepatic ALA synthetase, heme oxygenase, and cytochrome P-450 levels as well as aminopyrine N-demethylase activity were examined, some interesting results were observed. All of the enzyme activities were significantly altered by the administration of either Re glycolipid or free lipid A. Other degradation products, including free lipid A derived from more drastic acid hydrolysates, and a de-O-acylated glycolipid lost their ability to depress aminopyrine N-demethylase activity. Their effect on cytochrome P-450 levels and on heme oxygenase activity remained unchanged. This suggests that the activities of the parent compound can be partly dissociated (Yoshida et al 1982).

Recently, Kasai et al (1984) examined the effect of several synthetic lipid A analogs on hepatic microsomal cytochrome P-450 levels, aminopyrine N-demethylase, ALA synthetase and heme oxygenase activity. As shown in Table 1, these analogs were O- and/or N-fatty acyl derivatives of β-1,6-glucosamine disaccharide, carrying phosphoryl residues at C-1 and/or C-4'. All of the tested compounds failed to show significant depression of aminopyrine N-demethylase activity. However, all the analogs carrying 3-hydroxytetradecanoyl residues at C-3,4,6' and/or C-2,2' showed a significant increase in heme oxygenase activity. A similar effect was seen in the cytochrome P-450 levels, although the activity differed with the position and nature of the linkage of fatty acyl residues. In the assay of ALA synthetase activity, it appeared that the substitution of 3-hydroxytetradecanoyl groups at the position of C-2,3,4,2' and C-6' was required for depressor activity. These results suggest that the nature of the fatty acid linkages, especially the 3-hydroxy fatty acid, as well as the phosphate group in the glucosamine disaccharide, play an important role in the biological activity of product. The results also suggest that their activities can partly be dissociated. However, the required dose of synthetic analogs is much higher than that of natural lipid A.

The reason why some degradation products of LPS as well as the endotoxin-induced serum factor fail to diminish aminopyrine N-demethylase activity and yet lower cytochrome P-450 levels, is not known. If the endotoxin effect on hepatic MFO activity is due to indirect action via a cellular or humoral mediator, it may

TABLE 1   *Structures of synthetic lipid A analogs*

Sample number	R^1	R^2	R^3	R^4
301	C_{14}	C_{14}	H	H
302	C_{14}	C_{14}–OH	H	H
307	H	C_{14}–OH	H	H
311	C_{14}–OH	C_{14}–OH	H	H
314	C_{14}–OH	C_{14}–O–(C_{14})	H	H
304	C_{14}	C_{14}	P	H
312	C_{14}–OH	C_{14}–OH	P	H
315	C_{14}–OH	C_{14}–O–(C_{14})	P	H
303	C_{14}	C_{14}	H	P
316	C_{14}	C_{14}–OH	H	P
305	C_{14}	C_{14}	P	P
317	C_{14}	C_{14}–OH	P	P

C_{14} : tetradecanoyl
C_{14}–OH   : (R)-3-hydroxytetradecanoyl
C_{14}–O–(C_{14}) : (R)-3-tetradecanoyloxytetradecanoyl
P   : $\overset{O}{\underset{\parallel}{P}}(OH)_2$

be that two or more mediators combine to produce the multiple responses. If this is valid, then a degradation product might release only a single mediator. Alternatively, endotoxin might alter, selectively, the levels of some of the multiple forms of microsomal cytochrome P-450.

Endotoxin has the ability to induce tolerance in hepatic MFO activity (Egawa et al 1980, 1984; Williams and Szentivanyi 1982). Egawa et al (1980) studied the response in mice pretreated with endotoxin (Re glycolipid) by a single or by repeated daily injections. Essentially the same pattern of the host reactivity was observed in these animals. The altered reactivity was manifested as two refractory states. There was an early transient period of tolerance demonstrable at day 3 after the administration of a single injection and at day 2 in mice that had been treated daily with glycolipid. A late tolerance was evident about 11 days after a single injection and about at day 5 after daily injections. To define the active principle of the Re glycolipid that was required for the induction of the early refractory state, reciprocal cross-tolerance tests were performed using *Salmonella* free lipid A, Re glycolipid and LPS from *Pseudomonas aeruginosa* ATCC7700. These results suggested that the lipid A moiety represents the major determinant involved in the early refractory state. The two phases of tolerance resemble the pattern seen in tolerance to the lethal effect of LPS (Greer and Rietschel 1978).

Much work is needed before a reasonable explanation of the biochemical significance, the mechanism of action, and the structural component of LPS that are responsible for the host responses described in this chapter, becomes available.

ACKNOWLEDGMENTS

We are most grateful to Dr. L. Joe Berry for reviewing and correcting this chapter. We also wish to thank Mrs. M. Matsumoto for assistance in the preparation of this manuscript.

REFERENCES

Azhary R, Mannering GJ (1979) Effect of interferon inducing agents (polyriboinosinic acid-polyribocytidylic acid, tilorone) on hepatic hemoproteins (cytochrome P-450, catalase, tryptophan 2,3-dioxygenase, mitochondrial cytochrome), heme metabolism and cytochrome P-450-linked monooxygenase systems. *Mol. Pharmacol. 15*, 698.

Berry LJ (1964) Effect of endotoxins on the level of selected enzymes and metabolites. In: Landy M, Braun W (Eds), *Bacterial Endotoxins*, p 151. Rutgers University Press, New Brunswick, NJ.

Berry LJ (1977) Bacterial toxins. *CRC Rev. Toxicol. 5*, 239.

Bissell DM, Hammaker LE (1976a) Cytochrome p-450 heme and the regulation of hepatic heme oxygenase activity. *Arch. Biochem. Biophys. 176*, 91.

Bissell DM, Hammaker LE (1976b) Cytochrome P-450 heme and the regulation of δ-aminolevulinic acid synthetase in the liver. *Arch. Biochem. Biophys. 176*, 103.

Bissell DM, Hammaker LE (1977) Effect of endotoxin on tryptophan pyrrolase and δ-aminolaevulinate synthase: evidence for an endogenous regulatory haem fraction in rat liver. *Biochem. J. 166*, 301.

Carswell EA, Old LJ, Kassel RL, Green S, Fiore N, Williamson B (1975) An endotoxin-induced serum factor that causes necrosis of tumors. *Proc. Natl Acad. Sci. USA 72*, 3666.

Conney AH (1967) Pharmacological implications of microsomal enzyme induction. *Pharmacol. Rev. 19*, 317.

Cooper DY, Levin S, Narashimhulu S, Resenthal O, Estabrook RW (1965) Photochemical action spectrum of the terminal oxidase of mixed-function oxidase systems. *Science 147*, 400.

Dixon RL, Shultice RW, Fouts JR (1960) Factors affecting drug metabolism by liver microsomes. IV. Starvation. *Proc. Soc. Exp. Biol. Med. 103*, 333.

Eddington CL, Kampfschmidt RF (1968) Bilirubin production in endotoxin-treated or tumor-bearing rats. *Proc. Soc. Exp. Biol. Med. 129*, 580.

Egawa K, Kasai N (1978) Effect of administration of various bacterial cells on hepatic drug-metabolizing activity in mice (in Japanese). *Jpn. J. Bacteriol. 33*, 93.

Egawa K, Kasai N (1979) Endotoxic glycolipid as a potent depressor of the hepatic drug-metabolizing enzyme systems in mice. *Microbiol. Immunol. 23*, 87.

Egawa K, Yoshida M, Kasai N (1980) Studies on bacterial components and their inhibitory activities to the hepatic drug-metabolizing enzyme systems in mice. *Jpn. J. Med. Sci. Biol. 33*, 31.

Egawa K, Yoshida M, Kasai N (1981) An endotoxin-induced serum factor that depresses hepatic δ-aminolevulinic acid synthetase activity and cytochrome P-450 levels in mice. *Microbiol. Immunol. 25*, 1091.

Egawa K, Yoshida M, Sakaino R, Kasai N (1984) Hepatic drug-metabolizing enzyme system and endotoxin tolerance: structural requirement of LPS in induction of an early tolerance. *Microbiol. Immunol. 28*, 1181.

Farquhar D, Loo TL, Gutterman JU, Hersh EM, Luna MA (1976) Inhibition of drug-metabolizing enzymes in the rat after Bacillus Calmette-Guérin treatment. *Biochem. Pharmacol. 25*, 1529.

Gemsa D, Woo CH, Fudenberg HH, Schmid R (1974) Stimulation of heme oxygenase in macrophages and liver by endotoxin. *J. Clin. Invest. 53*, 647.

Gorodischer R, Krasner J, McDevitt JJ, Nolan JP, Yaffe SJ (1976) Hepatic microsomal drug metabolism after administration of endotoxin in rats. *Biochem. Pharmacol. 25*, 351.

Greer GG, Rietschel ET (1978) Inverse relationship between the susceptibility of lipopolysaccharide (lipid A)-pretreated mice to the hypothermic and lethal effect of lipopolysaccharide. *Infect. Immun. 20*, 366.

Higuchi T, Nakamura T, Uchino H (1978) Mechanism of enhanced toxicity of 6-mercaptopurine with endotoxin. *Biochem. Pharmacol. 27*, 2507.

Hojo H, Hashimoto Y (1977) Inhibition of drug-metabolizing enzymes in the mouse after treatment with host-mediating antitumor drugs. *Toxicol. Lett. 1*, 89.

Hojo H, Suzuki Y, Konishi Y, Uchiyama M (1976) Effect of zymosan on hepatic drug metabolism in mice. *Chem. Pharm. Bull. 24*, 10.

Hojo H, Suzuki Y, Uchiyama M (1976) Effect of zymosan on hepatic drug metabolism in mice. II. Identification of active component of yeast cell walls. *Chem. Pharm. Bull. 24*, 352.

Ichikawa Y (1981) Cytochrome P-450-linked mixed function oxidase systems (monooxygenase) of microsomal and mitochondrial types and their substrate specificities (in Japanese). *Seikagaku 53*, 221.

Imai Y, Sato R (1982) Multiplicity of hepatic microsomal cytochrome p-450 – molecular species and substrate specificity (in Japanese). *Metab. Dis. 19*, 1723.

Karp RD, Bradley SG (1968) Synergistic toxicity of endotoxin with pactamycin or sparsomycin. *Proc. Soc. Exp. Biol. Med. 128*, 1075.

Kasai N (1976) Structure and biological activity of endotoxin (in Japanese). *Jpn. J. Bacteriol. 31*, 44.

Kasai N, Egawa K, Mashimo J, Yoshida M, Sasaki M, Shiba T, Kusumoto S (1984) Shwartzman activities of synthetic lipid A analogues and their effects on hepatic enzyme. In: Homma JY, Kanegasaki S, Lüderitz O, Shiba T, Westphal O (Eds), *Bacterial Endotoxin*, p 81. Verlag Chemie, Weinheim.

Kato R (1977) Drug metabolism under pathological and abnormal physiological states in animals and man. *Xenobiotica 7*, 25.

Kato R, Gillette JR (1965) Effect of starvation on NADPH-dependent enzymes in liver microsomes of male and female rat. *J. Pharmacol. Exp. Ther. 150*, 279.

Lu AYH, West SB (1980) Multiplicity of mammalian microsomal cytochrome P-450. *Pharmacol. Rev. 31*, 277.

196

Marecki NM, Bradley SG (1974) Enhanced toxicity for mice of 6-mercaptopurine with bacterial endotoxin. *Antimicrob. Agents Chemother. 5*, 413.

Moore RN, Goodrum KJ, Couch RE Jr, Berry LJ (1978) Elicitation of endotoxemic effects in C3H/HeJ mice with glucocorticoid antagonizing factor and partial characterization of the factor. *Infect. Immun. 19*, 79.

Nebert DW (1979) Multiple forms of inducible drug-metabolizing enzymes: a reasonable mechanism by which any organism can cope with adversity. *Mol. Cell. Biochem. 27*, 27.

Omura T, Sato R, Cooper DY, Rosenthal O, Estabrook RW (1965) Function of cytochrome P-450 of microsomes. *Fed. Proc. 24*, 1181.

Quesenberry P, Morley A, Stohlman F Jr, Rickard K, Howard D, Smith M (1972) Effect of endotoxin on granulopoiesis and colony stimulating factor. *N. Engl. J. Med. 286*, 227.

Renton KW, Mannering GJ (1976) Depression of hepatic cytochrome P-450-dependent monooxygenase systems with administered interferon inducing agents. *Biochem. Biophys. Res. Commun. 73*, 343.

Rios A, Farquhar D, Loo TL (1977) Effect of immunotherapy with intravenous *C. parvum* on antipyrine metabolism. *Proc. Am. Soc. Clin. Oncol. 18*, C-332.

Rose WC, Bradley SG (1971) Enhanced toxicity for mice of combinations of antibiotics with *Escherichia coli* cells or *Salmonella typhosa* endotoxin. *Infect. Immun. 4*, 550.

Rose WC, Bradley SG, Lee IP (1972) Enhanced toxicity for mice of vincristine and other chemotherapeutic agents with *Salmonella typhosa* endotoxin and *Pseudomonas aeruginosa*. *Antimicrob. Agents Chemother. 1*, 489.

Sasaki K, Saitoh M, Takayanagi G (1979) Prevention of lipopolysaccharide-induced depression of drug-metabolizing enzyme activity by glutathione (in Japanese). *Iyakuhin Sogosayo Kenkyu (Drug Interaction Res.) 4*, 69.

Sasaki K, Saitoh M, Takayanagi G (1981) Effects of lipopolysaccharide (*E. coli*) and OK-432, a Streptococcus preparation, on the hepatic drug-metabolizing system in mice (in Japanese). *Yakugaku Zasshi 101*, 932.

Sato R, Omura T (1978) *Cytochrome P-450*. Kodansha and Academic Press.

Schacter BA, Marver HS, Meyer UA (1973) Heme and hemoprotein catabolism during stimulation of microsomal lipid peroxidation. *Drug Metab. Dispos. 1*, 286.

Singh G, Renton KW (1981) Interferon-mediated depression of cytochrome P-450-dependent drug biotransformation. *Mol. Pharmacol. 20*, 681.

Sipe JD, Vogel SN, Ryan JL, McAdam KPWJ, Rosenstreich DL (1979) Detection of a mediator derived from endotoxin-stimulated macrophages that induces the acute phase serum amyloid A response in mice. *J. Exp. Med. 150*, 597.

Sonawane BR, Yaffe SJ (1980) Gram-negative endotoxin administration decreases hepatic drug-metabolizing enzymes during development in rats. *Pediatr. Res. 14*, 939.

Soyka LF, Hunt WG, Knight SE, Foster RS Jr (1976) Decreased liver and lung drug-metabolizing activity in mice treated with *Corynebacterium parvum*. *Cancer Res. 36*, 4425.

Soyka LF, Stephens CC, MacPherson BR, Foster RS Jr (1979) Role of mononuclear phagocytes in decreased hepatic drug metabolism following administration of *Corynebacterium parvum*. *Int. J. Immunopharmacol. 1*, 101.

Toida I (1974) Effects of cord factor on microsomal enzymes. *Am. Rev. Respir. Dis. 110*, 641.

Vainio H (1973) Defective drug metabolism in rat liver in endotoxin shock. *Ann. Med. Exp. Biol. Fenn. 51*, 65.

Williams JF, Szentivanyi A (1975) Effect of *Bordetella pertussis* vaccine on the drug-metabolizing enzyme system of mouse liver. *Fed. Proc. 34*, 1001.

Williams JF, Szentivanyi A (1977) Depression of hepatic drug-metabolizing enzyme activity by *B. pertussis* vaccination. *Eur. J. Pharmacol. 43*, 281.

Williams JF, Szentivanyi A (1982) Possible involvement of α-adrenergic mechanisms and reticuloendothelial activation in the effects of bacterial lipopolysaccharide on hepatic enzyme activities. *Fed. Proc. 41*, 1722.

Williams JF, Lowitt S, Szentivanyi A (1979) Effect of endotoxin and phenobarbital on heme enzymes of rat liver. *Pharmacologist 21*, 232.

Williams JF, Lowitt S, Szentivanyi A (1980) Endotoxin depression of hepatic mixed function oxidase system in C3H/HeJ and C3H/HeN mice. *Immunopharmacology 2*, 285.

Yaffe SJ, Sonawane BR (1978) Inhibition of rat hepatic microsomal drug metabolizing enzymes by endotoxin. *Fed. Proc. 37*, 304.

Yoshida M, Egawa K, Kasai N (1982) Effect of endotoxin and its degradation products on hepatic mixed-function oxidase and heme enzyme systems in mice. *Toxicol. Lett. 12*, 185.

Handbook of Endotoxin, Vol. 3: Cellular Biology of Endotoxin
L.J. Berry, editor
© Elsevier Science Publishers B.V., 1985

CHAPTER 9

Effect of endotoxin on tryptophan metabolism

ROBERT J. MOON

1. INTRODUCTION

The effects of bacterial endotoxin on normal metabolic homeostasis are numerous and diverse (Berry 1975). In 1959, Berry et al (1959a) showed that endotoxin-poisoned animals become virtually depleted of carbohydrate. Cortisone not only antagonized carbohydrate depletion but also protected mice from the lethal actions of endotoxin. These observations form the cornerstone which has been the basis for much of the research on the metabolic toxicities of endotoxin. Subsequent studies established that when endotoxin was administered to cortisone-treated mice, the balance between protein catabolism and total carbohydrate deposition was disrupted (Berry et al 1959a, 1959b; Berry and Smythe 1961a, 1961b). The decrease in carbohydrate deposition was presumably due to inhibition of gluconeogenesis. While these general observations were well defined and reproducible, the metabolic mechanisms underlying these changes remained obscure. Throughout these studies, a constant theme has been that glucocorticoid hormones not only protect against endotoxin poisoning but also reverse or substantially antagonize the untoward metabolic effects of the bacterial poison.

The discovery (Feigelson et al 1962; Knox and Auerbach 1955) that cortisone could influence the activity of selected liver enzymes provided an important clue as to the direction subsequent studies should take. The first adaptive liver enzyme shown to be responsive to cortisone was tryptophan pyrrolase (subsequently named tryptophan oxygenase (tryptophan 2,3-dioxygenase, E.C. 1.13.11.11.)). The observations that tryptophan oxygenase activity is depressed in endotoxin-poisoned mice and that cortisone induction of the enzyme is also inhibited (Berry and Smythe 1963; Berry et al 1968a, 1968b) launched a series of studies on the effects of endotoxin on a whole range of adaptive liver enzymes. The two enzymes most extensively studied in this context are tryptophan oxygenase and phosphoenolpyruvate carboxykinase.

The major objective of this chapter is to discuss the development and current status of this field, particularly as it relates to tryptophan oxygenase and tryptophan metabolism. Changes in other adaptive enzyme systems will be discussed only as they are pertinent to the general theme under review.

2. EFFECTS OF ENDOTOXIN ON TRYPTOPHAN OXYGENASE

2.1. Selected factors involved in tryptophan oxygenase regulation

Tryptophan oxygenase is an iron porphyrin enzyme which exists in both the ferrous and ferric states (Tanaka and Knox 1959), as indicated by its absorption spectrum and inhibition reactions characteristic of iron porphyrin enzymes, as well as reactions such as light-reversible carbon monoxide inhibition and inhibition by cyanide or ferricyanide.

Isolation and purification of tryptophan oxygenase has shown the existence of 3 forms: an apoenzyme (Greengard and Feigelson 1962), as well as an oxidized and reduced holoenzyme (Tanaka and Knox 1959). Both the apoenzyme and oxidized holoenzyme require further activation to become catalytically active as the reduced holoenzyme. The overall in vitro activation process requires the presence of the substrate L-tryptophan, the prosthetic group hematin, and a reducing agent, such as ascorbic acid or H_2O_2 (Greengard and Feigelson 1961, 1962; Knox 1966; Knox et al 1966; Piras and Knox 1967). A sequential series of events occurs in the substrate-mediated activation process (Piras and Knox 1967).

Inactive apotryptophan oxygenase formed from de novo synthesis is conjugated with its prosthetic group hematin to form the oxidized holoenzyme. This step requires the presence of L-tryptophan or certain analogs, and can be inhibited by globin or thiol reagents. A second reaction involves the reduction of the oxidized holoenzyme to the catalytically active reduced holoenzyme. The reduction process involves ascorbic acid or H_2O_2 and has a specific requirement for L-tryptophan. Reversible loss of the active, reduced holotryptophan oxygenase to the inactive, oxidized holoenzyme can occur by oxidation in air following removal of L-tryptophan.

Tryptophan oxygenase from tryptophan-treated rats is more saturated with hematin than the enzyme from untreated or hydrocortisone-treated rats (Feigelson et al 1962). This increased saturation with hematin significantly increases the ratio of holo- to apoenzyme following tryptophan treatment (Greengard and Feigelson 1961); this ratio is relatively unchanged by hydrocortisone treatment (Feigelson et al 1962). Elevation of tryptophan oxygenase activity beyond activation of latent enzyme occurs following administration of corticosteroids or tryptophan as a result of either increased de novo apoenzyme synthesis or decreased apoenzyme breakdown, respectively (Civen and Knox 1959, 1960; Feigelson et al 1962; Feigelson and Greengard 1962, 1963; Goldstein et al 1962; Greengard et al 1963; Knox 1951; Schimke 1970; Schimke et al 1964, 1965a, 1965b).

Both corticosteroids and tryptophan increase the amount of apoenzyme present. This has been determined by increased radioactive label incorporation into proteins and RNA (Feigelson et al 1962), by the effects of inhibitors of protein synthesis (Goldstein et al 1962; Greengard and Feigelson 1962; Schimke et al 1965a), and by specific immunological reactions (Feigelson and Greengard 1962, 1963). Both cortisone and tryptophan increase ^{14}C-glycine incorporation into liver proteins which corresponds with changing tryptophan oxygenase activities, but only cortisone stimulates ^{32}P-orthophosphate incorporation into RNA at a time corresponding to its effects on tryptophan oxygenase activity. Following tryptophan administration, there is no increase in RNA labeling while the enzyme activity is rising. These results indicate that the mechanism of cortisone-mediated elevation of tryptophan oxygenase activity involves stimulation of RNA synthesis, whereas tryptophan-mediated elevation involves a different process, namely, decreased enzyme degradation. These conclusions were confirmed by the action of puromycin and actinomycin D on cortisone-mediated and tryptophan-mediated increases of tryptophan oxygenase activity. Both types of induction were greatly diminished by puromycin, an inhibitor of protein synthesis at the translation level. Only hormonal induction was inhibited by actinomycin D, a compound which inhibits DNA-dependent RNA polymerase, thus inhibiting RNA synthesis.

The decreased rate of enzyme degradation caused by tryptophan, as well as some of its analogs, is effective in vivo (Civen and Knox 1960; Feigelson et al 1962; Schimke et al 1965a), as well as in vitro (Schimke 1970; Schimke et al 1965a, 1965b). In vitro stabilization can be measured by comparing the effects of L-tryptophan or its analogs on the loss of tryptophan oxygenase activity normally occurring upon incubation without tryptophan or on loss of activity following inactivation with heat, ethanol, urea or trypsin. In vivo stabilization of tryptophan oxygenase in rats was indicated by retention of pre-labeled tryptophan oxygenase levels following tryptophan administration; whereas in control rats without tryptophan, there was a rapid loss of total counts precipitated by specific anti-tryptophan oxygenase antiserum.

Significant influence of corticosteroids on tryptophan oxygenase activity is also implied since the control levels of tryptophan oxygenase activity are significantly lower in adrenalectomized mice (Berry and Smythe 1963) and rats (Knox 1951, 1966; Knox and Auerbach 1955) than in intact animals. Additional factors which may influence tryptophan oxygenase activity include the presence of tryptophan analogs (Civen and Knox 1959; Frieden et al 1961; Moon and Berry 1968a; Schimke et al 1965b) and tryptophan metabolites including NAD (Cho-Chung and Pitot 1967; Frieden et al 1961).

Previous reports have noted increases in tryptophan oxygenase activity by purines (Cho-Chung and Pitot 1968; Chytil et al 1966; Julian and Chytil 1970) and by 3′,5′-cyclic AMP (Chytil 1968). The probable action of purines is thought to be through their conversion to hypoxanthine. This is indicated because blocking the conversion of cAMP to hypoxanthine with theophylline prevents activation of tryptophan oxygenase by cAMP (Chytil 1968). It has been suggested that xanthine oxidase may be operative in the regulation of tryptophan oxygenase activity (Chytil 1968; Julian and Chytil 1970). Xanthine oxidase, present in soluble (rat) liver preparations, activates tryptophan oxygenase when its substrate hypoxanthine is present. Removal of xanthine oxidase from the soluble fraction of rat liver preparations with specific antibody results in loss of tryptophan oxygenase by purines (Julian and Chytil 1970).

Because activation of purines was not dependent on the presence of hematin (Chytil 1968) or hemoglobin (Julian and Chytil 1970), xanthine oxidase probably activates tryptophan oxygenase by causing the reduction of the inactive holoenzyme, possibly due to the production of H_2O_2 (Chytil 1968).

Inhibition of xanthine oxidase with allopurinol inhibits tryptophan oxygenase activity in vitro (Becking and Johnson 1967; Chytil 1968; Julian and Chytil 1970). In vivo, allopurinol was found to be a potent inhibitor of tryptophan oxygenase in that 4 hours after an injection of allopurinol (20 mg/kg) to normal rats, tryptophan oxygenase activity was inhibited 90–95% (Becking and Johnson 1967; Julian and Chytil 1970). In mice, a similar inhibition was observed at lower doses (Becking and Johnson 1967). Allopurinol-mediated inhibition of tryptophan oxygenase activity is not due to a direct effect on the enzyme since allopurinol does not inhibit tryptophan oxygenase activity in a purified preparation of the apoenzyme. These data imply that allopurinol does not inhibit conjugation of the apoenzyme with hematin (Julian and Chytil 1970). Nor does allopurinol significantly inhibit hydrocortisone-mediated synthesis of tryptophan oxygenase, as measured by precipitation of the enzyme with specific anti-tryptophan oxygenase antibodies (Julian and Chytil 1970). Addition of ascorbic acid prevented inhibition of tryptophan oxygenase by allopurinol (Chytil 1968). These latter data further support the notion that allopurinol inhibits tryptophan oxygenase at the step of the reduction of the oxidized holoenzyme, possibly by the inhibition of xanthine oxidase activation.

2.2. Influence of endotoxin and cortisone on tryptophan oxygenase in vivo

Tryptophan oxygenase activity is depressed in endotoxin-poisoned mice (Agarwal, 1972, 1974, 1975; Agarwal and Berry 1966; Agarwal and Lazar 1977; Agarwal et al 1969; Berry and Smythe 1963; Berry et al 1968a, 1968b, 1968c; Moon and Berry 1968b). Because tryptophan oxygenase is the first enzyme in the pathway leading to the formation of pyridine nucleotides (see Section 3), its depression implies a block in the biosynthesis of such compounds, the consequence of which could be of biological significance since both NAD and nicotinamide protect mice against endotoxin lethality (Berry and Smythe 1963). A decrease in the levels of total oxidized pyridine nucleotide found in the liver of endotoxin-poisoned mice can be prevented by concurrent administration of nicotinamide. When given alone, nicotinamide approximately doubled the control levels of total oxidized pyridine nucleotides in 17 hours.

Cortisone, known to elevate tryptophan oxygenase activity (Feigelson and Greengard 1962, 1963; Knox and Auerbach 1955), protects mice against endotoxin lethality (Berry and Smythe 1963; Geller et al 1954) and also maintains the levels of total oxidized pyridine nucleotides (Berry and Smythe 1963). The protective role of cortisone, as well as its ability to elevate tryptophan oxygenase activity, is dependent on the time of its administration relative to the administration of endotoxin (Berry and Smythe 1963). Thus, 5 mg of cortisone given concurrent with 1 LD_{50} of endotoxin increased survival of mice and elevated tryptophan oxygenase activity; however, if cortisone was given more than 1 hour after endotoxin, cortisone failed to protect mice and failed to maintain tryptophan oxygenase activity at control levels.

An initial hypothesis early in studies on endotoxin and tryptophan oxygenase suggested that endotoxin might be a generalized inhibitor of protein synthesis. Certain inhibitors of protein synthesis (actinomycin D, ethionine, thiouracil, and 8-azaguanine) were found to potentiate endotoxin lethality and to prevent cortisone protection (Berry and Smythe 1968). These same compounds prevented cortisone elevation of tryptophan oxygenase activity in normal mice and prevented cortisone maintenance of tryptophan oxygenase activity at control levels in endotoxin-poisoned mice when endotoxin, cortisone and inhibitor were given concurrently. In the above instances, maintenance of tryptophan oxygenase activity at control levels during endotoxin-poisoning was associated with protection against the lethal effects of endotoxin.

Soon after work began on tryptophan oxygenase, a second adaptive liver enzyme, tyrosine α-ketoglutarate transaminase (TKT), was identified. TKT responded to cortisone in a way similar to tryptophan oxygenase. When tryptophan oxygenase and TKT activities were compared in endotoxin-poisoned animals, it was found that TKT remained inducible by cortisone while tryptophan oxygenase did not (Berry et al 1968a). Since induction of both enzymes was inhibited by actinomycin D and puromycin, it appeared that endotoxin was probably not a generalized inhibitor of protein synthesis. As more steroid-inducible hepatic enzymes were identified, it became clear that some behaved in a manner similar to tryptophan oxygenase and some did not (Berry 1973, 1975; Phillips and Berry 1970a, 1970b; Rippe and Berry 1972a). The two most thoroughly studied enzymes whose induction is blocked by endotoxin are tryptophan oxygenase and phosphoenolpyruvate carboxykinase (PEPCK). PEPCK studies have received most attention in recent years since this enzyme is a key regulator of gluconeogenesis and its inhibition by endotoxin could conceivably explain why gluconeogenesis is inhibited in poisoned animals. Simultaneous with studies on carbohydrate metabolism, the studies on tryptophan oxygenase and tryptophan metabolism continued.

2.3. Influence of tryptophan and selected analogs on tryptophan oxygenase activity in endotoxin-poisoned animals

Tryptophan and a number of its analogs have been tested for their ability to alter tryptophan oxygenase activity in vivo in endotoxin-poisoned mice. The primary objective of such experiments was to establish the relative importance of tryptophan oxygenase per se in the response of the animal to endotoxin. Because tryptophan is rapidly metabolized and excreted, substrate induction of tryptophan oxygenase in vivo is much more short-lived than hormonal induction. To maintain high levels of tryptophan oxygenase in vivo in mice, as much as 20 mg of L-tryptophan must be injected every 2 hours. Moon and Berry (1968b) showed that repeated substrate injections resulted in a steady increase in tryptophan oxygenase activity in both normal and endotoxin-poisoned animals. The rate of induction was about 50% slower in poisoned animals. One problem with this experimental approach was that endotoxin-poisoned mice which received large amounts of tryptophan frequently went into convulsion and died soon after tryptophan administration (see Section 3). To avoid this problem, structural analogs of tryptophan which were able to either increase or decrease tryptophan oxygenase activity were tested. The use of substrate analogs has allowed researchers to manipulate tryptophan oxygenase in vivo in a more specific manner than by using cortisone – which affects many other host physiologic and metabolic parameters in addition to tryptophan oxygenase – while avoiding the experimental problems of hyperreactivity in endotoxin-poisoned mice to tryptophan. The application of tryptophan analogs to endotoxin research allowed investigators to better address two particular issues, namely: is substrate induction of tryptophan oxygenase substantially impaired by endotoxin administration, and does tryptophan oxygenase activity per se correlate with survival of endotoxin-poisoned mice? α-Methyltryptophan, a nonmetabolizable analog of tryptophan, increases tryptophan oxygenase by mechanisms similar to tryptophan (Civen and Knox 1960; Moran and Sourkes 1963). The elevated enzyme activity lasts as long as 24 hours because of the in vivo persistence of the analog. α-Methyltryptophan injection resulted in elevated tryptophan oxygenase levels in endotoxin-poisoned mice although the activity is not as high as in normal mice. Despite the elevated tryptophan oxygenase activity, α-methyltryptophan did not protect mice from endotoxin (Moon 1971).

Depression of tryptophan oxygenase activity was accomplished with either 5-hydroxytryptophan (Moon and Berry 1968b) or allopurinol (Becking and Johnson 1967). Both of these components were able to enhance depression of tryptophan oxygenase in endotoxin-poisoned mice (Moon and Berry 1968b; Moon 1971). Treatment with 5-hydroxytryptophan or allopurinol did not enhance susceptibility to endotoxins. Cumulatively, these results imply that tryptophan oxygenase per se is not directly related to survival of endotoxin-poisoned mice.

2.4. Mechanisms of endotoxin-mediated inhibition of tryptophan oxygenase

The mechanism(s) by which endotoxin inhibits tryptophan oxygenase activity have received considerable study in a variety of in vivo and in vitro models. One early hypothesis to explain the inhibition was that endotoxin directly binds to tryptophan oxygenase, thereby inhibiting substrate binding. This has also been shown repeatedly *not* to be the case since endotoxin directly added to homogenates containing tryptophan oxygenase does not inhibit the activity of the enzyme in vitro. A second possibility was that endotoxin inhibits de novo synthesis of tryptophan oxygenase. Data which would support this notion are: cortisone induction of tryptophan oxygenase is inhibited by endotoxin, and substrate induction of the enzyme is depressed when compared to normal controls (Berry and Smythe 1963; Moon and Berry 1968b; Moon 1971). Further, it has been clearly demonstrated that depression of both tryptophan oxygenase and PEPCK results in a net loss of immunologically reactive enzyme protein (Rippe and Berry 1972b, 1973). This study was accomplished by purifying the respective enzymes, preparing antibody to them, and showing quantitatively less immunodiffusion reactivity in liver cell sap from endotoxin-poisoned mice than from normal mice.

The inhibition of induction does not reflect a generalized inhibition of protein synthesis since induction of tyrosine α-ketoglutarate transaminase, which is steroid-inducible, is not inhibited (Berry et al 1968a). Hence, if endotoxin does inhibit de novo synthesis of tryptophan oxygenase, a mechanistic model to explain this must include some dimension of specificity. Despite considerable effort such a mechanism has yet to be identified.

There have been two general experimental approaches attempting to ascertain the mechanism of endotoxin-mediated inhibition of enzyme activity and the inhibition of steroid-mediated induction. The first is that endotoxin has some type of direct toxicity on liver parenchymal cells; the second is that the action of endotoxin is mediated by molecules released from other points in the host. The first mechanism received some positive support when it was found that endotoxin could attach to and penetrate hepatic parenchymal cells and hepatoma tissue culture cells (Tavakoli and Moon 1982; Zlydaszyk and Moon 1976). Initial results suggested that the endotoxin was bound to parenchymal cell nuclei although more recent data using more critical techniques cast doubt on that assumption (Tavakoli and Moon 1982). The same authors did show that relatively high levels of endotoxin can directly interact with chromatin in vitro and also inhibit the ability of steroid receptor complexes to bind to DNA (Tavakoli and Moon 1981). Despite these data, no conclusive evidence is available which supports the notion that endotoxin associated with hepatic parenchymal cells does in fact lead to inhibition of steroid induction of tryptophan oxygenase. Attempts to incubate endotoxin with hepatoma tissue culture cells and inhibit induction have consistently failed (Moon, unpublished data). Cumulatively, the data indicate that a mechanism involving direct inhibition of induction by endotoxin is probably unlikely.

Direct studies implicating a mediator of toxicity rather than a direct action of endotoxin on tissues have been very promising (see Chapter 5). While most of these studies have involved the use of the PEPCK model, the data obtained in this system can be related to tryptophan oxygenase. Soon after administration of endotoxin, a soluble factor can be found in the blood which has been termed glucocorticoid antagonizing factor (GAF) (Moore et al 1976, 1978). This molecule is released from a variety of phagocytic cells. While it is clear that GAF blocks steroid induction of tryptophan oxygenase and PEPCK, the precise mechanism of action is unknown. Chemically, GAF is a relatively high molecular weight peptide which has been purified considerably but not to homogeneity.

Experiments have been performed which demonstrate the effectiveness of GAF both in vivo and in vitro. The in vivo experiments are relatively complex and involve the use of endotoxin-tolerant mice (Moore et al 1978). A group of mice are injected with endotoxin and bled, and their serum is administered to endotoxin-tolerant animals. The recipients of the endotoxin plasma or serum are then given glucocorticoid hormones to stimulate enzyme induction. The data show that mice which have received GAF have decreased steroid responsiveness. One must use tolerant animals in these studies to guarantee that the inhibition of induction is not due to a small amount of endotoxin which is being carried over in the serum obtained from endotoxin-poisoned animals. Application of serum directly to hepatoma tissue culture cells in vitro confirms the existence of GAF in vitro. This was accomplished by demonstrating that steroid induction was not as effective in the presence of such plasma as it is in its absence (Keitelman 1981). Keitelman has shown that as soon as 1 minute after administration of endotoxin, a substance can be detected in the blood which inhibits the ability of steroid hormones to induce PEPCK.

Mr. Bryon Petchow, working in the author's laboratory, has recently obtained in vivo data which support the notion that endotoxin action is mediated rather than direct. Mr. Petchow has injected Cr^{51}-labeled lipopolysaccharide both subcutaneously, intravenously and intraperitoneally into mice. The unpublished data show that while endotoxin is concentrated in the liver following intraperitoneal and intravenous injection, over 95% remains at the site of injection following subcutaneous administration. Despite this difference in in vivo localization, responsiveness to a variety of metabolic parameters is not significantly different if endotoxin is administered intravenously or subcutaneously.

While it is clear that the ability of endotoxin to inhibit enzyme induction involves the action of a mediator, what remains unclear is how the mediator is acting. For example, does GAF directly interact with steroid receptors preventing hormone association? Alternatively, does GAF bind to steroids thereby decreasing their availability for in vivo or in vitro action? Additional questions remain about GAF which deserve further consideration. For example, in recent years there have been a wide variety of mediators of endotoxin activity which have been described. What is the relationship between GAF and these other mediators of endotoxin activity?

Inherent in all of these studies is a general concern that particular lines of experimenting may be getting too narrow and defined and thus cause one to lose sight of the primary objectives of such studies: to define the basic mechanisms underlying the metabolic toxicities of endotoxin and to put into proper perspective the importance of these metabolic alterations in the overall pathophysiology of endotoxin shock. The relative importance of mediators of endotoxin to the overall pathophysiological alterations induced by endotoxin remains conjectural.

3. EFFECT OF ENDOTOXIN ON TRYPTOPHAN METABOLISM IN VIVO

While data directly linking the effects of endotoxin on tryptophan metabolism with the overall pathophysiology of endotoxin poisoning in normal animals do not exist, a number of curious observations have been made regarding the effects of endotoxin and tryptophan metabolism. One observation (Moon and Berry 1968b) was that when 15–20 mg of tryptophan was injected subcutaneously into endotoxin-poisoned animals, either concurrent with or 4 hours after endotoxin, within 1–2 hours following tryptophan the animals frequently went into convulsions and died. Since this amount of tryptophan had no apparent toxic effects on normal mice, the hyperreactivity of endotoxin-poisoned mice to tryptophan stimulated some interest.

A comprehensive review of tryptophan metabolism has been compiled by Meister (1965). It is of interest to note that tryptophan can be metabolized through several pathways in vivo, yielding numerically more metabolites than any other amino acid. General reviews of the two major pathways for tryptophan catabolism in mammals are shown in Figures 1 and 2. The first pathway leads to NAD biosynthesis (Figure 1) and is regulated by tryptophan oxygenase which converts tryptophan to formylkynurenine. The second pathway results in hydroxylation and decarboxylation of tryptophan to form 5-hydroxytryptamine or serotonin. A working hypothesis to explain the toxic effects of tryptophan in endotoxin-poisoned animals was developed. This hypothesis states: as a result of depressed tryptophan oxygenase activity, excess amounts of tryptophan are funneled into serotonin synthesis and excess serotonin production is the basis of the hyperreactivity. This hypothesis was supported by data showing that cyproheptadine, an antiserotonin drug, inhibits the hyperreactivity of endotoxin-poisoned animals to tryptophan (Moon and Berry 1968b). Since substrate induction of tryptophan oxygenase was far lower than in normal animals (see Section 2.3), the combination of these observations tends to support the notion that due to decreased tryptophan oxygenase activity, tryptophan was funneled in vivo into excessive serotonin biosynthesis.

Additional physiologic data were also obtained using temperature regulation and carbohydrate fluctuation as research tools. Following endotoxin poisoning, if mice were put into a warm environment, they experienced a rise in body temperature (Berry 1966). If they were put in the cold, they had a drop in body temperature. Administration of either serotonin or tryptophan greatly enhanced the drop in body temperature that was observed in endotoxin-poisoned mice housed at 15 °C (Moon and Berry 1968a). Cyproheptadine inhibits this effect. Further, endotoxin-poisoned mice became severely hypoglycemic following tryptophan load. It is well established that endotoxin alone results in decreased blood glucose and liver glycogen (Berry et al 1959a). Injection of tryptophan 4 hours after endotoxin dramatically enhanced these effects (Moon 1972). Cyproheptadine also prevented the depletion of blood glucose which in some animals reached as low as 100–300 mg/l. Hence, the immediate cause of hyperreactive death appears to be severe hypoglycemia. The mechanisms of this response are not clear. It is assumed that the additional stress of excessive serotonin production combined with endotoxin inhibition of gluconeogenesis resulted in this response.

A fourth way of implicating tryptophan oxygenase in the hyperreactivity of endotoxin-poisoned mice to tryptophan has been through the use of nonmetabolizable analogs of tryptophan which alter the activity of tryptophan oxygenase through substrate induction. The advantage of substrate regulation is that the multitude of secondary effects observed with cortisone are not encountered. Our working

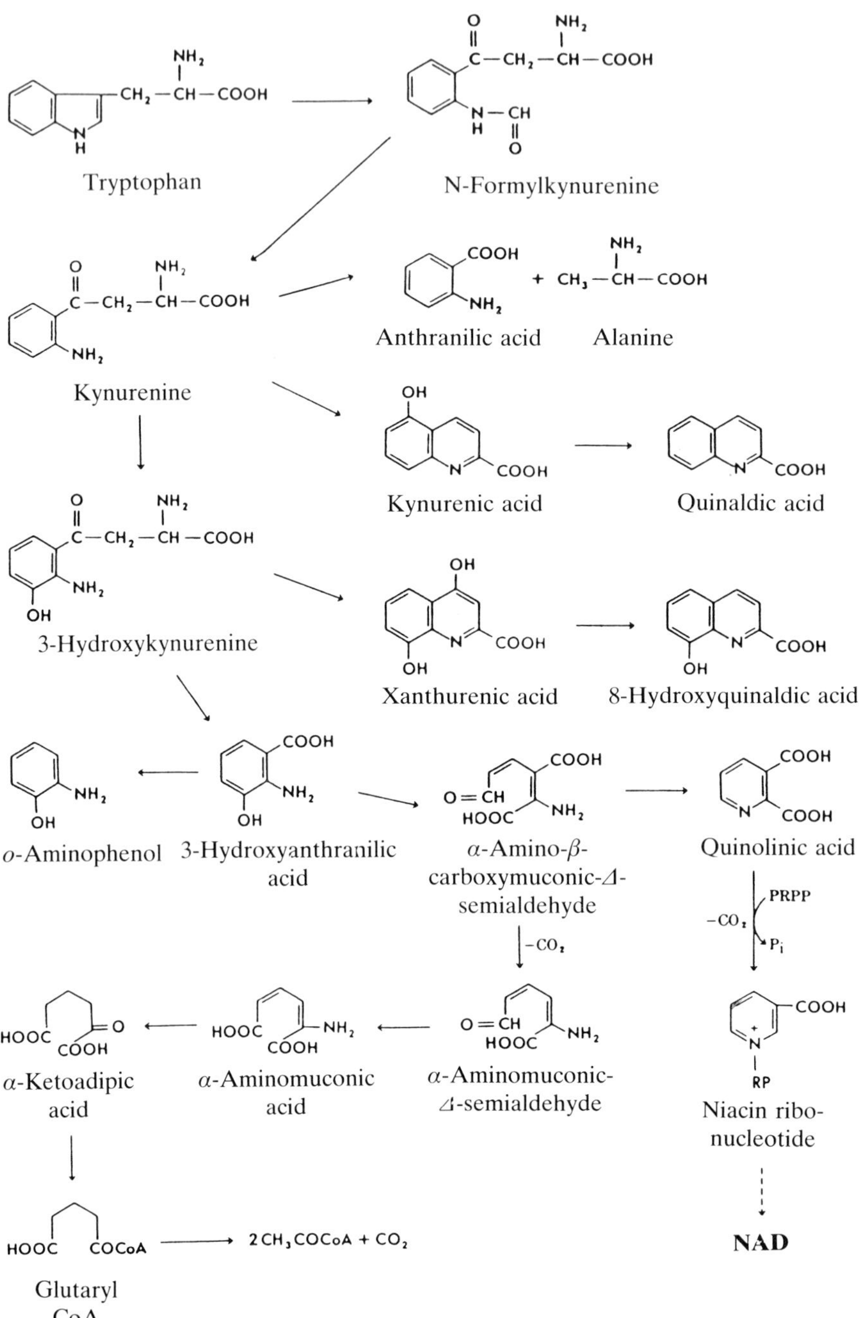

Fig. 1   *The kynurenine pathway.*

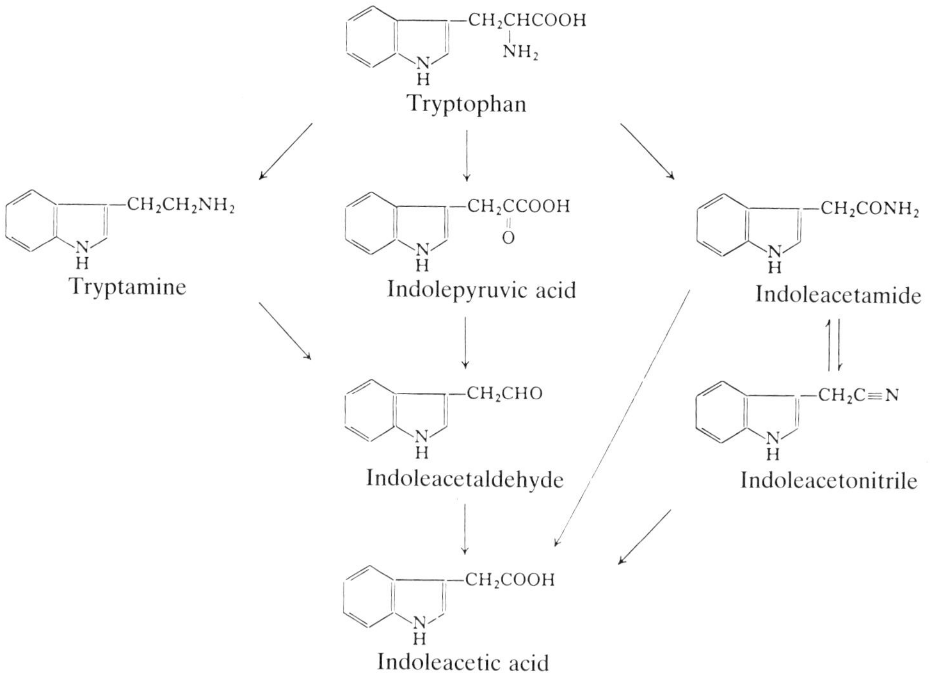
—CH₂CHCOOH
NH₂
N
H
Tryptophan
—CH₂CH₂NH₂
Tryptamine
—CH₂CCOOH
O
Indolepyruvic acid
—CH₂CONH₂
Indoleacetamide
—CH₂CHO
Indoleacetaldehyde
—CH₂C≡N
Indoleacetonitrile
—CH₂COOH
Indoleacetic acid

Fig. 2 *The serotonin pathway.*

hypothesis was that if tryptophan oxygenase per se was important in the hyperreactivity of endotoxin-poisoned mice to tryptophan, then elevation of the enzyme activity through α-methyltryptophan induction should not only prevent the endotoxin-mediated decrease in tryptophan oxygenase (see Section 2.3.) but also prevent the hyperreactivity of endotoxin-poisoned animals to tryptophan. Data published several years ago show that in fact both assumptions appear to be correct in that α-methyltryptophan increased tryptophan oxygenase activity in endotoxin-poisoned animals (although not as high as in normal animals) and mice showed no hyperreactivity to tryptophan (Moon 1971). Allopurinol inhibited tryptophan oxygenase activity in both endotoxin-poisoned animals and normal animals, but only endotoxin-poisoned animals were hyperreactive to tryptophan (Moon 1971). This latter data show that tryptophan oxygenase per se is not critical to the hyperreactivity of endotoxin-poisoned mice to tryptophan, but in order for the endotoxin-poisoned mice to exhibit hyperreactivity, tryptophan oxygenase must be depressed.

All of the data described to this point implicating tryptophan oxygenase depression in the funneling of tryptophan into serotonin synthesis are circumstantial. It was our wish to confirm these observations by more direct quantitative measures. The metabolic charts in Figures 1 and 2 show that only tryptophan which is metabolized through the tryptophan oxygenase pathway (Fig. 1) can release radiolabeled CO_2 from the benzene ring. Any ring-labeled tryptophan that is metabolized through the serotonin pathway (Fig. 2) will not be converted to CO_2 but instead the isotope will be excreted.

This information led to two separate experimental approaches designed to provide more quantitative data on the in vivo metabolism of tryptophan in endotoxin-poisoned animals. The experiments were designed to address two fundamental questions; namely, was tryptophan oxygenase as measured by in vitro enzyme assay in fact functionally decreased in hepatic cells? And, was exogenous tryptophan being funneled in excess into serotonin production in endotoxin-poisoned mice?

To address the first question, an experiment was designed using a perfused liver model. The working hypothesis was that if tryptophan oxygenase activity was functionally depressed in hepatic parenchymal cells, livers from endotoxin-poisoned mice should produce less $^{14}CO_2$ from benzene ring-labeled tryptophan than normal livers. Liver perfusion techniques to accomplish such a study are well described (Kim and Miller 1969; Moon et al 1975). Experimentally, a constant amount of benzene ring-labeled tryptophan was added to the perfusion medium and the amount of radioactivity released as $^{14}CO_2$ determined. The data obtained which have not been previously published are shown in Table 1.

In preliminary experiments (data not shown) we found that a relatively constant number of $^{14}CO_2$ counts per hour was obtained over the entire 6-hour perfusion period, which indicated that the livers remained viable and metabolically active. Table 1 shows that significantly fewer counts were obtained from livers of endotoxin-poisoned mice than normal mice. We interpret these data as supporting the notion that there is a functional decrease in tryptophan oxygenase activity in hepatic tissue from endotoxin-poisoned mice corresponding to the decrease in activity observed by in vitro assay.

To address the second question, benzene ring-labeled tryptophan was injected

TABLE 1 *Conversion of benzene ring-labeled tryptophan to $^{14}CO_2$ by perfused livers from normal and endotoxin-poisoned mice*

Experimental	Cumulative dpm of $^{14}CO_2$* with time		
	1 hr	2 hr	3 hr
Normal	2441 ± 978	4916 ± 1097	6490 ± 1074
Endotoxin	1477 ± 380	3074 ± 615	4393 ± 894

* 1×10^5 dpm of benzene ring-labeled tryptophan was added to the perfusion medium at time = 0, and the amounts of $^{14}CO_2$ evolved were monitored hourly for 3 hours. Data represent average dpm ± standard deviation from at least 6 separate experimental determinations.

intravenously into mice and the quantitative amount of tryptophan flowing into serotonin and kynurenine pathways determined by analysis of urinary metabolites.

The analysis of tryptophan metabolites in the urine is extremely complex in that there are well over fifteen different metabolites of tryptophan which can appear in the urine within 12 hours after exposure to the benzene ring-labeled amino acid. The studies by Morris and Moon (1974a, 1974b) show that the total quantity of radioactivity recovered from the urine was less in endotoxin-poisoned animals than in normal animals. Through a tedious separation of the various urinary metabolites on two-dimensional thin layer chromatography, these authors were able to quantitate the amount of metabolites flowing through the serotonin pathway and the NAD pathway. Cumulatively, the data show that there were significantly more metabolites of tryptophan appearing in the serotonin pathway of endotoxin-poisoned animals than in normal animals. For the kynurenine pathway metabolites, the results were reversed. These results are direct confirmation that tryptophan oxygenase activity in vivo is depressed in endotoxin-poisoned animals and support the hypothesis that this depression results in increased tryptophan catabolism towards serotonin.

Cumulatively, these studies support the overall hypothesis that the hyperreactivity of endotoxin-poisoned animals to tryptophan is a direct consequence of excess production of serotonin. Depression of tryptophan oxygenase per se with allopurinol did not result in hyperreactivity, which indicates that this response requires a combination of endotoxin and tryptophan before this abnormal biological response is observed. The ability of cyproheptidine to block the hypoglycemia, enhanced hypothermia, and hyperreactive death in endotoxin-poisoned animals given tryptophan leads to the assumption that serotonin is involved in this response. The direct physiological or biochemical data that have been obtained by monitoring urinary tryptophan metabolites and $^{14}CO_2$ further confirm this notion.

4. CURRENT STUDIES AND FUTURE DIRECTIONS

Since it became apparent that tryptophan oxygenase per se probably played little role in the overall response of experimental animals to endotoxin, research activity in this area has dropped off considerably. One interesting study which recently appeared has used isolated hepatic cells in culture to show that endotoxin does not directly antagonize the induction of tryptophan oxygenase (Lowitt et al 1981). Instead, the inhibition of tryptophan oxygenase induction by endotoxin is a mediated event involving the interaction of endotoxin with cellular constituents of hepatic nonparenchymal cells. The ability to use centrifugal elutriation to separate nonparenchymal cells into endothelial and Kupffer cells (Knook and Sleyster 1976) should now allow the definitive determination of the hepatic nonparenchymal cell type involved, although in all likelihood, on the basis of the mediator studies performed with PEPCK, it would be the Kupffer cells.

There have also been some interesting advances in understanding tryptophan metabolism in recent years which suggest further experiments on how endotoxin can alter this system. Ideas of particular note are the increased ability to grow hepatocytes in tissue culture and the observation that freshly isolated cells which have been exposed to collagenase have very low activities of various liver functions, including dexamethasone responsiveness, but that responsive levels returned to normal after 24 hours of culture (Ichihara et al 1982). Isolated cells are responsive to a number of hormones besides dexamethasone and might provide a potential for more critical work on how endotoxin can influence metabolism on a cellular level.

A poorly understood area of tryptophan oxygenase regulation has always been the mechanism by which the rate of enzyme degradation in tissue is controlled. Grinde and Jahnsen (1982) have recently shown that the protease inhibitor chymostatin inhibited TKT and tryptophan oxygenase inactivation, while a second protease inhibitor (leupeptin) did not. The authors suggest that a nonlysosomal protease is involved. It is conceivable that endotoxin might somehow enhance degradation of tryptophan oxygenase in vivo.

The studies on tryptophan oxygenase and tryptophan metabolism have contributed significantly to our knowledge of how endotoxin can influence normal metabolic homeostasis. The precise importance of metabolic alterations in the overall pathophysiology of gram-negative sepsis and endotoxin shock remains an open question. What does appear to be clear is that before the basic biological toxicity of endotoxin can be understood, the initial molecular interactions of endotoxin with host tissues and organs must be clarified. The studies of endotoxin effects on metabolism, which include the studies on tryptophan metabolism, appear to hold important experimental keys to unravelling these cellular and molecular interactions and as such deserve continued research attention.

REFERENCES

Agarwal MK (1972) Liver enzyme induction and endotoxicosis in mice. *Pathol. Microbiol.* *38*, 73-82.

Agarwal MK (1974) Analysis of the influence of selected neurotropic agents on hepatic metabolism in relation to endotoxicosis. *Biochem. Pharmacol. 23*, 2577-2584.

Agarwal MK (1975 Further characterization of hepatic tryptophan pyrrolase inhibition by zymosan. *Res. Commun. Chem. Pathol. Pharmacol. 10*, 387-390.

Agarwal MK, Berry LJ (1966) Effect of RES-active agent on tryptophan pyrrolase activity and endotoxin lethality. *J. Reticuloendothel. Soc. 3*, 223-235.

Agarwal MK, Lazar G (1977) Metabolic basis of endotoxicosis. *Microbios. 20*, 183-214.

Agarwal MK, Hoffman WW, Rosen F (1969) The effect of endotoxin and thorotrast on inducible enzymes in the isolated, perfused rat liver. *Biochim. Biophys. Acta. 177*, 250-258.

Becking GC, Johnson WJ (1967) The inhibition of tryptophan pyrrolase by allopurinol, an inhibitor of xanthine oxidase. *Can. J. Biochem. 45*, 1667-1672.

Berry LJ (1966) Effect of environmental temperature on lethality of endotoxin and its effects on body temperature in mice. *Fed. Proc. 25*, 1264-1270.

Berry LJ (1973) Effect of endotoxin on metabolic regulation in mice. *Ann. Immunolog. Hung. 17* (Suppl), 9-19.

Berry LJ (1975) Bacterial toxins. *Lloydia 38*, 8-20.

Berry LJ, Smythe DS (1961a) Effects of bacterial endotoxins on metabolism III. Nitrogen excretion after ACTH as an assay for endotoxin. *J. Exp. Med. 113*, 83-94.

Berry LJ, Smythe DS (1961b) Effects of bacterial endotoxins on metabolism IV. Renal function and adrenocortical activity as factors in the nitrogen excretion assay for endotoxin. *J. Exp. Med. 114*, 761-778.

Berry LJ, Smythe DS (1963) Effects of bacterial endotoxins on metabolism. VI. The role of tryptophan pyrrolase in response of mice to endotoxin. *J. Exp. Med. 118*, 587-603.

Berry LJ, Smythe DS, Young LG (1959a) Effects of bacterial endotoxin on metabolism I. Carbohydrate depletion and the protective role of cortisone. *J. Exp. Med. 110*, 389-405.

Berry LJ, Smythe DS, Young LG (1959b) Effects of bacterial endotoxin on metabolism II. Protein-carbohydrate balance following cortisone inhibition of intestinal absorption and adrenal response to ACTH. *J. Exp. Med. 110*, 407-413.

Berry LJ, Smythe DS, Colwell LS (1968a) Inhibition of inducible liver enzymes by endotoxin and actinomycin D. *J. Bacteriol. 91*, 107-115.

Berry LJ, Smythe DS, Colwell LS (1968b) Inhibition of hepatic enzyme induction as a sensitive assay for endotoxin. *J Bacteriol. 96*, 1191-1199.

Berry LJ, Smythe DS, Colwell LS, Chu PH (1968c) Influence of hypoxia, glucocorticoid, and endotoxin on hepatic enzyme induction. *Am. J. Physiol. 215*, 587-592.

Cho-Chung YS, Pitot HC (1967) Feedback control of rat liver tryptophan pyrrolase. I. Endproduct inhibition of tryptophan pyrrolase activity. *J. Biol. Chem. 242*, 1192-1198.

Cho-Chung YS, Pitot HC (1968) Regulatory effects of nicotinamide on tryptophan pyrrolase synthesis in rat liver in vivo. *Eur. J. Biochem. 3*, 401-406.

Chytil F (1968) Activation of liver tryptophan oxygenase by adenosine-3',5'-phosphate and by other purine derivatives. *J. Biol. Chem. 243*, 893-899.

Chytil F, Skrivanova J, Brana H (1966) Activating effect of purines on the liver tryptophan pyrrolase system. *Can. J. Biochem. 44*, 283-286.

Civen M, Knox WE (1959) The independence of hydrocortisone and tryptophan inductions of tryptophan pyrrolase. *J. Biol. Chem. 234*, 1787-1790.

Civen M, Knox WE (1960) The specificity of tryptophan and analogues as inducer, substrates, inhibitors, and stabilizers of liver tryptophan pyrrolase. *J. Biol. Chem. 235*, 1716-1718.

Feigelson P, Greengard O (1962) Regulation of liver tryptophan pyrrolase activity. *J. Biol. Chem. 237*, 1908-1913.

Feigelson P, Greengard O (1963) Immunologic and enzymatic studies on rat liver tryptophan pyrrolase during substrate and cortisone induction. *Proc. Natl Acad. Sci. USA 103*, 1075-1082.

Feigelson P, Feigelson M, Greengard O (1962) Comparison of the mechanisms of hormonal and substrate induction of rat liver tryptophan pyrrolase. *Recent Prog. Horm. Res. 18*, 491-512.

Frieden E, Westmark GW, Schor JM (1961) Inhibition of tryptophan pyrrolase by serotonin, epinephrine, and tryptophan analogs. *Arch. Biochem. Biophys. 92*, 176-182.

Geller P, Merrill ER, Jawetz E (1954) Effects of cortisone and antibiotics on lethal action of endotoxin in mice. *Proc. Soc. Exp. Biol. Med. 86*, 716-719.

Goldstein L, Stella EJ, Knox WE (1962) The effect of hydrocortisone on tyrosine-oc-ketoglutarate transaminase and tryptophan pyrrolase activities in the isolated, perfused rat liver. *J. Biol. Chem. 237*, 1723-1726.

Greengard O, Feigelson P (1961) The activation and induction of rat liver tryptophan pyrrolase in vivo by its substrate. *J. Biol. Chem. 236*, 158-161.

Greengard O, Feigelson P (1962) The purification and properties of liver tryptophan pyrrolase. *J. Biol. Chem. 237*, 1903-1907.

Greengard O, Smith MA, Acs G (1963) Relation of cortisone and synthesis of ribonucleic acid to induced and developmental enzyme formation. *J. Biol. Chem. 238*, 1548-1551.

Grinde B, Jahnsen R (1982) Effects of protein-degradation inhibitors on the inactivation of tyrosine aminotransferase, tryptophan oxygenase and benzopyrene hydroxylase in isolated rat hepatocytes. *Biochem. J. 202*, 191-196.

Ichihara H, Nakamura T, Tanaka K (1982) Use of hepatocytes in primary culture for biochemical studies on liver functions. *Mol. Cell. Biochem. 43*, 145-160.

Julian J, Chytil F (1970) Participation of xanthine oxidase in the activation of liver tryptophan pyrrolase. *J. Biol. Chem. 245*, 1161-1168.

Keitelman EL (1981) *The Effects of Endotoxin and Humoral Mediator(s) Derived from Endotoxin-treated Rats on the Humoral Induction of Phosphoenolpyruvate Carboxykinase in Reuber H-35 Hepatoma Cells.* Masters Thesis, Michigan State University, East Lansing, MI.

Kim JH, Miller LL (1969) The functional significance of changes in activity of the enzymes, tryptophan pyrrolase and tyrosine transaminase, after induction in intact rats and in the isolated, perfused rat liver. *J. Biol. Chem. 244*, 1410-1416.

Knook DL, Sleyster ECH (1976) Separation of Kupffer and endothelial cells of the rat liver by centrifugal elutration. *Exp. Cell. Res. 99*, 444-449.

Knox WE (1951) Two mechanisms which increase in vivo the liver tryptophan peroxidase system: specific enzyme adaptation and stimulation of the pituitary-adrenal system. *Br. J. Exp. Pathol. 32*, 462-469.

Knox WE (1966) The regulation of tryptophan pyrrolase activity by tryptophan. *Adv. Enzyme Regul. 4*, 287-297.

Knox WE, Auerbach VH (1955) The hormonal control of tryptophan peroxidase in the rat. *J. Biol. Chem. 214*, 307-313.

Knox WE, Piras MM, Tokuyama K (1966) Tryptophan pyrrolase of liver I. Activation and assay in soluble extracts of rat liver. *J. Biol. Chem. 241*, 297-303.

Lowitt S, Szentivanyi A, Williams JF (1981) Endotoxin inhibition of dexamethasone induction of tryptophan oxygenase EC- 1.13.11. 11 in suspension culture of isolated rat parenchymal cells involvement of the hepatic nonparenchymal cell fraction. *Biochem. Pharmacol. 30*, 1999-2006.

Meister A (Ed) (1965) *Biochemistry of the Amino Acids, Vol 2.* Academic Press, New York.

Moon RJ (1971) Tryptophan oxygenase and tryptophan metabolism in endotoxin-poisoned and allopurinol-treated mice. *Biochim. Biophys. Acta 230*, 342-348.

Moon RJ (1972) Carbohydrate metabolism and survival of endotoxin-poisoned mice given tryptophan. *Infect. Immun. 5*, 288-294.

Moon RJ, Berry LJ (1968a) Effect of tryptophan and selected analogues on body temperature of endotoxin-poisoned mice. *J. Bacteriol. 95*, 764-770.

Moon RJ, Berry LJ (1968b) Role of tryptophan pyrrolase in endotoxin poisoning. *J. Bacteriol. 95*, 1247-1253.

Moon RJ, Vrable RA, Broka JA (1975) In situ separation of bacterial trapping and killing functions of the perfused liver. *Infect. Immun. 12*, 411-418.

Moore RN, Goodrum KJ, Berry LJ (1976) Mediation of an endotoxin effect by macrophages. *J. Reticuloendothel. Soc. 19*, 187-197.

Moore RN, Goodrum KJ, Gouch R Jr, Berry LJ (1978) Factors affecting macrophage function: Glucocorticoid antagonizing factor. *J. Reticuloendothel. Soc. 23*, 321-332.

Moran JF, Sourkes TL (1963) Induction of tryptophan pyrrolase by α-methyltryptophan and its metabolic significance *in vivo. J. Biol. Chem. 238*, 3006-3008.

Morris KM, Moon RJ (1974a) Microanalysis of tryptophan metabolites in mice. *Anal. Biochem. 61*, 313-327.

Morris KM, Moon RJ (1974b) Quantitative analysis of serotonin biosynthesis in endotoxemia. *Infect. Immun. 10*, 340-346.

Phillips JL, Berry LJ (1970a) Circadian rhythm of mouse liver phosphoenolpyruvate carboxykinase. *Am. J. Physiol. 218*, 1440-1443.

Phillips JL, Berry LJ (1970b) Hormonal control of mouse liver phosphoenolpyruvate carboxykinase rhythm. *Am. J. Physiol. 219*, 697-701.

Piras MM, Knox WE (1967) Tryptophan pyrrolase of liver. II. The activating reaction in crude preparations from rat liver. *J. Biol. Chem. 242*, 2952-2958.

Rippe DF, Berry LJ (1972a) Effect of endotoxin on the activation of phosphoenolpyruvate carboxykinase by tryptophan. *Infect. Immun. 6*, 97-98.

Rippe DF, Berry LJ (1972b) Study of inhibition of induction of phosphoenolpyruvate carboxykinase by endotoxin with radial immunodiffusion. *Infect. Immun. 6*, 766-772.

Rippe DF, Berry LJ (1973) Immunological quantitation of hepatic tryptophan oxygenase in endotoxin-poisoned mice. *Infect. Immun. 8*, 534-539.

Schimke RT (1970) Protein turnover and the control of enzyme levels in animal tissues. *Acc. Chem. Res. 3*, 113-120.

Schimke RT, Sweeney EW, Berlin GM (1964) An analysis of the kinetics of rat liver tryptophan pyrrolase induction: the significance of both enzyme synthesis and degradation. *Biochem. Biophys. Res. Comm. 15*, 215-219.

Schimke RT, Sweeney EW, Berlin GM (1965a) The roles of synthesis and degradation in the control of rat liver tryptophan pyrrolase. *J. Biol. Chem. 240*, 322-331.

Schimke RT, Sweeney EW, Berlin GM (1965b) Studies of the stability in vivo and in vitro of rat liver tryptophan pyrrolase. *J. Biol. Chem. 240*, 4609-4620.

Tanaka T, Knox WE (1959) The nature and mechanism of tryptophan pyrrolase (peroxidase-oxidase) reaction of *Pseudomonas* and of rat liver. *J. Biol. Chem. 234*, 1162-1170.

Tavakoli H, Moon RJ (1981) In vitro interactions of endotoxin, chromatin, DNA, and steroid hormone receptors. In: *Pathophysiological Effects of Endotoxins at the Cellular Level*, 33-43. Alan R. Liss, New York.

Tavakoli H, Moon RJ (1982) In vitro and in vivo association and subcellular distribution of toxic and experimentally modified endotoxin. *Can. J. Microbiol. 28*, 822-829.

Zlydaszyk JC, Moon RJ (1976) Fate of ^{51}Cr-labeled lipopolysaccharide in tissue culture cells and livers of normal mice. *Infect. Immun. 14*, 100-105.

Handbook of Endotoxin, Vol. 3: Cellular Biology of Endotoxin
L.J. Berry, editor
© Elsevier Science Publishers B.V., 1985

CHAPTER 10

Modulation of antibody synthesis by bacterial endotoxins

ARTHUR G. JOHNSON

1. INTRODUCTION

The ability of the endotoxic lipopolysaccharide (LPS) isolated from the outer membrane of gram-negative bacteria to elevate antibody formation to unrelated antigens is now well known and readily demonstrable. On injection of microgram amounts of LPS together with any of a large variety of antigens into many species, one is usually favored with a pronounced increase in antibody titer. However, factors affecting the magnitude of the response have proved to be many, and the mechanism(s) involved are complex and as yet elusive. It appears that a multiplicity of cells and their functions, either directly or indirectly through secreted molecular mediators, are affected by this extraordinary molecular complex. In addition, the host has a variety of interacting pathways available for immune expression which also depend on factors as yet escaping our definition. Since multiple reviews covering various aspects of the adjuvant action of endotoxin have appeared recently (Jacobs 1983; Johnson 1983; Morrison and Ryan 1979; Nakano and Uchiyama 1983), this chapter will highlight the major findings, including those which appeared recently.

2. STRUCTURE–FUNCTION RELATIONSHIPS

LPS has proved to be a very complex molecule, whose structure is slowly being unraveled (Rietschel et al 1982; see also Volume 1 of this series). As originally isolated in earlier studies, it contained lipid, polysaccharide and protein moieties. It was found that the protein portion could be eliminated without loss of the endotoxic properties, and the protein-free lipopolysaccharide proved to be a very powerful adjuvant to the immune response (Johnson et al 1956). Nonetheless, the isolated protein portion proved later to have adjuvant properties in its own right (Izui et al 1980; Sultzer and Goodman 1976; Sultzer 1983). This entity, termed endotoxin protein, activated murine B cells directly. In addition, the polysaccharide moiety could also be removed, and the resulting lipid (termed lipid A) was a powerful adjuvant when solubilized correctly (Abdelnoor 1969; Chiller et al

216

1973). However, again the isolated polysaccharide alone too was capable of exerting a positive influence on antibody levels (Behling and Nowotny 1977; Friedman et al 1983; Nowotny et al 1975). Thus, one must conclude that all three of the components of this outer membrane protein–lipid–polysaccharide complex, as well as the whole can contribute to its adjuvant qualities. A broader range of testing (multiple species, etc.) however, needs to be carried out with the isolated components to determine their significance as adjuvant entities.

The actual stimulatory ligands on any of the moieties have defied attempts at their structural definition. Recently, Ribi et al (1982), starting with the heptoseless LPS from the Re mutant, removed 3-deoxy-D-manno-2-octulosonic acid (KDO), and isolated and purified a fully toxic diphosphoryl glucosamine disaccharide which was active as adjuvant (Johnson A., to be published) and reduced tumor growth. However, further treatment with weak acid removed one of the phosphate groups with a resultant loss of toxicity, while the adjuvant properties and inhibition of tumor growth were retained. In addition, Yasuda et al (1982) have synthesized lipid A-like structures including 2 of 5 synthetic analogs which showed adjuvant activity when incorporated into liposomes. Prior to this, Behling et al (1976) reported the adjuvant capacity of several synthetic N-acetylated D-glucosamine derivatives. Such defined molecules offer hope for chemical definition of the active adjuvant sites.

Obviously, detoxification of LPS with retention of its adjuvant activity has been a goal long strived for. In early studies, Johnson and Nowotny (1964) reported that treatment with boron trifluoride, potassium methylate or pyridinium formate diminished the toxicity of endotoxins with respect to lethality, pyrogenicity and the Shwartzman reaction. These endotoxoids, nevertheless, still possessed the ability to increase antibody formation to protein antigens. In addition, Friedman and Sultzer (1962) reported that acetylation of endotoxin reduced toxicity, yet retained the capacity to stimulate resistance to challenge with microorganisms. McIntire et al (1976) found LPS treated with o-phthalic anhydride lost much of its toxicity without loss of the adjuvant activity. This was confirmed by Chedid et al (1975). Thus, chemically detoxified endotoxins which retain adjuvant action are feasible. The greatest stumbling block to their utility appears to be the unique susceptibility of human beings to trace amounts of endotoxin which might not become manifest in standard safety tests of preparations with reduced toxicity.

3. IMMUNOMODULATING EFFECTS

The most important condition for the adjuvant action in vivo has proved to be the time of exposure to LPS relative to antigen injection. Thus, a striking change occurs in the antibody-forming potential of spleen cells when antigen injection is delayed for 1–2 days following LPS. For example, when antigen is injected together with LPS or within a few hours after LPS administration, enhancement of antibody levels measured 4–7 days later is seen. On the other hand, injection of antigen 1–2 days after LPS results in a profound shutdown of antibody synthesis (Behling and Nowotny 1977; Franzl and McMaster 1966; Jacobs 1983; Kind and Johnson 1959). An explanatory working hypothesis might be that injection of LPS alone fosters first the formation of nonspecific T helper cells in the spleen within hours. Injection of antigen at this time amplifies and directs the activity of such cells towards specific antibody formation. However, LPS also stimulates suppressor cells, which become functionally effective only after 1–2 days following LPS injection. Injection of antigen at this time after LPS then dictates amplification of the suppressive parameter of immunological reactivity. Whether or not inducement is dependent on the helper activity of the initial T helper cells engendered is not known. It is also not known as yet whether the helper cells per se wane in number at 1–2 days post-LPS, or whether their activity is overwhelmed by a stronger suppressive signal. Although both enhancement and suppression of antibody synthesis by LPS can also be demonstrated in vitro, the time relationships with respect to exposure to antigen vary from those found necessary in vivo (Hoffmann et al 1975; Kishimoto et al 1976).

Characterization of the suppressive phenomenon has revealed evidence for LPS-induced suppression mediated by B cells, T cells and macrophages. Persson (1977) initially deleted macrophages and T cells from suppressive murine spleen cell populations without losing suppression. Thus, the B cell was indicted as the responsible suppressing cell. From our laboratories, supportive evidence has also come from which it appeared that suppression induced by a glycolipid derived from LPS in C3H/HeJ mice was attributable to a B cell (Odean 1981). Kempf and Rubin (1979) extended implication of this cell by isolating a suppressor factor of 24 000 daltons from LPS-stimulated B cells. Koenig and Hoffmann (1979) attributed LPS-induced suppression in culture to an Ig^+ B cell, independent of T cells and macrophages.

Evidence suggesting multiple pathways of suppression was garnered by Uchiyama and Jacobs (1978) when they reported that a radiation-sensitive T suppressor cell followed LPS injection. This supported the findings of Walker and Weigle (1978), who demonstrated the capability of a B cell-depleted population to suppress the secondary response to turkey gamma globulin in culture. Nakano and Uchiyama (1983), on the other hand, have recorded data suggesting that the presence of macrophages under certain conditions could decrease LPS-induced factors and adjuvanticity. This indicates the macrophage could be a source of suppressive factors, and under appropriate conditions the equilibrium between helper and suppressor substances released by LPS would favor suppression. An intriguing hypothesis on an in vivo role for LPS-induced suppression has been offered by Gollahan et al (1983). They recorded the fact that germfree mice had higher IgA levels than conventional mice and reasoned that LPS contained in gut bacteria induced the formation of suppressor cells in vivo in gut-associated lymphoid tissue.

4. CELLS AFFECTED

Bacterial endotoxins bind to and affect the function of a myriad of cells. Thus, the final outcome of an increased antibody titer probably reflects the composite or equilibrated balance of this extraordinary compound on macrophages, B cells and the various subsets of regulatory T cells. A single target cell appears to be unlikely, and an 'initial' triggering may be exerted on several cells. Initial studies in rabbits designed to uncover tissue responses associated with the rise in antibody titers, revealed a marked increase in size and number of splenic germinal centers (Ward et al 1959). A decade later these organelles were defined as a source of B cells and with the finding of the selective mitogenic ability of LPS for this cell (Moller 1972; Peavey et al 1970), the logical hypothesis emerged that the endotoxin enhanced antibody formation by directly causing a rapid proliferation of B cells.

Indeed, Campbell and Kind (1979) found LPS enhanced B cell function in genetically deficient mice, and Skidmore et al (1975, 1976) demonstrated a correlation between the mitogenic activity of LPS for B cells and the adjuvant action. Others, however, have offered evidence for enhanced antibody levels in the absence of mitogenicity (Hoffmann et al 1977).

That effects on B cells may, in certain instances, reflect only the end result and be but a partial explanation, was suggested by the finding of Allison and Davies (1971), who demonstrated that removal of T cells by adult thymectomy or antilymphocyte serum abrogated the ability of LPS to increase anti-bovine serum albumin titers. Infusion of syngeneic thymus cells restored this capacity. Armerding and Katz (1974) obtained additional evidence in this regard in experiments which showed that LPS acts on carrier-specific T cells. A thymus dependency of LPS action was also implicated by the data of Kagnoff et al (1974) and Nakashima et al (1980). The latter directed attention to the importance of defining experimental conditions as a basis for obtaining and interpreting different results. Some of the most compelling data for T cell involvement in the adjuvant action of LPS have come from the experiments of McGhee et al (1979). Using cell transfer experiments between the unique C3H/HeJ mouse strain, which is not susceptible to the adjuvant action of endotoxin, and its genetically compatible susceptible counterpart, the C3H/HeN strain, they found both the T cell and the macrophage needed to be derived from the LPS-sensitive C3H/HeN strain in order for the adjuvant action to be expressed. Thus, LPS-induced T cell activities may be responsible for the increased B cell proliferation. Important conditions in this regard were described by Zabala and Lipsky (1982), who found LPS synergized with pokeweed mitogen in inducing Ig synthesis by human peripheral blood leukocytes without affecting the proliferative response. This LPS activity required relatively low numbers of T cells (10^4/well). However, larger numbers ($10^5 - 5 \times 10^5$) resulted in suppression. The number of macrophages was also critical with regard to enhancement or suppression. Thus, once again the ratio of T cells to B cells and macrophages determined the effect observed.

Additional evidence for T cell involvement has come from histological changes

ment of the T cell dependent production of antibody to SRBC in vitro by bacterial lipopolysaccharide. *J. Immunol. 114*, 738-741.

Hoffmann MK, Galanos C, Koenig S, Oettgen HF (1977) B cell activation by lipopolysaccharide. Distinct pathways for induction of mitosis and antibody production. *J. Exp. Med. 146*, 1640-1647.

Izui A, Morrison DC, Curry B, Dixon FJ (1980) Effect of lipid-A associated protein and lipid A on the expression of lipopolysaccharide activity. *Immunology 40*, 473-82.

Jacobs DM (1983) Immunomodulatory effects of endotoxin. In: Nowotny A (Ed), *Beneficial Effects of Endotoxins*, pp 307-326. Plenum Press, New York.

Johnson AG (1983) Adjuvant action of bacterial endotoxins on antibody formation: a historial perspective. In: Nowotny A (Ed), *Beneficial Effects of Endotoxins*, pp 249-255. Plenum Press, New York.

Johnson G, Nowotny A (1964) Relationship of structure to function in bacterial O antigens. III. Biological properties of endotoxoids. *J. Bacteriol. 87*, 809-814.

Johnson AG, Gaines S, Landy M (1956) Studies on the O antigen of *Salmonella typhosa*. V. Enhancement of antibody response to protein antigens by the purified lipopolysaccharide. *J. Exp. Med. 103*, 225-246.

Kagnoff MF, Billing P, Cohn M (1974) Functional characteristics of Peyer's patch lymphoid cells. II. Lipopolysaccharide is thymus dependent. *J. Exp. Med. 139*, 407-413.

Kempf KE, Rubin AS (1979) Generation by lipopolysaccharide of a late acting soluble suppressor of antibody synthesis. *Cell. Immunol. 43*, 30-40.

Kind PD, Johnson AG (1959) Studies on the adjuvant action of bacterial endotoxins. I. Time limitation of enhancing effect and restoration of antibody formation in x-irradiated rabbits. *J. Immunol. 92*, 415-427.

Kishimoto S, Takahama T, Mizumachi H (1976) In vitro immune response to the 2,4,5-trinitrophenyl determinant in aged C57/B1/6J mice: changes in the humoral immune response to avidity for the TNP determinant and responsiveness to LPS effect with aging. *J. Immunol. 116*, 294-300.

Koenig J, Hoffmann MK (1979) Generation by lipopolysaccharide of a late acting soluble suppressor of antibody synthesis. *Proc. Natl Acad. Sci. USA 76*, 4608-4612.

Lachman LB (1983) Interleukin 1. Release from LPS-stimulated mononuclear phagocytes. In: Nowotny A (Ed), *Beneficial Effects of Endotoxins*, pp 283-305. Plenum Press, New York.

McGhee JR, Farrar JJ, Michalek SM, Mergenhagen SE, Rosenstreich DL (1979) Cellular requirements for lipopolysaccharide adjuvanticity: a role for both T lymphocytes and macrophages for in vitro responses to particulate antigens. *J. Exp. Med. 149*, 793-807.

McIntire FC, Hargie MP, Schenck JR, Finley RA, Sievert HW, Rietschel ET, Rosenstreich DL (1976) Biological properties of non-toxic derivatives of a lipopolysaccharide from *Escherichia coli* K235. *J. Immunol. 117*, 674-678.

McMaster PD, Franzl RE (1968) The primary immune response in mice. II. Cellular responses of lymphoid tissue accompanying the enhancement or complete suppression of antibody formation by a bacterial endotoxin. *J. Exp. Med. 127*, 1109-1126.

Mizel SG (1980) Lymphocyte-activating factor (Interleukin 1), a possible mediator of endotoxin adjuvancy. In: Schlessinger D (Ed), *Microbiology – 1980*, pp 40-43. American Society for Microbiology, Washington DC.

Moatamed F, Karnowsky MJ, Unanue EA (1975) Early cellular responses to mitogens and adjuvants in the mouse spleen. *Lab. Invest. 32*, 303-312.

Moller G (1972) Lymphocyte activation by mitogens. *Transpl. Rev. 11.*

Morrison DC, Ryan JL (1979) Bacterial endotoxins and host immune responses. *Adv. Immunol. 28*, 293-450.

Nakano M, Uchiyama T (1983) In vitro adjuvant effect of endotoxin: review and experiments. In: Nowotny A (Ed), *Beneficial Effects of Endotoxins*, pp 307-326. Plenum Press, New York.

Nakashima I, Nagase F, Matsuura A, Yokochi T, Kato N (1980) Adjuvant actions of polyclonal lymphocyte activators. III. Two distinct types of T-initiating adjuvant action demonstrated under different experimental conditions. *Cell. Immunol. 52*, 429-437.

Nowotny A, Behling UH, Chang HL (1975) Relation of structure to function in bacterial endotoxins. VIII. Biological activities in a polysaccharide-rich fraction. *J. Immunol. 115*, 199-203.

Odean MJ (1981) *Characterization of the Cells Mediating LPS-induced Suppression.* M.Sc. Thesis, University of Minnesota.

Peavy DL, Adler WH, Smith RT (1970) The mitogenic effects of endotoxin and staphylococcal enterotoxin B on mouse spleen cells and human peripheral lymphocytes. *J. Immunol. 105*, 1453-1458.

Persson U (1977) Lipopolysaccharide induced suppression of the primary immune response to a thymus dependent antigen. *J. Immunol. 118*, 789-796.

Ribi E, Amano K, Cantrell J, Schwartzman S, Parker R, Takayama K (1982) Preparation and anti-tumor activity of nontoxic lipid A. *Cancer Immunol. Immunother. 12*, 91-96.

Rietschel ET, Galanos C, Lüderitz O, Westphal O (1982) The chemistry and biology of lipopolysaccharides and their lipid A component. In: Webb DR (Ed), *Immunopharmacology and the Regulation of Leukocyte Function*, pp 183-229. Academic Press, New York.

Rowlands DT, Claman HT, Kind PD (1965) The effect of endotoxin on the thymus of young mice. *Am. J. Pathol. 46*, 165-171.

Skidmore BJ, Chiller JM, Morrison DC, Weigle WO (1975) Immunologic properties of bacterial lipopolysaccharide (LPS): correlation between the mitogenic, adjuvant and immunogenic activities. *J. Immunol. 114*, 770-775.

Skidmore BJ, Morrison DC, Chiller JM, Weigle WO (1976) Immunologic properties of bacterial lipopolysaccharide (LPS). II. The unresponsiveness of C3H/HeJ mouse spleen cells to LPS-induced mitogenesis is dependent on the method used to extract LPS. *J. Exp. Med. 142*, 1488-1508.

Sultzer BM (1983) Lymphocyte activation by endotoxin protein. The role of the C3H/HeJ mouse. In: Nowotny A (Ed), *Beneficial Effects of Endotoxins*, pp 227-248. Plenum Press, New York.

Sultzer BM, Goodman GW (1976) Endotoxin proteins A and B cell mitogen and polyclonal activation of C3H/HeJ lymphocytes. *J. Exp. Med. 144*, 821-827.

Uccini S, Ruco L, Saravito G, Aderini L, Doria G, Barani CD (1976) Cell selection in the thymus of mice treated with *Escherichia coli* lipopolysaccharide (LPS). *Adv. Exp. Med. Biol. 66*, 95-99.

Uchiyama T, Jacobs DM (1978) Modulation of immune response by bacterial lipopolysaccharide (LPS): cellular bases of stimulatory and inhibitory effects of LPS on the in vitro IgM antibody response to a T dependent antigen. *J. Immunol. 121*, 2345-2351.

Walker SM, Weigle WO (1978) Effect of bacterial lipopolysaccharide on the in vitro secondary antibody response in mice. I. Description of the suppressive capacity of lipopolysaccharide. *Cell. Immunol. 36*, 170-179.

Ward PA, Johnson AG, Abell MR (1959) Studies on the adjuvant action of bacterial endotoxins on antibody formation. III. Histologic response of the rabbit spleen to a single injection of a purified protein antigen. *J. Exp. Med. 109*, 463-474.

White RG (1976) The adjuvant effect of microbial products on the immune response. *Annu. Rev. Microbiol. 30*, 579-600.

Yasuda T, Kanesasaki S, Tsumito T, Todakuma T, Homma JY, Inase M, Kusumoto S, Shiba T (1982) Biological activity of chemically synthesized analogues of lipid A. Demonstration of adjuvant effect in hapten-sensitized liposomal system. *Eur. J. Biochem. 124*, 405-407.

Zabala C, Lipsky PE (1982) Immunomodulation by bacterial endotoxin and other bacterial products in man. In: Yamamura Y, Kotani S, Azuma I, Koda A, Shiba T (Eds), *Immunomodulation by Microbial Products and Related Synthetic Compounds*. Excerpta Medica, Tokyo.

Handbook of Endotoxin, Vol. 3: Cellular Biology of Endotoxin
L.J. Berry, editor
© Elsevier Science Publishers B.V., 1985

CHAPTER 11

Endotoxin interactions with platelets

RICHARD I. WALKER AND LARRY C. CASEY

1. INTRODUCTION

Approximately 500 million years ago the horseshoe crab (*Limulus polyphemus*) evoked an antimicrobial defensive system consisting of a single cell type – the amebocyte – which degranulates to trap bacteria in a clot and subsequently destroy them. The antimicrobial system has become more specialized in mammals, but the teleologic descendant of the amebocyte – the platelet – is still important in the inflammatory response of the host to microorganisms. For example, platelets interact with pathogens and release mediators of inflammation, participate in trapping microorganisms for phagocytes and, although not bactericidal themselves, enhance the bactericidal effects of leukocytes (Clawson 1974; Clawson et al 1975; Smith 1972; Wilder and Lubin 1973).

The potentially beneficial effect of platelets on infection may become detrimental if thrombosis becomes too extensive to be resolved effectively. The host response to microorganisms can also be induced to self-destructive levels by challenge with purified endotoxin, a lipid-polysaccharide component of the cell walls of gram-negative bacteria. Within minutes after injection, most endotoxin in the bloodstream is associated with platelets (Braude 1964; Evans 1972). These cells have a profound influence on the rate of transport of endotoxin (Das et al 1973) and other particulate material (Donald 1972; Noyes et al 1959; Van Aken and Vreeken 1969) to the reticuloendothelial system. For example, plasma clearance of endotoxin is significantly slower in thrombocytopenic rabbits than in normal animals (Das et al 1973). In achieving this purpose, platelet-endotoxin complexes must pass through the capillary network of nonreticuloendothelial organs, a process which depends on maintenance of a balance between aggregation and disaggregation of platelets (Van Aken and Vreeken 1969).

Uncontrolled thrombosis in the microcirculation of major organs may be a key event during endotoxemia which could contribute to the diversity of host responses to endotoxin. This concept is consistent with microcinematographic observations by Urbaschek (1971). Within minutes after injection of endotoxin into rabbits, blood flow begins to slow and mast cells degranulate. Rouleaux formation of erythrocytes occurs, granulocytes adhere to the vessel walls and roll along the endothelium, and large platelet aggregates form and are partly disaggregated again. Soon irreversible platelet thrombi form, an event which could be significant to the further course of endotoxemia. Endothelial cells swell and together with wall-adhering blood cells, intensify the narrowing of the vessel lumen. Blood flow continues to decrease, first in the venules and capillaries and later in the arterioles. By 1 hour after challenge stasis occurs, capillary wall permeability increases and microbleeding begins.

2. GENERAL CHARACTERISTICS OF THE PLATELET RESPONSE TO ENDOTOXIN

The platelet response to endotoxin was recognized in the 1950s. Following in vivo administration of endotoxin there is platelet aggregation (Stetson, 1951) and increased plasma serotonin and histamine (Shimamoto 1958). Temperature affected both the aggregation and serotonin release (Des Prez et al 1961). At 4 °C no change was observed in either aggregation or serotonin release, whereas at 37 °C there was time-dependent serotonin release and aggregation. Furthermore, the clotting time of platelet-rich plasma was shortened following the addition of endotoxin (Des Prez 1964), and this could be related to the fact that platelets contribute a phospholipid called platelet factor 3 to the coagulation system (Horowitz and Spaet 1961). The clotting times of both platelet-rich plasma and whole blood were significantly shortened by the addition of endotoxin, but the phenomenon can be reversed by removing the platelets (Horowitz et al 1962; McKay et al 1958). Horowitz demonstrated that by lysing the platelets with detergent and then adding phospholipid (brain cephalin), the coagulation time could be shortened. Adding endotoxin did not further shorten the coagulation time once the platelets had been lysed or cephalin had been added.

Preincubation of antigen with immune plasma prior to addition of platelets and anticoagulant initiated effects similar to the endotoxin–platelet interaction (Des Prez 1964). The antigen-antibody reaction probably activates complement, and this complex could then react with platelets as a particle might. Movat et al (1965) and Spielvogel (1967) suggest that the antigen-antibody–platelet and platelet–endotoxin complexes behave as particles and that platelet injury is associated with phagocytosis. Movat showed that platelets ingest latex particles, ferritin–anti-ferritin complexes, and Spielvogel showed the ingestion of endotoxin by platelets. These particle ingestions by the platelets released serotonin into the plasma.

The similarity in platelet-induced injury by endotoxin and that induced by antigen-antibody reactions led Des Prez and Bryant (1966) to study the divalent cation requirements for these two types of reactions. To document that platelet injuries were similar to immune reactions, they compared endotoxin-induced release of serotonin with serotonin release induced by antigen added to platelets derived from immunized animals. With various concentrations of anticoagulants in platelet-rich plasma (heparin, 100 µg/ml; sodium citrate, 0.38%; sodium EDTA, 0.1% or 0.17%), Des Prez and Bryant (1966) showed that endotoxin caused release of serotonin when 0.38% sodium citrate or 0.1% sodium EDTA was present, while antigen-antibody did not. When the concentration of EDTA was increased to

0.17%, neither endotoxin nor antigen-antibody caused serotonin release. This difference was not observed when antigen and platelet-poor plasma were allowed to interact in the absence of chelating agents. When platelets and citrate were added following the incubation, the release of serotonin was similar to that induced by endotoxin. Des Prez and Bryant (1966) showed that the reason for the initial differences between endotoxin and antigen-antibody complex interactions with platelets were related to the divalent cation concentrations (Ca^{2+} and Mg^{2+}), and that more divalent cations were necessary for the antigen-antibody reactions than for endotoxin reactions in the presence of the same concentration of chelating agents. More recently, the noncytotoxic secretory activity of platelets has been shown to have a specific requirement for calcium (Morrison et al 1980).

Differences were also noted between platelet injury caused by thrombin and that induced by endotoxin (Des Prez 1964). The lag period of serotonin release and of aggregation led Des Prez to think that the effect of endotoxin on platelets was indirect, possibly secondary to the action of a proteolytic enzyme, because platelet-rich plasma incubated with trypsin and no calcium did not decrease the fibrinogen level, whereas with trypsin and calcium, lower fibrinogen values and increased serotonin levels in the plasma were seen, even though no aggregation occurred. Using thrombin-induced platelet injury, Des Prez compared it to endotoxin-induced platelet injury. Very small amounts of thrombin induced platelet injury, and this reaction was immediate and non-progressive as was induced clot formation. In contrast to endotoxin, concentrations of thrombin sufficient to produce gross clot formation in citrated plasma were less than those required to induce serotonin release from the platelets. Small doses of heparin prevented thrombin injury to platelets, whereas very large amounts of heparin were required to prevent endotoxin-induced platelet injury.

3. COMPONENTS OF THE PLATELET–ENDOTOXIN INTERACTION

Since the early studies of platelet aggregation and release responses, a variety of humoral and cellular elements have been identified as important to the platelet–endotoxin interaction.

3.1. Plasma factors

Des Prez et al (1961) showed that heating plasma to 56 °C for 30 minutes before adding the platelets and endotoxin, inactivated the serotonin release and the aggregation. This suggested that there is a heat-labile plasma factor necessary to mediate the effects of endotoxin on platelets. This factor, which had earlier been detected by McKay et al (1958), was of considerable interest because of its possible role in the immune mechanism and the coagulation system. With an intricate set of experiments, Ream et al (1965) showed not only that plasma and platelets were required for endotoxin to induce platelet injury, but also that the plasma factor could be better characterized. Heating the endotoxin to 56 °C did not alter its effect on platelets, but heating the plasma for 2 minutes at 56 °C inactivated the factor necessary for the platelet–endotoxin interaction.

Ream et al (1965) found evidence that a plasma factor necessary for the aggregation of platelets might be Factor V. By removing platelet-rich plasma from humans with known individual factor deficiencies, aggregation was observed after the addition of endotoxin in every case except with the plasma deficient in Factor V. If Factor V was added, the aggregation occurred. However, if the Factor V was allowed to be below 30% of normal (by dilution technique), aggregation did not occur following exposure to endotoxin (100 µg/ml). To simulate the lipopolysaccharide of endotoxin, several different fatty acids were added individually to platelet-rich plasma (normal), but no aggregation was observed. With rabbit platelet-rich plasma, endotoxin caused aggregation predictably, but with human platelet-rich plasma and endotoxin, the aggregation was variable. Additionally, Ream showed that contact Factor XII, believed to initiate the clotting mechanism, was not involved in the platelet–endotoxin interaction since platelet-rich plasma deficient in Factor XII allowed aggregation to occur when exposed to endotoxin. Müller-Berghaus and Schneberger (1971) confirmed that Factor XII (Hageman factor) was not involved in the platelet–endotoxin interaction because a single dose of endotoxin (10–20 µg/ml) did not decrease the amount of this factor. Since pretreatment with one of the coumarin drugs prevented the fall of the level of Hageman factor after the injection of thorotrast or endotoxin, Müller-Berghaus concluded that the contact factor was not the trigger mechanism in coagulation from the endotoxin–platelet interaction.

3.2. Complement

Even though Ream et al (1965) had demonstrated that the heat-labile plasma factor necessary for the platelet-endotoxin reaction could be Factor V, Des Prez (1967) was not satisfied. He demonstrated (a) that whole-blood complement titers were not decreased during the course of endotoxin-platelet injury in recalcified citrated plasma and (b) that plasma adsorbed with zymosan at 16 °C in the presence of EDTA did not lower the complement titers and it was furthermore similar to heated plasma (56 °C for 30 minutes) in failing to support the platelet–endotoxin interaction, and (c) the factor or factors removed or inactivated by zymosan and heating the plasma appeared to be similar in the amounts necessary to support the platelet–endotoxin interaction. Since Boyden (1966) had demonstrated that macroglobulin antibodies, reactive with the polysaccharide determinants of endotoxin of gram-negative bacteria, exist in all species virtually from birth, Des Prez speculated that the heat-labile plasma factor was a naturally occurring antibody.

Support for the fact that complement is involved in the platelet–endotoxin interaction was obtained by Gilbert and Braude (1962) who showed a decrease in serum complement after endotoxin injection into rabbits. Spielvogel (1967) demonstrated that the effects of endotoxin on rabbit platelets in vitro were due to the immune adherence phenomenon and the heat-labile plasma factor involved was complement. He found that measures known to inactivate some complement components (heating, NH_4OH, zymosan) also prevented degranulation of rabbit platelets in citrated platelet-rich plasma. It was also shown that loss of granules in the platelets was associated with adherence of endotoxin particles to the platelet membrane. In primate blood (in which immune adherence receptor sites are on the red blood cells rather than on the platelets), platelet damage was not observed.

Gewurz et al (1968) investigated the interaction of lipopolysaccharides with C3–C9 of the complement system and found that each of the six components was consumed during these incubations, even with small amounts (10–25 µg/ml) of endotoxin. Interestingly, he showed that lipopolysaccharide consumed greater quantities of C3–C9 than did anti-human gamma globulin. Lipopolysaccharide reacting with hyperimmune rabbit serum leads to a complement consumption profile similar to that induced by the immune complexes. When highly purified preparations of any of the six terminal complement components were incubated with lipopolysaccharide, no consumption occurred. This indicated other serum factors were needed. Gewurz concluded that the complement system was involved in the biological actions of lipopolysaccharide.

Kane et al (1973) studied the interactions of the classical and alternative complement pathways with endotoxin lipopolysaccharide with special emphasis on platelets and coagulation. Using C4-deficient guinea pigs, known to have a complete block in the activity of the classical complement pathway, but with the alternative pathway intact, he demonstrated that there was no fall in the platelet count as seen in the normal animals following endotoxin injection. The alternative pathway was activated in C4-deficient guinea pigs since the C3–C9 titers did decrease. When C4 was restored in the C4-deficient animals, thrombocytopenia did occur as in the controls. Kane concluded that the classical and alternative complement pathways were required for thrombocytopenia and shortening of the clotting time.

Brown and Lachmann (1973) studied the interactions of complement and platelets in lethal endotoxin shock in rabbits. With radioactive chromium-labeled platelets in normal, in C6-deficient, and C3–C9-depleted rabbits, they were able to study the effects of the potentially lethal dose of endotoxin on survival and platelet sequestration. In the normal rabbits, the platelets were acutely destroyed with only a small portion returning to the circulation and all animals died. In the C3–C9-depleted animals, platelets survived normally following endotoxin injection and none of the animals died. In C6-deficient rabbits, platelets were acutely lost from the circulation, but were destroyed only variably. C6-deficient rabbits in which the platelet recovery was greater than 65% lived, whereas when the platelet recovery was less than 45%, the animals died. Platelet factor 3 activation by endotoxin was also studied in these animals both in vitro and in vivo. Marked activation of platelet factor 3 occurred in normal rabbits, minor activation in C6-depleted rabbits, and no activation in C3–C9-depleted rabbits. Brown and Lachman concluded that complement depletion markedly alters the platelet response to endotoxin. Depletion of C3–C9 appears to be absolutely protective and correlates well with absence of mortality, normal platelet survival, and failure to activate platelet factor 3.

Morrison and Oades (1979) showed that different portions of the endotoxin molecule activate different complement pathways. Activation of the alternative pathway is essential for platelet lysis, and preparations containing lipid A activate the classical pathway which does not cause lysis. Up to 5 µg of lipid A per ml of rabbit platelet-rich plasma did not induce aggregation in vitro, although aggregation of the same platelet preparation could be accomplished with a comparable amount of whole endotoxin (Walker and Beasley 1980). Lipid A attaches the endotoxin to the platelet membrane, and this attachment is enhanced by lipid A protein (Morrison and Oades 1979). The alternative pathway requires the polysaccharide region of the molecule, but lysis is not obtained if the lipid A portion is not available for platelet–endotoxin binding. Perhaps lipid A attachment to platelet membranes is necessary to bring the carbohydrate portion of the molecule into proximity with the membrane surface, where the alternative complement pathway is initiated.

Evidence that both lipid A and polysaccharide components of endotoxin are necessary for lysis or aggregation of platelets is shown by endotoxin detoxification experiments. Polymyxin B binds to lipid A, the toxic moiety of the endotoxin complex, thereby detoxifying the endotoxin preparation (Morrison and Jacobs 1976). This detoxified preparation also loses its platelet lytic capability. Similarily, megarad doses of gamma radiation ^{60}Co destroy fatty acid groups on lipid A (L. Bertok, personal communication) and thereby reduce the toxicity of the molecule (Previte et al 1967). Irradiated endotoxin no longer induces platelet aggregation (Walker et al 1983).

3.3. Specific antibodies

Lethal amounts of endotoxin may disturb the disaggregation process in non-reticuloendothelial organs, thereby causing abnormal deposition of platelet-endotoxin complexes which could have locally damaging effects. Persistence of endotoxin-platelet aggregates in the vasculature of nonreticuloendothelial organs previously has been associated with toxicity (Carey et al 1958; Noyes et al 1959), and animals made tolerant to the lethal effects of endotoxin demonstrate improved reticuloendothelial clearance of the toxin (Carey et al 1958; Herring et al 1963). Platelet responses to endotoxin may be an important factor in such improved clearance by the reticuloendothelial system.

We tested the hypothesis that platelets from tolerant rabbits have different aggregation characteristics than platelets of animals that are not resistant to the lethal effects of endotoxin (Walker et al 1978). Adult New Zealand white rabbits (2.5 kg) were made tolerant to *Salmonella typhi* endotoxin (lipopolysaccharide W, Difco) by receiving 5 daily intraperitoneal injections of the toxin (0.2, 0.3, 0.4, 0.5 and 0.5 mg). The effectiveness of this regimen was determined by intravenous challenge of the rabbits with a lethal dose of endotoxin (0.15 mg/kg). Animals that had been made tolerant survived this dose of toxin.

Aggregation of platelets was determined on 1 ml samples of platelet-rich plasma

230

in plastic cuvettes equipped with plastic-covered magnetic stirring disks. Increased light transmission (aggregation) was recorded from a Beckman DU-2 spectrophotometer equipped with a magnetic stirrer and 37 °C water bath. The pattern of the aggregation was followed. When the pen never returned toward the base line, aggregation was termed 'secondary' or 'irreversible'. Tracings in which the recorder pen returned toward the base line after reaching a maximum were termed 'primary' or 'reversible' aggregations.

Platelets from rabbits made tolerant to the lethal effects of endotoxin have aggregation characteristics different from those of platelets from nontolerant animals. Platelets from tolerant rabbits responded more rapidly to the aggregating effects of endotoxin than did platelets from nontolerant animals. Furthermore, unlike platelets from nontolerant rabbits, platelets from tolerant rabbits aggregated reversibly. These characteristics of platelets from tolerant rabbits may enhance reticuloendothelial clearance of endotoxin and thereby promote survival.

Since platelet-free plasma from tolerant rabbits could confer tolerant characteristics on platelets from nontolerant animals, a humoral factor must be principally responsible for altered platelet responsiveness to endotoxin. The specificity of the humoral factor was suggested by the observation that induction of 'tolerant' responses to endotoxin did not affect the aggregation responses of rabbit platelets exposed to ADP or collagen (Walker et al 1980). The time at which this humoral factor appeared in the plasma (1 week after the first injection used to induce tolerance) correlated with the appearance of antipolysaccharide (but not anti-lipid A) antibodies in the plasma (Walker and Beasley 1980). Furthermore, when specific anti-O antisera were added to normal platelet rich plasma, tolerant-like responses were observed following addition of endotoxin. Addition of anti-lipid A did not accelerate aggregation or disaggregation after exposure of platelets to complete endotoxin.

Greisman reported that antisera induced by smooth endotoxin are directed against the O-specific side chains (Greisman et al 1973). This is consistent with our finding (Walker and Beasley 1980) that plasma from tolerant rabbits or specific antisera reacted only with the smooth endotoxin and not with purified lipid A. Likewise, antisera against lipid A did not interact immunologically with smooth endotoxin.

Bult and Herman (1979) obtained responses with guinea pig platelets which were similar to those described above for rabbits made tolerant to the toxin. They found that pretreatment of guinea pigs with endotoxin caused a more rapid onset of the aggregation response when endotoxin was subsequently added to platelet-rich plasma from these animals. This effect was due to the presence of a humoral factor directed against the polysaccharide component of the endotoxin.

Lipid A did not initiate platelet aggregation unless its specific antibody was also present (Walker and Beasley 1980). This observation, which is consistent with a report that platelet lysis is not initiated by lipid A alone (Morrison and Oades 1979), provides further insight into the platelet–endotoxin interaction. Platelets have specific membrane receptor sites (Ausprunk and Das 1978; Hawiger et al

1975; Washida 1978) to which lipid A can attach. However, complement is also involved in the rabbit platelet–endotoxin interaction (Spielvogel 1967), and depletion of these components correlates well with reduction of mortality of the animal and enhanced platelet survival. It now appears that lipid A serves as an attachment mechanism which brings the polysaccharide portion of the molecule into proximity with the platelet membrane where alternative complement components are assembled, thereby inducing visible platelet reactivity (Morrison and Oades 1979). In the tolerant animal the polysaccharide portion of the endotoxin is apparently modified by antibody, and this process could account for the altered platelet responsiveness to endotoxin seen in tolerant rabbits.

3.4. Prostaglandins

Prostaglandins are important mediators of platelet function and may contribute to the fate of these cells during endotoxemia. Recent work has demonstrated that the two endoperoxides derived from arachidonic acid, PGG_2 and PGH_2, are very potent in inducing rapid and irreversible aggregation of human platelets (Hamberg and Samuelsson 1974; Hamberg et al 1975). Acetylsalicylic acid, an inhibitor of cyclooxygenase and thus an inhibitor of endoperoxide formation, blocks secondary platelet aggregation, which suggests that the endoperoxides play a role in the platelet release reaction (Hamberg et al 1975). The addition of PGH_2 to platelet-rich plasma leads to rapid changes in shape, aggregation, and secretion of 5-hydroxytryptamine from the platelets. This phenomenon appears to be partly due to isomerization of the endoperoxides to thromboxane A_2 (TXA_2) by the enzyme thromboxane synthetase (Hamberg et al 1975).

The relative potencies of the prostaglandin endoperoxides and TXA_2 as platelet-aggregating agents are still unknown, but there is some evidence to suggest that TXA_2 is more potent (Needleman et al 1976). In contrast to the effects of endoperoxide on platelets, TXA_2 is at least 100 times more potent than PGH_2, as measured by the induction of contraction of aortic smooth muscle (Needleman et al 1976). The generation of TXA_2 during platelet aggregation in vivo may lead to amplification of the aggregation response as well as to local vasoconstriction. Butler et al (1982) demonstrated that a selective thromboxane synthetase inhibitor (N[7-carboxyheptyl]imidazole) failed to inhibit endotoxin-induced platelet aggregation, but did prolong the bleeding time. The prolongations of the bleeding time may be the result of the isomerization of the unconverted endoperoxides into prostacyclin (PGI_2) (Nijkamp et al 1977; Vermylen et al 1981), possibly by endothelial cells (Gryglewski et al 1976). These results may suggest that the most important role of TXA_2 in hemostasis is not as an amplifier of platelet aggregation, but rather as a potent vasoconstrictor which, by making the vessel smaller, facilitates hemostasis at the site of vessel injury.

Thromboxane may contribute to the pathophysiology of endotoxin shock. Cook et al (1981a) provided strong evidence for the importance of thromboxane in rat endotoxin shock by demonstrating that essential fatty acid-deficiency in rats, as well as the pretreatment of rats with either a cyclooxygenase inhibitor or a thromboxane receptor antagonist, all improved survival in a lethal model of endotoxin shock.

Furthermore, prostacyclin, which has the opposite effects of thromboxane (inhibits platelet aggregation and is a vasodilator), is beneficial in treating endotoxin shock (Fletcher and Ramwell 1980; Lefer et al 1980). For example, in dogs challenged with *Escherichia coli* endotoxin (1 mg/kg) and treated with a continuous infusion of prostacyclin (20 ng/kg/min) from 15 minutes before to 4 hours after the administration of the endotoxin, survival was increased from 42% to 83%. This effect must be due to factors other than the inhibition of platelet aggregation since prostacyclin has no effect on direct endotoxin-induced platelet aggregation and does not prevent the in vivo endotoxin-induced thrombocytopenia (Fletcher and Ramwell 1980).

The importance of thromboxane in primate endotoxin shock appears to be different than in rat endotoxin shock, as the pretreatment of baboons with either OKY 1581 (a selective thromboxane synthetase inhibitor) or imidazole fails to improve survival (Casey et al 1982). In addition, these drugs do not prevent the endotoxin-induced thrombocytopenia. However, it is clear that endotoxin-induced pulmonary hypertension is mediated by increased TXA_2 formation (Casey et al 1982), and not by the mechanical occlusion of pulmonary vessels by either platelet or white blood cell aggregates as previously suggested (Pennington et al 1973).

Since a large number of tissues and cells are capable of producing thromboxane, the source of the increased plasma thromboxane after intravenous endotoxin administration remains unclear. When endotoxin is injected into rabbits, their platelets develop an increased capacity to synthesize thromboxane (Prancan et al 1981). All thromboxane may not come from the platelets, however, because in sheep or baboons, the increase in plasma thromboxane occurs immediately after the injection of the endotoxin, whereas the decrease in platelet count occurs over several hours (Harris et al 1980; Smith et al 1981). The incubation of neutrophils in vitro with endotoxin will stimulate them to produce thromboxane which can then cause platelet aggregation (Spagnuolo et al 1980). In addition, neutropenia decreases endotoxin-induced increases in plasma thromboxane (Hüttemeier et al 1982).

3.5. Other cell types affecting platelets

Direct and indirect effects of endotoxin on platelets may be altered by responses of other cell types which interact with the toxin. These cells include macrophages, polymorphonuclear leukocytes and endothelial cells.

Macrophages are important in the detoxification of endotoxin (Filkins 1971) and regulation of many host responses to the toxin (Berry 1972). The importance of these cells to survival during endotoxemia is indicated by the increased resistance to the lethal effects of endotoxin obtained in normal strains of mice receiving bone marrow transplants from genetically resistant animals (Michalek et al 1980). Macrophages from genetically resistant mice (Sultzer and Goodman 1977) or tolerant mice (Snyder et al 1979) release lower amounts of mediators during phagocytosis, and this reduced responsiveness to endotoxin stimuli may diminish the degree of inflammatory responses obtained. For example, cultured mouse peritoneal macrophages secrete factors which cause rabbit platelets to aggregate and release serotonin (Blumenthal et al 1980). The nature of these factors is unknown, but macrophage preparations are known to secrete a variety of prostaglandins in response to inflammatory stimuli (Humes et al 1977) which include prostacyclin and TXA_2 (Krausz et al 1981; Cook et al 1981b). Reduction of such release could reduce microcirculatory damage. Additionally, resistant macrophages may secrete endotoxin-detoxifying substances or regulators of thrombosis, which would also reduce damage in the microcirculation.

Polymorphonuclear leukocytes could help reduce platelet aggregation. Tolerant animals mobilize granulocytes more effectively than normal animals (Fruhman 1972), and these cells may work with platelets in the delivery of the toxin to the reticuloendothelial system. Granulocytes are attracted to platelet thrombi and participate in their subsequent resolution through phagocytosis (Shirasawa and Chandler 1971) and release of mediators (Canoso et al 1974; Levine et al 1976; Rodvien et al 1976; Wautier et al 1976). Hydrogen peroxide is a leukocytic mediator released during phagocytosis (Iyer et al 1961; Root et al 1975) which reduces aggregation induced by ADP and other nonmicrobial substances (Canoso et al 1974; Levine et al 1976; Rodvien et al 1976). However, this substance has no effect on platelet aggregation induced by direct interaction with endotoxin (Walker et al 1980). This type of system seems well suited for elimination of microorganisms with simultaneous reduction of the dangerous additional aggregation induced by exposed collagen, ADP and other host factors.

Endotoxin can directly injure endothelial cell membranes with resulting intracellular edema and formation of gaps between cells which permit extrusion of erythrocytes (Baris et al 1980). In vivo endotoxin-dependent granulocyte and complement-mediated events could also damage endothelial cells (Sacks et al 1978). Subendothelium exposed by endothelial cell detachment permits platelet adhesion and various degrees of activation which may be important in the pathogenesis of the thrombotic complications of endotoxemia.

4. ENDOTOXIN INTERACTION WITH PRIMATE PLATELETS

Direct effects of endotoxin on primate platelets have been difficult to demonstrate in vitro. For example, aggregation responses readily observed in rabbit or dog platelet-rich plasma exposed to endotoxin are not obtained in primate preparations. However, thrombocytopenia does occur in primate models challenged with endotoxin (J.R. Fletcher, unpublished data). Male baboons weighing 20–30 kg were restrained and subjected to an LD_{50} dose of *E. coli* endotoxin (3 mg/kg given as an intravenous bolus via a peripheral vein). By 15 minutes after challenge, circulating platelets in those animals dropped to approximately 50% of normal values. Furthermore, when the responsiveness of the remaining platelets to ADP (2 µg/ml) was tested, they were found to be refractory to the induction of aggregation.

In 1965 Ream et al reported aggregation of washed human platelets in the presence of endotoxin and a heat-labile plasma factor. In general, however, relatively little success has been obtained in studies of platelet–endotoxin interactions in humans and other primates.

Das et al (1973) used the Limulus test to show that human platelets do interact with bacterial endotoxin. The sensitive component of the human platelet membrane interacts with endotoxin with subsequent release of adenine nucleotides and platelet factor 3 (Hawiger et al 1977). This receptor is apparently specific for endotoxin because zymosan does not release platelet factors as endotoxin does (Hawiger et al 1975). These authors go on to suggest that there is not only a specific endotoxin-sensitive component on human platelets, but also an indirect pathway of activation in which microbial antigens complexed with antibody and/or complement interact with platelets through receptors for the IgG Fc fragment and complement.

As Ream's work indicated (1965), plasma proteins are very important in the primate platelet–endotoxin interaction. In a plasma-free medium endotoxin, complexed with copper, adhered to human platelet surfaces and caused breaks in the membrane, pseudopod formation and centralization of platelet organelles (Ausprunk and Das 1978). Normal plasma interferes with this reaction. An interesting observation is that the copper-containing protein ceruloplasmin, which becomes elevated prior to thrombocytopenia associated with gram-negative infection, may potentiate adsorption of blood-borne endotoxin by human platelets.

Platelet interaction with endotoxin may initiate a signal to the host that bacteria are present, thereby inducing an inflammatory response. With this in mind, we tested (Sheil and Walker 1979) the hypothesis that human platelets may influence endotoxin-induced inflammation through alteration of the chemotactic responsiveness of polymorphonuclear leukocytes. Leukocyte-rich preparations were placed on filters in chemotaxis chambers and exposed to platelet-rich or platelet-free human plasma before and after activation of the plasma with *Salmonella typhi* (Westphal preparation – Difco) endotoxin.

With the addition of 25 ng – 0.5 µg of endotoxin to a milliliter of human platelet-rich plasma we observed an increase in leukocyte chemotaxis of 50–85% above that seen in unactivated platelet-rich plasma. However, activation of platelet-rich plasma with 1 and 2 µg of endotoxin resulted in an increased polymorphonuclear leukocyte chemotactic response of only 30–45%. These responses were significantly lower ($P < 0.01$) than the responses obtained with 1 and 2 µg of endotoxin in platelet-poor plasma. With the addition of 5, 10 and 100 µg of endotoxin, chemotaxis in response to activated platelet-rich plasma was increased by 75–100%.

Although caution must be exercised in interpreting these in vitro findings, observations made in vivo by others indicate that impairment of chemotactic responsiveness is a phenomenon associated with endotoxemia (Berthrong and Cluff 1953; Cluff 1953; Territo and Golde 1976). Furthermore, impairment of chemotaxis is associated with larger doses of endotoxin rather than with smaller doses of the toxin (Fruhman 1972). We can also make an association between platelet effects on leukocyte chemotaxis and survival from endotoxemia. Treatment of endotoxin with $CrCl_3$ reduces some toxic properties of the endotoxin molecule (Prigal et al 1973; Snyder et al 1978). Therefore, it is of interest that the reduction in chemotaxis seen in response to platelet-rich plasma activated with 1 and 2 μg of untreated endotoxin does not occur with Cr-treated endotoxin (Walker et al 1980).

5. POSSIBLE SIGNIFICANCE OF PLATELET INJURY DURING ENDOTOXEMIA

Microcirculatory failure may be a critical event during endotoxemia. The importance of events at this site is suggested by studies of the resolution of endotoxin-induced damage in the microcirculation of C3HeB/FeJ mice and the closely related, but endotoxin-resistant, C3H/HeJ mice (Walker and Parker 1980). These animals were challenged intraperitoneally with 1 mg of *Salmonella typhi* endotoxin, and early neutrophil sequestration in the pulmonary vasculature observed microscopically. In the C3H/HeJ group this neutrophil sequestration was apparent at 1 hour after challenge; it began to decline in incidence by 3 hours and was back to nearly normal levels by 6 hours. Although initial neutrophil sequestration in C3Heb/FeJ mice was similar to that of the C3H/HeJ mice, by 3 hours neutrophil sequestration became extensive and persisted at 6 hours in the C3Heb/FeJ mice in contrast to the near-absence of sequestration in the lungs of C3H/HeJ mice at this time. This difference in resolution of inflammation in the microcirculation may be associated with the finding that the challenge dose killed most of the sensitive but none of the resistant mice.

In later studies (Walker et al 1982), lung parenchymal tissue was excised from resistant and sensitive (this time C3H/HeN) mice 6 hours after challenge with 0.8 mg of endotoxin, and the in vitro production of TXA_2 or PGI_2 over a 30-minute period was determined by radioimmunoassay. TXA_2 and PGI_2 released by lung fragments taken 6 hours after challenge with endotoxin did not differ from normal levels of these substances in C3H/HeN mice. In contrast, TXA_2 release was normal at 6 hours in C3H/HeJ mice, but PGI_2 was significantly increased. The ratio of PGI_2/TXA_2 in resistant mice was about twice that seen in sensitive mice. Since PGI_2 can reduce leukocyte adhesion, an increase in the ratio of this substance to TXA_2 may be associated with resolution of polymorphonuclear leukocytes from the pulmonary microcirculation of endotoxin-resistant mice.

These microcirculatory disturbances can both affect platelets and be affected by them. Direct injury to platelets may be important with large doses of endotoxin. Cellular targets of endotoxin such as platelets can be expected to respond to the toxin in a dose-dependent manner. This type of response has been obtained in vitro with dog platelets which aggregate in greater numbers as the concentration of *Serratia marcescens* endotoxin is increased from 25 ng to 5 μg/ml (Fletcher and Ramwell 1980). In vivo dose effects of endotoxin on dog platelets have also been observed (Walker 1980). Small (4 μg/kg) doses of *S. typhi* endotoxin administered intravenously produce no thrombocytopenia, but large, sublethal (1.8 mg/kg) doses of endotoxin induced marked thrombocytopenia.

That effects due to platelet injury can be critical during endotoxemia is indicated by studies in several models. Ream (1965) found a positive correlation between toxicity of different endotoxin preparations (as measured by chick embryo lethality) and their ability to induce aggregation of washed human platelets. Further correlation of the degree of endotoxin interaction with platelets and lethality was obtained by Washida (1978). He found that the LD_{50} dose of endotoxin is 100 times less for newborn rabbits than for adult animals (50 μg/kg *vs* 5000 μg/kg). This increased sensitivity was associated with increased numbers of receptor sites for endotoxin on platelet membranes of newborn rabbits.

A significant contribution of platelets to host responses to endotoxemia is also indicated by studies which show that reduction of platelet aggregation increases survival from endotoxemia. A splenic extract which inhibits mouse platelet aggregation in vivo and in vitro also protects these animals from endotoxin-induced lethality (Spillert et al 1980). Lewis and Mustard (1969) found that inhibition of platelet aggregation during endotoxin shock prevents acute effects in adult rabbits. Furthermore, they obtained platelet aggregation and death similar to that seen during endotoxin shock infusion of ADP.

Studies in other species of animals indicate that platelet–endotoxin interactions may not be determinative during endotoxin shock. For example, in dogs made thrombocytopenic by estrogen pretreatment, mortality and gross necroscopy findings following challenge with endotoxin were similar to animals with normal platelet levels that were also challenged with endotoxin administered intravenously (From et al 1976). Levin and Cluff (1965) showed that platelets do not contribute significantly to tissue damage induced by endotoxins; they selectively depleted rabbits of circulating platelets by treatment with guinea pig anti-rabbit platelet antibody. This thrombocytopenia did not prevent the cutaneous hemorrhagic lesion of the localized Shwartzman reaction following administration of endotoxin.

Further evidence that direct endotoxin-platelet interactions are not determinative in survival from endotoxin shock has been reported by Morrison et al (1980). They found that platelets from both endotoxin-resistant C3H/HeJ mice and responsive C3H/St mice manifest a similar time-dependent increase in calcium-dependent secretion following incubation with endotoxin.

Petschow (1983) showed that over 85% of endotoxin administered subcutaneously to mice remained at the site of injection, but lethality was still obtained. Unfortunately, no study of thrombocytopenia or other microcirculatory events accompanied these data. However, the data indicate that mediated effects are important in endotoxemia, and if thrombocytopenia is involved, it must be due to indirect effects of host-generated factors. This may be one reason why factors such as PGI$_2$ which do not alter endotoxin-induced platelet aggregation are effective in enhancing survival following endotoxin challenge.

6. CONCLUSIONS

Endotoxin interacts rapidly with platelet membranes and, within minutes after experimental inoculation into animals, most is associated with the platelet fraction of the buffy coat. The endotoxin–platelet interaction contributes to inflammation by formation of aggregates and release of factors such as platelet factor 3, nucleotides, and serotonin. Also platelets may facilitate delivery of the toxin to the liver and spleen.

Most visible endotoxin effects are seen on platelets from subprimate animals which have immune adherence receptor sites. The initial part of this reaction involves the lipid region of the molecule and is followed by assembly of alternative complement pathway components on the polysaccharide portion of the molecule. This latter interaction causes platelet lysis. The presence of antibody to the O-polysaccharide region of the toxin alters the rabbit platelet response, so that aggregation occurs more rapidly and reversibly.

Primate platelets lack immune adherence receptor sites found in other animals. These platelets do interact with endotoxin and, through this process, can affect the inflammatory process.

Host factors generated in response to endotoxin can significantly affect platelet function and survival. For example, responses of endothelial cells, leukocytes and macrophages during endotoxemia are responsible for increased levels of factors such as thromboxane, prostacyclin, hydrogen peroxide and other substances which can enhance or reduce platelet aggregation.

There is considerable evidence that microvascular injury is significant to the outcome of endotoxemia. However, it is questionable that, particularly in primates, direct endotoxin–platelet interactions are essential to this process.

REFERENCES

Ausprunk DH, Das J (1978) Endotoxin-induced changes in human platelet membranes: morphologic evidence. *Blood 51*, 487-495.
Baris C, Guest MM, Frazer ME (1980) Direct effects of endotoxin on the microcirculation. *Adv. Shock Res. 4*, 153-160.
Berry LJ (1972) Factors influencing the function of the reticuloendothelial system in infections. *J. Reticuloendothel. Soc. 11*, 450-458.
Berthrong M, Cluff LE (1953) Studies of the effect of bacterial endotoxins on rabbit leukocytes. I. Effects of intravenous injection of the substances with and without induction of the local Shwartzman reaction. *J. Exp. Med. 98*, 331-347.
Blumenthal KM, Rourke FJ, Wilder MS (1980) Platelet activation by cultured mouse

peritoneal macrophages. *J. Reticuloendothel. Soc. 27*, 247-257.

Boyden SV (1966) Natural antibodies and the immune response. *Adv. Immunol. 5*, 1-28.

Braude AI (1964) Absorption, distribution and elimination of endotoxins and their derivatives. In: Landy M, Braun W (Eds), *Bacterial Endotoxins*, pp. 98-109. Rutgers University Press, New Jersey.

Brown DL, Lachmann PJ (1973) The behaviour of complement and platelets in lethal endotoxin shock in rabbits. *Int. Arch. Allergy 45*, 193-205.

Bult H, Herman AG (1979) Thromboxane A_2 biosynthesis during endotoxin-induced aggregation of platelets from normal and sensitized guinea pigs. *Agents Actions 9*, 560-565.

Butler KD, Maguire ED, Smith JR, Turnbull AA, Wattis RB, White AM (1982) Prolongation of rat bleeding time caused by oral doses of a thromboxane synthetase inhibitor which have little effect on platelet aggregation. *Thromb. Haemostasis 47*, 46-49.

Canoso RT, Rodvien R, Scoon K, Levine PH (1974) Hydrogen peroxide and platelet function. *Blood 43*, 645-656.

Carey FJ, Braude AI, Zalesky M (1958) Studies with radioactive endotoxin. III. The effect of tolerance on the distribution of radioactivity after intra-venous injection of *Escherichia coli* endotoxin labeled with Cr^{51}. *J. Clin. Invest. 37*, 441-457.

Casey LC, Fletcher JR, Zmudha MI, Ramwell PW (1982) Prevention of endotoxin-induced pulmonary hypertension in primates by the use of a selective thromboxane synthetase inhibitor, OKY 1581. *J. Pharmacol. Exp. Ther. 222*, 441-446.

Clawson CC (1974) Modification of neutrophil function by platelets. In: Baldini MG, Ebbe S (Eds), *Platelets: Production, Function, Transfusion, and Storage*, pp 287-296. Grune and Stratton, New York.

Clawson CC, Rao GHR, White JG (1975) Platelet interaction with bacteria. IV. Stimulation of the release reaction. *Am. J. Pathol. 81*, 411-420.

Cluff LE (1953) Studies of the effect of bacterial endotoxin on rabbit leukocytes. II. Development of acquired resistance. *J. Exp. Med. 98*, 349-362.

Cook JA, Wise WC, Knapp DR, Halushka PV (1981a) Sensitization of essential fatty acid-deficient rats to endotoxin by arachidonate pretreatment: role of thromboxane A_2. *Circ. Shock 8*, 69-76

Cook JA, Wise WC, Halushka PV (1981b) Thromboxane A_2 and prostacyclin production by lipopolysaccharide-stimulated peritoneal macrophages. *J. Reticuloendothel. Soc. 30*, 445-450.

Das J, Schwartz AA, Folkman J (1973) Clearance of endotoxin by platelets: role in increasing the accuracy of the limulus gelation test and in combating experimental endotoxemia. *Surgery 74*, 235-240.

Des Prez RM (1964) Effects of bacterial endotoxin on rabbit platelets. III. Comparison of platelet injury induced by thrombin and by endotoxin. *J. Exp. Med. 120*, 305-313.

Des Prez RM (1967) The effects of bacterial endotoxin on rabbit platelets. V. Heat labile plasma factor requirements of endotoxin-induced platelet injury. *J. Immunol. 99*, 966-973.

Des Prez RM, Bryant RE (1966) Effects of bacterial endotoxin on rabbit platelets. IV. The divalent ion requirements of endotoxin-induced and immunologically induced platelet injury. *J. Exp. Med. 124*, 971-982.

Des Prez RM, Horowitz HI, Hook EW (1961) Effects of bacterial endotoxin on rabbit platelets. I. Platelet aggregation and release of platelet factors in vitro. *J. Exp. Med. 114*, 857-874.

Donald KJ (1972) The role of platelets in the clearance of colloidal carbon from blood in

Rest RF, Cooney MH, Spitznagel JK (1978) Bactericidal activity of specific and azurophic granules from human neutrophils: studies with outer-membrane mutants of *Salmonella typhimurium* LT-2. *Infect. Immun. 19*, 131-137.

Ruscetti FW, Chervenick P (1975) Release of colony-stimulating activity from thymus-derived lymphocytes. *J. Clin. Invest. 55*, 520-527.

Salari SH, Devoe IW, Powell WS (1982) Inhibition of leukotriene B4 synthesis in human polymorphonuclear leukocytes after exposure to meningococcal lipopolysaccharide. *Biochem. Biophys. Res. Commun. 104*, 1517-1524.

Segal AW, Levi AJ (1973) The mechanism of the entry of dye into neutrophils in the nitroblue tetrazolium (NBT) test. *Clin. Sci. Mol. Med. 45*, 817-826.

Segal AW, Levi AJ (1975) Cell damage and dye reduction in the quantitative nitroblue tetrazolium (NBT) test. *Clin. Exp. Immunol. 19*, 309-318.

Shaefer WM, Martin LE, Yakrus M, Spitznagel JK (1983) Fractionation of bactericidal activity from human granulocytes under conditions of serine protease inhibition. In: Neidhardt FC (Ed), *Abstracts of the 83rd Annual Meeting of the American Society for Microbiology*, No B161, p. 50. American Society of Microbiology, Washington DC.

Shands JW Jr, Peavy DL, Gormus BJ, McGraw J (1974) *In vitro* and *in vivo* effects of endotoxin on mouse peritoneal cells. *Infect. Immun. 9*, 106-112.

Showell HJ, Freer RJ, Zigmond SH, Schiffmann E, Aswanikumar S, Corcoran B, Becker EL (1976) The structive-activity relations of synthetic peptides as chemotactic factors and inducers of lysosomal enzyme secretion for neutrophils. *J. Exp. Med. 143*, 1154-1169.

Snyderman R, Gewurz H, Mergenhagen SE (1968) Interactions of the complement system with endotoxic lipopolysaccharide: generation of a factor chemotactic for polymorphonuclear leukocytes. *J. Exp. Med. 128*, 259-275.

Snyderman R, Shin HS, Philips JK, Gewurz H, Mergenhagen SE (1969) A neutrophilic chemotactic factor derived from C5 upon interaction of guinea pig serum with endotoxin. *J. Immunol. 103*, 413-422.

Spagnuolo PJ, Ellner JJ, Hassid A, Dunn MJ (1980) Thromboxane A_2 mediates augmented polymorphonuclear leukocyte adhesiveness. *J. Clin. Invest. 66*, 406-414.

Springer GF, Adye JC (1975) Endotoxin-binding substances from human leukocytes and platelets. *Infect. Immun. 12*, 978-986.

Staber FG, Habild WF, Morrison DC, Tarcsay L (1981) Hemopoietic effects in mice of a lipid A-associated protein. *Exp. Hematol. 9*, 264-273.

Stendahl O, Edebo L (1972) Phagocytosis of mutants of *Salmonella typhimurium* by rabbit polymorphonuclear cells. *Acta Pathol. Microbiol. Scand. Sect. B 80*, 481-488.

Strauss BS, Stetson Jr CA (1960) Studies on the effect of certain macromolecular substances on the respiratory activity of the leucocytes on peripheral blood. *J. Exp. Med. 112*, 653-669.

Sveen K (1977) Rabbit polymorphonuclear leukocyte migration *in vitro* in response to lipopolysaccharides from bacteroides, fusobacterium and veillonella. *Acta Pathol. Microbiol. Scand. Sect. B 85*, 374-380.

Sveen K (1979) Chemotaxis or migration inhibition of rabbit peritoneal polymorphonuclear leukocytes caused by chemotractants at various concentrations. *Acta Pathol. Microbiol. Scand. Sect B 87*, 285-290.

Symon DNK, McKay IC, Wilkinson PC (1972) Plasma-dependent chemotaxis of macrophages towards *Mycobacterium tuberculosis* and other organisms. *Immunology 22*, 267-276.

Tagesson C, Stendahl O (1973) Influence of the cell surface lipopolysaccharide structure of *Salmonella typhimurium* on resistance to intracellular bactericidal systems. *Acta Pathol.*

Microbiol. Scand. Sect. B 81, 473-480.

Tenney SR, Rafter GW (1968) Leukocyte adenosine triphosphatases and the effect of endotoxin on their activity. *Arch. Biochem. Biophys. 126*, 53-58.

Territo MC, Golde DW (1976) Granulocyte function in experimental human endotoxemia. *Blood 47*, 539-544,

Thomas L, Good RA (1952) Studies on the generalized Shwartzman reaction. *J. Exp. Med. 96*, 605-623.

Toh K, Yamamoto N, Suzuki T, Kikuchi K (1979) Effect of concanavalin A and lipopolysaccharide on the cytotoxicity of macrophages and neutrophils. *J. Reticuloendothel. Soc. 26*, 273-282.

Tsan MF, Newman B, Chusid MJ, Wolff SM, McIntyre PA (1976) Surface sulphydryl groups and hexose monophosphate pathway activity in resting human polymorphonuclear leukocytes. *Br. J. Haematol. 33*, 205-211.

Walker RI, Sheil JM (1976) Contribution of granulocytopenia to endotoxin sensitivity of mice irradiated or undergoing graft-versus-host reaction. *Exp. Hematol. 4*, 329-333.

Walker RI, French JE, Walden DA, MacVittie TJ, Parker GA, Sobocinski PZ, Applebaum FR (1980) Protection of dogs from lethal consequences of endotoxemia with plasma or leukocyte transfusions. *Adv. Shock Res. 4*, 89-101.

Ward PA, Becker EL (1970) Biochemical demonstration of the activatable esterase of the rabbit neutrophil involved in the chemotactic response. *J. Immunol. 105*, 1057-1067.

Wehinger H, Hofacker M (1976) Latex phagocytosis by polymorphonuclear leukocytes. *In vitro* and *in vivo* studies with a simple screening test. *Eur. J. Pediatr. 123*, 125-132.

Weiss L (1970) Transmural cellular passage in vascular sinuses of rat bone marrow. *Blood 36*, 189-208.

Welch WD, Martin WH, Stevens P, Young LS (1979) Relative opsonic and protective activities of antibodies against K1, O and lipid A antigens of *Escherichia coli. Scand. J. Infect. Dis. 11*, 291-301.

Wilkinson PC (1980) Leukocyte locomotion and chemotaxis: effects of bacteria and viruses. *Rev. Inf. Dis. 2*, 293-318.

Wilson ME, Munkenbeck P, Morrison DC (1981) Influence of bacterial endotoxins on neutrophilic leukocytes: lack of a correlation between *in vivo* and *in vitro* responses. In: Majde JA, Person RJ (Eds), *Pathophysiological Effects of Endotoxins at the Cellular Level*, pp 157-172. Alan R. Liss, New York.

Wilson ME, Jones DP, Munkenbeck P, Morrison DC (1982) Serum-dependent and -independent effects of bacterial lipopolysaccharides on human neutrophil oxidative capacity *in vitro. J. Reticuloendothel. Soc. 31*, 43-57.

Woods MW, Whitby JL, Burk D (1961) Symposium on bacterial endotoxins. III. Metabolic effects of endotoxins on mammalian cells. *Bacteriol. Rev. 25*, 447-456.

Wozniak JT, Orchard TL, Mackie PH, Mistry DK, Stuart J (1978) Lysosomal enzyme cytochemistry of blood neutrophils. *J. Clin. Pathol. 31*, 648-653.

Yamada O, Moldow CF, Sacks T, Craddock PR, Boogaerts MA, Jacob HS (1981) Deleterious effects of endotoxin on cultured endothelial cells: an *in vitro* model of vascular injury. *Inflammation 5*, 115-126.

Yunis AA, Harrington WJ (1960) Metabolic effects of bacterial pyrogens upon human leukocytes. I. Studies on leukocyte respiration, glycolysis, and nucleic acid synthesis. *J. Lab. Clin. Med. 55*, 584-592.

Handbook of Endotoxin, Vol. 3: Cellular Biology of Endotoxin
L.J. Berry, editor
© Elsevier Science Publishers B.V., 1985

CHAPTER 13

The local Shwartzman reaction: endotoxin-mediated inflammatory and thrombo-hemorrhagic lesions*

HENRY Z. MOVAT AND CLEMENT E. BURROWES

1. HISTORICAL INTRODUCTION

Thromboembolism and acute inflammation are two distinct, yet closely related phenomena. The close relationship becomes obvious in a reaction known as the 'local Shwartzman phenomenon' (Shwartzman 1937). This reaction is characterized by thrombosis and hemorrhage in the dermis induced by an intracutaneous, followed next day by an intravenous injection of endotoxin. The local phenomenon was described independently by Shwartzman (1928a, 1928b, 1928c, 1937) and by Hanger (1928, 1930).

Unfortunately the term 'generalized Shwartzman phenomenon' brought some confusion. In this reaction two intravenous injections of endotoxin are given 24 hours after each other. For this process the term 'Sanarelli-Shwartzman phenomenon' has also been used. Furthermore, there are some unprecise quotations in the literature. In a review Sanarelli (1939) refers to one of his earlier publications (Sanarelli 1924) in which he described experimental cholera and stressed 'septicémie vibrionienne' (vibrionic septicemia) and varying degrees of immunity to the organism. Even in the cases in which the rabbits had what he described as a shock-like state, his attention was directed to the gastrointestinal tract, particularly the intestines. He stressed many characteristic clinical symptoms of human cholera, 'mais en général pauvre de faits anatomo-pathologiques' and no disseminated intravascular coagulation. With a culture filtrate of *Escherichia coli*, given intravenously, Sanarelli also observed in the intestines a mortal hemorrhagic congestion. It should be stressed that Sanarelli had no cause for indignation in his 1939 review, since Shwartzman never described a generalized phenomenon. He states 'De ce fait, un assez grand nombre d'auteurs ont tendance à attribuer, soit partiellement, soit entièrement à G. Shwartzman, la découverte d'un phénomène que nous avons signalé et nettement décrit douze années avant cet auteur.' In the Preface to Shwartzman's monograph (Shwartzman 1937), Jules Bordet refers to Sanarelli's

*   The published and unpublished experiments cited in this chapter were supported by grant T-1-15 of the Ontario Heart Foundation.

260

findings in the intestines and states that in view of Shwartzman's data those of Sanarelli are easier to interpret.

The confusion arose from the publications of Apitz (1933, 1934, 1935). In a paper published in English, Apitz (1935) writes about a 'generalized Shwartzman phenomenon', yet, as stated above, Shwartzman had observed and described a cutaneous or local phenomenon. Apitz had earlier described in German both the local (1933) and the generalized (1934) reactions without referring to the latter as a 'Shwartzman phenomenon'. However, in describing the generalized reaction due to two intravenous injections of *E. coli* filtrate, he correctly writes about 'hämorrhagische Diathese und glomerulär-vasculäre Nierenerkrankung' and about 'Blutungsreaktionen von Shwartzman-Typ', (1934), characteristics of disseminated intravascular coagulation.

To elicit a classical local Shwartzman reaction it is essential to inject endotoxin first intradermally (preparative injection), followed the next day (18–24 hours) by an intravenous challenging or provocative injection. Shwartzman (1928a) described the process as follows:

A phenomenon of local reactivity to *B. typhosus* culture filtrate is described. The work was done on rabbits. The reactivity was induced by skin injections of the filtrate 24 hours prior to the intravenous injection of the same filtrate. The local reactions appeared at the site of previous skin injections; they were fully developed in 4 to 5 hours after the intravenous injection; they were extremely severe and microscopically showed very pronounced necrosis of tissue with rupture of blood vessels and extensive local hemorrhage. The reactions in different rabbits varied in size from 1 × 1 cm to 4 × 4 cm.

For the reproduction of the phenomenon there was necessary an interval of a few hours between the skin and intravenous injections. An incubation period of 2 hours was insufficient. An interval of 24 hours was invariably sufficient. The ability to react disappeared in 48 hours after the preliminary skin injection.

Repeated injections of the filtrate into the same area of the skin with an interval of 24 hours between the injections did not result in reactions similar to the above described hemorrhagic necrosis. The second skin injection was followed by reddening, some swelling and migration of polymorphonuclear neutrophil leucocytes which showed no signs of necrobiosis. There was no rupture of blood vessels. Skin injections followed after a suitable interval by intravenous injections were necessary for the reproduction of severe local response.

In several experiments the preparatory skin injection of *B. typhosus* culture filtrate was substituted by local injections of turpentine in various doses and filtrates of culture of various strains of streptococci. No local response followed the intravenous injection of *B. typhosus* culture filtrate into animals prepared in this manner.

Shwartzman (1937) maintained that the intensity of the reaction after the intravenous challenge was not influenced by the intensity of the intradermal reaction. In the writers' experience the intensity of the thrombohemorrhagic reaction after intravenous endotoxin was more marked when the preparative injection induced an intense inflammatory reaction in the skin.

The local Shwartzman reaction serves as a model for the generalized Shwartzman

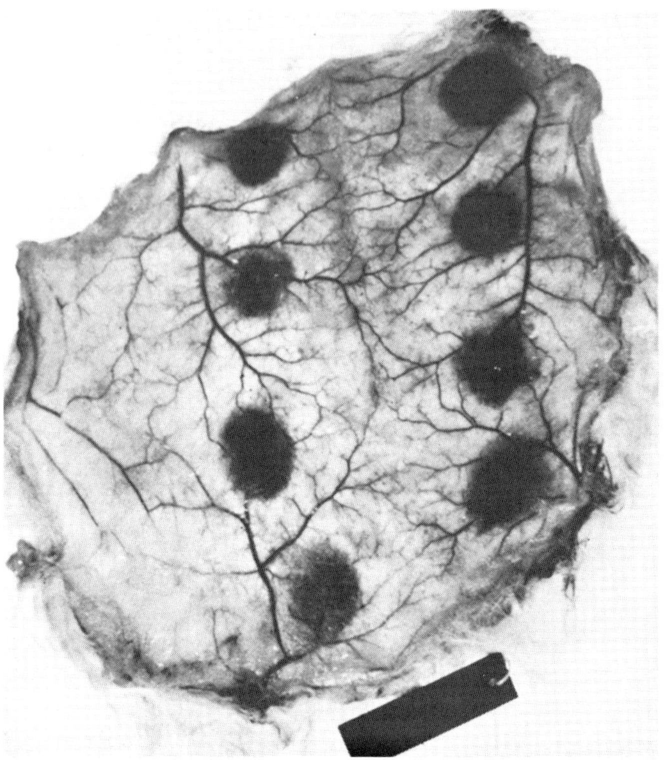

Fig. 2 *Macroscopic appearance of local Shwartzman reactions. Reproduced from Shwartzman (1937), with permission.*

as described below. The occlusion affects venules and small veins exclusively. Small arteries and arterioles are not affected. Some of the vessels are occluded mainly by leukocytes, consisting mostly of PMN and some monocytes (Fig. 6). Other vessels are occluded by mixed thrombi in which platelets and fibrin predominate (Figs 7 and 8), and in yet other vessels there are numerous erythrocytes. The red blood cells are found also extracellularly, often in very large numbers.

As indicated above, leukocytes accumulate in the dermis after the preparative injection. The majority of these are PMN. In the early lesions these appear normal, but in older lesions regressive changes are encountered, consisting of a loss in the normal distribution of lysosomes within the cytoplasm, loss of the limiting membrane of the PMN and nuclear pyknosis – that is, signs of cell necrosis. In some of the PMN, material is demonstrable in vacuoles with the ultrastructural characteristics of endotoxin (Fig. 9). Macrophages tend to phagocytose these degenerated PMN, as described before (Movat et al 1980). They often show evidence of phagocytosed endotoxin (Fig. 10).

Local Shwartzman reactions have been elicited in the liver and kidney. In the

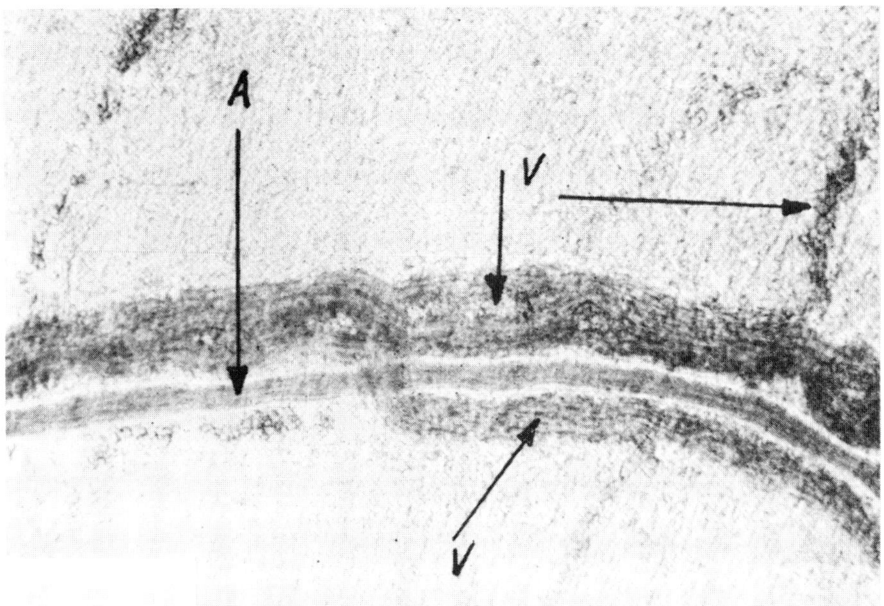

Fig. 3 *Part of a rabbit ear chamber showing the appearance of the microcirculation in vivo 18–20 hours after local application of endotoxin. A small artery (A), two small veins (V) and a venule (along the right edge of the illustration) are illustrated. Reproduced from Ebert and Koch-Weser (1958), with permission.*

latter the kidneys were perfused free of blood, and a glucose-sodium chloride solution containing endotoxin was perfused through the left kidney, the endotoxin being omitted in the control right kidney (Raiji et al 1977). The venous effluent was allowed to drain. Hence endotoxin came in contact only with the vasculature of one kidney. Twenty-four hours later 250 µg of endotoxin was administered intravenously, and the rabbits developed fibrin thrombi in the left (experimental) kidneys. Attempts to repeat these experiments were unsuccessful (N.A. Seniuk and H.Z. Movat, unpublished results). However, we were able to induce hepatic necrosis in the generalized Shwartzman reaction with two spaced intravenous injections of endotoxin (10 µg/kg) given 24 hours apart (N.A. Seniuk and H.Z. Movat, unpublished results). These lesions resembled those elicited exclusively in the liver, by giving the preparative injection (5 µg) into the bile duct and the provocative injection (50 µg) intravenously (Shiga and Mori 1980). The lesions were characterized by Kupffer cell necrosis after the preparative injection, and marked PMN infiltration and hepatic necrosis after the challenging dose of endotoxin. Fibrin was present in many sinusoids.

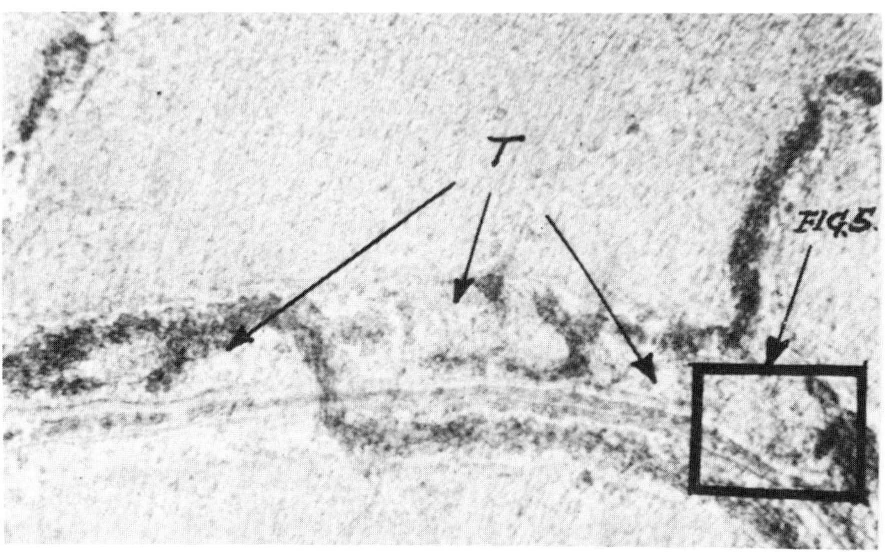

Fig. 4   *Thrombosis of the small veins (T) shown in Fig. 3, 90 minutes after an intravenous challenging injection of endotoxin. The column of erythrocytes in the small artery is narrower than in Fig. 3. From the observation in vivo the authors concluded that the thrombi consisted mainly of platelets and leukocytes. Reproduced from Ebert and Koch-Weser (1958), with permission.*

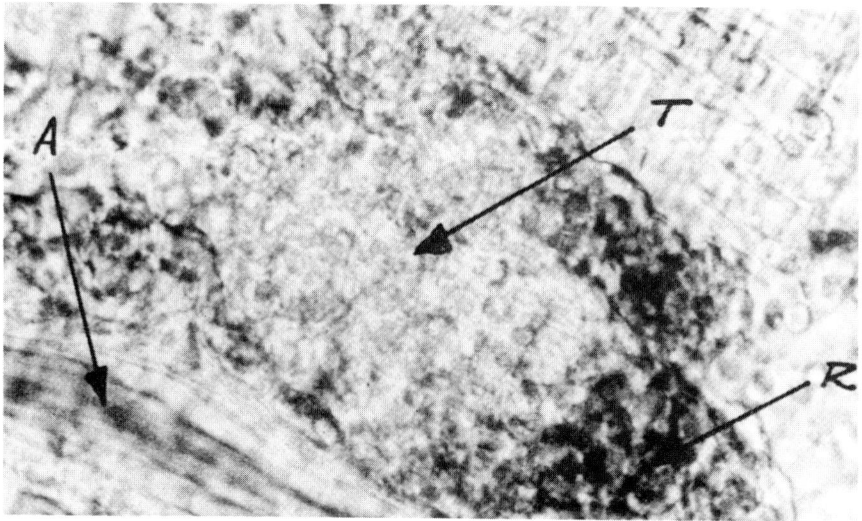

Fig. 5 *Higher magnification of the area in the rectangle in the lower right hand corner of Fig. 4. The thrombus (T) is composed of platelets and leukocytes, with red blood cells (R) at one edge of the thrombus. The small artery (A) is seen adjacent to the thrombosed small vein. Reproduced from Ebert and Koch-Weser (1958), with permission.*

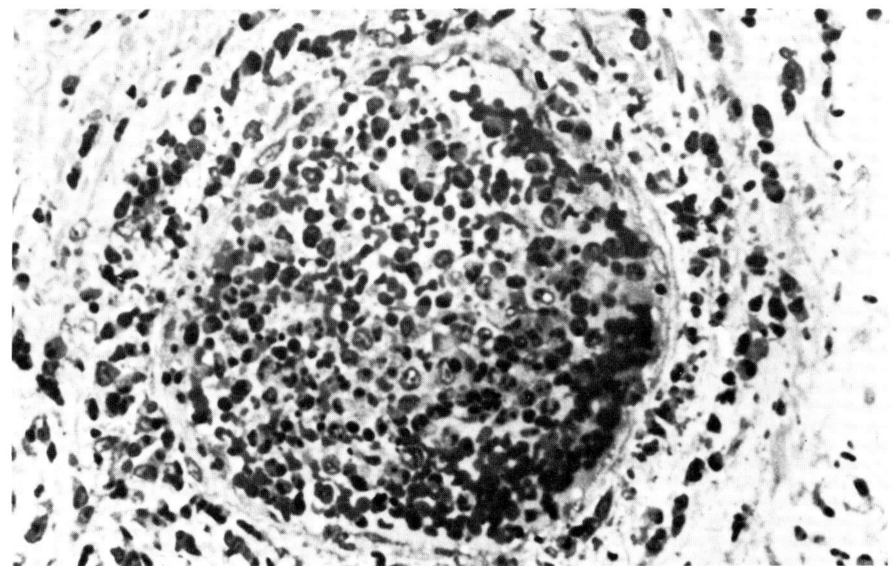

Fig. 6 *Four hours old local Shwartzman reaction showing a thrombotic occlusion of a small vein. The thrombus consists primarily of leukocytes. Such lesions were seen as early as 10 minutes after intravenous challenge.*

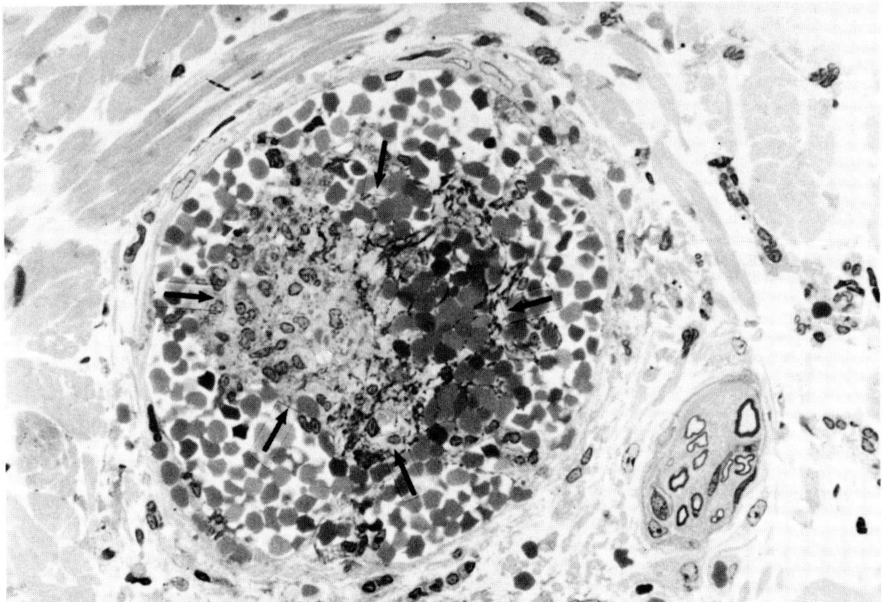

Fig. 7 *A lesion similar to Fig. 6, but the partial occlusion is caused by leukocytes, platelets and fibrin. The thrombus (or embolus) seems to 'float' in the lumen, rather than being attached to the vessel wall (arrows).*

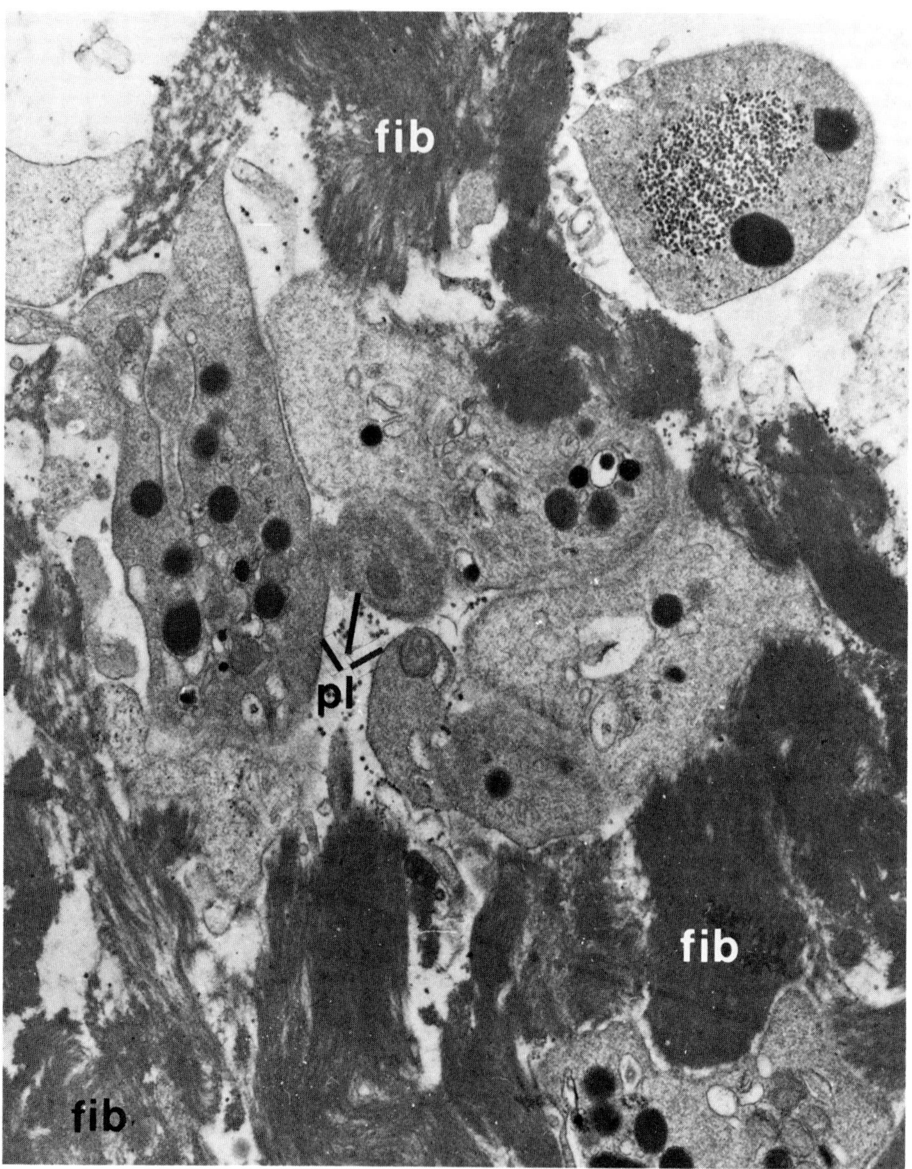

Fig. 8   *Electron micrograph of a thrombus, similar in composition to the one in Fig. 7, showing the platelets (pl) and fibrin (fib) in detail. Reproduced from Jeynes et al (1980), with permission.*

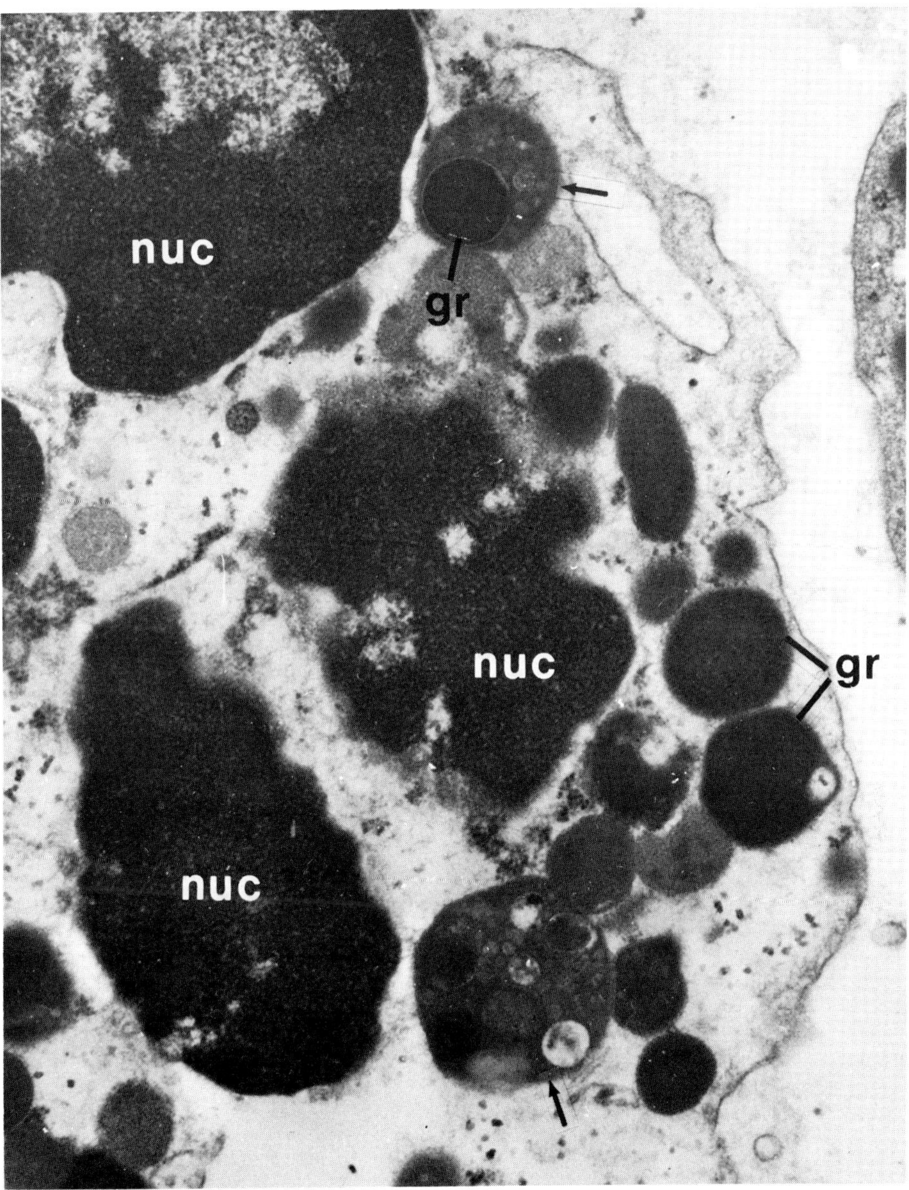

Fig. 9   *Part of a PMN with pycnotic nuclei (nuc), coalescent lysosomes or granules (gr) and ingested endotoxin-like material (arrows). In the phago-lysosome in the left upper corner there is in addition to the endotoxin-like material a granule (gr) preserved by the fixative just before lysis.*

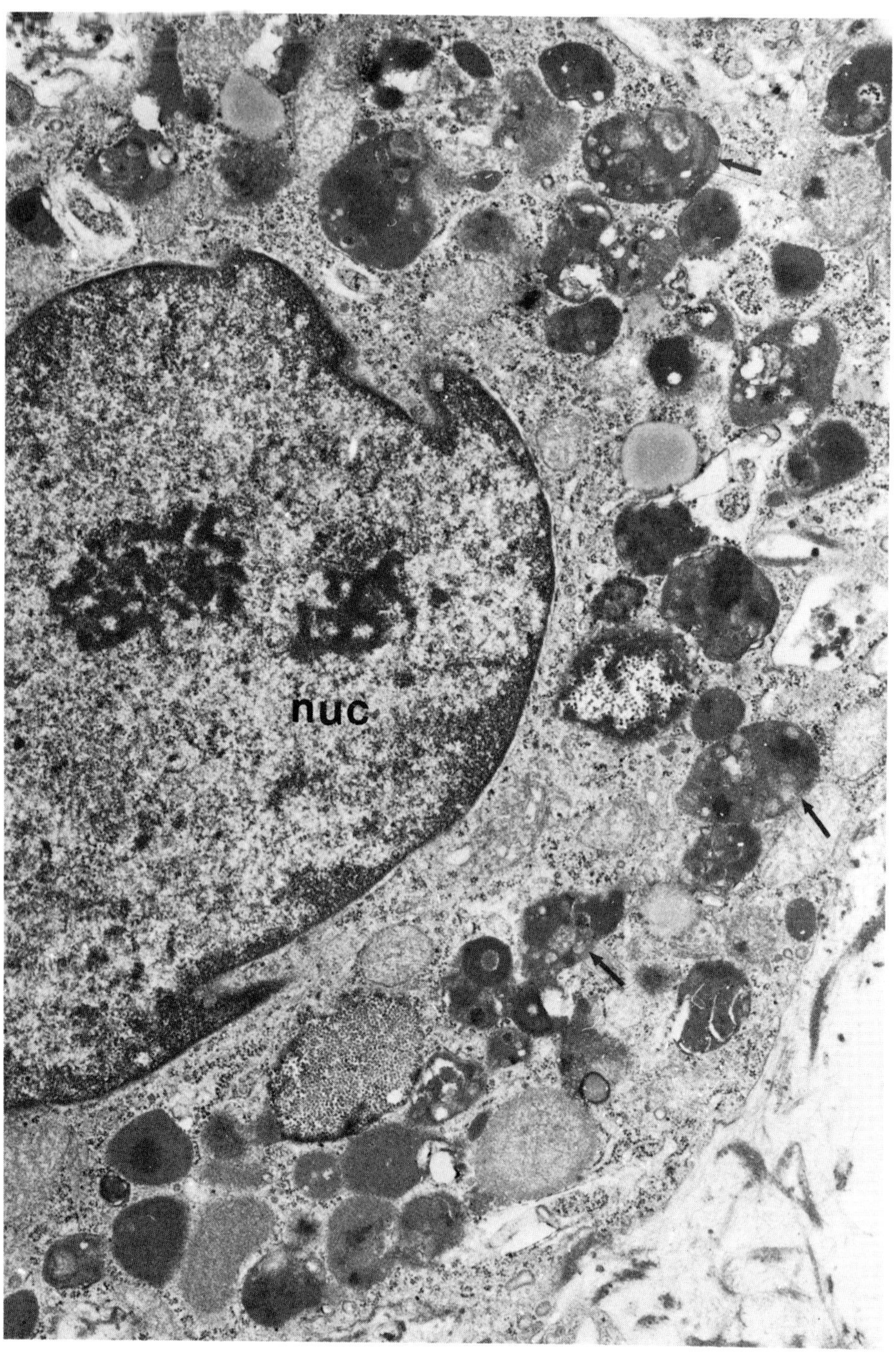

nuc

4. QUANTITATION AND KINETICS

Recently developed methods enable the quantitation of the various processes occurring in the skin. These methods make use of radiolabeling, and the administration of the radiolabeled cells and proteins as a pulse enables determination of rates – for example, rate of platelet or leukocyte accumulation. The following can be labeled and quantitated: increase in vascular permeability with the aid of ^{125}I- or ^{131}I-serum albumin (Udaka et al 1970); blood flow with ^{57}Co- or ^{85}Sr-microspheres administered intraarterially (Hay et al 1975); accumulation of leukocytes, either PMN (Issekutz and Movat 1980) or monocytes (Issekutz et al 1981) labeled with ^{51}Cr in vitro and administered intravenously; reversible or irreversible accumulation of ^{111}In-labeled platelets (Jeynes et al 1980); hemorrhage with ^{59}Fe-erythrocytes (Kopaniak et al 1980a); formation of fibrin clots with simultaneously administered ^{125}I-fibrinogen and ^{131}I-serum albumin (Movat et al 1980). Up to 3 isotopes can be given to one rabbit (Kopaniak et al 1980b).

It is of interest that following the intradermal injection there is a transient accumulation of leukocytes and platelets (Fig. 11). By the time the intravenous challenging endotoxin is administered (16–24 hours after the preparative injection), leukocytes are still demonstrable histologically (Fig. 1), but from the functional quantitative studies it is now known that the majority of these cells accumulated during the first 4 hours and, as shown in Figure 12, no cells accumulated in the lesions beyond 7–8 hours. In this respect the changes after an intradermal injection of endotoxin represent an acute inflammatory reaction, whose kinetics resemble those after an intradermal injection of formalin-killed *E. coli* (Kopaniak et al 1980b). Interestingly, when the endotoxin in the *E. coli* is inactivated by treating the organism with 0.05 N HCl, their phlogistic potency is considerably reduced (M.M. Kopaniak and H.Z. Movat, unpublished results). Not shown in Figure 12 are the transient hyperemia, increase in vascular permeability and hemorrhage, which also occur after an intradermal injection of endotoxin. However, even the largest doses of endotoxin that can be administered locally without causing endotoxin shock do not induce as intense as inflammatory reaction as the microorganisms (unpublished). An interesting observation was the transient accumulation of platelets, observed also in the inflammatory reaction of a reversed passive Arthus reaction (Crawford et al 1982). Issekutz (1983) elaborated on this in a recent publication. Because in relation to PMN the number of infiltrating monocytes is small and because of the small number of monocytes in the population of cells labeled with ^{51}Cr, Figure 12 probably does not fully illustrate the events. These cells, unlike the PMN, continue to accumulate at an inflammatory site for at least 24 hours (Issekutz et al 1981). This process may be important in relation to pathogenesis, as described below. The inflammation in response to endotoxin was reviewed in detail by Urbaschek and Urbaschek (1979).

Fig. 10 *Part of a macrophage, containing phagocytosed material, some with the ultrastructural characteristics of endotoxin (arrows).*

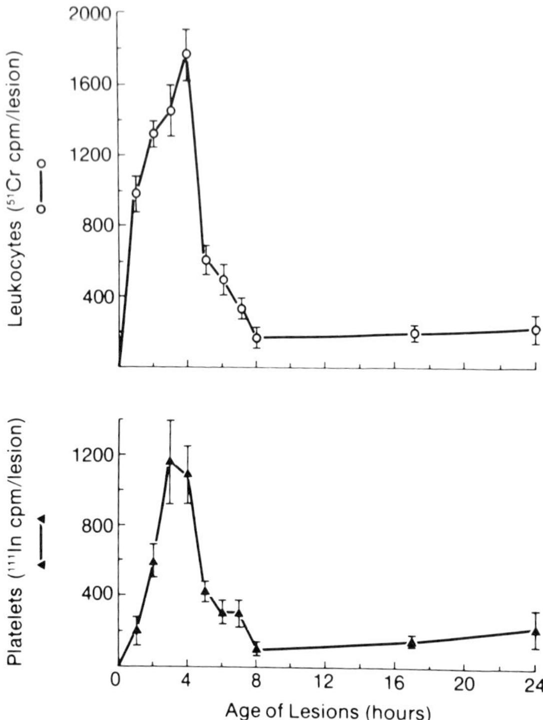

Fig. 11   *Kinetics of transient ^{51}Cr-leukocyte and ^{111}In-platelet accumulation after a preparative injection of E. coli O55:B5 endotoxin. Reproduced from Movat et al (1980), with permission.*

After the challenging intravenous injection of endotoxin, leukocytes and platelets accumulate in the lumen of dermal vessels where the preparative endotoxin has been deposited earlier. The ^{111}In- and ^{51}Cr-radioactivities increase parallel with the amount of endotoxin deposited into the dermis (Fig. 12). When accumulation (polymerization) of ^{125}I-fibrinogen was measured together with ^{131}I-

Fig. 12   *Dose response of ^{111}In-platelet and ^{51}Cr-leukocyte aggregation with increasing amounts of S. marcescens endotoxin. The accumulation of radiolabeled platelets and leukocytes occurred over a period of 1 hour after the intravenous challenging dose of endotoxin. Reproduced from Movat (1983), with permission.*

Fig. 13   *Kinetics of accumulation of ^{125}I-fibrinogen (fibrin thrombi) and ^{59}Fe-erythrocytes (hemorrhage) in a thrombo-hemorrhagic local Shwartzman reaction. Simultaneously injected ^{131}I-albumin remained constant in lesions of all ages. In lesions of all ages the radiolabeled proteins and erythrocytes circulated for the last 60 minutes before sacrifice. Reproduced from Movat et al (1980), with permission.*

5000
4000
3000
2000
1000
Platelets (^{111}In cpm/lesion)
Leukocytes (^{51}Cr cpm/lesion)
5000
4000
3000
2000
1000
0 2 4 6 8 10 12 14 16 18 20
µg Endotoxin/site

Fig. 12

4000
3000
2000
1000
^{125}I-fibrinogen ☐–☐ ^{131}I-albumin
(cpm/lesion)
2000
1000
^{59}Fe-erythrocytes
(cpm/lesion)
0 1 2 3 4 5 6
Age of Lesion (hours post provocative)

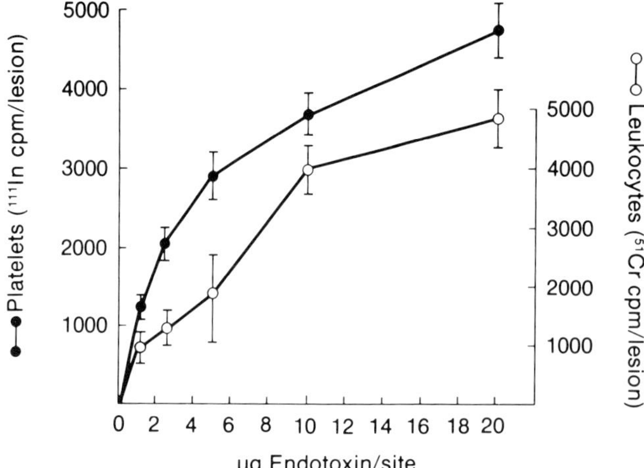

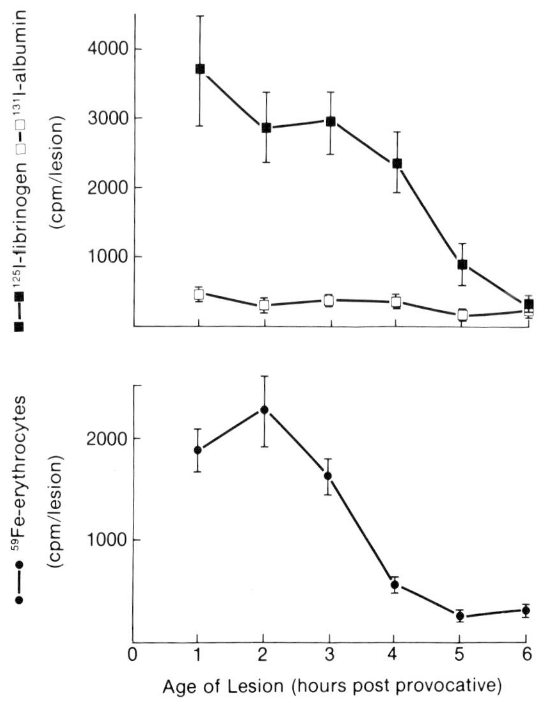

Fig. 13

serum albumin and ^{59}Fe-labeled erythrocytes, fibrinogen (fibrin) and hemorrhage peaked within 1–2 hours after the intravenous challenge and ^{131}I-albumin did not accumulate (Fig. 13). The latter is in keeping with earlier observations, using Evans blue as a vascular permeability tracer, that the vessels are not hyperpermeable in the thrombohemorrhagic lesion of the dermal Shwartzman reaction. Neither thrombus formation nor hemorrhage progress beyond 6 hours after challenge.

Prostaglandins of the E class enhance blood flow in the microcirculation and through this process increase vascular permeability induced by a number of mediators (Johnston et al 1976; Kopaniak et al 1978; Williams and Peck 1977) and infiltration of PMN with subsequent hemorrhage induced by zymosan-activated plasma (Issekutz and Movat 1982). Advantage was taken of this property of prostaglandin E_1: when it was injected into prepared sites at the time of challenge, the accumulation of both fibrinogen and of erythrocytes was increased compared with that in similar lesions injected with saline (Fig. 14). This means that increase in blood flow through the microcirculation (probably due to both vasodilatation and opening up of more vessels) enabled the formation of more fibrin-containing microthrombi, and that whatever caused the severe vascular injury, which resulted in hemorrhage, could injure more numerous vessels and hence induce more intense hemorrhage.

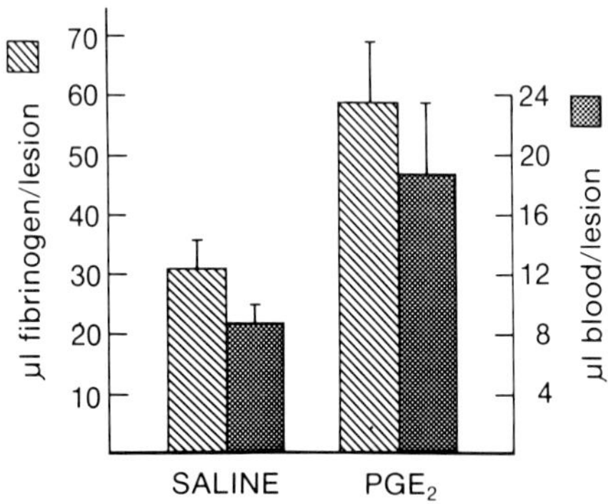

Fig. 14 *Elicitation of local Shwartzman reactions in animals which received at the time of the challenging intravenous injection into the prepared skin sites 1.0 µg of PGE₂ or saline. Both ^{125}I-fibrinogen and ^{59}Fe-erythrocytes were administered intravenously as a pulse. A blood sample was taken prior to sacrifice, and ^{125}I and ^{59}Fe were measured in the blood; on the basis of these data the amount of ^{125}I-fibrinogen and ^{59}Fe-erythrocytes could be expressed as microliters per site, accumulating in 1 hour. Reproduced from Movat (1983), with permission.*

5. POTENTIAL MECHANISMS WHICH OPERATE IN THE LOCAL SHWARTZMAN REACTION

5.1. The intrinsic clotting pathway

There are three lines of evidence that the contact phase of blood coagulation and the intrinsic clotting pathway are activated in the Shwartzman reaction: (a) activation of Factor XII (Hageman factor) in vitro, (b) suppression of the reaction by inhibitors of contact factors (Factor XII, high molecular weight kininogen) and (c) activation of the Factor XII-dependent kallikrein-kinin system.

Indirect evidence that Factor XII can be activated in vitro by endotoxin was gathered by several investigators (see Morrison and Ulevitch 1978; Müller-Berghaus and Lasch 1975). However, it was not until Morrison and Cochrane (1974) showed that highly purified Factor XII can be activated by highly purified LPS that a direct activation of the clotting factor was possible.

In vivo, it was shown that Factor XII may be activated when endotoxin is present in the circulation because it was associated with increased plasma levels of bradykinin (Kimball et al 1972; Nies et al 1968) and decreased levels of kininogen (Herman et al 1974). Further indirect evidence was provided for the activation of Factor XII by Erdös and Miva (1968). They demonstrated that glass beads released bradykinin from pre-shock rabbit plasma, but could not release the peptide from post-endotoxin plasma. With a decrease in blood pressure, a decrease of plasma prekallikrein and a decrease in 'immediate' antiplasmin activity in endotoxin shock of dogs was reported (Aasen et al 1978a, 1978b). Ellagic acid activates Factor XII (Ratnoff and Crum 1964), and it has been shown that the infusion of this substance also causes a DIC-like syndrome (McKay et al 1969, 1971). Infusion of ellagic acid together with norepinephrine (α-adrenergic stimulation) induced a DIC-like picture. Lysozyme was reported to inhibit the DIC-like picture induced by liquoid (an activator of Factor XII), but Müller-Berghaus and Schneberger (1971) could not inhibit the endotoxin-induced DIC with lysozyme. Although using a different model (a single injection of endotoxin in pregnant rats), Latour et al (1979) reported an inhibition by lysozyme of the systemic Shwartzman reaction of the rat. These investigators showed furthermore that in their model there was a reduction of Factor XII and prekallikrein (the latter, when activated by Factor XII cleaves bradykinin from kininogen). These authors also demonstrated that bromelain, which depletes rats of high molecular weight(HMW)-kininogen (Katori et al 1978), also inhibited their model of Shwartzman reaction. HMW-kininogen acts as a cofactor in the contact activation of Factor XII. In its absence the rate of activation of Factor XII by contact is markedly reduced (Chan et al 1976). In earlier experiments of Latour et al (1974), hepatic vein thrombosis was induced with endotoxin in the pregnant rat. These rats showed a decrease in Factor XII 2 hours after injection of endotoxin. The hepatic vein thrombosis was induced also by a slow injection of ellagic acid. Crucial experiments were performed by Shen and coworkers (1973). In these experiments the intrinsic clotting pathway was interrupted with the aid of an anti-human Factor VIII antibody which cross-reacted with rabbit Factor VIII. Shen et al noted that in the rabbits depleted of Factor VIII the parameters measured (^{125}I-fibrinogen survival, fibrinogen levels, blood clotting Factors VIII, VII and V, platelets, leukocytes, hematrocrit and histological deposition of fibrin in the kidneys) did not differ from those of control rabbits with intact Factor VIII. These studies strongly suggest that DIC can develop in the absence of the intrinsic clotting pathway. These experiments, however, would have to be repeated with the classical preparative, followed by the provocative endotoxin. Finally, the experiments by Skjørten and Evensen (1973) lend support to an important role of the intrinsic clotting pathway and Factor XII; they failed to induce DIC in fowl (which lack Factor XII) with endotoxin, although they did demonstrate accumulation of fibrin in pulmonary vessels when they injected tissue thromboplastin.

Indirect evidence for the role of the intrinsic clotting pathway in DIC comes also from clinical studies. Mason and Colman (1971) reported decreased levels of Factor XII and prekallikrein in patients with endotoxemia, who had DIC. Out of a group of patients that underwent transurethral resection or cytoscopy, those with gram-negative bacteremia and positive endotoxin assays had decreased levels of plasma prekallikrein (Robinson et al 1975).

5.2. The extrinsic clotting pathway

5.2.1. *Role of leukocytes*

Although some investigators used the noncommittal term 'leukocytes', the majority referred to PMN – that is, granulocytes or neutrophils. In this description a distinction will be made between leukocytes, meaning all white blood cells (WBC); PMN, referring to neutrophil leukocytes; mononuclear cells, a mixture of monocytes and lymphocytes (as is obtained by Percoll or Hypaque-Ficol centrifugation); and monocytes. The latter refers to these cells in the blood or after isolation from the blood (monocytes) and after emigration during an inflammatory reaction into the tissues (monocytes/macrophages). The term macrophages will be used in the next subheading in conjunction with certain organs – for example, pulmonary alveolar macrophages or other more or less 'fixed' locations. Monocytes can be separated first from PMN (e.g. Percoll) and thereafter from lymphocytes, through adherence (lymphocytes do not adhere). Finally, a negative effect can be achieved by making mononuclears free of monocytes, by removing these with a magnet, after they have phagocytosed iron filings, or by using blood from animals in which the PMN and monocytes are temporarily marginating or sequestering (Issekutz et al 1981).

The dramatic effect of endotoxin on leukocytes can be demonstrated by examining the blood of rabbits before and at various times after an intravenous dose of endotoxin (Fig. 15). When ^{51}Cr-labeled leukocytes are injected into rabbits and

276

their lungs examined after the intravenous dose of endotoxin, a temporary accumulation of the labeled leukocytes is demonstrable in the lungs (Fig. 16) (Cybulsky and Movat 1982, 1983).

Experiments with increasing doses of endotoxin pertaining to WBC show that there is a correlation between the number of WBC at the injection sites (^{51}Cr) and the amount of ^{125}I-fibrinogen which accumulates at these sites after an intravenous provocative injection of endotoxin (Movat et al 1980).

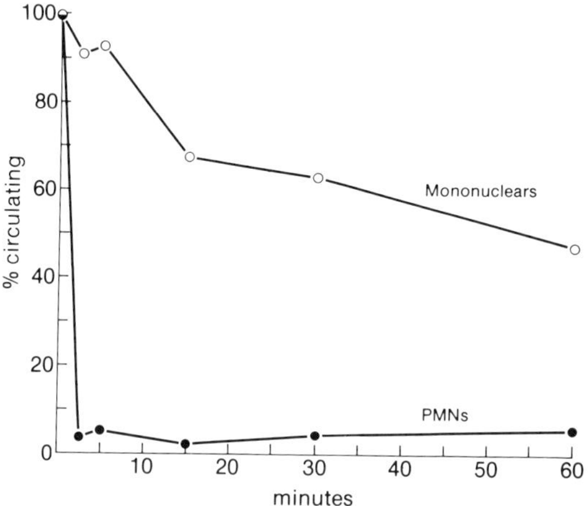

Fig. 15   *Almost instantaneous effect of intravenously administered endotoxin on blood PMN and a more gradual effect on mononuclear cells. Reproduced from Cybulsky and Movat (1982), with permission.*

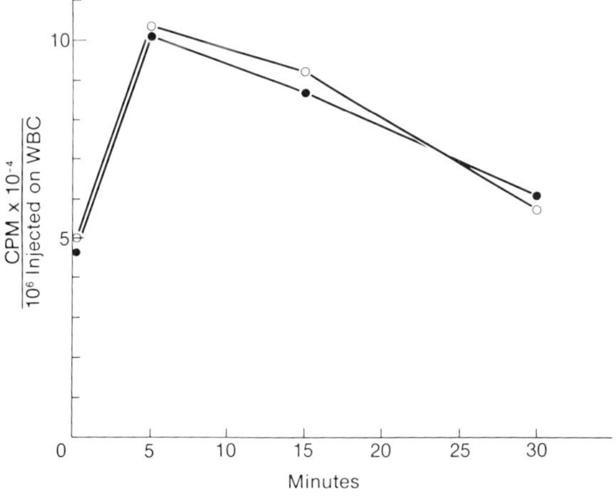

Fig. 16   *Accumulation of ^{51}Cr-leukocytes in the lung of rabbits after an intravenous injection of endotoxin. Reproduced from Cybulsky and Movat (1982), with permission.*

The depletion experiments date back about 35 years (Becker 1948; Stetson and Good 1951; Thomas and Good 1952). Stetson and Good (1951) referred to PMN and demonstrated a correlation between the degree of neutropenia and that of the suppression of the local Shwartzman reaction (Fig. 17). In all these studies nitrogen mustard was used to render the rabbits leukopenic. Margaretten and McKay (1969a) used a goat anti-rabbit leukocyte (peritoneal exudate) serum. They were able to induce a short-lived leukopenia and neutropenia, lasting only 4 hours. With this procedure the generalized Shwartzman reaction could not be prevented, whether the leukopenic state was induced before the preparative or before the provocative injection of endotoxin. Margaretten and McKay went on to demonstrate that the antileukocyte serum could be substituted with antigen (bovine serum albumin) in animals with circulating anti-bovine serum albumin. What they, in fact, had done was to elicit a so-called aggregate anaphylaxis, in which leukocytes and platelets aggregate as a sequence to the formation of antigen-antibody precipi-

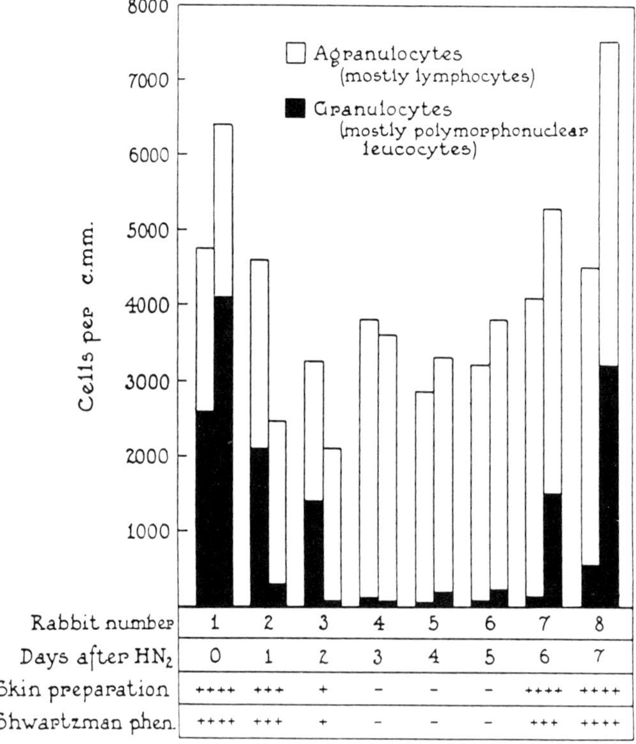

Rabbit number	1	2	3	4	5	6	7	8
Days after HN$_2$	0	1	2	3	4	5	6	7
Skin preparation	++++	+++	+	–	–	–	++++	++++
Shwartzman phen.	++++	+++	+	–	–	–	+++	++++

Fig. 17 *Correlation between leukocytes (mostly lymphocytes and PMN; monocytes not differentiated) in the blood after nitrogen mustard injection and the intensity of lesions after preparative (inflammatory) and challenging (thrombo-hemorrhagic) injections of endotoxin. Reproduced from Stetson and Good (1951), with permission.*

tates (Movat et al 1968). It had been demonstrated before that antigen-antibody precipitates can substitute for the provocative intravenous injection in the local Shwartzman reaction (Taichman et al 1966). More recently, Bohn and Müller-Berghaus (1976) depleted rabbits of their leukocytes and platelets with busulfan, but they referred to their experimental animals as 'granulocytopenic', not 'leukopenic'. Simultaneously, they produced a peritoneal exudate with glycogen and harvested the cells, which were 90–95% granulocytes. After induction of leuko- and thrombocytopenia, these investigators could not elicit the generalized Shwartzman reaction when the animals were transfused with platelets or a lysate of leukocytes, but could induce the reaction (measured as microclots in the glomeruli), when they transfused the rabbits with peritoneal exudate leukocytes (1×10^9 cells/kg) before the second or provocative injection of endotoxin. Bohn and Müller-Berghaus did not consider the possibility that the peritoneal exudate contains also extravasated monocytes. Even 4-hour-old peritoneal exudates contain at least 3–5% monocytes/macrophages and the number of these cells increases when the injections of glycogen-saline are administered several times ('priming'), as done by Bohn and Müller-Berghaus, to obtain a higher yield. If the exudate is collected after 3–5 days, as many as 60% of the exudate cells may be macrophages (Vadas et al 1981). The more macrophages there are in the exudate, the higher is its procoagulant activity (Robinson et al 1978). Despite the large number of exudate leukocytes injected by Bohn and Müller-Berghaus, a rise in circulating leukocytes could not be achieved. Unlike ^{51}Cr-labeled blood leukocytes (Issekutz and Movat 1980), only a very small percentage of labeled peritoneal exudate leukocytes recirculate (Jones et al 1977; Perper et al 1974). Furthermore, as already indicated, a large number of these ^{51}Cr-labeled leukocytes accumulate in the lungs, due to margination or sequestration, followed by demargination. Some of this (at least the sequestration) occurs also in the liver and spleen (Cybulsky and Movat 1982).

For each PMN or monocyte in the circulating pool there is one cell in the marginal pool, but in addition to these cells, there are about 15 band form PMN and about 5 monocytes in the bone marrow pool (Van Furth and Willemze 1979). These can be mobilized after an injection of endotoxin and are demonstrable in the organs in which they marginate (Cybulski and Movat 1983). It is very important to take this process into consideration when describing the changes which occur in conjunction with a local Shwartzman reaction – that is, after the intravenous challenge (Stetson and Good 1951). A rapid shift (approx. 2-fold) of PMN from the marginal to the circulating pool could not be demonstrated morphologically but only with ^{51}Cr-leukocytes. Hence the appearance of numerous PMN 90 minutes (or more) after an intravenous dose of endotoxin was attributed to a shift from the marrow pool to the lungs (Cybulsky and Movat 1982) and this was subsequently confirmed by autoradiography with ^{3}H-labeled cells (Cybulsky and Movat 1983). This observation is in keeping with earlier ones by Athens et al (1961).

The mechanisms which induce the neutropenia have not been elucidated. Although it has been demonstrated that complement-derived factors produce both in vitro granulocyte adherence and in vivo margination (Fehr and Jacob 1977), in vivo studies in which hemolytic complement and C3 were depleted could not confirm that an intact complement system is essential for the induction of neutropenia by endotoxin (Ulevitch et al 1975, 1978).

Some studies using radiolabeled endotoxin also indicate that the neutrophil interacts with endotoxin. Brunning and coworkers (1964) demonstrated by radioautography that injected ^{3}H-endotoxin localized in granulocytes in blood smears and that these cells could also be demonstrated in vessels of the lung, liver, spleen and kidney. Schrader et al (1964) examined the local Shwartzman reaction with the aid of similarly labeled endotoxin. They found radiolabel in the locally accumulating PMN when the ^{3}H-endotoxin was injected as the preparative intradermal injection and in the same cells in the lung when the ^{3}H-endotoxin was given as the challenging intravenous injection.

The in vivo studies described here which implicate PMN in the pathogenesis of DIC were supplemented by some in vitro findings. Peritoneal exudate leukocytes, which are predominantly neutrophils, contain some coagulant activity, but Niemetz and coworkers demonstrated a marked increase in the coagulant activity of rabbit peritoneal leukocytes from endotoxin-treated rabbits (Niemetz and Fani 1971; Niemetz 1972), which was enhanced in the presence of platelet membranes (Niemetz and Marcus 1974). A similar activity could be demonstrated also in leukemic cells (Garg and Niemetz 1973). The coagulant activity accelerated the clotting of normal plasma and activated Factor X in the presence of Factor VII and calcium, indicating a tissue factor or thromboplastin-like activity.

There is one more aspect that has to be considered in conjunction with PMN. In the local Shwartzman reaction there is evidence of phagocytosis of endotoxin both by neutrophils and macrophages. Of these two cells, the neutrophils show ultrastructurally regressive changes and cell death (Movat et al 1980). Perhaps the neutrophils interact with endotoxin, but, unlike macrophages, they cannot 'cope' with the LPS molecule (Filkins 1971). There is, in fact, some circumstantial evidence that upon interaction with gram-negative microorganisms, the neutrophil may induce bacteriolysis, but in this process releases endotoxin. De Voe (1976) incubated in vitro mouse PMN-leukocytes and ^{14}C-labeled meningococci. The cells were found to excrete radioactive products. Ultrastructurally, there was exocytosis of the membranous vesicles by the polymorphs. When leukocyte-egested material, obtained in a similar manner from rabbits, was injected into animals, followed by meningococcal endotoxin, changes resembling DIC induced in the generalized Shwartzman reaction and endotoxemia were observed (De Voe and Gilka 1976; De Voe et al 1977).

Although probably PMN, monocytes and tissue macrophages play some role in the pathogenesis of the local Shwartzman reaction, the PMN seemingly play no role in the thrombotic process, but only in the hemorrhage. With a single exception (Lerner et al 1977), all in vitro studies carried out in the past few years implicate the monocyte rather than the neutrophil in the production of a tissue factor or thromboplastin-like material after stimulation with endotoxin (Hiller et al 1977; Prydz and Allison 1978; Rickles et al 1977; Rivers et al 1975; Van Ginkel et al

1977). Tissue factor activity has been reported after incubation of a blood monocyte-rich preparation with immune complexes and lectins (Prydz and Allison, 1978; Prydz et al 1979; Rothberger et al 1977). Procoagulant activity was demonstrable also in peritoneal exudate cells stimulated with endotoxin and it increased in parallel with the aging of the exudate, which went hand in hand with disappearance of neutrophils and an increase in monocytes/macrophages (Robinson et al 1978). In all these studies, the mononuclear cells were separated from granulocytes by differential centrifugation. In various media, the monocytes cannot be separated from lymphocytes by centrifugation. It was possible to show that while the mononuclear cells are capable of generating tissue factor, granulocytes or lymphocytes devoid of monocytes cannot (Van Ginkel et al 1977). Furthermore, culture monocytes become adherent in tissue and lymphocytes do not, and the addition of endotoxin to the adherent cells brings about generation of tissue factor (Fig. 18). Edwards et al (1979) demonstrated that addition of nonadherent small lymphocytes enhances the generation of tissue factor by the monocytes when stimulated with endotoxin. Edwards et al showed furthermore that mere adherence, as reported by Van Ginkel et al (1977), did not induce the formation of the tissue factor. These investigators failed, however, to demonstrate a correlation between tissue factor production in vitro and delayed-type hypersensitivity in vivo (skin test), nor was there a correlation between generation of tissue factor and in vitro lymphocyte transformation, following addition of mitogens or antigens. Rothberger et al (1977) demonstrated first a stimulation of tissue factor generation by soluble immune complexes formed with IgG and subsequently showed that blood mononuclear cells form tissue factor after allogeneic stimulation of the cells (Rothberger et al 1978).

The studies of Edgington and coworkers demonstrated procoagulant activity in human and murine monocytes. First a cooperation between lymphocytes and monocytes was demonstrated when stimulated with LPS, immune complexes or plasma lipoproteins (Levy and Edgington 1980; Levy et al 1981a; Schwartz and Edgington 1981). A tissue factor or thromboplastin and a prothrombin-cleaving enzyme were demonstrated with human peripheral blood mononuclear cells (Levy et al 1981a; Schwartz et al 1981). The kinetics and metabolic requirements for direct lymphocyte induction of human procoagulant monokines by bacterial LPS was shown by Levy et al (1981b) and Schwartz et al (1982) provided evidence that exposure of murine mononuclear cells to LPS resulted in expression of surface-bound prothrombinase activity which was distinct from Factor Xa.

It is difficult to elicit in rats a local or generalized Shwartzman reaction with 2 spaced injections of endotoxin. Therefore modified reactions were elicited in this species, such as hepatic vein thrombosis in hyperlipemic rats (Latour et al 1974) or the generalized reaction in pregnant rats (Latour et al 1979). In both these reactions Factor XII and the intrinsic clotting pathway, catecholamines and prostaglandins have been implicated in the pathogenesis of thrombosis in the microcirculation. Recently evidence was presented that in the rat, blood leukocytes do not seem to play a role in DIC. According to Semeraro et al (1981), whereas rabbit

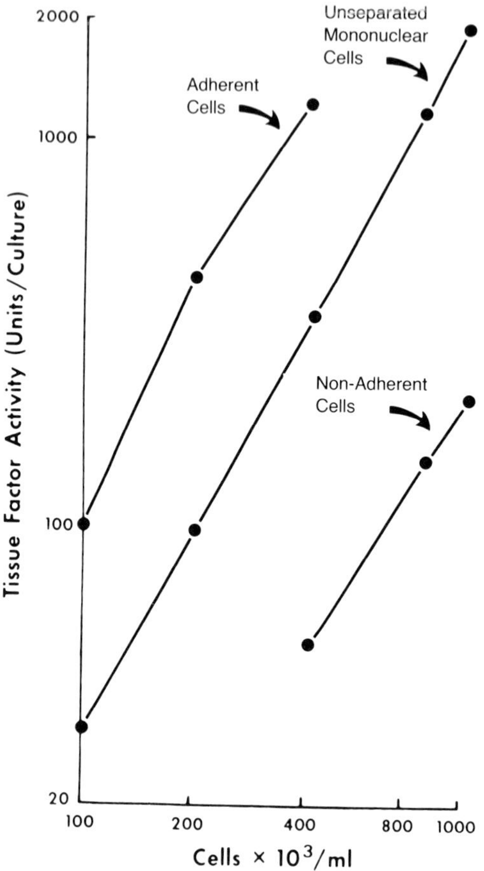

Fig. 18 *Effect of mononuclear cell concentration on tissue-factor generation. Each curve plots the amount of tissue factor generated by cultures of specific mononuclear cell subpopulations as a function of increasing cell concentration. Reproduced from Edwards et al (1979), with permission.*

leukocytes exhibited procoagulant activity when exposed to endotoxin, no such activity could be detected with rat leukocytes in vitro or in vivo. The procoagulant was believed to be a product of monocytes.

5.2.2. Role of fixed macrophages

The fact that the Shwartzman reaction, local or generalized, cannot be elicited in animals rendered leukopenic implies that the formed elements of the blood play an important role in the pathogenesis of these reactions.

Two relatively recent observations, implicating tissue macrophages, appear both suggestive and challenging with respect to formation of a substance or substances which render the animal exposed to endotoxin susceptible to intravascular coagulation.

One of the studies implicating liver macrophages or Kupffer cells was conducted by Mathison and Ulevitch (1979). Their findings have relevance mostly to the challenging or provocative injection of the Shwartzman reaction, since they followed the fate of 250 µg ^{125}I-labeled LPS injected intravenously in rabbits. Rough LPS (*S. minnesota* R595) and a smooth LPS (*E. coli* O111:B4PII) were used. The bulk of the native LPS was cleared with a half-life of 30 minutes. Most of the uptake occurred in the liver, spleen and to a lesser extent in the kidneys and

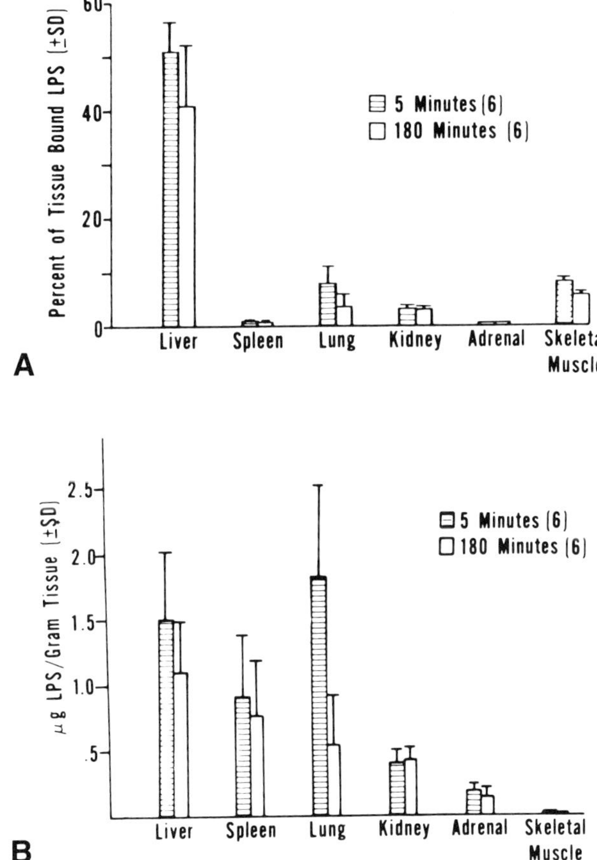

Fig. 19   *A: the distribution of tissue-bound LPS 5 and 180 minutes after intravenous injection of 250 µg of R 595 (rough) LPS in 2.5 kg rabbits. The number of rabbits used is shown in parenthesis. B: as A, but showing the concentration of tissue-bound LPS. Reproduced from Mathison and Ulevitch (1979), with permission.*

adrenals. By electron-autoradiography it could be observed that 5 minutes after injection, endotoxin was concentrated in phagocytic vacuoles of Kupffer cells and splenic macrophages. A few labeled neutrophils were encountered in the liver, spleen and lungs. The LPS remaining in the plasma beyond 30 minutes was gradually converted into low density LPS (buoyant density of less than 1.2 g/cm^2). This LPS disappeared with a half-life of 12 hours. The organ distribution at 180 minutes resembled that at 5 minutes (Fig. 19). The reduction in the buoyant density of LPS required high density lipoprotein (Ulevitch et al 1979). In a later study (Mathison et al 1980), the distribution of ^{125}I-R595 in Rhesus monkeys was found to be similar to that in rabbits, and decomplementation with cobra venom factor (C3 less than 5% of normal) had no effect in either rabbits or monkeys. The significance of these findings remains to be ascertained. In earlier studies, it was reported that blocking the reticuloendothelial system with thorotrast, trypan blue or denatured albumin substituted for the first injection of endotoxin, required for the elicitation of the generalized Shwartzman reaction (Good and Thomas 1952; Lee 1962). All these findings strongly implicate an important role for fixed macrophages in the host's response to endotoxin. More recently, Maier and Ulevitch (1981a, 1981b) reported production by cultured hepatic macrophages of a procoagulant activity after stimulation with endotoxin which was DFP-sensitive, heat-labile (56°C) and capable of activating Factor X directly (Fig. 20). Acid hydrolases were only slightly increased, but there was a marked increase in plasminogen activator, and lactic dehydrogenase. There was evidence of marked Kupffer cell injury by the endotoxin.

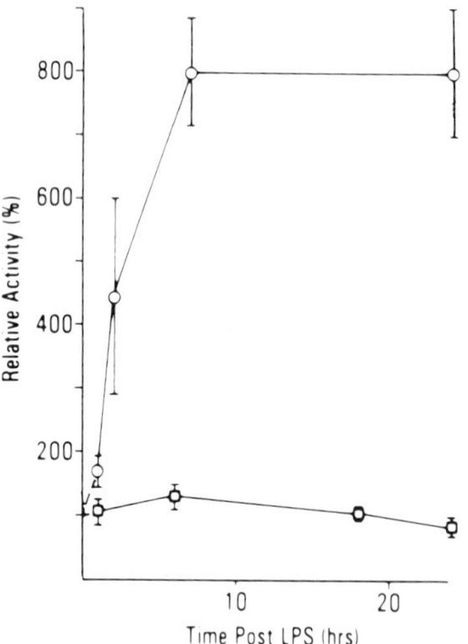

Fig. 20 *Induction of procoagulant factor by LPS in cultured fixed tissue macrophages of the liver (Kupffer cells). The LPS was added to 0–48 hours-old cell cultures. The relative activity is based on initial untreated culture values which were 100%. Reproduced from Maier and Ulevitch (1981a), with permission.*

Procoagulant activity assayed with the two-stage assay was encountered in alveo lar macrophages, which also generated plasminogen activator (Wasi et al 1983). The macrophages were separated by unit gravity sedimentation to a purity of more than 95%. The procoagulant was present in the culture supernatant and in the membrane fraction of the disrupted cells. It was not inactivated by DFP, but lost its activity upon incubation with phospholipase C, it required Factor VII and was therefore considered to represent tissue thromboplastin.

The second observation implicating fixed macrophages pertains to an endotoxin-resistant strain of mice (C3H/HeJ). Studies with these mice suggest a central role of macrophages in the host's response to endotoxin (Rosenstreich and Vogel 1980). First, there is the reversal of the endotoxin-resistant state of the mice by adoptive transfer of bone marrow cells (Michalek et al 1978). When endotoxin-resistant mice are lethally irradiated and adoptively transfused with bone marrow cells from histocompatible, but endotoxin-sensitive mice (C3H/HeN), they become sensitive to a number of biological effects of endotoxin (antibody formation, adjuvancy, production of colony stimulating factor and lethality). The marked insensitivity of C3H/HeJ mice to endotoxin can be overcome by pretreating these mice with viable BCG (*Bacillus Calmette Guérin*) (Vogel et al 1980). Musson et al (1978) compared the uptake of radiolabeled endotoxin-sensitive (C3H/St) mice with the uptake in endotoxin-resistant (C3H/HeJ) mice and found considerably less uptake by various organs in the endotoxin-resistant strain. The most recent observation on the C3H/ HeJ mice (that have also other defects omitted in this discussion) is that their lack of responsiveness to endotoxin can be corrected by an interferon-like molecule (Vogel et al 1982).

What seems pertinent to the local and in particular the generalized Shwartzman reaction is the capacity of endotoxins to interact directly and functionally with host cells and tissues, rather than their ability to localize in such tissues, which seems to be the critical factor in determining host responses to endotoxin. Part the defect in the C3H/HeJ mice has been attributed to a failure of endotoxin to stimulate phagocytic cells to release leukocyte endogenous mediator or LEM (now recognized as endogenous pyrogen or interleukin 1). While an endotoxin-responsive strain of mice responded when stimulated with endotoxin (e.g. increase in number of neutrophils, decrease in plasma iron concentration, formation of acute phase proteins), the endotoxin-resistant C3H/HeJ strain of mice did not respond. These responses are attributable to LEM (interleukin 1). However, both strains responded with these parameters measured when LEM (interleukin 1) was injected (Kampschmidt and Upchurch 1980). (For a discussion of systemic PMN-responses to endotoxin, interleukin 1, activated complement and inflammatory reactions see Cybulsky and Movat 1983).

5.3. Platelets: complement-dependent and -independent mechanisms

It can be stated with relative certainty that platelets probably do not play a primary role in the local Shwartzman reaction. However, they may play a secondary role in aggravating the occlusive changes. This is very prominent morphologically, which may be the reason why several pathologists have written in favor of a role for platelets in the local and the generalized Shwartzman phenomena (e.g. Stetson 1951a; Margaretten and McKay 1969b; Mustard and Packham 1979).

Platelets react with aggregation and release of platelet constituents to a number of stimuli (Mustard and Packham 1979). Only the highlights of platelet–endotoxin interactions will be described here, and the reader is referred to the reviews of Müller-Berghaus and Lasch (1975), Morrison and Ulevitch (1978) and Morrison and Ryan (1979), and also to Chapter 11. Recent data on the possible role of complement in endotoxin–platelet interactions will be described in some detail.

Endotoxins seem to have a profound effect, both in vivo and in vitro, on the platelets of some species. This effect has been demonstrated with the platelets of rabbits, rats, guinea pigs and dogs. Platelets of all these animals have immune adherence receptor sites on their membranes. Little or no effect of endotoxin on platelets of primates has been demonstrated and these do not have immune adherence receptor sites (Morrison and Ulevitch 1978). Complement has been shown to play a critical role in the responses of platelets to endotoxin. When rabbits were depleted of C3 with cobra venom factor, the marked decrease in the number of circulating platelets observed in normal controls after a lethal dose of endotoxin, was abrogated (Brown and Lachmann 1973). Morrison and coworkers (Morrison et al 1978; Morrison and Oades 1979) demonstrated that only those preparations of endotoxin which had the capacity to activate the alternative pathway of complement were able to initiate platelet lysis in vitro. Both granule-associated serotonin and cytoplasmic lactic dehydrogenase were released from the platelets. Isolated lipid A, which activates the classical complement pathway by interacting, in the absence of antibody, with the first component of complement, did not initiate lysis of rabbit platelets. However, lipid A did play a significant role, being instrumental in the attachment of endotoxin to the platelet. This was facilitated by the lipid-associated protein. Having attached itself to the platelet membrane, endotoxin can now, via its polysaccharide portion, provide the appropriate surface for the assembly of the alternative pathway of complement, the terminal components of which can lyse the platelet membrane. More recently, Morrison et al (1980) described that both isolated lipid A and endotoxin from polysaccharide-deficient (rough) mutants had the capacity to elicit a complement-independent secretory (nonlytic) response of platelets. Their experiments suggest that endotoxin is perturbing the platelet membrane in such a manner that it acquires the capacity to bind and/or translocate Ca^{2+}. Furthermore, after that a method for isolating murine platelets had been developed (Oades and Morrison 1980), it was demonstrated that the platelets of the endotoxin-resistant C3H/HeJ mice described above respond (secreted serotonin) to stimulation with endotoxin just like the platelets of normal mice. In another recent study it was shown that both endotoxin- and collagen-aggregated platelets require complement, but whereas collagen-induced aggregation shows 'fusion' of platelets with degranulation and release of serotonin and platelet factor 3, endotoxin induces a loose platelet aggregation without degranulation and release of serotonin and platelet factor 3 (Murphy et al 1980).

286

In the studies of Mathison and Ulevitch (1979) quoted above, a rough and a smooth LPS were used. Lipid A (and rough LPS) activate the classical complement pathway, in the absence of antibody, by activating C1 (Cooper and Morrison 1978). The O antigens of some smooth LPS induce complement activation via the alternative pathway (Mathison and Ulevitch 1983). Furthermore, the O antigen of smooth or semismooth LPS modulates the anticomplementary activity of the lipid A portion of the LPS molecule so that activation of C1 is markedly diminished (Morrison and Kline 1977). Therefore, smooth and semismooth LPS with insufficient O antigen for the activation of the alternative pathway have little anticomplementary activity (Morrison and Ulevitch 1978).

Mathison and Ulevitch (1981) carried out further studies by examining in vivo the interaction between LPS and rabbit platelets and by modulating complement and high density lipoproteins. Between 0.5 and 2 minutes after injection of 250 µg of rough LPS (some radiolabeled with ^{125}I), 35% of the LPS was in the buffy coat of the blood. Radioautographs of blood smears showed 86% of the grains over platelets and 14% over PMN and monocytes, but not lymphocytes or erythrocytes. These studies were extended to demonstrate that binding of LPS to platelets occurred only with the rough anticomplementary (R 595) LPS. Binding of the smooth LPS was noted with the leukocytes, but not with the platelets (Mathison and Ulevitch 1983). Labeled leukocytes or platelets were not observed between 5 and 500 minutes after injection. Within 5 minutes after injection of LPS there was thrombocytopenia and neutropenia. Return of ^{51}Cr-labeled platelets (which had been injected) to normal levels occurred 30 minutes after injection of the LPS. However, ^{125}I-LPS-containing platelets were not observed in the blood samples up to 500 minutes, suggesting that the platelets which had bound LPS were removed from the circulation. Interestingly, none of the interactions between platelets and rough LPS occurred in rabbits treated with cobra venom factor to deplete their C3 and terminal complement components. However, the complement depletion did not affect the binding of LPS to PMN and the development of neutropenia. These observations are in keeping with earlier ones by the same group of investigators on endotoxemia as a model for gram-negative sepsis (Ulevitch et al 1975, 1978; Ulevitch and Cochrane 1978).

Also of interest was the fact that the low density LPS described above, with a half-life of 12 hours, did not bind to any elements of the blood. Furthermore, low density LPS prepared in vitro produced a slowly developing thrombocytopenia, hypotension, fibrin deposition in small vessels and death in all the experimental animals. This picture resembled the course in the C3-depleted rabbits.

One approach in studies with platelets, as with leukocytes, was to deplete the experimental animals of these blood elements. In the studies of Bohn and Müller-Berghaus (1976) cited above, reconstitution of thrombo- and leukopenic rabbits with infused platelets did not restore their capacity to elicit a generalized Shwartzman reaction. Furthermore, in one experiment the reaction could be readily elicited in rabbits rendered thrombocytopenic with antiplatelet antibody (Kramer and Müller-Berghaus 1977).

5.4. Vascular endothelium

It has been known for over a century that vessels lined by normal endothelium are 'nonthrombogenic', and the vessel wall injury, primarily endothelial, is part of the triad of Virchow (Mustard and Packham 1979).

About 50 years ago Apitz studied the local (Apitz 1933) and the generalized (Apitz 1934) Shwartzman reactions and postulated that endotoxin 'altered' or 'stimulated' the endothelium and that in the generalized reaction a 'systemic alteration' (i.e. throughout the vascular system) takes place. Since then, a number of investigators have attempted to demonstrate changes in vascular endothelium after injection of endotoxin. Such 'changes', if they exist as a causative factor, need not necessarily be demonstrable morphologically, as most investigators attempted to show. They could, in fact, represent not the cause, but the effect of the changes taking place in the blood and the microcirculation.

In an electron microscopic study, sequestration and aggregation of platelets in the pulmonary microcirculation was noted 15 minutes, aggregates of leukocytes within 1 hour, and fibrin formation, as evidence of intravascular coagulation, not before 2 hours after the injection of endotoxin (McKay et al 1966). McKay et al interpreted the degenerative changes in the endothelium which they observed as secondary to platelet aggregation and fibrin formation. They interpreted the fibrin deposits as embolic in nature. Hoff et al (1967) observed, by electron microscopy, electron dense particles in endothelial vacuoles, which they interpreted to be endotoxin. McGrath and Stewart (1969) prepared and examined sheets of endothelium of mesenteric vessels, which, after injection of endotoxin, showed 'distorted nuclei, apparent nuclear vacuolization, and missing nuclei'. Gaynor et al (1970) presented not only evidence of ultrastructural changes, but also circulating endothelial cells in the blood. Heparin administration did not prevent development of the lesions, indicating that the endothelial changes may be independent of the abnormalities in the clotting system. Unlike McKay et al (1966), these investigators interpreted the DIC as a thrombotic phenomenon developing in situ at sites of endothelial damage. Gaynor (1973) went on to show that leukopenia, induced by nitrogen mustard, did not prevent the development of the endothelial changes. In search of a more sensitive technique for demonstrating endothelial injury, endothelial turnover was studied and found enhanced after endotoxin (Gaynor 1971; Gerrity et al 1975). Evensen et al (1975), who were interested in the role of complement in endotoxin injury, found no difference in the number of ^{3}H-labeled endothelial cells of the aorta of normal rats and C3-depleted rats, after administration of endotoxin.

One report already quoted above (Raiji et al 1977) strongly implicates an important, although not necessarily primary, role of endothelial cells in the development of fibrin thrombi in the microcirculation. Raiji et al (1977) cannulated the renal artery and vein of the left kidney in rabbits. Through the arterial cannula they perfused the kidney free of blood (0.145 M NaCl containing 0.2 M glucose) and then perfused slowly with 50 µg endotoxin, followed by saline-glucose and restitution of blood flow. The endotoxin was therefore in contact with renal endothelium and since it was collected via the renal vein, it was in contact presumably with the endothelium only for a few minutes. Next day, the animals received 250 µg endotoxin intravenously and were killed 4–8 hours after the intravenous injection. All animals whose left kidney was perfused for about 7 minutes with endotoxin developed renal cortical necrosis with fibrin thrombi in their glomeruli and focal accumulation of PMN-leukocytes in the cortex. No such changes were seen in the right kidneys and only 1 out of 6 rabbits perfused with saline instead of endotoxin showed a few focal necroses and fibrin thrombi. Raiji and coworkers stated that their results were 'compatible with the hypothesis that endotoxin mediates its action at least in part through a local mechanism in which fixation of endotoxin to the endothelial surface may play an important role'. As already stated, in the writer's laboratory the studies of Raiji et al could not be repeated (N.A. Seniuk and H.Z. Movat, unpublished results).

In vitro studies are difficult to interpret. For the barrier function of endothelium and for in vivo and in vitro adhesion of leukocytes, endothelial cells grown in culture are a relatively good model (Beesley et al 1978, 1979; Gimbrone and Buchanan 1983; Harlan et al 1981; Hoover et al 1980; Lackie and De Bono 1977; Yamada et al 1981). However, the in vitro endothelial culture model is unsatisfactory in attempts to separate between a primary injury of the blood or of the endothelial lining of blood vessels.

In a recent study, Reidy and Schwartz (1983) injected slowly a massive dose (3 mg/kg) of endotoxin intravenously and followed this by perfusion-fixation at 0.5–48 hours after the injection of the endotoxin. The aorta was prepared for scanning and transmission electron microscopy. Examination of the endothelium revealed no endothelial denudation and no exposure of subendothelium. However, 18 hours after the intravenous endotoxin, several endothelial cells were raised up into the lumen, above the monolayer; by 24 hours there was some loss of cell membrane interdigitation, and by 48 hours occasional cells were lying above the monolayer, to which several small cells (identified as monocytes by nonspecific esterase staining) were adhering. Autoradiographs revealed a marked increase in endothelial cell replication between 24 and 96 hours. Weibel-Palade bodies were not increased in number compared to saline-injected controls. Because there was no change in the number of cells per unit area, the authors concluded that the endotoxin produced desquamation with simultaneous replacement by cell movement and cell replication. These results were stated to imply 'that small injuries can be healed without exposure of the endothelium and that an increase in endothelial cell replication does not necessarily imply that endothelial cell denudation had occurred'.

6. PATHOGENETIC CONSIDERATIONS: FACTS AND HYPOTHESES

6.1. Initiating events

Data collected over the last 3 decades show quite conclusively that endotoxin injected into the dermis elicits an acute inflammatory reaction. This has been substantiated by classic histologic, vital microscopic and ultrastructural observations and by quantitating various parameters of the inflammatory reaction.

While the local inflammatory reaction is undeniable, we do not know precisely what else happens when 10–50 µg of endotoxin is injected into one or more sites of dermal connective tissue. It is known from other models of acute inflammation, that the 'inflammatory process' spreads very rapidly along afferent lymph vessels to the lymph nodes draining that particular region and beyond these lymph nodes along efferent lymphatic vessels (Johnston et al 1979). Hence endotoxin injected locally can gain access via the thoracic duct to blood, and with the blood it can be carried in the formed elements or in plasma to all parts of the body. In this process changes occur that are probably similar, if not identical, to those in a generalized Shwartzman reaction when the endotoxin is administered intravenously. It is true that these changes are mild compared to a bolus of 250 µg given intravenously, but the changes described after such a bolus (Mathison and Ulevitch 1983) give us a good guideline as to what to expect in a 'mini-endotoxemia.' Without repeating what has been described in Section 5 ('potential mechanisms'), a few important events should be stressed which are pertinent to what follows after a provoking or challenging intravenous injection in a local Shwartzman reaction. There is probably binding of endotoxin to platelets, PMN and monocytes, and there is stimulation of the bone marrow (direct or via interleukin 1) to deliver more of these cells, which can in fact be detected at least in one of the marginating organs, the lungs. Native or plasma-bound low density endotoxin can bind to and stimulate fixed macrophages in the liver, spleen, lung and perhaps other organs. Among the many changes occurring in these activated macrophages, as well as in blood monocytes, one should be stressed: that of generation of procoagulant activity. In addition to tissue factor or thromboplastin, this comprises probably also other, less well-defined coagulation factors. One more item to stress is that monocytes emigrate together with PMN to the site of the inflammatory reaction taking place in the dermis. Endotoxin has been demonstrated at the inflammatory site in both PMN and monocytes, by autoradiography and ultrastructurally. Hence generation of procoagulant activity could take place locally at the injected site. Among the effects of endotoxin is the generation of interleukin 1 and one of its effects is the enhanced formation of acute phase proteins, which includes fibrinogen (Kampschmidt 1982).

Despite all this knowledge and all these assumptions one would wish to know more about the prepared site. For example, if a preparative injection or injections would be deposited in the dermis of a circumscribed area, and the endotoxin, which normally resorbs, would not be permitted to disseminate but would be

290

drained *via* cannulated draining lymph vessels, would a 'preparation' take place? In the accumulation of monocytes/macrophages and PMN that have 'seen' endotoxin in the local inflammatory lesion, sufficient to induce 'preparation' – that is, a state in which after the intravenous challenging injection microemboli which form in the circulating blood tend to impact in the terminal vascular bed? Why is it that these microemboli form in the postcapillary venules and small veins – that is, sites in the microcirculation where exudation of plasma, transient aggregation of platelets and emigration of leukocytes takes place after the deposition of endotoxin? $C5a_{des\ Arg}$ causes an inflammatory reaction with vascular injury (increased vascular permeability and hemorrhage) which is leukocyte-dependent (Issekutz et al 1980a, 1980b). While endotoxin may induce inflammation directly, probably most of the chemotactic attraction of PMN and monocytes occurs through the in situ formation of $C5a_{des\ Arg}$, after injection of endotoxin. In this connection it would be important to know whether an anticomplementary rough endotoxin is more potent than a smooth or semismooth preparation in preparing for a local Shwartzman reaction. Could $C5a_{des\ Arg}$ chemoattract and 'prepare' a dermal site for a subsequent intravenous challenge with endotoxin?

6.2. Formation of microthrombi in the microcirculation and the possible causes of hemorrhage

The various processes described in Section 5 trigger enhanced coagulation, and this is what happens when an intravenous challenging dose of endotoxin is administered 18–24 hours after the preparative injection. As described above (Sections 3 and 4), the microthrombi in the local Shwartzman reaction consist of formed elements of the blood and fibrin aggregated directly or indirectly by endotoxin. Fibrin is a very prominent feature, and some investigators, particularly those interested in the generalized Shwartzman reaction, consider fibrin the quintessence of the reaction (Müller-Berghaus and Lasch 1975). In the glomeruli one encounters almost exclusively fibrin.

Several factors have been considered instrumental in the formation of fibrin clots, which form in the blood and embolize into several organs and to the prepared site in the local Shwartzman reaction. The local Shwartzman reaction can be inhibited by massive doses of heparin (Table 1). The fibrin in the microthrombi is closely associated with platelets and leukocytes and both have been thought to influence the formation of fibrin. However, conversion of fibrinogen or soluble fibrin to polymerized fibrin (fibrin precipitates) occurs both in platelet-depleted (Müller-Berghaus and Kramer 1976) and leukocyte-depleted rabbits (Müller-Berghaus and Eckhardt 1975). Other factors include vasomotor changes, due to the effect of catecholamines and glucocorticoids. Fibrinolysis also plays a role, albeit a secondary one.

A number of models have been described, particularly in the older literature, in which a generalized Shwartzman-like reaction has been induced by various forms of manipulation, ranging from the injection of acid polymers and snake

Table 1 Effect of massive heparinization on the local Shwartzman reaction

Group	Rabbits	Hemorrhage (6 hr)	Necrosis (24 hr)
Treated	31	0*	0
Control	53	38 (71%)	24 (62%)

*Eight of the 31 treated rabbits showed some petechial hemorrhage at the injection site.

Serratia marcescens endotoxin was used, 50–100 µg given as a preparative intradermal injection and 24 hours later a provocative intravenous injection of 125–200 µg. Five mg of heparin was given every 30–40 minutes, beginning 5–15 minutes before and at least 4 hours after the provocative injection. Summarized from Cluff and Berthrong (1953).

venoms, to activation of Factor XII with elagic acid and injection of thrombin with simultaneous inhibition of fibrinolysis (reviewed by Müller-Berghaus and Lasch 1975).

Vasomotor changes have been implicated in the precipitation of fibrin, and catecholamines, which stimulate α-receptors of the adrenergic system, induce vasomotor changes in the microcirculation (McKay et al 1971). Lambert et al (1969) used another approach to demonstrate the importance of vasomotor reactions. When sympathetic denervation was performed on one ear, the ensuing local Shwartzman reactions were markedly attenuated. With the aid of both histologic and microangiographic examination, it was demonstrated that in the denervated tissue arterial and arteriolar vessels did not react maximally, as they did in the undenervated controls. It was also observed that in the denervated rabbits there was much less hemorrhage. These investigators also showed that denervated ears failed to respond to nonspecific vasodilator stimuli. As described in Section 4 ('Quantitation and Kinetics') locally injected prostaglandins of the E class enhance the blood flow, which can be quantitated (Johnston et al 1976; Kopaniak et al 1978; Williams and Peck 1977). When PGE_1 is injected into the prepared site of a Shwartzman lesion at the time of the intravenous challenge, both the ^{125}I-fibrinogen and the ^{59}Fe-erythrocytes are found increased compared to saline injected controls (Fig. 14).

Corticosteroids also seem to have an effect on fibrin deposition, at least in the glomeruli, and may do so by influencing the effect of catecholamines on blood vessels. Adrenalectomy performed between the preparative and the provocative injection of a generalized Shwartzman reaction prevented localization of fibrin clots in the glomeruli, an effect which could be reversed by corticosteroids (Latour and McKay 1970). This effect of hydrocortisone was dose-dependent and could be partially blocked by the α-adrenergic blocking agent phenoxybenzamine. Norepinephrine infused to the adrenalectomized animals induced fibrin thrombi in 60% of the adrenalectomized animals and the control adrenalectomized had none.

Finally, fibrinolysis probably plays some role, at least in the formation of fibrin thrombi in glomeruli, but perhaps also in the local Shwartzman reaction. Fibrinolysis is impaired in pregnancy (Margaretten et al 1964; Müller-Berghaus and Schmidt-Ehry 1972). Perfusion of normal rats with thrombin for various lengths of time does not lead to glomerular fibrin deposition, but in pregnant rats with extended perfusion, the fibrin persists in the glomeruli. Microthrombi persist – that is, they are not lysed – when normal rats are given endotoxin or thrombin and the fibrinolytic inhibitor aminocaproic acid (Lee 1962; Margaretten et al 1964).

Because the generalized Shwartzman reaction and prolonged endotoxemia are models for human DIC, and because coagulation and occlusion of small vessels by thrombi dominate the picture in these conditions, most studies conducted clinically, in experimental animals and in vitro, were directed at the formation of these microthrombi. In several of the affected organs in a generalized reaction there is hemorrhage. However, in the local Shwartzman reaction the hemorrhage dominates the picture, as described already by Shwartzman (1928a, 1928b, 1928c, 1937) and in more detail by Stetson (1951a). However, Stetson did not fail to recognize the importance of vascular occlusion, since he felt that tissue necrosis preceded the hemorrhage and that it was tissue death, including that of vessels, which led to the hemorrhage. In the studies in which the rate of formation of fibrin microthrombi and hemorrhage were quantitated (Movat et al 1980), both radiolabeled fibrinogen and erythrocytes were at peak levels in local Shwartzman lesion within 1 hour (Fig. 13). This indicates that the two events occur simultaneously. With radiolabels of higher specifity the events occurring earlier than 1 hour should be detectable. *Intra vitam* observations indicate formation of thrombi and of hemorrhage within minutes after the intravenous challenging injection of endotoxin (Ebert and Koch-Weser 1958).

The mechanism underlying the hemorrhage has not been demonstrated with certainty. However, since the studies of Stetson and Good (1951), evidence has accumulated that PMN may play a critical role. In an earlier section (5.2.) the evidence in favor of monocytes was outlined with respect to the coagulation process, and the pitfalls in ascribing this role to PMN discussed. However, there is little evidence implicating monocytes in the production of acute hemorrhagic lesions. Stetson's (1951b) observations on the Arthus phenomenon and the crucial role of PMN in the elicitation of hemorrhage, without the thrombotic component, has been well documented in subsequent years (Cochrane and Janoff 1974; Wasi and Movat 1979). A lysate of PMN-lysosomes and some proteases (including elastase) isolated from the lysate, induce a marked increase in vascular permeability and hemorrhage which can be quantitated. These lesions develop rapidly and persist for several hours (Wasi and Movat 1981). PMN-elastase can be released during blood coagulation. In serum the level of elastase-related antigen is increased 11.5 fold (288±125 ng/ml) above plasma levels (Plow 1982).

REFERENCES

Aasen AO, Frølich W, Saugstad OD, Amundsen E (1978a) Plasma kallikrein activity and prekallikrein levels during endotoxin shock in dogs. *Eur. Surg. Res. 10*, 50-62.

Aasen AO, Ohlsson K, Larsbraaten M, Amundsen E (1978b) Changes in plasminogen levels, plasmin activity and activity of anti-plasmins during endotoxin shock in dogs. *Eur. Surg. Res. 10*, 63-72.

Apitz K (1933) Über hämorrhagische Hautreaktionen nach örtlicher Umstimmung. *Z. Gesamte Exp. Med. 89*, 699-764.

Apitz K (1934) Die Wirkung bakterieller Kulturfiltrate nach Umstimmung des gesamten Endothels beim Kaninchen. *Virchows Arch. Pathol. Anat. 293*, 1-33.

Apitz K (1935) A study of the generalized Shwartzman phenomenon. *J. Immunol. 29*, 255-266.

Athens JW, Haab OP, Raab SO, Maurer AM, Ashenbruckner H, Cartwright GE, Wintrobe MM (1961) Leukokinetic studies. IV. The total blood, circulating and marginating granulocyte pools and the granulocyte turnover rate in normal subjects. *J. Clin. Invest. 40*, 989-995.

Becker RM (1948) Suppression of local tissue reactivity (Shwartzman phenomena) by nitrogen mustard, benzoyl or x-ray irradiation. *Proc. Soc. Exp. Biol. Med. 69*, 247-250.

Beesley JE, Pearson JD, Carleton JS, Hutchings A, Gordon JL (1978) Interaction of leukocytes with vascular cells in culture. *J. Cell Sci. 33*, 85-101.

Beesley JE, Pearson JD, Hutchings A, Gordon JL (1979) Granulocyte migration through endothelium in culture. *J. Cell Sci. 38*, 237-248.

Bohn EG, Müller-Berghaus G (1976) The effect of leukocyte and platelet transfusion on the activation of intravascular coagulation by endotoxin in granulocytopenic and thrombocytopenic rabbits. *Am. J. Pathol. 84*, 239-258.

Brown DL, Lachman PJ (1973) The behaviour of complement and platelets in lethal endotoxin shock in rabbits. *Int. Arch. Allergy Appl. Immunol. 45*, 193-205.

Brunning RD, Woolfrey BF, Schrader WH (1964) Studies with tritiated endotoxin. II. Endotoxin localization in the formed elements of the blood. *Am. J. Pathol. 44*, 401-409.

Chan JYC, Habal FM, Burrowes CE, Movat HZ (1976) Interaction between factor XII (Hageman factor), high molecular weight kininogen and prekallikrein. *Thromb. Res. 9*, 423-433.

Cluff LE, Berthrong M (1953) The inhibition of the local Shwartzman reaction by heparin. *Bull. Johns Hopkins Hosp. 92*, 353-369.

Cochrane CG, Janoff A (1974) The Arthus reaction. A model of neutrophil- and complement-mediated injury. In: Zweifach BW, Grant L, McCluskey RT (Eds), *The Inflammatory Process Vol 3*, pp 86-162. Academic Press, London–New York.

Cooper NR, Morrison DC (1978) Binding and activation of the first component of human complement by the lipid A region of lipopolysaccharide. *J. Immunol. 120*, 1862-1868.

Crawford JP, Movat HZ, Ranadive NS, Hay JB (1982) Pathways to inflammation induced by immune complexes: development of the Arthus reaction. *Fed. Proc. 41*, 2583-2587.

Cybulsky MI, Movat HZ (1982) Experimental bacterial pneumonia in rabbits: polymorphonuclear leukocyte margination and sequestration in the lungs of rabbits and quantitation and kinetics of ^{51}Cr-labelled polymorphonuclear leukocytes in *E. coli* induced lung lesions. *Exp. Lung Res. 4*, 47-66.

Cybulsky MI, Movat HZ (1983) The application of ^{51}Cr-labelled PMN-leukocytes in quantitating PMN kinetics in systemic and local inflammatory mediated processes: effects of endotoxin, complement and interleukin I. *Surv. Synth. Pathol. Res. 1*, 208-228.

De Voe IW (1976) Egestion of degraded meningococci by polymorphonuclear leukocytes. *J. Bacteriol. 125*, 258-266.

De Voe IW, Gilka F (1976) Disseminated intravascular coagulation in rabbits: synergistic activity of meningococcal endotoxin and materials egested from leucocytes containing meningococci. *J. Med. Microbiol. 9*, 451-458.

De Voe IW, Gilka F, Gilchrist JE, Yu E (1977) Pathology of rabbits treated with leukocyte-degraded meningococci in combination with meningococcal endotoxin. *Infect. Immun. 16*, 271-279.

Ebert RH, Koch-Weser D (1958) *In vivo* observations of the Shwartzman phenomenon. *Trans. Am. Clin. Climatol. Assoc. 70*, 103-114.

Edwards RL, Rickles FR, Bobrove AM (1979) Mononuclear cell tissue factor: cell of origin and requirements for activation. *Blood 54*, 359-370.

Erdös EG, Miva I (1968) Effect of endotoxin shock on the plasma kallikrein-kinin system of the rabbit. *Fed. Proc. 27*, 92-95.

Evensen SA, Pickering RJ, Batbouta J, Shepro D (1975) Endothelial injury induced by bacterial endotoxin: effect of complement depletion. *Eur. J. Clin. Invest. 5*, 463-469.

Fehr J, Jacob HS (1977) *In vitro* granulocyte adherence and *in vivo* margination: two associated complement-dependent functions. *J. Exp. Med. 146*, 641-652.

Filkins JP (1971) Comparison of endotoxin detoxification by leukocytes and macrophages. *Proc. Soc. Exp. Biol. Med. 137*, 1396-1400.

Galanos C, Lüderitz O, Rietschel ET, Westphal O (1977) Newer aspects of the chemistry and biology of bacterial lipopolysaccharides, with special reference to their lipid A component. In: Goodwin TW (Ed), *International Review of Biochemistry Vol 14: Biochemistry of Lipids II*, pp 239-335. University Park Press, Baltimore.

Garg SK, Niemetz J (1973) Tissue factor activity of normal and leukemic cells. *Blood 42*, 729-735.

Gaynor E (1971) Increased mitotic activity in rabbit endothelium after endotoxin: an autoradiographic study. *Lab. Invest. 24*, 318-320.

Gaynor E (1973) The role of granulocytes in endotoxin-induced vascular injury. *Blood 41*, 797-808.

Gaynor E, Bouvier C, Spaet TH (1970) Vascular lesions: possible pathogenetic basis of the generalized Shwartzman reaction. *Science 170*, 986-988.

Gerber IE (1936) Morphologic aspects of the local Shwartzman phenomenon. *Arch. Pathol. 21*, 331-350.

Gerrity RG, Caplan BA, Richardson M, Cade JF, Hirsh J, Schwartz CJ (1975) Endotoxin-induced endothelial injury and repair. I. Endothelial cell turnover in the aorta of the rabbit. *Exp. Mol. Pathol. 23*, 379-385.

Gimbrone MA Jr, Buchanan MR (1983) Interaction of platelets and leukocytes with vascular endothelium. *Ann. N.Y. Acad. Sci. 401*, 171-183.

Good RA, Thomas L (1952) Studies on the generalized Schwartzman reaction. II. The production of bilateral cortical necrosis of the kidneys by a single injection of bacterial toxin in rabbit previously treated with Thorotrast or Trypan blue. *J. Exp. Med. 96*, 625-641.

Hanger FM (1928) Effect of intravenous bacterial filtrates on skin tests and local infections. *Proc. Soc. Exp. Biol. Med. 25*, 775-777.

Hanger FM (1930) The effect of inflammatory reactions on tissue immunity. *J. Exp. Med. 52*, 485-491.

Harlan JM, Killen DD, Harker LA, Striker GE (1981) Neutrophil-mediated endothelial injury *in vitro*. Mechanisms of cell detachment. *J. Clin. Invest. 68*, 1394-1403.

Hay JB, Johnston MG, Hobbs BB, Movat HZ (1975) The use of radioactive microspheres to quantitate hyperemia in dermal inflammatory sites. *Proc. Soc. Exp. Biol. Med. 150*, 641-644.

Herman CM, Oshima G, Erdös EG (1974) The effect of adrenocorticosteroid pretreatment on kinin system and coagulation response in septic shock in the baboon. *J. Lab. Clin. Med. 84*, 731-739.

Hiller E, Saal JG, Ostendorf P, Griffith GW (1977) The procoagulant activity of human granulocytes, lymphocytes and monocytes stimulated by endotoxin. *Klin. Wochenschr. 55*, 751-756.

Hoff HF, Gottlob R, Blümel G (1967) Ultrastructural changes in arteries induced by endotoxin. *Naturwissenschaften 54*, 287-288.

Hoover RL, Folger R, Haering WA, Ware BR, Karnovsky MJ (1980) Adhesion of leukocytes to endothelium: roles of divalent cations, surface charge, chemotactic agents and substrate. *J. Cell Sci. 45*, 73-86.

Issekutz AC (1983) Quantitation and kinetics on the inflammatory reaction induced in the skin of rabbits. *Surv. Syn. Pathol. Res. 1*, 89-110.

Issekutz AC, Movat HZ (1980) The *in vivo* quantitation and kinetics of rabbit neutrophil leukocyte accumulation in the skin in response to chemotactic agents and *Escherichia coli. Lab. Invest. 42*, 310-317.

Issekutz AC, Movat HZ (1982) The effect of vasodilatory prostaglandins on polymorphonuclear leukocyte infiltration and vascular injury. *Am. J. Pathol. 107*, 300-309.

Issekutz AC, Movat KW, Movat HZ (1980a) Enhanced vascular permeability and haemorrhage-inducing activity of zymosan-activated plasma. *Clin. Exp. Immunol. 41*, 505-511.

Issekutz AC, Movat KW, Movat HZ (1980b) Enhanced vascular permeability and haemorrhage-inducing activity of C5a$_{\text{des Arg}}$: probable role of polymorphonuclear leukocyte lysosomes. *Clin. Exp. Immunol. 41*, 512-520.

Issekutz TB, Issekutz AC, Movat HZ (1981) The *in vivo* quantitation of monocyte migration into acute inflammatory tissue. *Am. J. Pathol. 103*, 47-55.

Jeynes BJ, Issekutz AC, Issekutz TB, Movat HZ (1980) Quantitation of platelets in the microcirculation. Measurement of Indium-111 in microthrombi induced in rabbits by inflammatory lesions and related phenomena. *Proc. Soc. Exp. Biol. Med. 165*, 445-452.

Johnston MG, Hay JB, Movat HZ (1976) The modulation of enhanced vascular permeability by prostaglandins through alterations in blood flow (hyperemia). *Agents Actions 6*, 705-711.

Johnston MG, Hay JB, Movat HZ (1979) Kinetics of prostaglandin production in various inflammatory lesions measured in draining lymph. *Am. J. Pathol. 95*, 225-236.

Jones DG, Richardson DL, Kay AB (1977) Neutrophil accumulation *in vivo* following the administration of chemotactic factors. *Br. J. Haematol. 35*, 19-2.

Kampschmidt RF (1982) Leukocytic endogenous mediator/endogenous pyrogen. In: Powanda MC, Canonico PG (Eds), *Infection: the Physiologic and Metabolic Response of the Host*, pp 55-74. Elsevier, Amsterdam–New York.

Kampschmidt RF, Upchurch HF (1980) Neutrophil release after injection of endotoxin or leukocytic endogenous mediator in rats. *J. Reticuloendothel. Soc. 28*, 191-201.

Karsner HT, Moritz AR (1934) Pathologic histology of the Shwartzman phenomenon with interpretative comments. *J. Exp. Med. 60*, 37-45.

Katori M, Ikeda K, Harada Y, Uchida Y, Tamaka K, Ohishi S (1978) A possible role of prostaglandins and bradykinin as a trigger of exudation in carrageenin-induced rat pleurisy. *Agents Actions 8*, 108-112.

Kimball HR, Melmon KL, Wolff SM (1972) Endotoxin-induced kinin production in man. *Proc. Soc. Exp. Biol. Med. 139*, 1078-1082.

Kopaniak MM, Hay JB, Movat HZ (1978) The effect of hyperemia on vascular permeability. *Microvasc. Res. 15*, 77-82.

Kopaniak MM, Issekutz AC, Movat HZ (1980a) The quantitation of hemorrhage in the skin. Measurement of hemorrhage in the microcirculation in inflammatory lesions and related phenomena. *Proc. Soc. Exp. Biol. Med. 163*, 126-131.

Kopaniak MM, Issekutz AC, Movat HZ (1980b) Kinetics of acute inflammation induced by *E. coli* in rabbits. Quantitation of blood flow, enhanced vascular permeability, hemorrhage, and leukocyte accumulation. *Am. J. Pathol. 98*, 485-498.

Kramer W, Müller-Berghaus M (1977) Effect of platelet antiserum on the activation of intravascular coagulation by endotoxin. *Thromb. Res. 10*, 47-70.

Lackie JM, De Bono D (1977) Interactions of neutrophil granulocytes (PMNs) and endothelium *in vitro. Microvasc. Res. 13*, 107-112.

Lambert PB, Ming S-C, Palmerio C (1969) Vasomotor factors in the Shwartzman phenomenon. *Am. J. Pathol. 57*, 559-580.

Latour J-G, McKay DG (1970) Requirement of the adrenal glands for provocation of the generalized Shwartzman reaction. *Lab. Invest. 22*, 281-285.

Latour J-G, Léger C, Renault S (1974) Activation of Hageman factor and initiation of hepatic vein thrombosis in the hyperlipemic rat. *Am. J. Pathol. 76*, 179-192.

Latour J-G, Giroux C, Léger-Gauthier C (1979) Activation of the Hageman factor-prekallikrein system in the pathogenesis of the generalized Shwartzman phenomenon. *Lab. Invest. 41*, 237-246.

Lee L (1962) Reticuloendothelial clearance of circulating fibrin in the pathogenesis of the generalized Shwartzman reaction. *J. Exp. Med. 115*, 1065-1082.

Lerner RG, Goldstein R, Cummings G (1977) Endotoxin induced disseminated intravascular clotting: evidence that it is mediated by neutrophil production of tissue factor. *Thromb. Res. 11*, 253-261.

Levy GA, Edgington TS (1980) Lymphocyte cooperation is required for amplification of macrophage procoagulant activity. *J. Exp. Med. 151*, 1232-1239.

Levy GA, Schwartz BS, Curtiss LK, Edgington TS (1981a) Plasma lipoprotein induction and suppression of the generation of cellular procoagulant activity *in vitro*. Requirements for cellular collaboration. *J. Clin. Invest. 67*, 1614-1622.

Levy GA, Schwartz BS, Edgington TS (1981b) The kinetics and metabolic requirements for direct lymphocyte induction of human procoagulant monokines by bacterial lipopolysaccharide. *J. Immunol. 127*, 357-363.

Maier RV, Ulevitch RJ (1981a) The response of isolated rabbit hepatic macrophages (H-Mø) to lipopolysaccharide (LPS). *Circ. Shock 8*, 165-181.

Maier RV, Ulevitch RJ (1981b) The induction of a unique procoagulant activity in rabbit hepatic macrophages by bacterial lipopolysaccharides. *J. Immunol. 127*, 1596-1600.

Margaretten W, McKay DG (1969a) The effect of antiserum on the generalized Shwartzman reaction. *Am. J. Pathol. 57*, 299-305.

Margaretten W, McKay DG (1969b) The role of the platelet in the generalized Shwartzman reaction. *J. Exp. Med. 129*, 585-590.

Margaretten W, Zunker HO, McKay DG (1964) Production of the generalized Shwartzman reaction in pregnant rats by intravenous infusion of thrombin. *Lab. Invest. 13*, 552-559.

Mason JW, Colman RW (1971) The role of Hageman factor in disseminated intravascular coagulation induced by septicemia, neoplasia or liver disease. *Thromb. Diath. Haemorrh. 26*, 325-331.

Mathison JM, Ulevitch RJ (1979) The clearance, tissue distribution, and cellular localization of intravenously injected lipopolysaccharide in rabbits. *J. Immunol. 123*, 2133-2143.

Mathison JM, Ulevitch RJ (1981) *In vivo* interaction of bacterial lipopolysaccharide (LPS) with rabbit platelets: modulation by C3 and density lipoproteins. *J. Immunol. 126*, 1575-1580.

Mathison JC, Ulevitch RJ (1983) Mediators involved in the expression of endotoxic activity. *Surv. Syn. Pathol. Res. 1*, 34-48.

Mathison JM, Ulevitch RJ, Fletcher JR, Cochrane CG (1980) The distribution of lipopolysaccharide in normocomplementemic and C3-depleted rabbits and rhesus monkeys. *Am. J. Pathol. 101*, 245-264.

McGrath JM, Stewart GJ (1969) The effects of endotoxin on vascular endothelium. *J. Exp. Med. 129*, 833-848.

McKay DG (1965) *Disseminated Intravascular Coagulation. An Intermediary Mechanism of Disease.* Hoeber Medical Division, Harper and Row, New York.

McKay DG, Margaretten W, Csavossy I (1966) An electron microscope study of the effects of bacterial endotoxin on the blood-vascular system. *Lab. Invest. 15*, 1815-1829.

McKay DG, Müller-Berghaus G, Cruse V (1969) Activation of Hageman factor by elagic acid and the generalized Shwartzman reaction. *Am. J. Pathol. 54*, 393-420.

McKay DG, Latour J-G, Lopez AM (1971) Production of the generalized Shwartzman reaction by activated Hageman factor and adrenergic stimulation. *Thromb. Diath. Haemorrh. 26*, 71-76.

Michalek SM, Moore RN, McGhee JR, Rosenstreich DL, Mergenhagen SM (1978) The primary role of lymphoreticular cells in the mediation of host responses to bacterial endotoxins. *Infect. Immun. 2*, 448-457.

Morrison DC, Cochrane CG (1974) Direct evidence for Hageman factor (factor XII) activation by bacterial lipopolysaccharides (endotoxins). *J. Exp. Med. 140*, 797-811.

Morrison DC, Kline LF (1977) Activation of the classical and properdin pathways of complement by bacterial lipopolysaccharide (LPS). *J. Immunol. 118*, 362-368.

Morrison DC, Oades ZG (1979) Mechanisms of lipopolysaccharide-initiated rabbit platelet responses. II. Evidence that lipid A is responsible for binding of lipopolysaccharide to the platelet. *J. Immunol. 122*, 753-758.

Morrison DC, Ryan JL (1979) A review – bacterial endotoxins and host immune function. *Adv. Immunol. 28*, 293-450.

Morrison DC, Ulevitch RJ (1978) The effects of bacterial endotoxins on host mediation systems. *Am. J. Pathol. 93*, 527-604.

Morrison DC, Kline LF, Oades ZG, Henson PM (1978) Mechanisms of lipopolysaccharide-initiated rabbit platelet responses. I. Alternative complement pathway dependence of the lytic response. *Infect. Immun. 20*, 744-751.

Morrison DC, Oades ZG, Di Pietro D (1980) Endotoxin-initiated membrane changes in rabbit platelets. In: Agarwal MK (Ed), *Bacterial Endotoxins: Unity and Diversity of Host Responses*, pp 311-326. Elsevier/North Holland, Amsterdam, New York.

Movat HZ (1983) The local Shwartzman reaction. Endotoxin-mediated inflammatory and thrombo-hemorrhagic lesions. *Surv. Syn. Pathol. Res. 1*, 241-250.

Movat HZ, Uriuhara T, Taichman NS, Rowsell HC, Mustard JF (1968) The role of PMN-leucocyte lysosomes in tissue injury, inflammation and hypersensitivity. VI. The participation of the PMN-leukocyte and the blood platelet in systemic aggregate anaphylaxis. *Immunology 14*, 637-648.

Movat HZ, Jeyens BJ, Kopaniakk MM, Wasi S (1980) Quantitation of the development and progression of the local Shwartzman reaction. In: Agarwal MK (Ed), *Bacterial Endotoxins: Unity or Diversity of Host Response*, pp 179-201. Elsevier/North Holland, Amsterdam, New York.

Müller-Berghaus G, Eckhardt T (1975) The role of granulocytes in the activation of intravascular coagulation and the precipitation of soluble fibrin by endotoxin. *Blood 45*, 631-641.

Müller-Berghaus G, Kramer W (1976) Effect of platelet antiserum on the precipitation of soluble fibrin by endotoxin. *Thromb. Haemostasis 35*, 237-248.

Müller-Berghaus G, Lasch H-G (1975) Microcirculatory disturbances induced by generalized intravascular coagulation. In: Schmier J, Eichler O (Eds), *Handbook of Experimental Pharmacology Vol XVI/3: Experimental Production of Diseases: Heart and Circulation*, pp 429-514. Springer-Verlag, New York–Berlin–Heidelberg.

Müller-Berghaus G, Schmidt-Ehry B (1972) The role of pregnancy in the induction of the generalized Shwartzman reaction. *Am. J. Obstet. Gynecol. 114*, 847-849.

Müller-Berghaus G, Schneberger R (1971) Hageman factor activation in the generalized Shwartzman reaction induced by endotoxin. *Br. J. Haematol. 21*, 513-527.

Murphy TL, Taylor FB III, Welsman IM, Gillian JM, Taylor FB Jr (1980) Complement-dependent activation of cannine platelets by endotoxin and collagen: *in vitro* studies. *Clin. Immunol. Immunopathol. 16*, 57-71.

Musson RA, Morrison DC, Ulevitch RJ (1978) Distribution of lipopolysaccharide in LPS-responsive and LPS-unresponsive mice. *Infect. Immun. 21*, 448-457.

Mustard JF, Packham MA (1979) The reaction of the blood to injury. In: Movat HZ (Ed), *Inflammation, Immunity and Hypersensitivity (2nd Ed.)*, pp 557-664. Harper and Row, Hagerstown–New York.

Niemetz J (1972) Coagulant activity of leukocytes. *J. Clin. Invest. 51*, 307-313.

Niemetz J, Fani K (1971) Role of leucocytes in blood coagulation and the generalized Shwartzman reaction. *Nature (New Biol.) 232*, 247-248.

Niemetz J, Marcus AJ (1974) The stimulatory effect of platelets and platelet membranes on the procoagulant activity of leukocytes. *J. Clin. Invest. 54*, 1437-1443.

Nies AS, Forsthy RP, Williams HE, Melmon KL (1968) Contribution of kinins to endotoxin shock in unanesthetized rhesus monkeys. *Circ. Res. 22*, 155-164.

Oades ZG, Morrison DC (1980) A simple and reproducible method to assess murine platelet responses *in vitro*. *J. Immunol. Methods 34*, 79-84.

Perper RJ, Sanda M, Chinea G, Oronsky AL (1974) Leukocyte chemotaxis *in vivo*: I. Description of a model of cell accumulation using adoptively transferred ^{51}Cr-labelled cells. *J. Lab. Clin. Med. 84*, 378-393.

Plow EF (1982) Leukocyte elastase release during coagulation. *J. Clin. Invest. 69*, 564-572.

Prydz H, Allison AC (1978) Tissue thromboplastin activity of isolated human monocytes. *Thromb. Haemostasis 39*, 582-589.

Prydz H, Lyberg T, Deteix P, Allison AC (1979) *In vitro* stimulation of tissue thromboplastin (factor III) activity in human monocytes by immune complexes and lectins. *Thromb. Res. 15*, 465-474.

Raiji L, Keane WF, Michael AF (1977) Unilateral Shwartzman reaction: cortical necrosis in one kidney following *in vivo* perfusion with endotoxins. *Kidney Int. 12*, 91-95.

Ratnoff OD, Crum JD (1964) Activation of Hageman factor by solutions of elagic acid. *J. Lab. Clin. Med. 63*, 359-365.

Reidy MA, Schwartz SM (1983) Endothelial injury and regeneration. IV. Endotoxin: a nondenuding injury to aortic endothelium. *Lab. Invest. 48*, 25-34.

Rickles FR, Levin J, Hardin JA, Barr CF, Coward ME Jr (1977) Tissue factor generation by human mononuclear cells: effects of endotoxin and dissociation of tissue factor generation from mitogenic response. *J. Lab. Clin. Med. 89*, 792-803.

Rivers RPA, Hathaway WE, Weston WL (1975) The endotoxin-induced coagulant activity of human monocytes. *Br. J. Haematol. 30*, 311-316.

Robinson JA, Klodmycky ML, Loeb HS, Racic MR, Gunnar RM (1975) Endotoxin, prekallikrein, complement and systemic vascular resistance: sequential measurements in man. *Am. J. Med. 59*, 61-67.

Robinson JA, Rapaport SI, Brown SF (1978) Procoagulant activity of peritoneal leukocytes: effects of cortisone and endotoxin. *Am. J. Physiol. 235*, H333-H337.

Rosenstreich DL, Vogel SN (1980) Central role of macrophages in the hosts response to endotoxin. In: Schlesinger D (Ed), *Microbiology – 1980*, pp 11-15. American Society for Microbiology, Washington D.C.

Rothberger H, Zimmerman TS, Spiegelberg HL, Vaughan JH (1977) Leukocyte procoagulant activity: enhancement of production *in vitro* by IgG antigen-antibody complexes. *J. Clin. Invest. 59*, 549-557.

Rothberger H, Zimmerman TS, Vaughan JH (1978) Increased production and expression of tissue thromboplastin-like procoagulant activity *in vitro* by allogeneically stimulated human leukocytes. *J. Clin. Invest. 60*, 649-655.

Sanarelli G (1924) De la pathogénie du choléra (neuvième mémoire). Le choléra expérimental. *Ann. Inst. Pasteur 38*, 11-72.

Sanarelli G (1939) L'allergie salivaire hémorrhagique. *Ann. Inst. Pasteur 63*, 105-144.

Schrader WH, Woolfrey BF, Brunning RD (1964) Studies with tritiated endotoxin. III. The local Shwartzman reaction. *Am. J. Pathol. 44*, 597-611.

Schwartz BS, Edgington TS (1981) Lymphocyte collaboration is required for induction of murine monocyte procoagulant activity by immune complexes. *J. Immunol. 127*, 438-443.

Schwartz BS, Levy GA, Curtis LA, Fair DS, Edgington TS (1981) Plasma lipoprotein induction and suppression of the generation of cellular procoagulant activity *in vitro*. I. Two procoagulant activities are produced by peripheral blood mononuclear cells. *J. Clin. Invest. 67*, 1650-1658.

Schwartz BS, Levy GA, Fair DS, Edgington TS (1982) The murine lymphoid procoagulant activity induced by bacterial lipopolysaccharide and immune complexes is a monocyte prothrombinase. *J. Exp. Med. 155*, 1464-1479.

Semeraro N, Colucci M, Mussoni L, Donati MB (1981) Rat blood leucocytes, unlike rabbit leucocytes, do not generate procoagulant activity on exposure to endotoxin. *Br. J. Exp. Pathol. 62*, 638-642.

Shen SM-C, Rapaport SI, Feinstein DI (1973) Intravascular clotting after endotoxin in rabbits with impaired intrinsic clotting produced by a factor VIII antibody. *Blood 42*, 523-534.

Shiga J, Mori W (1980) A light and electron-microscopic study of the so-called uni-visceral Shwartzman reaction. *Acta Pathol. Jpn. 30*, 705-712.

Shwartzman G (1928a) A new phenomenon of local skin reactivity to *B. typhosus* culture filtrate. *Proc. Soc. Exp. Biol. Med. 25*, 560-561.

Shwartzman G (1928b) Phenomenon of local skin reactivity to culture filtrates of various organisms. *Proc. Soc. Exp. Biol. Med. 26*, 207-208.

Shwartzman G (1928c) Studies on bacillus typhosus toxic substances. I. Phenomenon of local reactivity to *B. typhosus* culture filtrate. *J. Exp. Med. 48*, 247-268.

Shwartzman G (1937) *Phenomenon of Local Tissue Reactivity and its Immunological, Pathological and Clinical Significance*. Paul B. Hoeber, New York.

Skjørten F, Evensen SA (1973) Induction of disseminated intravascular coagulation in factor XII-deficient fowl: morphological effects of liquoid, bacterial endotoxin and tissue thromboplastin in the normal anticoagulated fowl. *Thromb. Diath. Haemorrh. 30*, 25-35.

Stetson CA Jr (1951a) Studies on the mechanism of the Shwartzman phenomenon. Similarities between reactions to endotoxins and certain reactions of bacterial allergy. *J. Exp. Med. 101*, 421-436.

Stetson CA Jr (1951b) Similarities in the mechanisms determining the Arthus and Shwartzman phenomenon. *J. Exp. Med. 94*, 347-358.

Stetson CA, Good RA (1951) Studies on the mechanisms of the Shwartzman phenomenon. Evidence for the participation of polymorphonuclear leucocytes in the phenomenon. *J. Exp. Med. 93*, 49-64.

Taichman NS, Uriuhara T, Movat HZ (1966) Ultrastructural alterations in the local Shwartzman reaction. *Lab. Invest. 14*, 2160-2176.

Thomas L, Good RA (1952) Studies on the generalized Shwartzman reaction. I. General observations concerning the phenomenon. *J. Exp. Med. 96*, 605-624.

Udaka K, Takeuchi Y, Movat HZ (1970) Simple method for quantitation of enhanced vascular permeability. *Proc. Soc. Exp. Biol. Med. 133*, 1384-1387.

Ulevitch RJ, Cochrane CG (1978) Role of complement in lethal bacterial lipopolysaccharide-induced hypotensive and coagulative changes. *Infect. Immun. 19*, 204-211.

Ulevitch RJ, Cochrane CG, Henson PM, Morrison DC, Doe WF (1975) Mediation systems in bacterial lipopolysaccharide-induced hypotension and disseminated intravascular coagulation. I. The role of complement. *J. Exp. Med. 142*, 1570-1590.

Ulevitch RJ, Cochrane CG, Bangs K, Herman CM, Fletcher JR, Rice CL (1978) The effect of complement depletion on bacterial lipopolysaccharide (LPS)-induced hemodynamic and hematologic changes in the rhesus monkey. *Am. J. Pathol. 92*, 227-240.

Ulevitch RJ, Johnston AR, Weinstein DB (1979) New function for high density lipoproteins. Their participation in intravascular reactions of bacterial lipopolysaccharides. *J. Clin. Invest. 64*, 1516-1524.

Urbaschek B, Urbaschek R (1979) The inflammatory response to endotoxins. In: Hauck G, Irwin JW (Eds), *Microcirculation in Inflammation*, pp 74-104. Bibliotheca Anatomica No 17. S. Karger, Basel.

Vadas P, Wasi S, Movat HZ, Hay JB (1981) Extracellular phospholipase A_2 mediates inflammatory hyperaemia. *Nature (London) 293*, 583-585.

Van Furth R, Willemze R (1979) Phagocytic cells during an acute inflammatory reaction. *Curr. Top. Pathol. 68*, 179-212.

Van Ginkel CJW, Van Aken WG, Oh JIH, Vreeken J (1977) Stimulation of monocyte procoagulant activity by adherence to different surfaces. *Br. J. Haematol. 37*, 35-45.

Vogel SN, Moore RN, Sipe SD, Rosenstreich DL (1980) BCG-induced enhancement of endotoxin sensitivity in C3H/HeJ mice. *J. Immunol. 124*, 2004-2009.

Vogel SN, Weedon LL, Moore RN, Rosenstreich DL (1982) Correction of defective macrophage differentiation in C3H/HeJ mice by an interferon-like molecule. *J. Immunol. 128*, 380-387.

Wasi S, Movat HZ (1979) Phlogistic substances in neutrophil leukocyte lysosomes: their possible role *in vivo* and their *in vitro* properties. *Curr. Top. Pathol. 68*, 213-238.

Wasi S, Burrowes CE, Hay JB, Movat HZ (1983) Plasminogen activator and thromboplastin activity from sheep alveolar macrophages. *Thromb. Res. 30*, 27-45.

Wasi S, Movat HZ (1981) Hemorrhage and exudation in the microcirculation induced in rabbits with human neutrophil lysosomal enzymes. *Thromb. Haemost. 46*, 59.

Williams TJ, Peck MJ (1977) Role of prostaglandin-mediated vasodilatation in inflammation. *Nature (London) 270*, 530-532.

Yamada O, Moldow CF, Sacks T, Craddock PR, Boogaerts MA, Jacob HS (1981) Deleterious effects of endotoxin on cultured endothelial cells: an *in vitro* model of vascular injury. *Inflammation 5*, 115-126.

Handbook of Endotoxin, Vol. 3: Cellular Biology of Endotoxin
L.J. Berry, editor
© Elsevier Science Publishers B.V., 1985

CHAPTER 14

Agents which modulate reticuloendothelial function and affect endotoxin sensitivity*

J.A. COOK, T.S. ROGERS, W.C. WISE AND P.V. HALUSHKA

1. INTRODUCTION

Injection of lipopolysaccharide from gram-negative bacteria in laboratory animals induces a complex sequence of hemodynamic, hematologic, and metabolic changes frequently culminating in severe circulatory shock and death (Cook et al 1980c; Gilbert 1960; Ulevitch and Cochrane 1978). The mediation of the host response involves numerous cell types (for example, tissue macrophages, blood monocytes, neutrophils and platelets) as well as activation of the complement, arachidonic acid, and coagulation systems (Morrison and Ulevitch 1978). The mechanisms by which these various cell types and humoral factors interact in vivo to induce primary or secondary sequelae and ultimately decompensated circulatory collapse remain unclear.

2. ENDOCYTIC ASPECTS OF ENDOTOXIN SENSITIVITY

From a historical perspective, beginning with the observations of Metchnikoff (1905), the reticuloendothelial system (RES) has been regarded as playing a key role in phagocytosis and intracellular digestion. More than two decades ago, studies by Fine et al (1959) and Zweifach (1958) were the first to establish potential links between the functional status of reticuloendothelial cells and tolerance to circulatory shock. The importance of the RES has been predicated upon several different lines of evidence. First, procedures using radiolabeling (Braude et al 1955; Carey et al 1958; Chedid et al 1966; Noyes et al 1959), immunofluorescence, or specific antibodies (Cremer and Watson 1957; Rubenstein et al 1962; Tanaka et al 1959) have demonstrated that most endotoxin is rapidly sequestered in the RES component of liver and spleen. More recent autoradiographic procedures employing radioiodinated endotoxin have shown that endotoxin is concentrated in

*Work presented in this chapter was supported in part by National Institutes of Health grants GM27673 and HL29566, and by the Drug Science Foundation.

phagocytic vacuoles of hepatic Kupffer cells, splenic macrophages, and leukocytes (Mathison and Ulevitch 1979). Secondly, macrophages of the liver and spleen have been reported to be of crucial importance in endotoxin detoxification (Palmerio and Fine 1969; Rutenburg et al 1967). These observations led to the hypothesis that shock susceptibility centers on a basic etiology of splanchnic ischemia and associated RES failure (Palmerio and Fine 1969). Thirdly, induction of endotoxin tolerance by prior exposure to sublethal doses of the lipopolysaccharide was shown to be temporally related with stimulation of RES phagocytic function (Beeson 1947a, 1947b; Benacerraf and Sebestyen 1957). Endotoxin-tolerant animals are not only more resistant to bacterial endotoxin from other gram-negative bacteria but exhibit enhanced host resistance to heterologous bacterial infections (Chedid et al 1976). In a classical series of experiments, Beeson (1947b) demonstrated that endotoxin-tolerant rabbits could be resensitized to lethal endotoxin by administration of RES-'blockading' doses of colloidal material. It was subsequently shown that potentiated endotoxin lethality occurs following treatment with a variety of RES-blockading agents, for example colloidal carbon, lead acetate, saccharated iron oxide, thorotrast and trypan blue (Berry and Smythe 1965; Cooper and Stuart 1961; Filkins 1977; Janoff et al 1962, 1963; Noyes et al 1959; Selye et al 1966).

3. DISSOCIATION OF ENDOCYTIC ACTIVITY AND ENDOTOXIN SENSITIVITY

While macrophages may contribute significantly to host defense during certain types of systemic stress, these cells may also play a pivotal role in the pathogenesis of endotoxin shock. Suter et al (1958) and Halpern (1959) provided the first evidence that the link between RE phagocytic function and resistance to endotoxemia was more complex than originally thought. These early studies demonstrated that infection of mice with *Mycobacterium tuberculosis* (strain *Bacillus Calmette-Guérin*; BCG) led to an intense, long-lasting stimulation of RES function. In view of these actions, it appeared quite feasible that BCG vaccination might induce tolerance to bacterial endotoxin. Contrary to anticipated results, the BCG-infected mice were rendered 100–1000 times more susceptible to lethal endotoxemia (Suter et al 1958). These mice exhibited a markedly abbreviated clinical course with augmented shock manifestations compared to normal mice given much higher doses of endotoxin. This phenomenon of enhancement of lethal endotoxemia was not restricted simply to mycobacterium infections in mice; it could be induced with other infectious or killed microbes including *Corynebacterium parvum*, *Vibrio cholerae, Bacillus anthracis*, Variola vaccine, *Bordetella pertussis* vaccine, *Eperythrozoom coccoides*, and *Salmonella typhi* infections (Adlam 1973; Suter 1962). Similarly, cell wall-derived products from mycobacteria, for example N-acetyl-muramyl-alanyl-(1-seryl)-D-isoglutamine (MDP) (Ribi et al 1979) and trehalose dimycolate (Parant et al 1977), both RES stimulants, have been reported to induce endotoxin hyperreactivity in mice.

This increased susceptibility to endotoxin is not restricted to bacteria or bacterial products but can also be demonstrated with other potent RE-phagocytic stimulants, such as zymosan (Benacerraf et al 1959), glucan (Lemperle 1966) and pyran copolymer (Regelson et al 1973). The dissociation between RE-phagocytic activity and endotoxin lethality is further supported by the pharmacologic action of certain lipids (DiLuzio 1972). Stuart and Cooper (1962) investigated the effect of glycerol triolin and ethyl stearate on RES function and endotoxin susceptibility in mice. Administration of glycerol triolate activated the RES clearance of colloidal carbon and ^{51}Cr-labeled endotoxin. Despite the marked stimulation of RES function, this agent rendered mice extremely sensitive to endotoxin lethality. In contrast, ethyl stearate, which induced a hypophagocytic state, did not appreciably enhance endotoxin susceptibility (Stuart and Cooper 1962). In subsequent studies, DiLuzio and Crafton (1970) evaluated the effect of RE depression induced by methyl palmitate on endotoxin-induced lethality in rats. The administration of methyl palmitate resulted in a marked degree of resistance to lethal and even supralethal doses of endotoxin (see Section 6.1). In contrast to methyl palmitate, the RE-blockading agent trypan blue sensitizes animals to endotoxin shock (Noyes et al 1959). This agent did not, however, alter hepatic localization of ^{51}Cr-labeled endotoxin, and it was concluded that the potentiated endotoxin lethality was not related to alterations in phagocytic clearance of the toxin. It is apparent from the aforementioned studies that there are no clear discernable relationships between RE phagocytic function per se and host response to endotoxin. These observations suggest that the pharmacologic actions of these agents on other RE functional or metabolic parameters must be considered.

4. EXOCYTIC ASPECTS OF ENDOTOXIN SENSITIVITY

Over the past few years, the concept of the macrophage as a secretory cell has emerged. Through various exocytic functions, these phagocytes participate in complex inflammatory interactions with components of the cellular and humoral immune system (for reviews see Nathan et al 1980; Stenson and Parker 1980; Unanue 1976). Thus, in counterpoint to what is known about the endocytic aspects of the relationship between RES function and shock, there is increasing evidence that the RES is a source of a considerable number of mediators that may contribute to local and systemic injuries during endotoxemia. The multiplicity of endotoxin effects may, therefore, be partly due to the variety of mediators that are released by endotoxin-stimulated mononuclear phagocytes. These include endogenous pyrogen, colony stimulating activity, procoagulant factor, interleukin 1, glucocorticoid antagonizing factor, macrophage insulin-like activity, plasminogen activator, collagenase, and others (Filkins 1982; Maier and Ulevitch 1981; Moore et al 1978; Morrison and Ulevitch 1979; Rosenstreich and Vogel 1980; Unanue 1976). A deleterious contribution of lymphoreticular cells to the pathophysiology of endotoxic shock is also suggested from previous studies in which spleen cells (Glode et al 1976) or bone marrow cells (Michalek et al 1980) from C3H/HeN mice, which are normally responsive to endotoxin, were reconstituted with C3H/HeJ mice, a related but genetically endotoxin-resistant strain. The chimeric recipients of these endotoxin-responsive cells became more sensitive to lethal endotoxemia.

There is a considerable body of evidence that arachidonic acid metabolites may participate in the pathophysiology of shock resulting from either infection or the administration of endotoxin (for reviews see Bult and Herman 1982; Fletcher and Ramwell 1980; Rietschel et al 1980). Prophylactic or post-treatment regimens of nonsteroidal antiinflammatory agents, which block all arachidonic acid cyclo-oxygenase products, improve survival and/or circulatory function in experimental septic or endotoxin shock (Bult and Herman 1982; Fletcher and Ramwell 1980).

In addition to the classical prostaglandins of the E and F series, recent interest has focused on two other arachidonic acid metabolites: thromboxane (TX) A_2 and prostacyclin (PGI_2). Through their opposing effects on vasoactivity and platelet aggregation, TXA_2 and PGI_2 are emerging as potentially important controlling influences of vascular hemostasis and homeostasis (for review see Moncada and Vane 1978). PGI_2 is a potent endogenous inhibitor of platelet aggregation and as a vasodilator, PGI_2 is 4–8 times more potent than PGE_2 (Armstrong et al 1978; Weiss and Turitto 1979). TXA_2 is proposed to mediate the mechanism for the aggregation and release of platelets (Hamberg et al 1975) and has also been shown to be an extremely potent vasoconstrictor (Svensson and Fredholm 1977). Circulatory dysfunction characterized by tissue ischemia, systemic hypotension, and disseminated intravascular coagulation is a frequent manifestation of septic shock in patients and experimental animals. It has therefore been proposed that an unfavorable TXA_2/PGI_2 balance may contribute to these altered hemodynamic and hemostatic sequelae of shock (Butler et al 1983; Cook et al 1980c, 1981b; Demling et al 1981; Frolich et al 1980; Halushka et al 1983; Harris et al 1980).

Although it is clear that these arachidonic acid metabolites are present in blood during many pathophysiologic events in shock, their exact interrelationships and cellular sources remain to be clearly defined. As noted previously, the macrophage represents a major site of endotoxin sequestration and a major source of endotoxin-induced mediators. A potential role of macrophage-derived arachidonic acid metabolites in the sequelae of endotoxin shock is supported by studies demonstrating direct in vitro endotoxin stimulation of PGE_2 (Kurland and Bockman 1978; Moore et al 1979; Wahl et al 1977) and PGF_2 (Rietschel et al 1980) or TXB_2 and 6-keto-$PGF_{1\alpha}$ synthesis by peritoneal macrophages (Cook et al 1981b; Feuerstein et al 1981a, 1981b). Indeed, Feuerstein et al (1981b) proposed that TXB_2 and 6-keto-$PGF_{1\alpha}$ are the major endogenous arachidonic acid metabolites produced by rat peritoneal macrophages and the rate of their release is higher than that of $PGF_{2\alpha}$ or PGE. Such direct in vitro stimulatory actions of endotoxin on the synthesis of TXB_2 by pulmonary parenchyma (Feuerstein and Ramwell 1981) or of PGI_2 by vascular endothelium could not be demonstrated (Villa et al 1981). Previous studies have also noted a correlation between a lack of prostaglandin release by macrophages and endotoxin resistance in nonresponder C3H/HeJ mice (Wahl et al 1979). From these observations, we have postulated that the RES may be a significant source of arachidonic acid metabolites in endotoxin shock. If this hypothesis is tenable, then agents which modulate reticuloendothelial function and endotoxin susceptibility should also affect arachidonic acid metabolism. The following is a review of certain pathophysiologic features produced by four factors which alter reticuloendothelial function. Two of these agents, glucan and lead acetate, both sensitize animals to endotoxin shock and yet have divergent effects upon reticuloendothelial phagocytic function. In a similar fashion, two other factors, methyl palmitate and endotoxin tolerance, have opposing effects on phagocytic function but are both protective.

5. FACTORS WHICH SENSITIZE TO ENDOTOXIN

5.1. Glucan

Glucan is a β-1,3-glucopyranose derivative of the yeast cell wall and comprises the active polysaccharide fraction of zymosan (DiLuzio 1976). Intravenous injection of glucan induces a hyperphagocytic state that is associated with a significant hyperplasia and hypertrophy of the major RE organs (DiLuzio 1976). This increased functional status of the RES induced by glucan has been shown to enhance nonspecific or specific host resistance to certain bacterial (Delville and Jacque 1977; Kokoshis et al 1978; Williams et al 1983), fungal (Stevens et al 1976; Williams et al 1978), protozoan (Cook et al 1980b; Holbrook et al 1981) or viral infections (Reynolds et al 1980; Williams and DiLuzio 1980). Like bacterial immunopotentiators (e.g. BCG), however, glucan renders laboratory animals highly susceptible to lethal endotoxemia (Lemperle 1966). Potential mechanisms whereby glucan induces endotoxin sensitivity were initially investigated by Crafton and DiLuzio (1969) and DiLuzio and Crafton (1970). Vascular clearance and tissue distribution of ^{51}Cr-labeled *Salmonella enteritidis* endotoxin as well as the ^{131}I RE-test lipid emulsion were markedly enhanced by glucan treatment. However, the endotoxin sensitivity did not appear to be due directly to the hyperphagocytic state since glucan-treated rats injected with methyl palmitate demonstrated a normal RES clearance while still maintaining a degree of endotoxin susceptibility. It was concluded, therefore, that the endotoxin reactivity was not due to the phagocytic event but reflected some other enzymatic or functional change that is related to a massive RE cell population.

Clearance studies in our laboratory suggest that the RES in glucan-treated rats may be particularly vulnerable to endotoxin injury. Pretreatment with glucan markedly accelerated intravascular clearance of the RE-test lipid emulsion, and clearance took place approximately four times more rapidly than in controls (Table 1). This rapid clearance was associated with a concomitant increase in hepatic localization of the emulsion. However, measurement of phagocytic function 2 hours

TABLE 1   *Clearance and tissue distribution of the ^{131}I RE-test lipid emulsion in glucan-pretreated and control rats during endotoxin shock*

		Tissue distribution					
		Liver		Lung		Spleen	
	T/2(min)	% ID/g	% ID/TO	% ID/g	% ID/TO	% ID/g	% ID/TO
Glucan control	2.2*	5.9	74.4	0.2	0.7	0.6*	1.4
	±0.4	±0.6	±3.3	±0.04	±0.1	±0.1	±0.2
Normal control	9.2	4.3	45.7	0.5	0.6	2.3	2.2
	±1.5	±0.7	±2.7	±0.1	±0.1	±0.4	±0.1
Glucan + endotoxin	12.3	3.8	35.2*	0.8	1.7	1.6	2.7
	±1.9	±0.5	±1.1	±0.1	±0.2	±0.3	±0.4
Control + endotoxin	9.5	3.4	38.6	1.2	1.5	3.6	2.6
	±1.5	±0.4	±2.7	±0.4	±0.6	±0.4	±0.2

Clearance and tissue distribution of the ^{131}I RE-test lipid emulsion was determined 2 hours after administration of *S. enteritidis* endotoxin (10 µg/100 g body wt.) or isovolumetric saline in glucan-pretreated or normal rats (N = 7/group). Data expressed as percent injected dose per gram (% ID/g) or per total organ (% ID/TO). * $p < 0.05$ compared to normal controls. Reproduced from Cook et al (1980), with permission.

after injection of a low dose of endotoxin revealed a markedly depressed clearance rate and reduced hepatic uptake of the lipid. This dose of endotoxin in normal rats produced minimal changes in RE phagocytic function. It is of interest that macrophages from BCG-infected mice are also more sensitive to in vitro toxicity of endotoxin (Peavy et al 1979). The investigators concluded that this sensitivity parallels the susceptibility of BCG-infected mice to endotoxin, and that release of vasoactive substances from their cells in vivo may in part be responsible for endotoxin lethality.

Alterations in lysosomal membrane permeability have been suspected as possible etiologic factors in glucan-induced endotoxin hypersensitivity (Lemperle 1966). Studies by DiLuzio and Crafton (1970) did demonstrate elevated serum levels of β-glucuronidase and acid phosphatase. This deterioration of lysosomal integrity has subsequently been corroborated by our laboratory as reflected by elevated serum lysosomal enzyme activities (Table 2) and ultrastructural changes (Fig. 1) (Cook et al 1980a). Comparable pathologic sequelae were observed in control rats given much larger doses (100 times) of endotoxin. Kupffer cells in glucan–endotoxin-treated rats exhibited pleomorphic and heterophagic vacuoles and their cellular integrity was frequently disrupted (Fig. 1). Fibrin and cellular debris were

TABLE 2   *Effects of glucan pretreatment on endotoxin (LPS)-induced lethality and associated sequelae*

Group	Serum β-glucuronidase (units/ml)	Blood glucose (mg/100 ml)	24-hr mortality (dead/total)
Unshocked control	46±8* (5)	125±16* (5)	0/5*
Control + LPS	117±20* (5)	72±16* (5)	0/5*
Glucan + LPS	336±46 (7)	34±7 (7)	13/14
Glucan + methyl-prednisolone + LPS	284±52* (7)	110±16* (7)	1/10*

Glucan (10 mg/kg) was administered intravenously for 5 consecutive days, and the rats were challenged with *S. enteritidis* LPS (0.1 mg/kg) 3 days later. Methylprednisolone was administered intravenously (60 mg/kg) 60 minutes before LPS. Blood glucose and serum β-glucuronidase levels were measured 4 hours after LPS. Number of animals given in parentheses. Values presented as mean±SEM. * $p < 0.05$ compared to glucan + LPS group. Reproduced from Cook et al (1980a), with permission.

also present in sinusoids. These extensive sinusoidal alterations, not seen in rats given glucan or low doses of endotoxin alone, would suggest that granulomatous foci, hypertrophic Kupffer cells and/or sequestered inflammatory cells are significant sources of lysosomal acid hydrolase release. These severe sinusoidal changes may also have contributed to the markedly depressed hepatic clearance observed 2 hours post-shock (Table 1). Similar increases in lysosomal enzyme activity and acute hepatic damage have been reported following endotoxin shock in mice treated with the macrophage stimulants BCG (Shands and Senterfitt 1972) and *Corynebacterium parvum* (Ferluga et al 1979).

We have observed that, in contrast to normal shocked rats given much higher doses (100 times) of endotoxin, glucan-pretreated rats succumbed to lethal shock before overt increases in certain hepatocyte enzyme markers (glutamic pyruvic acid transaminase or ornithine carbamyl transferase) were apparent (Cook et al 1980a). Nevertheless, hepatocellular dysfunction may be a significant factor contributing to shock pathogenesis. Shocked glucan-pretreated rats exhibited accelerated severe hypoglycemia (Table 2) and a rapid depletion of glycogen granules in the cytoplasmic matrix of hepatocytes (Fig. 1) (Cook et al 1980a). These findings suggest that severe metabolic disturbances of glucoregulation are prevalent. Pretreatment with methylprednisolone protected the glucan group from endotoxin shock (Table 2). Although glucocorticoids exert both membrane-stabilizing effects (Galvin et al 1978) and enhancement of gluconeogenic function (Berry 1971;

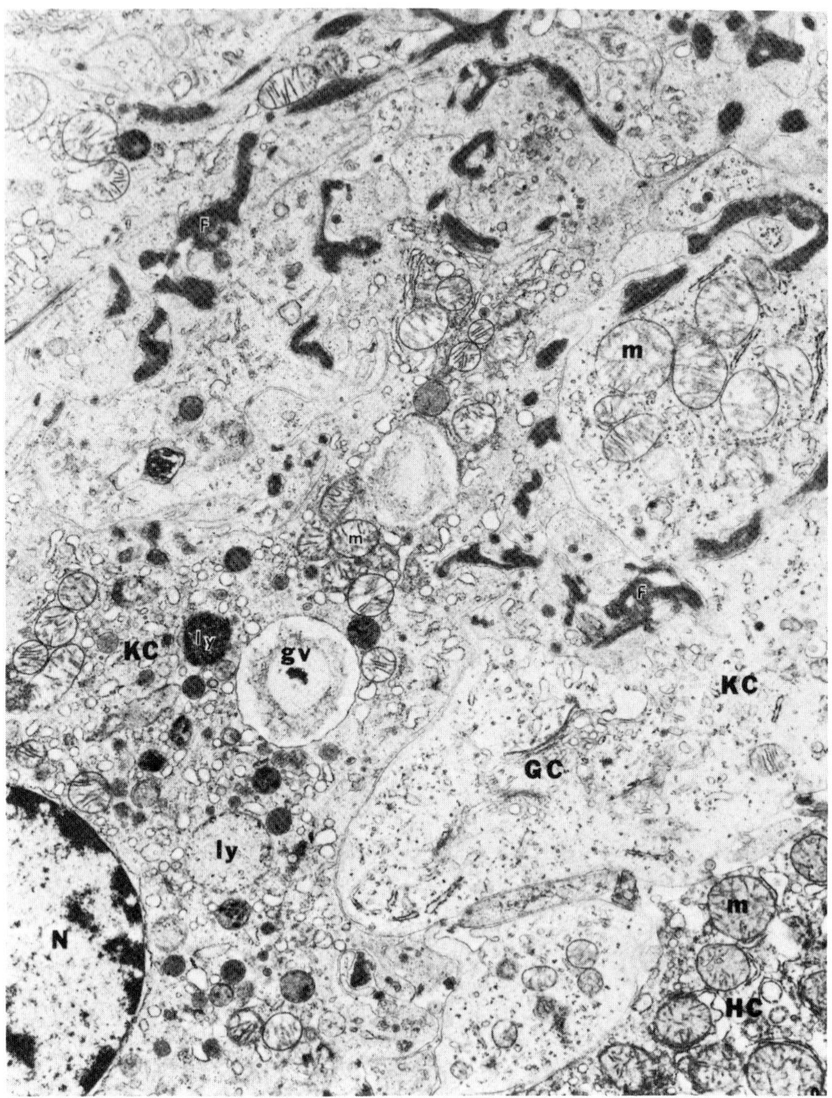

Fig. 1 *Liver tissue from rat injected with low dose of endotoxin (10 μg/100g body wt.)
following glucan stimulation of the RES. A portion of a granulomatous lesion is illustrated.
The granuloma consists of several hypertrophic Kupffer cells (KC) which contain
heterophagic vacuoles (Ly), rough endoplasmic reticulum, Golgi complexes (GC), and
mitochondria (m). One Kupffer cell also contains large vacuoles (gv), which are filled with
electron-lucent material that may be glucan. Deposits of fibrin (F) occupy the intercellular
space between adjacent Kupffer cells (magnification × 9900). HC = hepatocyte; N = nucleus.
Reproduced from Cook et al (1980a), with permission.*

310

Schuler et al 1976), our studies demonstrate that methylprednisolone was more effective in maintaining plasma glucose levels than in preventing disruption of lysosomal integrity. Alterations in carbohydrate homeostasis, including depressed gluconeogenic function, are hallmarks of endotoxin shock (Berry 1971; Filkins 1978). Additionally, Senterfitt and Shands (1978) have reported that the gluconeogenic pathway showed general rather than selective enzymatic malfunctions in mice rendered hyperactive to endotoxin by BCG.

Since there is a growing body of evidence that the macrophage is a central effector cell in host-mediated responses to endotoxin, it is reasonable to postulate that exocytic products from glucan-activated macrophages may play a role in the potentiated shock response. In view of the importance of arachidonic acid metabolites as mediators of certain sequelae of endotoxemia, studies were initiated to assess changes in these metabolites in glucan-stimulated rats. Plasma levels of TXA_2 and PGI_2 were assessed by radioimmunoassay of their stable immunoreactive metabolites TXB_2 and 6-keto-$PGF_{1\alpha}$. These metabolites were measured at 30 minutes and 4 hours after endotoxin administration. These time intervals were selected since our previous studies with normal rats have indicated peak TXB_2 and 6-keto-$PGF_{1\alpha}$ levels at 30 minutes and 4 hours post-endotoxin, respectively (see Section 6.2.) (Cook et al 1982). Control rats administered doses of endotoxin which are normally well tolerated (0.1 mg/kg) exhibited minimal elevations in TXB_2, whereas 6-keto-$PGF_{1\alpha}$ levels were nondetectable at both time intervals (Table 3). Similarly, minimal changes in activity were seen in glucan control groups. In contrast, endotoxin administration in glucan-pretreated rats produced synergistic elevations of both TXB_2 and 6-keto-$PGF_{1\alpha}$. Furthermore, unlike normal

TABLE 3   *Plasma immunoreactive (i)TXB_2 and 6-keto-$PGF_{1\alpha}$ in glucan-pretreated rats and control rats 30 minutes and 4 hours after endotoxin (LPS)*

Group	Time after LPS	(i)TXB_2 (pg/ml)	(i)6-keto-$PGF_{1\alpha}$ (pg/ml)
Normal (N = 5)	–	ND	ND
Control (N = 7)	30 min	243±48*	ND
Glucan (N = 5)	30 min	296±58*	ND
Glucan + LPS (N = 10)	30 min	1596±342	2457±250
Control (N = 5)	4 hrs	401±36*	ND
Glucan + LPS (N = 5)	4 hrs	3406±636	2345±540

Studies were conducted after an intravenous dose of 0.1 mg/kg *S. enteritidis* LPS. Values presented as mean±SEM. * $p < 0.01$ compared to glucan + LPS group. ND = nondetectable levels of (i)TXB_2 and (i)6-keto-$PGF_{1\alpha}$ (< 200 pg/ml). Reproduced from Cook et al (1982), with permission.

rats given higher doses of endotoxin (Cook et al 1980c), the group treated with glucan and endotoxin exhibited sustained increases in TXB_2 at 4 hours after endotoxin injection.

Since the alterations in plasma TXB_2 and 6-keto-$PGF_{1\alpha}$ could be reflecting the severity of shock rather than a primary event, we assessed the effect of pretreatment with glucan on endotoxin-stimulated in vitro synthesis of arachidonic acid metabolites by adherent peritoneal macrophages. Endotoxin incubated with adherent peritoneal macrophages significantly stimulated TXB_2 and 6-keto-$PGF_{1\alpha}$ synthesis (Table 4). However, peritoneal macrophages from rats pretreated with glucan intravenously or intraperitoneally exhibited further enhancement (2–3-fold) of endotoxin-stimulated synthesis of arachidonic acid metabolites (Table 4).

These results demonstrate that glucan-induced hypersensitivity to endotoxin is associated with synergistic increases in both plasma TXB_2 and 6-keto-$PGF_{1\alpha}$ in response to a normally well-tolerated dose of endotoxin. In a similar fashion, increases in basal and endotoxin-stimulated arachidonic acid metabolism were observed in vitro with glucan-activated macrophages. The latter results demonstrate that, in addition to altering phagocytic function and the total number of functional cells, glucan significantly altered the capacity of these cells to metabolize arachidonic acid. Additional studies are needed to elucidate the mechanisms underlying this altered cellular thromboxane and prostacyclin production. It appears, however, that macrophages at various stages of differentiation, that is, 'resident', 'elicited' or 'activated', have distinct functional and metabolic capacities including synthesis of arachidonic acid metabolites (Snider et al 1983). It was found that

TABLE 4   *Effect of glucan on (i)TXB_2 and (i)6-keto-$PGF_{1\alpha}$ synthesis by endotoxin(LPS)-stimulated adherent peritoneal macrophages*

Group	(i)TXB_2 (ng/ml)		(i)6-keto-$PGF_{1\alpha}$ (ng/ml)	
	−LPS	+LPS	−LPS	+LPS
Control (N = 10)	4.6±0.4	17.4±1.1	7.1±1.9	20.4±2.8
Glucan i.v. (N = 5)	16.9±1.8*	45.8±3.8*	17.2±2.3*	69.2±4.2*
Glucan i.p. (N = 5)	8.6±0.9*	40.5±2.3*	11.2±0.7	46.4±2.0*

Glucan was administered intravenously (i.v.) (10 mg/kg) for 5 consecutive days. Peritoneal cells were harvested 3 days after the last injection. In one group glucan was administered as an equivalent single dose intraperitoneally (i.p.) (50 mg/kg) 3 days prior to collection of cells. Adherent peritoneal macrophages were incubated (1×10^6 cells/ml) for 5 hours with or without *S. enteritidis* LPS (50 µg/ml). Values presented as mean±SEM. * $p < 0.001$ compared to control cells. Reproduced from Cook et al (1982), with permission.

312

resident peritoneal macrophages produce predominantly PGE_2 and PGI_2, although pyran copolymer- or BCG-activated macrophages also produce larger amounts of PGE_2 and the highest TXB_2 levels. Peptone- or thioglycolate-elicited macrophages generally synthesized prostaglandins and TXB_2 at a rate intermediate between resident and activated cells. Kato and Yamamoto (1982) and Grimm et al (1978) have likewise demonstrated increased arachidonic acid synthesis in macrophages activated by BCG and *C. parvum* respectively. These fatty acid cyclooxygenase products generated by the activated cells may play key roles in modulating immune responses, for example immunosuppression, as well as increasing the vulnerability of animals to lethal endotoxemia.

It appears that arachidonic acid metabolites mediate toxicity since we have observed that a fatty acid cyclooxygenase inhibitor, indometacin, attenuates the severity of endotoxin shock in glucan-stimulated rats (Cook et al 1982). In a similar fashion, Ferluga et al (1979) studied the effect of several pharmacologic agents including indometacin on hepatic function in mice rendered hyperreactive to endotoxin by *C. parvum*. They also concluded that prostaglandins are likely toxic mediators since indometacin ameliorated hepatic injury. Additionally, we have studied the response of essential fatty acid-deficient rats to glucan-induced hypersensitivity to endotoxin (Table 5). These rats lack the eicosanoid precursor arachidonic acid. The animals also exhibited reduced susceptibility to endotoxin following glucan pretreatment. Furthermore, the intrinsic resistance of essential fatty acid-deficient rats to glucan and endotoxin was prevented by prior administration of ethyl arachidonic acid before glucan injections were initiated. Similar

TABLE 5　*Resistance of essential fatty acid-deficient (EFAD) rats to glucan sensitization to endotoxin (LPS) shock*

Group	(i)TXB$_2$ (pg/ml)	(i)6-keto-PGF$_{1\alpha}$ (pg/ml)	24-hr mortality (dead/total)
Glucan + LPS (control diet)	1596±342 (10)	2457±250 (10)	13/14
Glucan + LPS (EFAD)	335±47* (5)	ND (5)	1/6*
Glucan + LPS (EFAD + ethyl arachidonate)	1273±121 (5)	3145±918 (5)	7/7

(i)TXB$_2$ and (i)6-keto-PGF$_{1\alpha}$ levels were measured 30 minutes after *S. enteritidis* LPS (0.1 mg/kg). One group of EFAD rats were given ethyl arachidonate (400 mg/rat intraperitoneally) on day 4 and day 1 before beginning glucan injections. Number of animals given in parentheses. Values presented as mean±SEM. *p < 0.01 compared to glucan + LPS (control diet). ND = nondetectable levels. Reproduced from Cook et al (1982), with permission.

results, therefore, have been obtained with two different experimental approaches, both of which decreased prostaglandin and thromboxane synthesis. Collectively these data implicate arachidonic acid metabolites as a component of glucan-potentiated susceptibility to endotoxin.

5.2. Lead acetate

Parenterally administered lead salts are effective phagocytic inhibitors. Trejo et al (1972) demonstrated impaired intravascular clearance of colloidal carbon in rats administered lead acetate (5 mg/rat). A depressed clearance half-time was evident within 6 hours following lead injection, and this phagocytic depression persisted for 72 hours. Similar suppressive actions of intravenous lead salts measured by clearance of colloidal carbon in rats and mice (Filkins and Buchanan 1973; Seyberth et al 1972) and of the ^{131}I RE-test lipid emulsion (Cook et al 1974; Trejo et al 1972) have been reported. This lead-induced phagocytic depression appears to be a cellular defect rather than an opsonic deficiency, since prior opsonization of colloids with normal serum did not alter the clearance kinetics (Trejo et al 1972). The depressed phagocytic activity following lead is associated with progressive ultrastructural changes in hepatic Kupffer cells (Cook et al 1975a; Hoffmann et al 1974, 1975). Granular electron-dense aggregates are seen in cytoplasmic vesicles of approximately 30% of Kupffer cells 5 hours after an injection of lead acetate (20 mg/kg) (Fig. 2). This material is predominantly associated with large vacuoles and smaller lysosome-like granules. It is probable that this electron-dense material represents a nondiffusable lead ligand since our studies have demonstrated a tissue distribution of ^{203}Pb, administered with a lead acetate carrier (20 mg/kg), that is characteristic of particulate clearance by the RES (Table 6). The rapid clearance of isotope was associated with an approximate 70% hepatic uptake of the total injected dose within 5 hours.

While investigating the pharmacologic action of a number of heavy metals, Selye et al (1966) observed that intravenous administration of well-tolerated doses of lead acetate (5 mg) to rats markedly sensitized these animals (100 000-fold) to the lethal effects of bacterial endotoxin. This increased susceptibility of lead-sensitized rats to endotoxin has been subsequently reported by several groups of investigators (Cook and DiLuzio 1973; Cook et al 1974, 1975a; Filkins 1973; Filkins and Buchanan 1973; Jones and Kiesow 1974; Shumer and Erve 1973; Trejo and DiLuzio 1971; Trejo et al 1972). The potentiating effect of lead acetate on endotoxin-induced lethality is not species-specific and can be demonstrated in a variety of other animals, including mice (DeClercq and Merigan 1969; Seyberth et al 1972), baboons (Holper et al 1973) and chicks (Truscott 1970).

Since the RES was thought to play an essential role in preventing systemic toxicity by sequestering intravascular endotoxin, Selye et al (1966) initially postulated that lead acetate may induce a blockade of the RES. Although intravascular clearance of certain colloidal agents is depressed by lead (Trejo et al 1972), it is

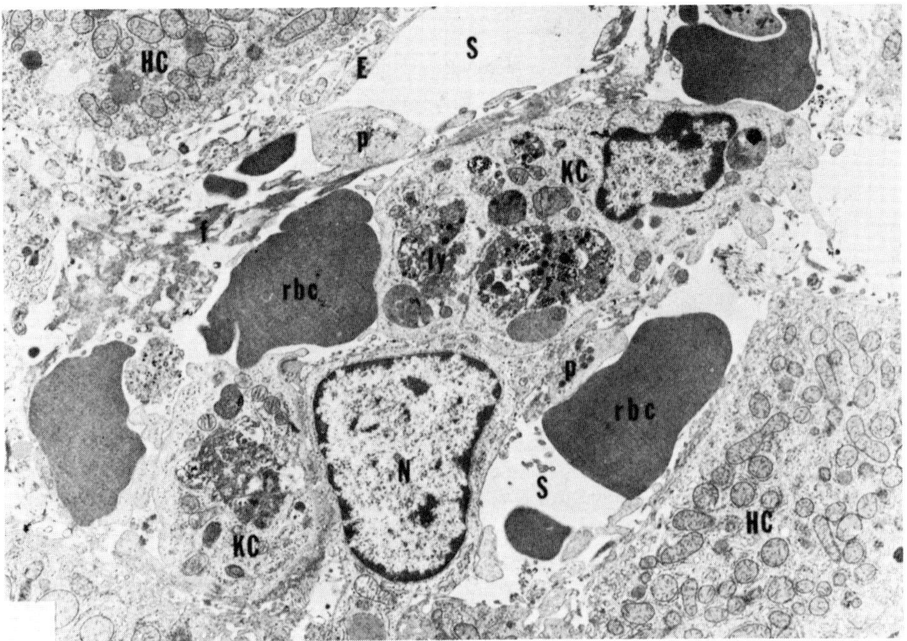

Fig. 2   *Liver tissue from a rat injected with Pb and endotoxin showing hepatic sinusoidal (S) engorgement with Kupffer cells (KC), red blood cells (rbc), blood platelets (p), and strands of fibrin (f). Kupffer cells are filled with pleomorphic lysosomes (ly) which contain remnants of red blood cells and electron-dense particulate material. Hepatocyte (HC) fine structure shows minor alterations including the lack of demonstrable glycogen particles and redistribution of rough and smooth endoplasmic reticulum. Endothelial (E) cell continuity is difficult to trace in these preparations. The nucleus (N) of a lipocyte occupies the lower central portion of the field. (× 2550). Reproduced from Cook et al (1981a), with permission.*

questionable whether phagocytic alterations per se contribute to the enhanced susceptibility of these animals to endotoxin. Increased sensitivity of lead-treated rats to endotoxin shock did not correlate well with periods of maximal phagocytic depression (Trejo et al 1972). Likewise, an actual depression of intravascular clearance of endotoxin following lead administration was either not apparent (Filkins and Buchanan 1973) or was quantitatively minor (Seyberth et al 1972). Studies conducted in our laboratory have also assessed the possibility that endotoxin alters the tissue distribution and thus toxicity of lead when these two agents are given conjointly (Cook et al 1981a). Tissue distribution of ^{203}Pb injected with a lead acetate carrier (20 mg/kg) was measured in rats with and without a concurrent dose of *S. enteritidis* endotoxin (Table 6). However, with the exception of lower renal localization of the isotope seen at 1 hour post-injection in the lead-endotoxin group, endotoxin did not significantly affect tissue distribution of the heavy metal.

TABLE 6 *Distribution of ^{203}Pb following acetate injection with and without concurrent endotoxin administration*

	Tissue distribution 1 hr (% ID/TO)		Tissue distribution 5 hrs (% ID/TO)	
	Pb	Pb + LPS	Pb	Pb + LPS
Liver	55.6 ±4.6	59.13 ±8.04	69.45 ±8.07	74.42 ±4.22
Spleen	0.98 ± .10	1.10 ±0.20	1.21 ±0.2	1.35 ±0.4
Kidney	0.85 ±0.04	0.57* ±0.07	1.68 ±0.29	1.64 ±0.14
Lung	0.58 ±0.06	0.67 ±0.06	0.39 ±0.06	1.08 ±0.32
Heart	0.25 ±0.02	0.21 ±0.02	0.073 ±0.010	0.10 ±0.01
Brain	0.038 ±0.008	0.03 ±0.006	0.05 ±0.004	0.05 ±0.006

^{203}Pb (0.5 Ci/100 g body wt.) was added to carrier lead acetate and injected intravenously (20 mg/kg) with or without *S. enteritidis* LPS (100 µg/kg). N = 8. Tissue distribution expressed as percentage of the injected dose, per total organ (% ID/TO). * $p < 0.05$ compared to the Pb group. Reproduced from Cook et al (1980a), with permission.

It is, therefore, unlikely that the toxic synergism between lead and endotoxin can be attributed to any unique clearance of these agents when administered in combination.

The high hepatic uptake of this intravenous dose of lead, that is, 60-70% of the injected doses, however, provides a basis for the hepatotoxic effect induced by heavy metal interaction with endotoxin (Table 7). The hepatic dysfunction was denoted by marked elevations in serum transaminase and ornithine carbamyl transferase activity (Table 7), or impaired sulfobromphthalein clearance (Cook and DiLuzio 1973; Cook et al 1974; Holper et al 1973). Ultrastructural observations also reveal extensive disruption of sinusoidal integrity involving endothelial and Kupffer cells in rats 5 hours after the administration of lead and endotoxin (Fig. 2) (Cook et al 1981a). These hepatic alterations appear to be progressive since sinusoidal microthrombosis and marked degenerative changes in hepatic parenchymal cells can be observed 24 hours after lead–endotoxin interaction in the baboon (Hoffman et al 1974).

TABLE 7 *Synergism between lead and endotoxin effect on hepatic integrity*

Group	Plasma glutamic pyruvic transaminase (SF units/ml plasma)	Plasma ornithine carbamyl transferase (sigma units/ml)	Mortality (dead/total)
Saline	19±5	343±30	–
Endotoxin	26±6	350±99	0/5
Pb	18±1	230±20	0/5
Pb + endotoxin	199±31*	34540±8283*	10/12
Pb + endotoxin + polysorbate	165±20*	27390±6217*	5/5
Pb + endotoxin + methyl palmitate	40±6**	2431±210**	0/11**
Pb + endotoxin + methylprednisolone	39±14**	1951±325**	0/10**
Pb + endotoxin + cysteine	70±29**	6235±1220**	2/12**

Rats were killed 5 hours after an intravenous dose of 20 mg/kg lead acetate and/or endotoxin (100 µg/kg); cysteine (150 mg/kg) or methylprednisolone (60 mg/kg) was administered concurrently with lead and endotoxin. Methyl palmitate was administered (1 µg/kg) for 2 days prior to lead-endotoxin injection. N = 7–8/group. Data expressed as mean±SEM. $*p <$ 0.01 compared to saline control group. $**p < 0.01$ compared to Pb + endotoxin group. Reproduced from Cook et al (1981a), with permission.

One sequela of hepatic dysfunction due to the toxic synergism between lead and endotoxin is severe hypoglycemia. Indeed, as with certain RE stimulants, the mechanism of lead sensitization to endotoxin has been attributed to lead-induced impairment of hepatic gluconeogenesis (Cornell and Filkins 1974; Filkins 1973; Rippe and Berry 1973). Studies by Hadbavny et al (1978) have also shown that lead acetate administration sensitized rats to lethal insulin-induced hypoglycemia, increased glucose tolerance and enhanced whole body oxidation of glucose. On the basis of these observations, it was postulated that the RES plays a modulatory role in insulin metabolism which is disrupted by lead poisoning.

It has been demonstrated that several agents protect against lead sensitization to endotoxin and reduce the severity of hepatic dysfunction and associated sequelae. These include diverse pharmacologic agents such as methylprednisolone (Cook and DiLuzio 1973), cysteine (Bertok 1968; Cook et al 1981a) and the RE phagocytic depressant methyl palmitate (Cook et al 1981a) (Table 7). When the protective agent cysteine is administered concurrently with lead acetate in the rat, there is a significant change in organ distribution of the heavy metal characterized by a several-fold increase in renal concentration and a concomitant 80% reduction in hepatic clearance of lead (Cook et al 1981a). This chelator may therefore prevent lead interaction with hepatic sulfhydryl proteins (Bertok 1968) or other metabolic factors essential in host resistance to endotoxin. Sakaguchi et al (1982) proposed that hepatic dysfunction secondary to lead-potentiated endotoxin shock was partly due to lipid peroxide formation and reduced sulfhydryl proteins. Hepatic peroxide levels markedly increased in livers of mice 6 hours post-intoxication. In contrast, there was a concomitant decrease in activity of the superoxide anion scavengers, glutathione peroxidase and superoxide dismutase.

Endoperoxide metabolites of arachidonic acid metabolism are also highly reactive intermediates that could mediate cellular damage. These observations, coupled with a previous report by Jones and Kiesow (1974) that lead acetate potentiates endotoxin-induced consumptive coagulopathies, prompted us to measure plasma levels of the arachidonic acid metabolites TXB_2 and 6-keto-$PGF_{1\alpha}$ in this shock model. In these studies, plasma from rats was collected at various intervals following an intravenous dose of lead acetate (20 mg/kg), with or without *S. enteritidis* endotoxin (100 µg/kg). When these agents were administered in combination, the plasma TXB_2 levels at 4 hours were higher than those produced by lead or endotoxin alone (Table 8). This level of TXB_2 may have contributed to systemic coagulopathies as there is a marked synergism between lead and endotoxin with respect to plasma levels of fibrin split products, an index of the severity

TABLE 8   *Plasma (i)TXB_2 levels during lead(Pb)-potentiated endotoxin (LPS) shock*

Group	Plasma (i)TXB_2 (pg/ml)	
	30 min post-injection	4 hrs post-injection
Control	ND (5)	ND (5)
Pb	183±59 (5)	496±157 (5)
LPS	345±44 (8)	528±99 (9)
Pb + LPS	625±120 (5)	1582±122* (8)

Lead acetate was administered intravenously (20 mg/kg) with or without a concurrent intravenous dose of 100 µg/kg *S. enteritidis* LPS. Number of animals in parentheses. ND = nondetectable levels (< 150 pg/ml). * $p < 0.01$ compared to Pb or LPS controls. Reproduced from Cook et al (1981a), with permission.

of disseminated intravascular coagulation (DIC) (Table 9). This marked elevation in fibrin split products in rats treated with lead and endotoxin was also associated with a severe thrombocytopenia at 4 hours following injections. There was a synergistic increase in plasma levels of 6-keto-PGF$_{1\alpha}$ in rats treated with lead and endotoxin compared with the lead-treated or endotoxin-treated group (Table 10). Elevated plasma PGI$_2$, therefore, did not prevent the severe systemic coagulopathies following combined administration of lead and endotoxin and may

TABLE 9   *Platelet counts and fibrin/fibrinogen degradation products following lead-potentiated endotoxin (LPS) shock*

Group	Platelet count ($\times 10^3$/mm^3)	Serum fibrinogen/fibrin degradation products (µg/ml)
Pb (N = 5)	520±18	0.3±0.1
LPS (N = 5)	420±35	5.3±3.7
Pb + LPS (N = 5)	130±51*	51.2±14*

Studies were conducted at 4 hours after lead acetate administration (20 mg/kg) with or without a concurrent dose of *S. enteritidis* LPS (100 µg/kg). * $p < 0.001$ compared to Pb or LPS control groups. Reproduced from Cook et al (1981a), with permission.

TABLE 10   *Plasma 6-keto-PGF$_{1\alpha}$ levels in rats following lead(Pb)-potentiated endotoxin (LPS) shock*

Group	30 min post-injection	4 hrs post-injection
Control	ND (5)	ND (5)
Pb	ND (5)	417±160 (5)
LPS	ND (5)	408±112 (9)
Pb + LPS	ND (5)	2350±496* (8)

Lead acetate (20 mg/kg) was administered intravenously with or without a concurrent dose of *S. enteritidis* LPS (100 mg/kg intravenously). Number of animals in parentheses. ND = nondetectable levels (< 200 pg/ml). * $p < 0.01$ compared to LPS or Pb controls. Reproduced from Cook et al (1981a), with permission.

be deleterious via its potent hypotensive actions (Bult and Herman 1982; Harris et al 1980).

Since arachidonic acid metabolites are increased secondary to lead–endotoxin interaction, studies were undertaken to assess whether either the fatty acid cyclooxygenase inhibitor indometacin, or essential fatty acid (EFA) deficiency would prevent lead sensitization to endotoxin shock. Rats were pretreated intravenously with indometacin (10 mg/kg) 30 minutes before administration of lead and endotoxin. Both the indometacin and EFA deficiency prolonged survival time ($p <$ 0.05) as indicated by respective mortalities of 40% and 18% compared to a 100% mortality in the shocked control group at 12 hours (Table 11). With indometacin, however, the overall 24-hour mortality was not changed. These data concur with the previous report of Reichgott and Engelman (1975) that administration of indometacin did not prevent the lethal lead–endotoxin interaction in the rat. Nevertheless, survival time was significantly improved, and such an observation suggests that arachidonic acid metabolites (e.g., TXA_2 and PGI_2) may contribute to the toxic synergism between lead and endotoxin. It is probable, however, that additional factors independent of arachidonic acid metabolites, for example lead inhibition of gluconeogenic function, may be the determining lethal factor in the extreme sensitization to endotoxin.

TABLE 11   *Effect of indometacin pretreatment or essential fatty acid deficiency (EFAD) on survival during lead(Pb)-potentiated endotoxin (LPS) shock*

Group	Mortality (dead/total) at			
	6 hrs	12 hrs	18 hrs	24 hrs
Pb	0/5	0/5	0/5	0/5
LPS	0/5	0/5	0/5	0/5
Pb + LPS	3/10	10/10	10/10	10/10
Pb + LPS (indometacin-pretreated)	0/10*	4/10*	8/10	10/10
Pb + LPS (EFAD)	0/11	2/11	6/11	6/11

Lead acetate (20 mg/kg) was administered intravenously with or without concurrent *S. enteritidis* LPS (100 µg/kg). Indometacin was injected intravenously (10 mg/kg) 30 minutes prior to lead-endotoxin administration. * $p <$ 0.05 compared to Pb + LPS group. Reproduced from Cook et al (1981a), with permission.

6. FACTORS WHICH PROTECT AGAINST ENDOTOXIN

6.1. Methyl palmitate

Artificial lipid emulsions have been used both to measure phagocytic function, for instance RE-test lipid emulsion, and to experimentally manipulate reticuloendothelial function (for review see DiLuzio 1972). Most studies with these emulsions have employed polysorbate 20, or less frequently, soya lecithin and poloxamer as dispersing agents (DiLuzio 1972). Pretreatment of rats with emulsions of glycerol triolate markedly stimulates RES phagocytic clearance whereas the glycerol ester of oleic acid produces mild but measurable RE stimulation (Stuart et al 1960; Stuart and Cooper 1963). In contrast, certain butyl, ethyl, and methyl esters containing 14 to 20 carbon units with varying degrees of saturation induced a depression of phagocytic clearance by the RES (DiLuzio and Blickens 1966). Wooles and DiLuzio (1963) stressed the employment of methyl palmitate as the lipid emulsion of choice to induce RE depression. RE cellular elements of spleen and liver of laboratory animals were histologically normal following injection of this emulsion, indicating that the depressed phagocytic function was not due to tissue destruction (Wooles and DiLuzio 1964). Hepatic parenchymal cell sulfobromphthalein clearance was also normal, thus indicating a selective functional change of Kupffer cells. Depression of phagocytic function was evident within 5 hours after administration of the emulsion and persisted for approximately 2 weeks (DiLuzio and Blickens 1966). Studies using ^{14}C-labeled methyl palmitate indicated that the phagocytic depression was not due to persistent storage of methyl palmitate in macrophages (Blickens and DiLuzio 1964). To test the possibility of opsonic deficiency, clearance studies were conducted in rats with and without serum-opsonized ^{131}I-labeled RE-test lipid emulsion (Saba and DiLuzio 1968). Vascular clearance of the opsonized RE-test lipid emulsion was enhanced. Depressed phagocytic uptake, however, in both liver and spleen persisted, thus indicating that the improved vascular clearance was by other tissues. These observations demonstrated that despite a mild opsonic deficiency, the major defect in clearance was due to hepatic and splenic phagocytosis.

The marked protective action of methyl palmitate against endotoxin shock (Crafton and DiLuzio 1969) is particularly intriguing since most RE-blockading agents sensitize to lethal endotoxemia. DiLuzio and Crafton (1970) were the first to demonstrate that rats pretreated with this lipid emulsion were refractory to endotoxin shock even when challenged with supra lethal doses of endotoxin (approximating $3 \times LD_{100}$) in polysorbate or normal control groups. The observation that prior administration of plasma from methyl palmitate-treated rats did not modulate host resistance of normal recipients suggested that no stable circulating plasma factor was involved. In contrast to its depressive action on clearance of colloidal carbon and the ^{131}I-labeled RE-test lipid emulsion, methyl palmitate did not alter the clearance of ^{51}Cr-labeled endotoxin as there was no change in the early rapid or slower exponential rate of clearance. The only major difference in tissue distribution between polysorbate-treated controls and methyl palmitate-treated rats was an approximate 80% reduction in splenic localization of the radioisotope (DiLuzio and Crafton 1970).

The effect of methyl palmitate on lysosomal integrity as originally reported by DiLuzio and Crafton (1970) is equally enigmatic. These investigators reported that an LD_{100} dose of endotoxin in normal rats produced no significant elevation in serum β-glucuronidase or acid phosphatase activity when measured at 3 hours post-shock. In contrast, this same dose of endotoxin, which was nonlethal in methyl palmitate-treated rats, caused either the same or greater elevation in plasma acid hydrolase activity. The authors concluded that lysosomal labilization may not be a causative factor in endotoxin shock and that loss of lysosomal integrity may be a result of endotoxin injury rather than its cause. Observations from our laboratory with regard to the effect of methyl palmitate on lysosomal stability in endotoxemia are at variance with these earlier studies (Cook et al 1982). We have also shown that methyl palmitate-pretreated rats are highly refractory to lethal endotoxemia. This improved survival, however, was associated with a markedly reduced activity of serum β-glucuronidase at 4 hours post-endotoxin challenge (Table 12). The protected animals also exhibited less severe hypoglycemia and reductions in serum fibrinogen/fibrin degradation products (Table 12). Additionally, as noted in the previous section, methyl palmitate was highly protective against lead-potentiated endotoxin shock in the rat (Cook et al 1981a). The animals treated with lead and endotoxin exhibited improved survival and preservation of hepatocyte integrity as denoted by reduced plasma elevations of hepatic transaminase and ornithine car-

TABLE 12 *Effect of methyl palmitate pretreatment on endotoxin(LPS)-induced lethality and associated sequelae*

Group	Serum β-glucuronidase (units/ml)	Blood glucose (mg/100 ml)	Serum fibrinogen/ fibrin degradation products (μg/ml)	24-hour mortality (dead/total)
Unshocked controls	39±8* (10)	120±20* (10)	< 0.5 (5)	–
Control + LPS	358±55 (12)	37±5 (12)	46±8.2 (12)	10/12
Methyl palmitate + LPS	55±12* (12)	110±18* (12)	16±4.4* (10)	1/14*

Methyl palmitate (1 μg/kg) was administered intravenously for 2 days prior to challenging with *S. enteritidis* LPS (20 mg/kg). Blood glucose, serum β-glucuronidase, and split products were measured 4 hours after *S. enteritidis* (15 mg/kg). Number of animals in parentheses. Values are presented as mean±SEM. * $p < 0.001$ compared to shocked control. Reproduced from Cook et al (1982), with permission.

bamyl transferase activity (Table 7). The latter protective action of methyl palmitate was not due to an altered clearance or distribution of lead acetate (Cook et al 1981a).

Since glucan-induced stimulation of RES function and hyperreactivity to endotoxin was associated with increased arachidonic acid metabolism, we were interested to see whether methyl palmitate would have opposing effects on endotoxin-stimulated synthesis of arachidonic acid metabolites (Cook et al 1982). Administration of an LD_{80} dose of endotoxin in polysorbate control rats induced a peak increase in plasma TXB_2 activity at 30 minutes with less pronounced increases in plasma 6-keto-$PGF_{1\alpha}$ (Table 13). By 4 hours there was reversal of the plasma TXB_2/6-keto-$PGF_{1\alpha}$ ratio with plasma 6-keto-$PGF_{1\alpha}$ levels increasing to more than 4000 pg/ml. Methyl palmitate-pretreated rats exhibited an approximately 80% reduction in plasma TXB_2 at 30 minutes after endotoxin compared with polysorbate controls, whereas plasma 6-keto-$PGF_{1\alpha}$ levels at this time interval were nondetectable. A similar degree of inhibition of the rise of plasma 6-keto-$PGF_{1\alpha}$ levels was seen at 4 hours after endotoxin in the methyl palmitate-pretreated group. As this inhibitory effect of methyl palmitate on arachidonic acid metabolism in endotoxin shock could be elicited by pharmacologic actions independent of its effects on the RES, the direct effect of this agent on in vitro macrophage arachidonic acid metabolism was assessed. As indicated in Table 14, peritoneal macrophages harvested from rats given intravenous or intraperitoneal pretreatment with methyl palmitate exhibited reductions in basal and/or endotoxin-stimulated synthesis of TXB_2 and 6-keto-$PGF_{1\alpha}$. Whether this inhibitory action of methyl palmitate is due to inhibition of membrane phospholipase or fatty acid cyclooxygenase, or represents interfer-

TABLE 13   *Plasma (i)TXB_2 and (i)6-keto-$PGF_{1\alpha}$ in methyl palmitate-pretreated rats and polysorbate control rats at 30 minutes and 4 hrs after endotoxin (LPS)*

Group	Time after LPS	(i)TXB_2 (pg/ml)	(i)6-keto-$PGF_{1\alpha}$ (pg/ml)
Normal (N = 10)	–	ND	ND
Control (N = 16)	30 min	2207±282	840±50
Methyl palmitate (N = 12)	30 min	484±50*	ND
Control (N = 6)	4 hours	581±142	4276±802
Methyl palmitate (N = 10)	4 hours	673±124	1080±168*

Studies were conducted after *S. enteritidis* LPS (15 mg/kg intravenously). ND = nondetectable levels of (i)TXB_2 and (i)6-keto-$PGF_{1\alpha}$ (< 200 pg/ml). Values presented as mean±SEM. * $p < 0.01$ compared to control. Reproduced from Cook et al (1982), with permission.

TABLE 14   *Effect of methyl palmitate on (i)TXB_2 and (i)6-keto-$PGF_{1\alpha}$ synthesis by control and endotoxin(LPS)-stimulated adherent peritoneal macrophages*

Group	(i)TXB_2 (ng/ml)		(i)6-keto-$PGF_{1\alpha}$ (ng/ml)	
	−LPS	+LPS	−LPS	+LPS
Control (N = 10)	4.6±0.4	17.4±1.1	7.1±1.9	20.4±2.8
Methyl palmitate i.v. (N = 5)	7.1±1.0*	13.4±0.8*	6.1±1.7	14.9±2.4
Methyl palmitate i.p. (N = 5)	1.3±0.1**	9.2±0.7**	1.5±0.5	2.3±0.3**

Methyl palmitate was injected either intraperitoneally or intravenously 2 days prior to the study. Adherent peritoneal macrophages were incubated (1×10^6 cells/ml) for 5 hours with or without *S. enteritidis* LPS (50 µg/ml). Values presented as mean±SEM. * $p < 0.05$ compared to control cells. ** $p < 0.001$ compared to control cells. Reproduced from Cook et al (1982), with permission.

ence at other sites in the arachidonic acid cascade remains to be delineated. These observations, however, further support the hypothesis that the RES plays a role in modulating arachidonic acid metabolism in endotoxemia.

6.2. Endotoxin tolerance

Animals are made tolerant to otherwise lethal doses of endotoxin by pretreatment with one or more low doses administered over a period of days prior to challenge with a larger dose. Tolerance is characterized by the attenuation of many of the classical symptoms of endotoxin shock including febrile response, hypotension, hypoglycemia, serum activity of lysosomal and hepatic enzymes, thrombocytopenia, systemic coagulopathies, increased levels of arachidonic acid metabolites, and lethality (Chedid and Parant 1971; for review see Rietschel et al 1980). While the phenomenon of tolerance has been studied for decades, the mechanisms responsible are not clearly understood. Several biochemical and physiologic conditions are associated with tolerance, including lysosomal stabilization, adrenal hypersecretion, increased degradation of endotoxin, antibody production, and hyperphagocytic activity of macrophages (for review see Chedid and Parant 1971). Past attempts to find one essential and consistent alteration that would represent a single mechanism of tolerance have been unsuccessful.

To investigate the participation of antibodies in the development of the tolerant state, Beeson (1947) attempted to transfer tolerance using plasma from tolerant animals, and was unsuccessful. He also found that the tolerance developed from a specific organism is maintained, although to a lesser degree, when animals are challenged with other, nonrelated organisms. Additionally, specific antibodies to endotoxin remain in circulation after tolerance is lost. These observations, taken together with results from later studies showing that thymectomized and agammaglobulinemic animals are capable of developing tolerance, strongly suggest that there is no direct relation between an immune response and endotoxin tolerance (Chedid and Parant 1971; for review see Snell 1971).

Other early work by Beeson (1946), using tolerant rabbits, strongly suggested that the RES was a major participant in the development of tolerance. Rabbits were made tolerant to *Salmonella typhosa* and reduced sensitivity to endotoxin was measured by a diminished febrile response. To investigate the role of the RES in this acquired resistance to endotoxin, the animals were administered the RE-blockading agent thorotrast. Tolerant rabbits given thorotrast showed a marked febrile response. Thus, the restoration of the febrile response suggested that the acquired tolerance had been abolished by the RE-blockading agent.

Additional studies by Greisman et al (1963) of tolerance and RE blockade, however, suggest that RE blockade does not abolish tolerance and that humoral factors at least contribute to tolerance. The investigators agreed that tolerance is associated with an increased ability to clear colloidal carbon from the circulation, but added that a general increased phagocytic activity of the RES is not necessary or sufficient to result in endotoxin tolerance. This evidence included the observation that tolerance (as measured by diminished febrile response) could be passively transferred, even to normal RE-blockaded animals. When compared to the controls after blockade, the tolerant animals still maintained tolerance, at least that measurable by reduced febrile response. It has been postulated that a highly specific factor acting as an opsonin prepares endotoxin for clearance by the RES. Studies by Jenkin and Rowley (1961) suggested that RE-blockading agents depress phagocytosis by inactivating these opsonins. Thus, a humoral opsonin that contributes to tolerance need not be totally inactivated by RE-blockading agents.

Greisman et al (1963) and others (Chedid and Parant 1971; for review see Snell 1971) have shown therefore that generalized enhanced phagocytosis of macrophages cannot account for acquired endotoxin tolerance. However, in light of the fact that macrophages release a battery of substances when stimulated with endotoxin, further consideration of the metabolism of these cells is warranted. Several groups have investigated the effect of the tolerant state on endotoxin-induced release of mediators from macrophages.

Snyder et al (1979) reported that peritoneal macrophages of tolerant rats release a smaller fraction of their β-glucuronidase before and after incubation with LPS in vitro than nontolerant controls. The results are consistent with the hypothesis that macrophage-derived mediators regulate the toxicity of endotoxin. The investigators suggested that tolerance is associated with a decreased propensity of macrophages to release lysosomal enzymes during phagocytosis. Recent in vivo studies of tolerant rats in our laboratory are consistent with these observations (Wise et al 1983). Serum β-glucuronidase and acid phosphatase levels were measured in tolerant and nontolerant rats after the injection of endotoxin. As is shown in Figure 3, levels of these lysosomal enzymes are decreased in tolerant animals as compared to nontolerant controls. In addition, measurements of SGOT and SGPT were significantly lower in tolerant animals, indicating preservation of hepatic integrity.

Another endotoxin-induced macrophage product, glucocorticoid-antagonizing factor (GAF), has been reported to contribute to the irreversibility of the shock state (Moore et al 1978). GAF is thought to make animals refractory to hydrocortisone-induced glucose and phosphoenol pyruvate carboxykinase synthesis and glycogen deposition. Tolerant mice were found to be unresponsive to endotoxin-induced GAF release (Goodrum and Berry 1979). Goodrum and Berry suggested that tolerance renders macrophages incapable of stimulation by endotoxin.

The onset of fever is an early symptom of endotoxin shock and results, in part, from the release of endogenous pyrogen. Macrophages are considered a significant

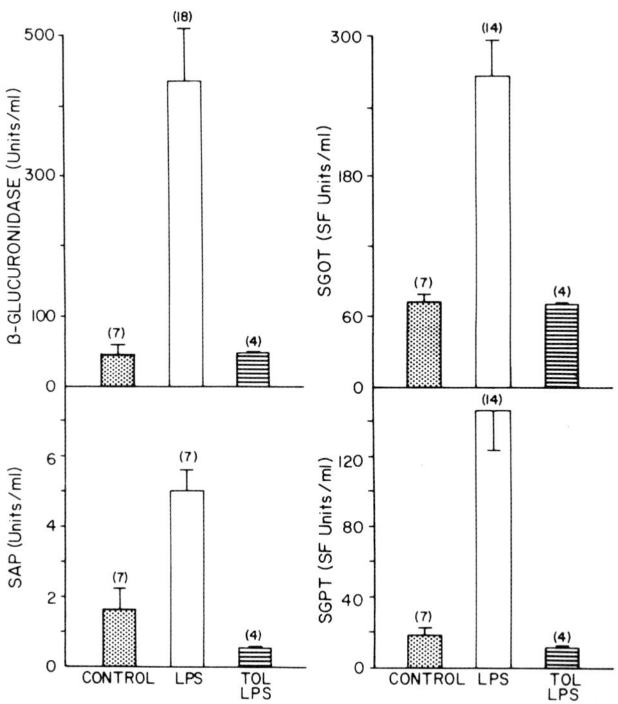

Fig. 3 *Lysosomal and hepatic enzymes during endotoxemia in tolerant (TOL) and nontolerant (control) Long-Evans rats. All enzyme levels were determined 4 hours after intravenous S. enteritidis endotoxin (15 mg/kg) (LPS). SAP = serum acid phosphatase; SGOT = serum glutamic oxaloacetic acid transaminase; SGPT = serum glutamic pyruvic transaminase. Values are presented as mean±SEM. Numbers in parentheses represent the number of animals per group. Only the LPS-treated group is significantly different (p < 0.01) from controls. Reproduced from Wise et al (1983), with permission.*

326

source of endotoxin-induced endogenous pyrogen (Dinarello et al 1968). As first reported by Beeson (1946), febrile response (and presumably the release of endogenous pyrogen) is diminished in the tolerant state. It has been shown that Kupffer cells rendered tolerant to endotoxin release a fraction of the amount of endogenous pyrogen released by normal cells. Other endogenous pyrogen-releasing cells do not exhibit this decreased production. Dinarello et al (1968) also showed that normal Kupffer cells, when incubated with endotoxin and plasma from tolerant animals, show decreased endogenous pyrogen release. This suggests that a plasma factor present in tolerance can inhibit endotoxin-induced endogenous pyrogen release.

Another important class of macrophage-derived endotoxin-induced mediators that are found decreased in the tolerant state are the eicosanoids. Rietschel et al (1980) reported a correlation between prostaglandin release and the susceptibility of mice to endotoxin. Mice pretreated with sublethal doses of endotoxin are much more susceptible (hyperreactive) to challenge 24 hours later. In contrast, mice challenged 4 days after pretreatment are tolerant to the endotoxin (Greer and Rietschel 1978). The investigators measured and compared PGE_2 and $PGF_{2\alpha}$ production by peritoneal macrophages from hyperreactive and tolerant mice (Schade and Rietschel 1979). It was found that cells from hyperreactive mice released PGE_2 and $PGF_{2\alpha}$ in greater amounts than did the non-pretreated control animals. Cells from the tolerant mice were completely refractory in their ability to release these prostaglandins upon endotoxin stimulation. In an attempt to correlate prostaglandin release to phagocytic activity, the investigators compared the uptake of zymosan by macrophages of hyperreactive and tolerant mice. They found that the phagocytic capacity of the tolerant macrophages was much greater than that of the hyperreactive cells and concluded that endotoxin-induced synthesis of arachidonic acid metabolites did not parallel phagocytic activity.

In our laboratory, two other eicosanoids, TXB_2, and 6-keto-$PGF_{1\alpha}$, were measured and compared in tolerant and nontolerant states (Wise et al 1983). We measured plasma levels of these metabolites in tolerant and normal rats after the injection of endotoxin at normally lethal or supralethal doses, 15 and 50 mg/kg, respectively. Nontolerant controls given 15 or 50 mg/kg endotoxin showed increased TXB_2 levels compared with unshocked controls (Fig. 4). Tolerant rats given either dose of endotoxin showed significantly lower TXB_2 levels and enhanced survival compared with nontolerant controls. To determine whether altered TXB_2 production in the tolerant rats was due to altered platelet TXB_2 synthesis, these cells were tested in vitro for their ability to synthesize TXB_2 (Wise et al 1983). As no significant difference was seen between tolerant and nontolerant rats, it was concluded that altered TXB_2 synthesis was occurring elsewhere. It is interesting that plasma TXB_2 levels in tolerant rats given an otherwise supralethal dose of 50 mg/kg increased to the same extent as those seen in nontolerant controls given the lethal 15 mg/kg dose. These findings raise the possibility that reduced TXB_2 production per se is either not essential in endotoxin tolerance and/or that potential aggregatory or vasopressor responses to TXA_2 are altered. Walker and Beasley

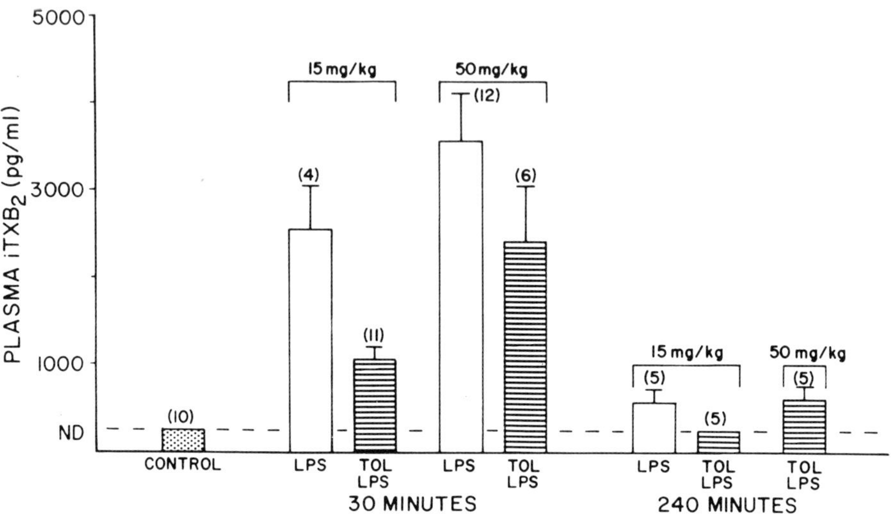

Fig. 4   *Plasma immunoreactive (i)TXB$_2$ (stable metabolite of TXA$_2$) levels 30 and 240 minutes after endotoxin (S. enteritidis 15 mg/kg intravenously) (LPS) in tolerant (TOL) and nontolerant (control) Long-Evans rats. Values are presented as mean±SEM. Numbers in parentheses represent the number of animals per group. All values are significantly (p < 0.01) different from control values. Reproduced from Wise et al (1983), with permission.*

(1980) reported that antibodies against the polysaccharide endotoxin antigen in tolerant rabbits appeared to accelerate the in vitro aggregation response of platelets to endotoxin, but in contrast with normal rabbits this aggregation was quickly reversible. Thus, potential deleterious aggregatory responses to TXA$_2$ could be attenuated in the tolerant animal.

Depressed plasma levels of 6-keto-PGF$_{1\alpha}$ were even more striking than those of TXB$_2$ in endotoxin-tolerant rats (Fig. 5). In contrast to a shocked control plasma level of approximately 4000 pg/ml at 4 hours post-endotoxin, plasma 6-keto-PGF$_{1\alpha}$ levels in tolerant rats were nondetectable or just above detectable limits (Fig. 5).

Rat peritoneal macrophages from tolerant and nontolerant controls were stimulated with endotoxin, and levels of TXB$_2$ and 6-keto-PGF$_{1\alpha}$ were measured. As is seen in Figure 6, levels of these eicosanoids are significantly lower in the tolerant cells compared with nontolerant controls both before and after endotoxin stimulation. Our finding that macrophages from tolerant cells are refractory to TXB$_2$ and 6-keto-PGF$_{1\alpha}$ synthesis extends the observations of Rietschel et al (1980) that these cells do not release PGE$_2$ and PGF$_{2\alpha}$ upon endotoxin stimulation. Additionally, Schade and Rietschel (1979) noted that, as production of PGE$_2$ and PGF$_{2\alpha}$

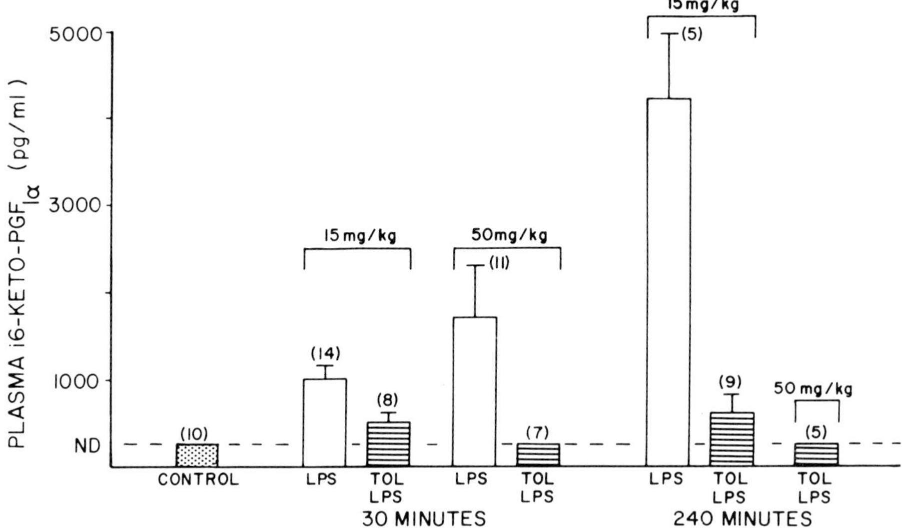

Fig. 5 *Plasma immunoreactive (i)6-keto-PGF$_{1\alpha}$ (stable metabolite of PGI$_2$) levels 30 and 240 minutes after endotoxin (S. enteritidis 15 mg/kg or 50 mg/kg intravenously) (LPS) in tolerant (TOL) and nontolerant (control) Long-Evans rats. Values are presented as mean±SEM. Numbers in parentheses represent the number of animals per group. All values are significantly (p < 0.01) different from control values except TOL + LPS at 30 and 240 minutes. Reproduced from Wise et al (1983), with permission.*

could be stimulated by incubation of the cells with zymosan, it appeared that the refractoriness of the tolerant state was endotoxin-specific and not due to altered capacity to release the prostaglandins. Collectively, these results raise the possibility that the RES metabolism of arachidonic acid is altered in endotoxin tolerance.

7. CONCLUSION

The material reviewed here lends credence to the concept that the RES plays a pivotal role in the pathogenesis of endotoxemia. No clear relationships, however, can be discerned with regard to predicting host response to endotoxin by measuring phagocytic function as the only end point. This is apparent since the presence of certain agents which depress or enhance RES phagocytic clearance of colloidal materials cannot consistently be related with endotoxin susceptibility. Additionally, there are frequent discrepancies in clearance and tissue distribution of colloidal materials used to measure RES function and the actual clearance and tissue localization of bacterial endotoxin.

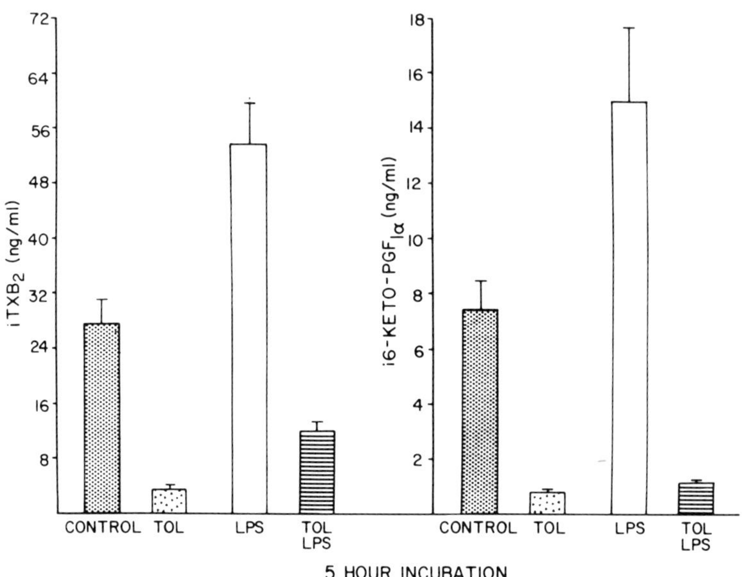

Fig. 6 *Culture medium levels of immunoreactive (i)TXB$_2$ and (i)6-keto-PGF$_{1\alpha}$ (ng/ml) 5 hours after in vitro incubation of adherent peritoneal cells (1 × 10^6/ml) from tolerant (TOL) and nontolerant (control) Long-Evans rats with and without S. enteritidis endotoxin (50 µg/ml) (LPS). Values are presented as mean±SEM with 5 animals per group. All values are significantly (p < 0.01) different from the control values. Reproduced from Wise et al (1983), with permission.*

In addition to phagocytic function, the potent secretory function of macrophages gives the phagocyte the potential to exert regulatory influences on its surrounding environment in control of inflammation and critical steps in the immune process. Of the plethora of mediators released following macrophage interaction with endotoxin, we have emphasized arachidonic acid metabolites and restricted this review to the pattern of release of eicosanoids with agents which modulate RES function and alter endotoxin sensitivity. The RE-phagocytic stimulant glucan, or the RE-phagocytic depressant lead acetate, both of which sensitize to lethal endotoxemia, elevated in vivo plasma levels of TXB$_2$ and 6-keto-PGF$_{1\alpha}$, and these factors appear to be related with shock severity. Parallel patterns of synthesis of these arachidonic acid metabolites were observed in vitro, at least with glucan-activated peritoneal macrophages. In contrast, RES phagocytic depression induced by methyl palmitate or RES phagocytic stimulation following endotoxin tolerance, both of which are protective, were associated with highly significant reductions in plasma TXB$_2$ and 6-keto-PGF$_{1\alpha}$. Similar refractory responses to in vitro endotoxin stimulation were observed with peritoneal macrophages. The question can be

330

posed as to cause-and-effect relationships between arachidonic acid metabolism and the severity of endotoxin shock. However, fatty acid cyclooxygenase inhibitors (Fletcher 1982) as well as essential fatty acid deficiency (Cook et al 1981c) have been previously demonstrated to be protective in experimental endotoxin shock. These separate experimental approaches also prevented or ameliorated certain shock sequelae in rats rendered hyperreactive to endotoxin.

The role of the RES in modulating the release of arachidonic acid metabolites during different types of circulatory stress merits further investigation. The relationship of these cyclooxygenase products to different stages of macrophage activation or heterogeneity of function within a given cell population (Goldyne and Stobo 1979; Picker et al 1980; Snider et al 1983) should be clarified. Likewise, the effect of other inflammatory mediators, for example complement (Hartung et al 1983; Rutherford and Schenkein 1983), and lipoxygenase products (Feuerstein et al 1981a) on macrophage synthesis of these metabolites needs to be more fully characterized.

Macrophages are not only a potent endogenous source of eicosanoid synthesis; these metabolites may function as important local hormones regulating many aspects of macrophage function. The latter include effects on phagocytosis (Razin et al 1978), spreading adhesion and migration (Cantarow et al 1978), regulating the release of other mediators (Kurland et al 1978; Schnyder et al 1982; Wahl et al 1979a), and expression of concanavalin A and Fc receptors (Razin and Globerson 1979). The E series prostaglandins have been shown to be potent inhibitors of lymphocyte function, for example mitogenesis and antibody or lymphokine production (Bray et al 1978; Goodwin et al 1977). Depending upon the state of macrophage activation or the nature of the inflammatory process, other arachidonic acid metabolites, for instance TXA_2, may be more prominent. Since the cyclooxygenase products, TXA_2 and PGI_2 have opposing functions on platelet aggregation and vascular smooth muscle, similar opposing actions of these and other metabolites may become evident with certain inflammatory and immune responses (Parker et al 1979).

These mediators may be of key importance in clinical septic shock or other forms of circulatory dysfunction. Studies from our laboratory (Reines et al 1982) and others (Fletcher 1982; Oettinger et al 1983; Parratt et al 1982) have demonstrated increases in plasma eicosanoids in patients during severe sepsis. More recently, Ninnemann et al (1983) have shown that endotoxin induces prostaglandin synthesis-dependent suppressor cells (monocytes) in septic patients with thermal injuries. Macrophage-derived prostaglandins also have been implicated in the pathophysiology of immunocompromised trauma patients (Miller 1982). It is evident that our present understanding of the immunology, biochemistry and cellular sources of arachidonic acid metabolites in circulatory homeostasis and the immune response must be clarified substantially.

ACKNOWLEDGEMENTS

We appreciate the technical assistance of Chance Kaplan, Sarah Ashton, Katherine Haines, and Marsha Black and the typing of Yvonne Mazzell and Barbara White.

REFERENCES

Adlam C (1973) Studies on the histamine sensitisation produced in mice by *Corynebacterium parvum. J. Med. Microbiol. 6*, 527.

Armstrong JM, Lattimer N, Moncada S, Vane JR (1978) Comparison of the vasodepressor effects of prostacyclin and 6-oxo-prostaglandin $F_{1\alpha}$ with those of prostaglandin E_2 in rats and rabbits. *Br. J. Pharmacol. 62*, 125.

Beeson PB (1946) Development of tolerance to typhoid bacterial pyrogen and its abolition by reticuloendothelial blockade. *Proc. Soc. Exp. Biol. Med. 61*, 248.

Beeson PB (1947a) Tolerance to bacterial pyrogens. *J. Exp. Med. 86*, 39.

Beeson PB (1947b) Effect of reticuloendothelial blockade on immunity to the Shwartzman phenomenon. *Proc. Soc. Exp. Biol. Med. 64*, 146.

Benacerraf B, Sebestyen M (1957) Effect of bacterial endotoxins on the reticuloendothelial system. *Fed. Proc. 16*, 860.

Benacerraf B, Thorbecke GJ, Jacoby D (1959) Effect of zymosan on endotoxin toxicity in mice. *Proc. Soc. Exp. Biol. Med. 100*, 796.

Berry LJ (1971) Metabolic effects of bacterial endotoxins. In: Kadis S, Weinbaum G, Ajl SJ (Eds), *Microbial Toxins Vol 5: Bacterial Endotoxins*, p 165. Academic Press, New York–London.

Berry LJ, Smythe DS (1965) Some metabolic aspects of tolerance to bacterial endotoxin. *J. Bacteriol. 90*, 970.

Bertok L (1968) Effect of sulfhydryl compound on the lead acetate-induced endotoxin hypersensitivity of rats. *J. Bacteriol. 95*, 1974.

Blickens DA, DiLuzio NR (1964) Effect of methyl cellulose on the reticuloendothelial system. *J. Reticuloendothel. Soc. 1*, 68.

Braude AI, Carey FJ, Zalesky M (1955) Studies with radioactive endotoxin. II. Correlation of physiologic effects and distribution of radioactivity in rabbits injected with lethal doses of *E. coli* endotoxin labelled with radioactive sodium chromate. *J. Clin. Invest. 34*, 858.

Bray MA, Gordon D, Morley J (1978) Prostaglandins as regulators in cellular immunity. *Prostaglandins Med. 1*, 183.

Bult H, Herman AG (1982) Prostaglandins in circulatory shock. In: Herman AG, Vanhoutte PM, Denolin H, Goossens A (Eds), *Cardiovascular Pharmacology and the Prostaglandins,* p 327. Raven Press, New York.

Butler RR, Wise WC, Halushka PV, Cook JA (1983) Gentamicin and indomethacin in the treatment of septic shock: effects on prostacyclin and thromboxane A_2 production. *J. Pharmacol. Exp. Ther. 225*, 94.

Carey FJ, Braude AI, Zalesky M (1958) Studies with radioactive endotoxin. III. The effect of tolerance on distribution of radioactivity after intravenous injection of *E. coli* endotoxin labeled with ^{51}Cr. *J. Clin. Invest. 37*, 441.

Cantarow WD, Cheung HT, Sundharadas C (1978) Effects of prostaglandins on spreading, adhesion and migration of mouse peritoneal macrophages. *Prostaglandins 16*, 39.

Chedid L, Parant M (1971) Hypersensitivity and tolerance in reactions to endotoxin. In:

Kadis S, Weinbaum G, Ajl SJ (Eds), *Microbial Toxins Vol 5: Bacterial Endotoxins*, p 415. Academic Press, New York–London.

Chedid L, Parant F, Parant M, Boyer F (1966) Localization and fate of ^{51}Cr-labeled somatic antigens of smooth and rough salmonellae. *Ann. N.Y. Acad. Sci. 133*, 712.

Chedid L, Parant M, Damais C, Parant F, Juy D, Galelli A (1976) Failure of endotoxin to increase nonspecific resistance to infection of lipopolysaccharide low-responder mice. *Infect. Immun. 13*, 722.

Cook JA, DiLuzio NR (1973) Protective effect of cysteine and methylprednisolone in lead acetate-endotoxin induced shock. *Exp. Mol. Pathol. 19*, 127.

Cook JA, Marconi EA, DiLuzio NR (1974) Lead, cadmium, endotoxin interaction: effect on mortality and hepatic function. *Toxicol. Appl. Pharmacol. 28*, 292.

Cook JA, Diluzio NR, Hoffmann EO (1975a) Factors modifying susceptibility to bacterial endotoxin: the effect of lead and cadmium. *CRC Crit. Rev. Toxicol. 3*, 201.

Cook JA, Hoffmann EO, DiLuzio NR (1975b) Influence of lead and cadmium on the susceptibility of rats to bacterial challenge. *Proc. Soc. Exp. Biol. Med. 150*, 741.

Cook JA, Dougherty WJ, Holt T (1980a) Enhanced sensitivity to endotoxin induced by the RE stimulant, glucan. *Circ. Shock. 7*, 225.

Cook JA, Holbrook TW, Parker BW (1980b) Visceral leishmaniasis in mice: protective effect of glucan. *J. Reticuloendothel. Soc. 27*, 567.

Cook JA, Wise WC, Halushka PV (1980c) Elevated thromboxane levels in the rat during endotoxic shock: protective effects of imidazole, 13-azaprostanoic acid, or essential fatty acid deficiency. *J. Clin. Invest. 65*, 227.

Cook JA, Dougherty WJ, Holt T (1981a) Distribution of [203]Pb during lead potentiated endotoxic shock. *Exp. Mol. Pathol. 34*, 253.

Cook JA, Wise WC, Halushka PV (1981b) Thromboxane A_2 and prostacyclin production by lipopolysaccharide stimulated peritoneal macrophages. *J. Reticuloendothel. Soc. 30*, 445.

Cook JA, Wise WC, Knapp DR, Halushka PV (1981c) Essential fatty acid deficient rats: a new model for evaluating arachidonate metabolism in shock. *Adv. Shock Res. 6*, 93.

Cook JA, Halushka PV, Wise WC (1982) Modulation of macrophage arachidonic acid metabolism: potential role in the susceptibility of rats to endotoxic shock. *Circ. Shock. 9*, 605.

Cooper GN, Stuart AE (1961) Sensitivity of mice to bacterial lipopolysaccharide following alterations of activity of the reticuloendothelial system. *Nature (London) 191*, 294.

Cornell RP, Filkins JP (1974) Depression of hepatic gluconeogenesis by acute lead administration. *Proc. Soc. Exp. Biol. Med. 147*, 371.

Crafton CG, DiLuzio NR (1969) Relationship of reticuloendothelial functional activity to endotoxin lethality. *Am. J. Physiol. 217*, 736.

Cremer N, Watson DW (1957) Influence of stress on distribution of endotoxin in RES determined by fluorescein antibody technique. *Proc. Soc. Exp. Biol. Med. 95*, 510.

DeClercq E, Merigan TC (1969) An active interferon inducer obtained from *Hemophilus influenza* type B. *J. Immunol. 103*, 899.

Delville J, Jacques PJ (1977) Therapeutic effect of yeast glucan in mice infected with *Mycobacterium leprae. Arch. Int. Physiol. Biochem. 85*, 965.

Demling RH, Smith M, Gunther R, Flynn JT, Gee M (1981) Pulmonary injury and prostaglandin production during endotoxemia in conscious sheep. *Am. J. Physiol. 240*, H348.

DiLuzio NR (1972) Employment of lipids in the measurement and modification of cellular, humoral and immune responses. *Adv. Lipid Res. 10*, 43.

DiLuzio NR (1976) Pharmacology of the reticuloendothelial system: accent on glucan. *Adv. Exp. Med. Biol. 73A*, 412.

DiLuzio NR, Blickens DA (1966) Influence of intravenously administered lipids on reticuloendothelial function. *J. Reticuloendothel. Soc. 3*, 250.

DiLuzio NR, Crafton CG (1969) Influence of altered reticuloendothelial function on vascular clearance and tissue distribution of *S. enteritidis* endotoxin. *Proc. Soc. Exp. Biol. Med. 132*, 686.

DiLuzio NR, Crafton CG (1970) A consideration of the role of the reticuloendothelial system (RES) in endotoxin shock. *Adv. Exp. Med. Biol. 9*, 27.

Dinarello CA, Bodel PT, Atkins E (1968) The role of the liver in the production of fever and in pyrogenic tolerance. *Trans. Assoc. Am. Physicians 81*, 334.

Ferluga J, Kaplun A, Allison AC (1979) Protection of mice against endotoxin-induced liver damage by antiinflammatory drugs. *Agents Actions 9*, 566.

Feuerstein N, Ramwell PW (1981) *In vitro* and *in vivo* effects of endotoxin on prostaglandin release from rat lung. *Br. J. Pharmacol. 73*, 511.

Feuerstein N, Bash JA, Woody JN, Ramwell PW (1981a) 3-Deazaadenosine, a transmethylase inhibitor, suppresses the effect of lipopolysaccharide on release of prostacyclin and thromboxane. *J. Pharm. Pharmacol. 33*, 401.

Feuerstein N, Foegh N, Ramwell PW (1981b) Leukotrienes C_4 and D_4 induce prostaglandin and thromboxane release from rat peritoneal macrophages. *Br. J. Pharmacol. 72*, 389.

Filkins JP (1973) Hypoglycemia and depressed hepatic gluconeogenesis during endotoxicosis in lead sensitized rats. *Proc. Soc. Exp. Biol. Med. 142*, 915.

Filkins JP (1977) Depression of RES function and sensitization to endotoxin shock and hypoglycemia. *J. Reticuloendothel. Soc. 22*, 461.

Filkins JP (1978) Phases of glucose dyshomeostasis in endotoxicosis. *Circ. Shock 5*, 347.

Filkins JP (1982) Role of the RES in the pathogenesis of endotoxic hypoglycemia. *Circ. Shock 9*, 269.

Filkins JP, Buchanan BJ (1973) Effects of lead acetate on sensitivity to shock, intravascular carbon and endotoxin clearances, and hepatic endotoxin detoxification. *Proc. Soc. Exp. Biol. Med. 142*, 471.

Fine J, Rutenberg S, Schweinburg FB (1959) The role of the reticuloendothelial system in hemorrhagic shock. *J. Exp. Med. 110*, 547.

Fletcher JR (1982) The role of prostaglandins in sepsis. *Scand. J. Infect. Dis. Suppl. 31*, 55.

Fletcher JR, Ramwell PW (1980) The role of prostaglandin synthetase inhibitors in shock and trauma. In: *Prostaglandin Synthetase Inhibitors: New Clinical Application*, p 257. Alan R. Liss, New York.

Frolich JC, Olgetree M, Peskar BA, Brigham KL (1980) Pulmonary hypertension correlated to pulmonary thromboxane synthesis. *Adv. Prostaglandin Thromboxane Res. 7*, 745.

Galvin MJ, Shupe K, Lefer AM (1978) Anti-endotoxin actions of methylprednisolone in isolated perfused cat liver. *Pharmacology 17*, 181.

Gilbert RP (1960) Mechanisms of the hemodynamic effects of endotoxin. *Physiol. Rev. 40*, 245.

Glode LM, Mergenhagen SE, Rosenstreich DL (1976) Significant contribution of spleen cells in mediating the lethal effects of endotoxin *in vivo*. *Infect. Immun. 14*, 626.

Goldyne ME, Stobo JD (1979) Synthesis of prostaglandins by subpopulations of human peripheral blood monocytes. *Prostaglandins 18*, 687.

Goodrum KJ, Berry LJ (1979) The use of Reuber hepatoma cells for the study of a lipopolysaccharide-induced macrophage factor: glucocorticoid antagonizing factor. *Lab.*

Invest. 41, 174.

Goodwin JS, Messner RP, Bankhurst AD, Peake GT, Saiki JH, Williams RC (1977) Prostaglandin producing suppressor cells in Hodgkin's disease. *N. Engl. J. Med. 297*, 963.

Greer GG, Rietschel ET (1978) Inverse relationship between the susceptibility of lipopolysaccharide (lipid A) pretreated mice to the hypothermic and lethal effect of lipopolysaccharide. *Infect. Immun. 20*, 366.

Greisman SE, Carozza FA, Hills JD (1963) Mechanisms of endotoxin tolerance. I. Relationship between tolerance and reticuloendothelial system phagocytic activity in the rabbit. *J. Exp. Med. 117*, 663.

Grimm W, Seitz M, Kirchner H, Gemsa D (1978) Prostaglandin synthesis in spleen cell cultures of mice injected with *Corynebacterium parvum. Cell. Immunol. 40*, 419.

Hadbavny AM, Buchanan BJ, Filkins JP (1978) Insulin and glucoregulatory alterations of RES depression. *J. Reticuloendothel. Soc. 24*, 57.

Halpern BN (1959) The role and function of the reticuloendothelial system in immunological processes. *J. Pharm. Pharmacol. 11*, 321.

Halushka PV, Cook JA, Wise WC (1983) Beneficial effects of UK37,248, a thromboxane synthetase inhibitor, in experimental endotoxic shock in the rat. *Br. J. Clin. Pharmacol. 15*, 133s.

Hamberg M, Svensson J, Samuelsson B (1975) Thromboxanes: a new group of biologically active compounds derived from prostaglandin endoperoxides. *Proc. Natl Acad. Sci. 72*, 2994.

Harris RH, Zmudka M, Moddox Y, Ramwell PW, Fletcher JR (1980) Relationships of TxB_2 and 6-keto-$PGF_{1\alpha}$ to hemodynamic changes during baboon endotoxic shock. *Adv. Prostaglandin Thromboxane Res. 7*, 843.

Hartung H, Hadding U, Bitter-Suermann D, Gemsa D (1983) Stimulation of prostaglandin E and thromboxane synthesis in macrophages by purified C3b. *J. Immunol. 130*, 2861.

Holbrook TW, Cook JA, Parker BW (1981) Glucan enhanced immunogenicity of killed erythrocytic stages of *Plasmodium berghei. Infect. Immun. 32*, 542.

Hoffmann EO, DiLuzio NR, Holper K, Brettschneider L, Coover J (1974) Ultrastructural changes in the liver of baboons following lead and endotoxin administration. *Lab. Invest. 30*, 311.

Hoffmann EO, Cook JA, DiLuzio NR, Coover JA (1975) The effects of acute cadmium administration in the liver and kidney of the rat – light and electron microscopic studies. *Lab. Invest. 32*, 655.

Holper K, Trejo RA, Brettschneider L, DiLuzio NR (1973) Enhancement of endotoxic shock in the lead sensitized subhuman primates. *Surg. Gynecol. Obstet. 136*, 593.

Janoff A, Zweifach BW (1963) Effect of endotoxin – tolerance, cortisone, and thorotrast on release of enzymes from subcellular particles of mouse liver. *Proc. Soc. Exp. Biol. Med. 114*, 695.

Janoff A, Weissmann G, Zweifach BW, Thomas L (1962) Pathogenesis of experimental shock. IV. Studies on lysosomes in normal and tolerant animals subjected to lethal trauma and endotoxemia. *J. Exp. Med. 116*, 451.

Jenkins CR, Rowley D (1961) The role of opsonins in the clearance of living and inert particles by cells of the reticuloendothelial system. *J. Exp. Med. 114*, 363,

Jones RB, Kiesow LA (1974) Potentiation of endotoxin-induced consumptive coagulopathy by lead acetate administration. *Infect. Immun. 10*, 1343.

Kato K, Yamamoto K (1982) Involvement of prostaglandin E_1 in delayed-type hypersensitivity suppression induced with live *Mycobacterium bovis* BCG. *Infect. Immun. 36*, 426.

Kokoshis PL, Williams DL, Cook JA, DiLuzio NR (1978) Increased resistance to *Staphylococcus aureus* infection and enhancement in serum lysozyme activity by glucan. *Science 199*, 1340.

Kurland JI, Bockman R (1978) Prostaglandin E production by human blood monocytes and mouse peritoneal macrophages. *J. Exp. Med. 147*, 952.

Kurland JI, Broxmeyer HE, Pelus LM, Bockman RS, Moore MAS (1978) Role of monocyte-macrophage-derived colony-stimulating factor and prostaglandin E in the positive and negative feedback control of myeloid stem cell proliferation. *Blood 52*, 388.

Lemperle G (1966) Effects of RES stimulation on endotoxin shock in mice. *Proc. Soc. Exp. Biol. Med. 122*, 1012.

Maier RV, Ulevitch RJ (1981) The response of isolated hepatic rabbit macrophages (H-MØ) to lipopolysaccharide (LPS). *Circ. Shock 8*, 165.

Mathison JC, Ulevitch RJ (1979) The clearance, tissue distribution, and cellular localization of intravenously injected lipopolysaccharide in rabbits. *J. Immunol. 123*, 2133.

Metchnikoff E (1905) *Immunity in Infective Disease.* Cambridge University Press, Boston.

Michalek SM, Moore RN, McGhee JR, Rosenstreich DL, Mergenhagen SE (1980) The primary role of lymphoreticular cells in the mediation of host responses to bacterial endotoxin. *J. Infect. Dis. 141*, 55.

Miller CL (1982) Infection in massively injured patients. *Prog. Clin. Biol. Res. 108*, 91.

Moore RN, Goodrum KJ, Couch RE, Berry LJ (1978) Elicitation of endotoxemic effects in C3H/HeJ mice with glucocorticoid antagonizing factor and partial characterization of the factor. *Infect. Immun. 19*, 79.

Moore RN, Urbaschek R, Wahl LM, Mergenhagen SE (1979) Prostaglandin regulation of colony-stimulating factor production by lipopolysaccharide-stimulated murine leukocytes. *Infect. Immun. 26*, 408.

Moncada S, Vane JR (1978) Pharmacology and endogenous roles of prostaglandin endoperoxides, thromboxane A_2 and prostacyclin. *Pharmacol. Rev. 30*, 293.

Morrison DC, Ulevitch RJ (1978) The effects of bacterial endotoxins on host mediation system. *Am. J. Pathol. 93*, 527.

Nathan CF, Murray HW, Cohn ZA (1980) The macrophage as an effector cell. *N. Engl. J. Med. 303*, 622.

Ninnemann JL, Stockland AE, Condie JT (1983) Induction of prostaglandin synthesis-dependent suppressor cells with endotoxin: occurrence in patients with thermal injuries. *J. Clin. Immunol. 3*, 142.

Noyes HE, McInturf CR, Blahuta GJ (1959) Studies on distribution of *Escherichia coli* endotoxin in mice. *Proc. Soc. Exp. Biol. Med. 100*, 65.

Oettinger WKE, Walter GO, Jensen UM, Beyer A, Peskar A (1983) Endogenous prostaglandin $F_{2\alpha}$ in the hyperdynamic state of severe sepsis in man. *Br. J. Surg. 70*, 237.

Palmerio C, Fine J (1969) The nature of resistance to shock. *Arch. Surg. 98*, 679.

Parker CW, Stenson WF, Huber MG, Kelly JP (1979) Formation of thromboxane B_2 and hydroxyarachidonic acids in purified human lymphocytes in the presence and absence of PHA. *J. Immunol. 122*, 1572.

Parant M, Parant F, Chedid L, Drapier JC, Petit JF, Wietzerbin J, Lederer E (1977) Enhancement of nonspecific immunity to bacterial infection by cord factor (6,6'-trehalose dimycolate). *J. Infect. Dis. 135*, 771.

Parratt JR, Coker SJ, Hughes B (1982) The possible role of prostaglandins and thromboxanes in the pulmonary consequences of experimental endotoxin shock and clinical sepsis. In: McConn R (Ed), *Role of Chemical Mediators in the Pathophysiology of Acute*

Illness and Injury. p 195. Raven Press, New York.

Peavy DL, Baughn RE, Musher DM (1979) Effects of BCG infection on the susceptibility of mouse macrophages to endotoxin. *Infect. Immun. 24*, 59.

Picker LJ, Raff HV, Goldyne ME, Stobo JD (1980) Metabolic heterogeneity among human monocytes and its modulation by PGE_2. *J. Immunol. 124*, 2557.

Razin E, Globerson A (1979) The effect of various prostaglandins on plasma membrane receptors and function of mouse macrophages. *Adv. Exp. Med. Biol. 114*, 415.

Razin E, Bauminger S, Globerson A (1978) Effect of prostaglandins on phagocytosis of sheep erythrocytes by mouse peritoneal macrophages. *J. Reticuloendothel. Soc. 23*, 237.

Regelson W, Morahan P, Kaplan AM, Baird LG, Munson JA (1973) Synthetic polyanions: molecular weight, macrophage activation and immunologic response. In: Wagner WH, Hahn H, Evans R (Eds), *Activation of Macrophages*, p 97. Excerpta Medica, Amsterdam.

Reichgott MJ, Engelman K (1975) Indomethacin: lack of effect on lethality of endotoxin in rats. *Circ. Shock 2*, 215.

Reines HD, Halushka PV, Cook JA, Wise WC, Rambo W (1982) Plasma thromboxane concentrations are raised in patients dying with septic shock. *Lancet 2*, 174.

Reynolds JA, Kastello MD, Harrington DG, Crabbs CL, Peters CJ, Jemski JV, Scott GH, DiLuzio NR (1980) Glucan-induced enhancement of host resistance to selected infectious diseases. *Infect. Immun. 30*, 51.

Ribi EE, Cantrell JL, Von Eschen KB, Schwartzman SM (1979) Enhancement of endotoxic shock by N-acetylmuramyl-l-alanyl-(l-seryl)-d-isoglutamine (Muramyl Dipeptide). *Cancer Res. 39*, 4756.

Rietschel ET, Schade U, Luderitz O, Fischer H, Peskar BA (1980) Prostaglandins in endotoxicosis. In: D Schlessinger (Ed), *Microbiology – 1980*, p 66. American Society for Microbiology, Washington, D.C.

Rippe DF, Berry LJ (1973) Metabolic manifestations of lead acetate sensitization to endotoxin in mice. *J. Reticuloendothel. Soc. 13*, 527.

Rosenstreich D, Vogel SN (1980) Central role of macrophages in the host response to endotoxin. In: D Schlessinger (Ed), *Microbiology – 1980*, p 11. American Society for Microbiology, Washington, D.C.

Rubenstein HS, Fine J, Coons AH (1962) Localization of endotoxin in the walls of the peripheral vascular system during lethal endotoxemia. *Proc. Soc. Exp. Biol. Med. 111*, 458.

Rutenburg S, Skarnes R, Palmerio C, Fine J (1967) Detoxification of endotoxin by perfusion of liver and spleen. *Proc. Soc. Exp. Biol. Med. 125*, 455.

Rutherford B, Schenkein H (1983) C3 cleavage products stimulate release of prostaglandins by human mononuclear phagocytes *in vitro. J. Immunol. 130*, 874.

Saba TM, DiLuzio NR (1968) Evaluation of humoral and cellular mechanisms of methyl palmitate induced reticuloendothelial depression. *Life Sci. 7(2)*, 337.

Sakaguchi O, Abe H, Sakaguchi S, Hsu C (1982) Effect of lead acetate on superoxide anion generation and its scavengers in mice given endotoxin. *Microbiol. Immunol. 26*, 767.

Schade V, Rietschel ET (1979) Differences in lipopolysaccharide-induced prostaglandin release and phagocytosis capacity of peritoneal macrophages from LPS-hyperreactive and tolerant mice. *Toxicon 17 (Suppl 1)*, 162.

Schnyder J, Dewald B, Baggiolini M (1982) Prostaglandin E_2 is a feedback regulator of macrophage activation. *Adv. Exp. Biol. Med. 155*, 535.

Selye H, Tuchweber B, Bertók L (1966) Effect of lead acetate on susceptibility of rats to bacterial endotoxins. *J. Bacteriol. 91*, 884.

Senterfitt VC, Shands JW (1978) Endotoxin induced metabolic alterations in BCG infected (hyperreactive) mice. *Proc. Soc. Exp. Biol. Med. 159*, 69.

Seyberth HW, Schmidt-Gayk H, Hackenthal E (1972) Toxicity clearance and distribution of endotoxin in mice as influenced by actinomycin D, cycloheximide, α-amanitin and lead acetate. *Toxicon 10*, 491.

Shands JW, Senterfitt VC (1972) Endotoxin induced hepatic damage in BCG-infected mice. *Am. J. Pathol. 67*, 23.

Schuler JJ, Erve PR, Schumer W (1976) Glucocorticoid effect on hepatic carbohydrate metabolism in the endotoxin-shocked monkey. *Ann. Surg. 183*, 345.

Shumer W, Erve PR (1973) Endotoxin sensitivity of adrenalectomized rats treated with lead acetate. *J. Reticuloendothel. Soc. 13*, 122.

Snell ES (1971) Endotoxin and the pathogenesis of fever. In: Kadis S, Weinbaum G, Ajl SJ (Eds), *Microbial Toxins Vol 5: Bacterial Endotoxins*, p 277. Academic Press, New York - London.

Snider ME, Fertel RH, Zwilling BS (1983) Production of arachidonic acid metabolites by operationally defined macrophage subsets. *Prostaglandins 25*, 491.

Snyder SL, Eklund SK, Walker RI (1979) Release of β-glucuronidase from peritoneal macrophages of normal and endotoxin-tolerant mice. *Can. J. Microbiol. 25*, 1245.

Stenson WF, Parker CW (1980) Prostaglandins, macrophages, and immunity. *J. Immunol. 125*, 1.

Stevens M, Steven P, Cook JA, Ichinose H, DiLuzio NR (1976) Protective effect of the immunostimulant glucan on sporotrichosis infections of mice. *J. Reticuloendothel. Soc. 20*, 66a.

Stuart AE, Cooper GN (1962) Susceptibility of mice to bacterial endotoxin after modification of reticuloendothelial function by simple lipids. *J. Pathol. Bacteriol. 83*, 245.

Stuart AE, Cooper GN (1963) Stimulation of reticuloendothelial phagocytic function by glyceryl tricaprate and 2-oleodistearin. *Exp. Mol. Pathol. 2*, 215.

Stuart AE, Biozzi G, Stiffel C, Halpern BN, Mouton D (1960) The stimulation and depression of reticuloendothelial phagocytic function by simple lipids. *Br. J. Exp. Pathol. 41*, 599.

Suter E (1962) Hyperreactivity to endotoxin in infection. *Trans. N.Y. Acad. Sci. Ser. II. 24*, 281.

Suter E, Ullman GE, Hoffman RG (1958) Sensitivity of mice to endotoxin after vaccination with BCG *(Bacillus Calmette-Guérin). Proc. Soc. Exp. Biol. Med. 99*, 167.

Svensson J, Fredholm B (1977) Vasoconstrictor effect of thromboxane A_2. *Acta Physiol. Scand. 101*, 366.

Tanaka N, Nishimura T, Yoshiyuki T (1959) Histochemical studies on the cellular distribution of endotoxin of *Salmonella enteritidis* in mouse tissues. *Jpn. J. Microbiol. 3*, 191.

Trejo RA, DiLuzio NR (1971) Impaired detoxification as a mechanism of lead acetate-induced hypersensitivity to endotoxin. *Proc. Soc. Exp. Biol. Med. 136*, 889.

Trejo RA, DiLuzio NR, Loose LD, Hoffman E (1972) Reticuloendothelial and hepatic function alterations following lead acetate administration. *Exp. Mol. Pathol. 17*, 145.

Truscott RB (1970) Endotoxin studies in chicks: effect of lead acetate. *Can. J. Comp. Med. 34*, 134.

Ulevitch RJ, Cochrane CG (1978) Role of complement in lethal bacterial lipopolysaccharide-induced hypotensive and coagulative changes. *Infect. Immun. 19*, 204.

Unanue ER (1976) Secretory function of mononuclear phagocytes. *Am. J. Pathol. 83*, 396.

Villa S, deGaetano G, Semeraro N (1981) Increased vascular prostacyclin activity in rats after endotoxin administration. *Experientia 37*, 494.

Wahl LM, Olsen CE, Sandberg AL, Mergenhagen SE (1977) Prostaglandin regulation of macrophage collagenase production. *Proc. Natl Acad. Sci. 74*, 4955.

Wahl LM, Rosenstreich DL, Glode LM, Sandberg AL, Mergenhagen SE (1979) Defective prostaglandin synthesis by C3H/HeJ mouse macrophages stimulated with endotoxin preparations. *Infect. Immun. 23*, 8.

Walker RI, Beasley WJ (1980) Evidence that antibodies to polysaccharide alter platelet response to endotoxin in tolerant rabbits. *Can. J. Microbiol. 26*, 1241.

Weiss HJ, Turitto VT (1979) Prostacyclin (prostaglandin I_2, PGI_2) inhibits platelet adhesion and thrombus formation on subendothelium. *Blood 53*, 244.

Williams DL, DiLuzio NR (1980) Glucan-induced modification of murine viral hepatitis. *Science 208*, 67.

Williams DL, Cook JA, Hoffmann EO, DiLuzio NR (1978) Protective effect of glucan in experimentally induced candidiasis. *J. Reticuloendothel. Soc. 23*, 479.

Williams DL, Browder IW, DiLuzio NR (1983) Immunotherapeutic modification of *Escherichia coli*-induced experimental peritonitis and bacteremia by glucan. *Surgery 93*, 448.

Wise WC, Cook JA, Halushka PV (1983) Arachidonic acid metabolism in endotoxin tolerance. *Adv. Shock Res. 10*, 131.

Wooles WR, DiLuzio NR (1963) Reticuloendothelial function and the immune response. *Science 142*, 1078.

Wooles WR, Diluzio NR (1964) Inhibition of homograft acceptance and homo- and heterograft rejection in chimeras by reticuloendothelial system stimulation. *Proc. Soc. Exp. Biol. Med. 115*, 756.

Zweifach BW (1958) Microcirculatory derangements as a basis for the lethal manifestations of experimental shock. *Br. J. Anaesth. 30*, 466.

Handbook of Endotoxin, Vol. 3: Cellular Biology of Endotoxin
L.J. Berry, editor
© Elsevier Science Publishers B.V., 1985

CHAPTER 15

Interactions between endotoxin and protein synthesis*

S.G. BRADLEY

1. INTRODUCTION

Gram-negative bacterial endotoxin or lipopolysaccharide (LPS) has the capability to elicit elevated enzyme activities and to lower other enzyme activities when administered to intact animals or when added to cells in culture and to subcellular fractions (Berry 1977: Bradley 1979; Urbaschek and Urbaschek 1977). Another characteristic feature of endotoxicosis resulting from gram-negative bacterial sepsis in man and laboratory animals is a marked loss of body protein (Beisel 1975; Clowes et al 1976; Garlick et al 1980; Tavakoli and Mela 1982). The loss of cellular constituents appears to be the result of enhanced degradation of intracellular proteins (Long et al 1981). The primary site of intracellular protein degradation is generally considered to be the lysosome. Consistent with this proposition, increased protein catabolism is usually associated with increased lysosomal enzyme activity (Ballard 1978; Hersko and Ciechanover 1982). Indeed, lysosomal enzyme activities are elevated in animals and cell cultures treated with LPS (Bradley 1979 and 1981).

Bacterial LPS persists in the blood for some time after intravenous injection before it is cleared by the liver. LPS in blood may be found in platelets and polymorphonuclear neutrophils, but not in erythrocytes. Circulating LPS is mainly in the plasma and may interact directly with complement and the coagulation system. In plasma, LPS is predominantly complexed with a high density lipoprotein (Freudenberg et al 1980). LPS is converted to a low density form by binding to plasma lipoproteins in the blood. The low density form has a lower molecular size than the injected LPS, reduced ability to produce acute neutropenia in rabbits, and reduced anticomplementary activity (Ulevitch and Johnston 1978). These alterations occur in the absence of extensive degradation of the glycolipid backbone of LPS. Mathison and Ulevitch (1979) found no evidence of major degradative changes in the subunit structure of LPS incubated with serum lipoprotein or liver

---

*   The ongoing research in immunotoxicology is supported in part by a contract from the National Institute of Environmental Health Sciences and by a Center of Excellence Award from the Commonwealth of Virginia.

homogenates which is in contrast to the conclusion of Skarnes (1966), who proposed that LPS is enzymatically cleaved during inactivation by serum proteins.

2. EFFECTS OF LPS ON ANIMALS

Endotoxin is involved in a complex array of metabolic processes in which LPS specifically perturbs protein synthesis and catabolism, and selected proteins interact with LPS, thereby modulating its activity. It is not known whether LPS exerts its primary toxic effects directly on cells or through release of soluble mediators (Tavakoli and Moon 1981). It is well established, however, that LPS attaches quickly to the surface of most mammalian cell types and becomes internalized subsequently (Silver 1981).

Endotoxin injected intravenously into rabbits is cleared rapidly from the blood and accumulates in the liver. According to Maier et al (1981), LPS is localized in the hepatic macrophages whereas the hepatocytes and endothelial cells of the rabbit liver are relatively free of LPS. Endotoxin in the phagocytic vacuoles of rabbit hepatic macrophages can be recovered as apparently unaltered LPS. Hepatic macrophages, treated with LPS in vitro, show elevated levels of acid phosphatase, lactate dehydrogenase, lysozyme, and plasminogen activator after 45 hours but decreased β-glucuronidase activity. Livers from LPS-treated mice have impaired glycogen synthesis and gluconeogenesis which are correlated with reduced activities of glycogen synthase *a*, phosphoenolpyruvate carboxykinase, glucose-6-phosphatase, and fructose 1,6-diphosphatase (McCallum 1981). Endotoxin alters gluconeogenesis when it is added directly to hepatocytes from normal fasted rats (Filkins and Cornell 1974). Kupffer cells from C3HeB/FeJ mice, cultured in the presence of LPS, release a substance that inhibits phosphoenolpyruvate carboxykinase activity in mouse hepatocytes from either the LPS-responsive C3HeB/FeJ or refractory C3H/HeJ mouse. The spleen is another target organ when data are based upon μg LPS per gram of tissue. Direct adrenal damage by LPS has also been proposed as contributing to the endotoxemic syndrome (Maier et al 1981). It is difficult to determine in studies performed in the presence of serum whether LPS is directly responsible for the observed effect or whether a mediator is produced, for example, by activation of the complement system (Wilson et al 1981). Endotoxin unquestionably exerts profound effects in vivo on a number of host mediator systems.

Endotoxin is responsible for a wide variety of pathophysiological effects, for example, fever, acidosis, decreased cardiac output, cardiac arrhythmia, vasodilatation, granulocytosis, thrombocytopenia, disseminated intravascular coagulation, hypoxia, and death. Many of the effects of LPS are biphasic: hyperglycemia followed by hypoglycemia; hypercalcemia followed by hypocalcemia; and hyperferremia followed by hypoferremia. LPS also acts as a mitogen and adjuvant, and stimulates resistance to a number of infections.

Table 1   *Substances that elicit enhanced lethality when administered in combination with gram-negative bacterial lipopolysaccharide*

Test substance	Treatment	Reference
Actinomycin D	with LPS	Berry 1964
Adriamycin	pre-LPS	Bradley 1974
α-Amanitin	with LPS	Seyberth et al 1972
8-Azaguanine	with LPS	Berry and Smythe 1964
5-[3,3-bis(2-chloroethyl)-1-triazeno]imidazole-4-carboxamide	post-LPS	Bradley et al 1975a
bis-DEAE-fluorenone(tilorone)	pre-LPS	Levine and Sowinski 1976
Cadmium acetate	with LPS	Cook et al 1974
Camptothecin	with LPS	Bradley 1974
Cannabinol	with LPS	Munson et al 1978
Chlorambucil	post-LPS	Bradley 1974
Colchicine	with LPS	Marecki and Bradley 1973
Cycloheximide	with LPS	Rose and Bradley 1971
Cyclophosphamide	with LPS	Rose et al 1972a
Cytosine arabinoside	with LPS	Bradley et al 1975a
Daunomycin	pre-LPS	Rose et al 1972b
Emetine	with LPS	Marecki and Bradley 1973
5-Fluorouracil	post-LPS	Marecki and Bradley 1973
5-Fluorouracil deoxyriboside	post-LPS	Bradley 1974
Glucan	pre-LPS	Crafton and Diluzio 1969
Insulin	with LPS	Pieroni and Levine 1969
Lead acetate	with LPS	Selye et al 1966
6-Mercaptopurine	with LPS	Marecki and Bradley 1973
Methotrexate	with LPS	Marecki and Bradley 1973
6-Methylmercaptopurine riboside	with LPS	Marecki and Bradley 1973
Mithramycin	with LPS	Bradley et al 1975a
Mycobacterium bovis BCG	pre-LPS	Suter et al 1958
Nogalamycin	with LPS	Bradley et al 1975a
Oxytetracycline	post-LPS	Floersheim and Logara-Kalantzis 1972
Pactamycin	with LPS	Rose and Bradley 1971
Polyinosinic polycytidylic acid	with LPS	Rose et al 1972a
Procarbazine	with LPS	Rose et al 1972b
Propranolol	with LPS	Filkins 1979
Pyran copolymer	pre-LPS	Munson and Regelson 1971
Sparsomycin	with LPS	Karp and Bradley 1968
Staphylococcal pyrogenic exotoxin	pre-LPS	Schlievert 1982
Streptococcal pyrogenic exotoxin	pre-LPS	Kim and Watson 1970
6-Sulfanilamidoindazole	pre-LPS	Miller et al 1979
Tenuazonic acid	post-LPS	Rose 1973
Tetracycline	post-LPS	Floersheim and Logara-Kalantzis 1972
Δ^8-Tetrahydrocannabinol	with LPS	Munson et al 1978
Δ^9-Tetrahydrocannabinol	with LPS	Bradley et al 1977

Table 1 (Continued)

Test substance	Treatment	Reference
Thorium dioxide	pre-LPS	Beeson 1947
Trehalose-6,6′-dimycolate	pre-LPS	Yarkoni and Rapp 1979
Trypan blue	pre-LPS	Beeson 1974
Vinblastine	with LPS	Marecki and Bradley 1973
Vincristine	with LPS	Rose et al 1972a
Zymosan	pre-LPS	Benacerraf et al 1959

3. ENHANCED TOXICITY OF LPS AND CHEMICALS

The susceptibility of animals to LPS is altered by a variety of chemicals (Bradley 1978; Rose 1973). The responsiveness of an animal to a simultaneous or sequential combination of a drug and LPS may be enhanced or decreased, reflecting the overall consequences of interactions involving the host and drug, the host and LPS, and the drug and LPS. Each of these pairwise interactions is actually a complex array of events; for example, the host may activate or detoxify a drug and, concurrently, the drug may stimulate or inhibit activity or synthesis of an enzyme. The end result of exposure to two agents may be the expected additive effect based upon the responses to each agent individually. Alternatively, the net effect of a combination may be markedly greater than expected (hyperadditive or synergistic) or markedly less than expected (hypoadditive or antagonistic) (Loewe 1957).

Drugs that inhibit protein synthesis or ribonucleic acid synthesis render animals unusually susceptible to LPS (Table 1). The bases of the enhanced toxicity of combinations of LPS and pactamycin, 6-mercaptopurine, vincristine or Δ^9-tetrahydrocannabinol are examined in detail later in this chapter. Drugs that inhibit protein synthesis or ribonucleic acid synthesis generally elicit enhanced lethality of LPS when the two agents are given simultaneously. Drugs that affect protein synthesis or ribonucleic acid synthesis potentiate the action of LPS dramatically, up to a millionfold (Bradley et al 1975a; Dowling and Feldman 1970; Pieroni et al 1970). Two anthracycline antibiotics, adriamycin and daunomycin, act hyperadditively with LPS when given before the microbial toxin, perhaps indicating that a metabolite of each drug is involved in the enhanced toxic effect rather than the unaltered antibiotic molecule (Table 1). Some drugs affecting deoxyribonucleic acid synthesis, structure, or function act hyperadditively with LPS when the microbial toxin is administered prior to the drug. The majority of the drugs that affect deoxyribonucleic acid structure, function, and activity fail to act hyperadditively with LPS (Table 2). The determinative factors responsible for the differences in the effects of combinations of LPS and drug acting on DNA are unknown. Several alkylating agents act hyperadditively with LPS (chlorambucil and 5-[3,3-bis(2-chloroethyl)-1-triazeno]imidazole-4-carboxamide) whereas other alkylating agents

Table 2 *Chemicals not found to elicit enhanced lethality when administered in combination with gram-negative bacterial lipopolysaccharide*

Chemical	Reference
Asparaginase	Rose et al 1972a
Azathioprine	Seyberth et al 1972
1,3-bis(2-chloroethyl)-1-nitrosourea	Bradley et al 1975a
Bleomycin	Bradley 1974
Chloramphenicol	Berry and Smythe 1964
1-(2-Chloroethyl)-3-cyclohexyl-1-nitrosourea	Bradley et al 1975a
Decoyinine	Bradley 1974
Dibromomannitol	Marecki and Bradley 1973
Diethylstilbestrol	Trejo et al 1972
Estradiol	Dobson and Kelly 1973
Hydroxyurea	Bradley et al 1975a
Methyltubercidin-5′-phosphate	Bradley 1974
Mitomycin C	Marecki and Bradley 1973
Nitrogen mustard	Marecki and Bradley 1973
Polyadenylic-polyuridylic acid	Bradley et al 1975a
Porfiromycin	Bradley 1974
Psicofuranine	Karp and Bradley 1968
Tris(1-aziridinyl)phosphine sulfide	Marecki and Bradley 1973
Uracil arabinoside	Bradley 1974

(1,3-bis(2-chloroethyl)-1-nitrosourea and 1-(2-chloroethyl)-3-cyclohexyl-1-nitrosourea) do not (Tables 1 and 2).

These interactions between LPS or the indigenous gram-negative bacterial flora and antitumor drugs have significance in different fields: (a) some of the adverse reactions experienced by patients receiving treatment for neoplastic diseases may represent enhanced toxicity as a consequence of an interaction in vivo involving the antitumor drug and LPS; (b) the synergistic toxicity of a combination of LPS with certain antitumor drugs may provide a means for improving the effectiveness of cancer chemotherapy; (c) LPS may be involved in certain adverse reactions to environmental hazards and pollutants; (d) means to manage patients in shock may be developed as a consequence of studies on the characterization of the toxophore of LPS and on the amelioration of the toxicity of drug–LPS combinations; and (e) toxicologic assessments using LPS-treated animals may be better indicators of what is to be expected in man than risk assessments based upon normal animals.

4. MANIFESTATIONS OF INTERACTIONS INVOLVING LPS AND CHEMICALS

The enhanced toxicity of LPS in combination with chemicals may be observed in any of a variety of manifestations ranging from retinal hemorrhage to death (Table 3). The range of toxic manifestations encompasses, in general, the spectrum of symptoms associated with either LPS or the chemical. Combinations involving drugs (pactamycin, sparsomycin, or vincristine) that appear to potentiate the action of LPS cause loss of regulation of body temperature, coagulopathies, liver damage, and kidney damage. A combination with a drug (5-fluorouracil) whose action appears to be potentiated by LPS results in hematopoietic damage. These changes can be provoked by a number of treatments; therefore, death remains the standard criterion for detecting enhanced toxicity of combinations of drugs with LPS.

5. SOURCE OF ENDOTOXIN IN ANIMALS

The source of endotoxin in an interaction with a chemical may be purified LPS, or killed or live gram-negative bacteria (Table 4). Gram-positive bacteria, in general, do not enhance the susceptibility of animals to chemicals. Endotoxic gram-negative bacteria are abundant in the environment of man. The sigmoid colon and

Table 3 *Manifestations of enhanced toxicity elicited by combinations of chemicals and gram-negative bacterial lipopolysaccharide*

Chemical	Manifestation	Reference
5-Fluorouracil	Elevated serum glutamate-pyruvate transaminase	Marecki et al 1975
	Granulocyte depletion	Marecki et al 1975
	Thrombocytopenia	Marecki et al 1975
Pactamycin	Elevated serum-urea-nitrogen	Bradley and Bond 1975
	Hypothermia	Bradley and Bond 1975
	Lysosomal fragility	Bradley and Bond 1975
	Retinal hemorrhage	Karp and Bradley 1968
Sparsomycin	Retinal hemorrhage	Karp and Bradley 1968
Δ^9-Tetrahydrocannabinol	Hypothermia	Munson et al 1978
Vincristine	Elevated serum-urea-nitrogen	Munson et al 1976
	Hypothermia	Munson et al 1976
	Leukocytosis	Munson et al 1976
	Low fibrinogen level	Munson et al 1976
	Low prothrombin activity	Munson et al 1976
	Thrombocytopenia	Munson et al 1976

rectum of man, for example, contain about 10^{11} bacteria per gram of fecal material. Bacteria account for 10% to 20% of the fecal mass. The anaerobic gram-negative *Bacteroides* may make up 95% to 99% of the colonic flora; the facultative enteric bacteria and enterococci make up 1% to 5% of the flora; and *Pseudomonas* contributes a smaller percentage. About 1% to 10% of the mass of a gram-negative bacterial cell is LPS; therefore, a gram of fecal material contains about 1 mg of LPS. Seepage of a few microliters of colonic contents into the circulation or

Table 4   *Enhanced lethality of combinations of bacterial cells and drugs*

Bacterium	Live/killed	Drug	Reference
Bacteroides fragilis	live	Vincristine	Munson et al 1976
Bacteroides melaninogenicus	live	Vincristine	Munson et al 1976
Bacteroides ovatus	killed	Δ^9-THC*	Bradley et al 1977
Bacteroides thetaiotaomicron	killed	Δ^9-THC	Bradley et al 1977
Branhamella catarrhalis	killed	6-Mercaptopurine	Marecki and Bradley 1974
'Clostridium leptum'	killed	Δ^9-THC	Bradley et al 1977
Enterobacter cloacae	killed	Δ^9-THC	Bradley et al 1977
Escherichia coli	live	Actinomycin D 6-Mercaptopurine	Bradley et al 1977 Marecki and Bradley 1973
	killed	Actinomycin D Pactamycin Δ^9-THC	Rose and Bradley 1971 Bradley and Rose 1973 Bradley et al 1977
Klebsiella pneumoniae	live	6-Mercaptopurine	Marecki and Bradley 1973
Listeria monocytogenes	killed	Vincristine	Munson et al 1976
Mycoplasma sp.	live	Pactamycin Sparsomycin	Gabridge 1974 Gabridge 1974
Proteus mirabilis	live killed	6-Mercaptopurine Pactamycin	Marecki and Bradley 1973 Bradley and Rose 1973
Pseudomonas aeruginosa	live killed	6-Mercaptopurine Δ^9-THC Vincristine	Marecki and Bradley 1973 Bradley et al 1977 Rose et al 1972a
Salmonella minnesota S	killed	6-Mercaptopurine	Marecki and Bradley 1974
S. minnesota Re595	killed	6-Mercaptopurine	Marecki and Bradley 1974
Serratia marcescens	killed	Δ^9-THC	Bradley et al 1977

* Δ^9-THC = Δ^9-tetrahydrocannabinol.

peritoneum of a patient undergoing chemotherapy might have disastrous consequences. One of the most remarkable synergistic drug–LPS combinations involves mithramycin (Bradley et al 1975a). As little as a dose of 0.1 µg/kg LPS kills about half of mice concurrently given mithramycin (0.5 mg/kg). The situation for man is even more dramatic because man is 1000-fold more sensitive to LPS than the mouse. A challenge with 10^4 gram-negative bacteria could cause adverse effects in a patients treated with cytotoxic drugs. This number of bacteria can be released into circulation by vigorous brushing of the teeth.

6. ENHANCED TOXICITY OF LIPID–PROTEIN COMPLEXES AND CHEMICALS

The toxophore of LPS, at least with respect to synergistic lethality, is lipid A. Lipid A complexed to bovine serum albumin or concanavalin A renders mice hyperreactive to 5-fluorouracil, mithramycin, pactamycin, Δ^9-tetrahydrocannabinol or vincristine (Table 5). A number of lipids or fatty acids complexed to proteins have less but definite endotoxic activity. The endotoxic activity of myristic acid–protein complexes has been confirmed by four criteria: (a) the complex elicits enhanced lethality in combination with mithramycin; (b) mice previously treated with the complex are resistant to a dose of LPS that kills normal mice; (c) mice previously treated with LPS are resistant to a lethal combination of complex and mithramycin (Bradley et al 1975b); and (d) mice are protected from the lethal action of LPS by simultaneously administered complex (Bradley 1978). The protein constituent of the complex serves as a solubilizing carrier and does not otherwise contribute to the specificity of the endotoxic activity of a lipid–protein complex. A wide variety of proteins can serve effectively as the carrier for lipid A (Table 5).

7. ATTENUATION OF THE ENHANCED TOXICITY OF COMBINATIONS OF LPS AND CHEMICALS

Animals can generally be protected from the adverse effects of combinations of LPS and drugs by prior treatment with LPS (Table 6). Mice treated with a combination of methotrexate and LPS, however, are not protected by prior treatment with LPS. In some instances, prior treatment with LPS alters the susceptibility of the animals to the drug alone. LPS-pretreated mice are more susceptible to vincristine than normal mice; this presumably reflects impaired drug detoxification in LPS-pretreated animals. Conversely, LPS-pretreated mice are less susceptible to 5-fluorouracil or 6-mercaptopurine than normal mice. The basis of this protection is not known. It is conceivable that LPS pretreatment stimulates the bone marrow to produce an abundance of the hematopoietic stem cells that are particularly vulnerable to 5-fluorouracil.

Animals can be protected from the adverse effects of particular combinations of

treated mice to the extent that the drug concentration is half that of normal mice at 1 minute after intravenous injection. This enhanced clearance results in non-detectable levels of 6-MP after 20 minutes whereas 6-MP can be detected in the serum of normal mice for a period of 45 minutes.

It is reasonable to propose that agents which protect against either 6-MP or LPS alone can protect against combinations of the two agents. This has been found to be true. Caffeine, an agent that inhibits phosphodiesterase activity, alters the response of mice to high doses of LPS and protects against combinations of 6-MP and LPS. Methylprednisolone protects against high doses of 6-MP and LPS as well as against combinations of these two agents (Table 8). It is possible that this steroid alters the response of the host to these two agents through different mechanisms of action but this is not known. Polymyxin B prevents LPS-induced leukopenia, thrombocytopenia, and disseminated intravascular coagulation in the rabbit (Corrigan and Bell 1971), and protects mice against high doses of LPS or 6-MP as well as against the synergistic combination. Although each of these three agents is capable of protecting against the lethality associated with combinations of 6-MP and LPS, there is no apparent common denominator for their action.

10. BASIS OF THE VINCRISTINE–LPS INTERACTION

Mice given vincristine plus LPS intravenously die 6–8 hours after treatment in contrast to mice challenged intraperitoneally, which die 18–24 hours later (Munson et al 1976). The time course and dose–effect relation of mice administered vincristine and LPS resemble those of mice given a lethal dose of LPS. Hemoglobin levels and hematocrit values are not markedly altered by vincristine, LPS, or the combination. A combination of vincristine and LPS elicits a slight leukocytosis. The combination markedly depresses the platelet count, fibrinogen level, and prothrombin activity; these changes are also observed in animals receiving a high dose of LPS. A high dose of vincristine also decreases the platelet count but has little effect on prothrombin activity and actually increases fibrinogen activity (Munson et al 1976). Serum glutamate transaminase activities are elevated and sulfobromphthalein excretion is retarded in mice given vincristine plus LPS, but these changes are not markedly more severe than those observed in mice given a sublethal dose of vincristine (Floersheim and Szeszak 1971). Liver damage per se does not seem to be the basis of the enhanced lethality of the vincristine–LPS combination.

Vincristine at 2 mg/kg impairs drug metabolism as measured by the duration of phenobarbital-induced sedation. Endotoxin at 2 mg/kg also impairs drug metabolism when given 1 day before, but not when given simultaneously with the phenobarbital used to induce sleep (Rose et al 1972a). Phenobarbital pretreatment, which induces liver mixed function oxidases, protects mice from the vincristine–LPS combination but not from vincristine alone (Table 7). The liver also participates in the neutralization of the toxicity of LPS. However, liver homogenates prepared from mice treated with vincristine (4 mg/kg) neutralize endotoxicity of LPS to the same extent as homogenates from untreated mice (Rose et al 1972a).

The SUN levels of mice receiving a combination of vincristine plus LPS rise dramatically, similar to the values observed in nephrectomized mice. Mice receiving the combination of vincristine plus LPS and nephrectomized mice die at about the same time (Munson et al 1976), indicating that renal shut-down can lead to death in rodents (Prior and Visek 1973). Histologic examination of sections of kidneys from mice receiving a high dose of LPS or a combination of vincristine plus LPS shows interstitial edema in and around the glomeruli and a marked widening of Bowman's space. In addition, leukocytic infiltration into the interstitial tissue is observed, and the vascular channels appear congested. There is slight hypercellularity at the center of the glomeruli, and some hemolyzed erythrocytes are seen within the capillary tuft. The changes indicate that there is intracapillary coagulation of erythrocytes (Munson et al 1976).

11. BASIS OF THE Δ^9-TETRAHYDROCANNABINOL–LPS INTERACTION

A combination of Δ^9-tetrahydrocannabinol (Δ^9-THC) and LPS results in enhanced mortality for mice (Bradley et al 1977). The hyperadditive action of combinations of Δ^9-THC and LPS may reflect effects of one or both substances on the action of, or host response to, the other agent. It is reasonable to propose that the interaction arises as a result of (a) altered metabolism of Δ^9-THC in LPS-treated animals (or the converse); (b) altered distribution of Δ^9-THC in LPS-treated animals (or the converse); (c) exacerbation of cardiovascular collapse; or (d) exaggeration of selected cellular processes.

Both hexobarbital and cannabinoids are metabolized by the liver (Siemens and Kalant 1975). Mantilla-Plata and Harbison (1976) claim that phenobarbital pre-treatment antagonizes, and proadifen-hydrochloride potentiates, Δ^9-THC-induced mortality in mice. These workers report that there is a significant increase in the plasma levels of Δ^9-THC metabolites in phenobarbital-pretreated mice. They suggest that the metabolites in phenobarbital-treated mice may be different from those in untreated mice (Mantilla-Plata and Harbison 1974). Rose et al (1972a) have found that LPS decreases the levels of the hepatic mixed function oxidase microsomal enzyme activities. Impaired metabolism of Δ^9-THC by drug-metabolizing enzymes of the liver can conceivably lead to higher and sustained blood levels of Δ^9-THC and thereby increase the levels of Δ^9-THC in the brain. Stimulation of mixed function oxidase activities by phenobarbital, however, fails to protect mice from Δ^9-THC or combinations of Δ^9-THC and LPS. For these and other reasons described below, it appears that altered metabolism of Δ^9-THC by LPS is not the probable basis of the interaction.

Tolerance to Δ^9-THC may develop rapidly but the duration of the effects of prior treatment is variable (McMillan et al 1971). Some investigators have suggested that tolerance to cannabinoids is the consequence of a change in their rate of metabolism but the mechanism of development of tolerance is unknown. Untreated and Δ^9-THC-tolerant mice (80 mg/kg intraperitoneally for 4 days) have equivalent amounts of cannabinoid in the plasma and brain 4 hours after acute Δ^9-THC injection as untreated mice. The tolerance-inducing regimen leads to an accumulation of Δ^9-THC in the brain: 2 μg per gram of brain on day 1, 3.2 μg/g on day 2, and 6 μg/g on day 4. These results indicate that tolerance has developed because the central nervous system is less sensitive to this agent. Accordingly, Dewey et al (1976) have concluded that tolerance to Δ^9-THC in pigeons or dogs cannot be accounted for by alterations in absorption or metabolism of the cannabinoid or altered distribution in peripheral tissues. Moreover, Martin et al (1978) have found that, although the level of radiolabeled Δ^9-THC in the brains of LPS-treated mice that died is higher than that in control mice, it is less than that in animals receiving a larger but nonlethal dose of Δ^9-THC alone. For these reasons, it appears that altered distribution of Δ^9-THC in LPS-treated mice is not the probable basis of the interaction. Unexpectedly, prior treatment with Δ^9-THC to elicit tolerance protects mice from lethality due to high doses of LPS alone as well as to high doses of Δ^9-THC alone or a combination of Δ^9-THC and LPS (Table 10).

One of the effects that may be critical in the pathophysiology of both Δ^9-THC and LPS is that they induce hypotension leading to circulatory stasis and ischemia. Both LPS and Δ^9-THC elicit either bradycardia or tachycardia, depending upon

Table 10   *Nonreciprocal cross resistance of mice to the lethal action of Δ^9-tetrahydrocannabinol (Δ^9-THC) and LPS elicited by prior treatment with Δ^9-THC*

Prior treatment	Δ^9-THC (mg/kg)[*]	LPS (mg/kg)	No. of mice	Mortality (%)[**]
DMSO	500	none	20	65
Δ^9-THC	500	none	20	35[***]
saline	500	none	20	70
LPS	500	none	30	80
DMSO	none	15	20	75
Δ^9-THC	none	15	20	5[***]
saline	none	15	30	70
LPS	none	15	20	15[***]
DMSO	150	1.5	20	85
Δ^9-THC	150	1.5	20	35[***]
saline	150	1.5	40	70
LPS	150	1.5	30	13[***]

[*] Δ^9-THC in the challenge dose was dissolved in Emulphor.
[**]   Mortality was scored on day 9.
[***]   Significantly different from diluent-treated controls at $p<0.05$.

the species and experimental design. Moreover, vagotomy and propranolol modify the cardiovascular responses to both Δ^9-THC and LPS. Endotoxin and Δ^9-THC perturb the levels of histamine, serotonin, and catecholamines, all of which can be related to exaggerated sympathoadrenal activity. Both LPS and Δ^9-THC affect the level of calcium-dependent ATPase activity in myocardial sarcoplasmic reticulum and it has been proposed by some workers that both Δ^9-THC and LPS alter calcium binding and calcium flux. Accordingly, it is reasonable to propose that the enhanced lethality of Δ^9-THC and LPS, in combination, reflects primary effects on the cardiovascular system. This synergy may be explained by the action of Δ^9-THC and LPS on different sites in the central nervous system, leading to decreased sympathetic outflow.

12. EFFECTS OF LPS ON CELLS

Endotoxin exerts a variety of effects on lymphocytes, in particular B cells. Many immature B lymphocyte lines, including the murine B cell leukemia line BCL_1 can be activated by LPS to secrete IgM (Knapp et al 1979). Stimulation of BCL_1 cells by LPS results in a 10-fold increase in the amount of mRNA for the μ chain of secreted IgM. The increase in mRNA for secreted μ chain is accompanied by a 3- to 4-fold increase in mRNA for membrane IgM; however, the rate of synthesis of μ chain of membrane IgM is reduced 2-fold (Yuan and Tucker 1982). LPS-stimulated normal murine B lymphocytes (Tsuda et al 1980), and LPS-stimulated 70Z/3 lymphoma cells – a murine cell line with the characteristics of immature B lymphocytes (Perry and Kelley 1979) – contain increased levels of mRNA for $\varkappa$ chain also. These results indicate that LPS acts through both transcriptional and translational regulatory events controlling secretion of IgM by B lymphocytes.

Murine B lymphocytes are polyclonally activated by LPS in low cell density cultures. IgM^+, IgD^+, IgG^- lymphocytes are the major LPS-responsive cell population in the low cell density cultures (Severinson-Gronowicz et al 1979a). The development of IgG secretion after polyclonal activation of murine spleen cells is a complex process requiring several cycles of splenocyte proliferation (Severinson-Gronowicz et al 1979b).

Endotoxin added to mouse spleen cell cultures stimulates production of colony stimulating factor, a humoral regulatory factor necessary to support the development of granulocyte/macrophage colonies in vitro in soft agar cultures. Purified splenic lymphocytes produce only small amounts of colony stimulating factor in response to LPS unless macrophages are added to the cultures. Supernatant fluid from LPS-treated macrophages of C3H/HeB mice, but not of C3H/HeJ mice, can replace the intact macrophages. Supernatant fluid from LPS-treated macrophages activates B lymphocytes from congenitally athymic nude mice, from C3H/HeB mice, or from C3H/HeJ mice to secrete colony stimulating factor (Apte et al 1980). In addition to colony stimulating factor, LPS has been shown to induce production of a number of mediators including: (a) glucocorticoid antagonizing factor that acts on hepatocytes (Goodrum and Berry 1978); (b) serum amyloid A inducer (Sipe et al 1979); (c) endogenous pyrogen (Haeseler et al 1977); and (d) plasminogen activator (Mathison and Ulevitch 1979).

Mouse macrophages can be activated by LPS leading to increased ability to ingest complement-coated erythrocytes in the absence of IgG and to decreased levels of 5'-nucleotidase in the plasma membrane. The defective activation of macrophages by LPS in C3H/HeJ mice appears to be secondary to the lymphocyte defect. Peritoneal macrophages of C3HeB/FeJ or C3H/HeJ mice can be activated by LPS, provided that C3HeB/FeJ lymphocytes are present. Moreover, lymphokines induced by LPS in C3HeB/FeJ spleen cells can activate resident macrophages in vitro. Serum components play a role in macrophage activation and the serum from C3H/HeJ mice appears to lack this function (Nowakowski et al 1980).

Macrophages of C3H/HeJ mice are refractory to most of the biological effects elicited by LPS or lipid A: (a) they are not killed by LPS in vitro (Glode et al 1977); (b) their phagocytic activity is not increased by in vivo injections of LPS (Ruco and Meltzer 1978a); (c) they exhibit defective Fc receptor-mediated phagocytosis when cultured in vitro over a 24-hour period (Vogel and Rosenstreich 1979); (d) they fail to respond to macrophage-activating factor to kill tumor cells (Ruco and Meltzer 1978b); and (e) they are refractory to macrophage migration inhibition factor (Tagliabue et al 1978) but their capability to phagocytize bacteria is not impaired and may even be enhanced (Cuffini et al 1980).

Endotoxin or lipid A activates peritoneal macrophages of C3H/HeSt mice but not of C3H/HeJ mice to become nonspecifically cytotoxic for tumor cells in vitro. This activation of macrophages results from a direct interaction between the protein-free LPS or lipid A and macrophages, and appears to be entirely independent of B lymphocytes. Endotoxin appears to interact directly with the macrophages of LPS-responsive mice to induce a lytic effect on tumor cells (Doe and Henson 1979). Similarly, LPS activates rat alveolar macrophages directly without an interaction with lymphocytes (Sone and Fidler 1980).

In C3HeB/FeJ mice treated with LPS, the maximum influx of polymorphonuclear neutrophils and macrophages occurs after 96 hours. In C3H/HeJ mice treated with LPS, the maximum influx of polymorphonuclear neutrophils occurs after 12 hours and the maximum for macrophages after 48 hours. The influx of inflammatory cells into the peritoneum is much greater in C3H/HeJ mice than in C3HeB/FeJ mice and may be integrally related to the resistance of C3H/HeJ mice to the lethal action of LPS (Moeller et al 1978).

Endotoxin elicits a variety of alterations in cellular activities and organellar function. Lysosomal enzyme activities are elevated in human embryonic lung (WI-38) cells in culture and in African green monkey kidney (Vero) cells treated with LPS (10 µg/ml of medium) (McGivney and Bradley 1978 and 1979a). The increase in lysosomal enzyme activities is progressive with time, achieving a twofold increase within 4 hours. The rise in activity is inhibited by cycloheximide or actinomycin D, is temperature-dependent, and occurs whether serum is present or

358

absent. Because the changes in lysosomal enzyme activities occur in the absence of added serum, the roles of antibody, complement, hormones, and lymphokines is minimized. Endotoxin does not enhance the level of lysosomal enzyme activities when added directly to the granular fraction. A number of investigators have proposed that the increased lysosomal activity elicited by LPS is effected by a direct action on the cell membrane or some subcellular constituent and is not an immunologic phenomenon (Evans 1974; Mørland 1979).

Endotoxin elicits enhanced lysosomal enzyme activities in established normal cell cultures and primary cultures of spleen cells or peritoneal exudate cells from normal mice (McGivney and Bradley 1977; Mørland 1979), but not in several transformed cell lines (McGivney and Bradley 1979a) or primary culture of spleen cells from LPS-tolerant mice or genetically refractory C3H/HeJ mice (McGivney and Bradley 1977). The proposition that lysosomes may be the target for the effects of LPS is supported by the observation that lysosomes from LPS-treated mice are more fragile than those isolated from untreated mice (Bradley and Bond 1975). A number of investigators have suggested that increased lysosomal enzyme activity is one of the important biological effects of LPS on a cell (Allison et al 1973; Mørland and Mørland 1978).

The activities of malate dehydrogenase, succinate dehydrogenase, and adenylate kinase are decreased in the granular fraction of Vero cells exposed to LPS at a dose of 10 μg/ml of medium for 2 hours. These changes in mitochondrial enzyme activities precede those for lysosomal enzymes and are elicited by lower amounts of LPS (McGivney and Bradley 1979a). Endotoxin acts directly on normal established cell cultures, causing early changes in mitochondria under conditions in which the activities of catalase and hexokinase are not perturbed. Mitochondrial enzyme activities leak extensively from the isolated granular fraction exposed to LPS for 45 minutes. Accelerated loss of mitochondrial enzyme activities from the granular fraction to the soluble fraction is dependent upon the LPS concentration and incubation time (McGivney and Bradley 1979a). Similar results have been obtained with primary cultures of mouse liver cells and subcellular fractions of mouse liver cells (McGivney and Bradley 1979b). The activities of mitochondrial enzymes are decreased in the granular fraction of mouse liver cells exposed to LPS and the mitochondrial enzymes leak into the supernatant fraction within 45 minutes after exposure to LPS. Loss of mitochondrial enzyme activity from the isolated granular fraction is accelerated by free lipid A as well as by diverse LPS preparations. Endotoxin causes isolated mitochondria from mouse liver to swell (McGivney and Bradley 1979b) and affects enzymes located in the inner membrane of the mitochondrion, the mitochondrial matrix, and the intermembrane space simultaneously. These results indicate that (a) lipid A is the toxic moiety of LPS, (b) LPS affects lysosomes, and (c) LPS disrupts the integrity of the mitochondrion.

13. EFFECTS OF LPS ON MITOCHONDRIA

Endotoxin, even at 100 µg/mg of mitochondrial protein, does not impair reduction of the cytochromes when reduced nicotinamide adenine dinucleotide (NADH) or glutamate is the electron donor. Accordingly, LPS does not inhibit the electron carriers of the respiratory chain directly. Endotoxin at 100 µg/mg of mitochondrial protein impairs reduction of the cytochromes when succinate is the electron donor, indicating that LPS can act at, or prior to, the linkage of flavin adenine dinucleotide-iron sulfur protein to coenzyme Q-cytochrome b → cytochrome c_1-cytochrome c (McGivney and Bradley 1979b).

Endotoxin inhibits oxygen consumption by isolated mouse liver mitochondria when glutamate plus malate is the substrate and adenosine-5'-diphosphate (ADP) is added (State 3 respiration) but has little effect when ADP is not added (State 4 respiration). Uncoupling does not seem to be the mechanism of action of LPS when glutamate is the substrate. Both LPS and lipid A cause a loss of dependency of the rate of respiration on ADP but the amount of ADP utilized to the amount of O_2 consumed (the ADP/O value) in isolated mouse liver mitochondria treated with LPS or lipid A is not altered significantly (McGivney and Bradley 1979b). Impaired coupling of oxidative phosphorylation by LPS, therefore, is secondary to other interactions between LPS and a mitochondrial constituent.

Endotoxin or lipid A stimulates State 4 respiration by isolated mouse liver mitochondria when succinate is the substrate but has little effect on State 3 respiration (McGivney and Bradley 1980a). Accordingly, the manner by which LPS alters the respiratory capabilities of isolated mitochondria differs with various substrates. Endotoxin appears to act as an uncoupling agent, in that State 4 respiration is stimulated, when succinate is the substrate. The ADP/O value, however, is not markedly altered by LPS with succinate as the substrate indicating that the coupling sites are functioning. In respiring mitochondria, the rate of oxygen consumption is determined by the concentration of ADP rather than by the concentration of the substrate. The dependence of the respiratory rate upon ADP, that is, the respiratory control ratio, is a useful measure of the structural integrity of isolated mitochondria. The deleterious effect of LPS or lipid A on respiratory control with either glutamate or succinate as the substrate may reflect damage to the integrity of the mitochondria rather than an effect on a particular site in electron transport or oxidative phosphorylation.

Mela (1981) has concluded that the amount of LPS required for direct inhibition of mitochondrial function is excessively high. Both Mela (1981), and McGivney and Bradley (1979b) used about 10 µg LPS per mg of mitochondrial protein. A mouse liver may be expected to contain about 5 mg of mitochondrial protein; therefore, 50 µg LPS would be adequate to produce a measurable injury. More than 80% of intravenously administered LPS is found in the liver, and therefore sufficient LPS is available to act on the mitochondria of normal mice administered 250 to 500 µg LPS (Bradley 1979). In mithramycin-treated mice, however, only minute amounts of LPS are required to kill the animals. Under these cir-

360

cumstances, LPS cannot act stoichiometrically to produce the lethal event but must provoke the production, release, or both, of mediator(s) that precipitate(s) the lethal sequelae. It is not known whether the inhibitor of protein synthesis renders mice exquisitely sensitive to LPS because it stimulates production or release of mediator, or because it prevents repair of cellular injury.

14. EFFECTS ON CELLS IN RELATION TO EFFECTS ON ANIMALS

The concentration of LPS that elicits a marked decrease in mitochondrial enzyme activities in cultured cells (10 µg/ml) is comparable to the LPS dose that kills 50% of treated mice (10–20 µg of endotoxin/g of mouse weight). Furthermore, LPS causes a substantial change in mitochondrial enzyme activities in cultured cells within 2–4 hours, which corresponds to the time when animals display loss of thermal regulation and pactamycin-treated mice die after intravenous administration of LPS (Bradley and Bond 1975). It is well established that LPS can affect cellular metabolic processes directly and that not all the cellular effects of LPS can be attributed to immunologic reactions, changes in hormone levels, or humoral mediators.

Gram-negative bacterial LPS impairs mitochondrial function when injected into animals, added to cells in culture, or added to subcellular fractions (Bradley 1979). Although LPS affects the activity of isolated mitochondria, it is not definitively established that the interaction of LPS with the mitochondria per se is the causal early event in endotoxicosis. It is conceivable that LPS acts first on the plasma membrane thereby triggering several diverse series of reactions. In order to identify the site of some of the early determinative alterations elicited by LPS, the effects of LPS on liver cells in primary cultures and subcellular fractions of liver from LPS-susceptible mice, from mice rendered resistant to LPS by prior treatment with endotoxin, and from genetically LPS-refractory mice have been examined.

Liver cells from normal LPS-susceptible mice have reduced activities of mitochondrial enzymes after having been exposed to LPS for a few hours. Liver cells from LPS-refractory mice, whether rendered resistant by prior treatment with endotoxin or from the genetically refractory C3H/HeJ mouse, exhibit normal mitochondrial enzyme activity after being exposed to LPS. Conversely, isolated mitochondria from LPS-susceptible and LPS-refractory mice are sensitive to endotoxin indicating that refractoriness is a cellular and not an organellar attribute (McGivney and Bradley 1980b). Several investigators have reported that cells of C3H/HeJ and C3H/HeN mice bind LPS equally well (Springer et al 1977). These observations indicate that, although binding of LPS should be a prerequisite for a response, binding is not by itself sufficient to cause a response. The mechanism by which LPS is bound to cell membranes is largely unknown. Any membrane constituent with a high affinity for the lipid moiety of LPS may bind it. Binding to such high-affinity, but nonspecific, sites on membranes may be sufficiently strong to obscure specific binding to a specific LPS receptor. This phenomenon may account for the failure of most, but not all, workers to detect differences in binding of LPS by cells from LPS-susceptible and LPS-refractory mice (Nygren et al 1979). The possibility that refractoriness in the C3H/HeJ mouse reflects altered internalization of LPS or impaired triggering of crucial responses on the plasma membrane must also be considered. Alternatively, refractoriness in the C3H/HeJ mouse may involve impaired translocation of the toxophore or effector through the cytoplasm to the ultimate target.

15. EFFECTS ON LIVER IN RELATION TO MOLECULAR EVENTS

Numerous metabolic activities in liver are altered after endotoxicosis. Such changes include depletion of carbohydrate reserves, inhibition of particular enzyme induction, and inhibition of gluconeogenesis (Berry 1977). Endotoxin induces a transient hyperglycemia followed by a marked hypoglycemia. The early hyperglycemia is the result of increased hepatic glucose output due to enhanced glycogenolysis. The hypoglycemia develops as a consequence of both elevated uptake of glucose by peripheral tissue and the failure of the liver to compensate for this by augmented glucose output (Wolfe et al 1977). Moreover, LPS increases production of macrophage insulin-like activity (Filkins 1980) as well as glucocorticoid antagonizing factor (Berry 1977). Thus gluconeogenesis is impaired by glucocorticoid antagonizing factor, and tissue utilization of glucose is augmented by macrophage insulin-like activity. The net result in endotoxicosis is severe hypoglycemia. Liver damage initiated by LPS also leads to markedly elevated levels of transaminase activities in serum within a few hours (Munson et al 1976), and to decreased levels of microsomal mixed function oxygenase activity (Rose et al 1972a).

The tissue localization of LPS has not been completely resolved. It is generally agreed that LPS administered intravenously is taken up by the liver almost immediately (Walker and Snyder 1978). Some workers have reported that LPS is taken up by hepatocytes (Willerson et al 1970; Zlydaszyk and Moon 1976) whereas other workers have reported that LPS is taken up by Kupffer cells (Cremer and Watson 1957; Golub et al 1968). Mathison and Ulevitch (1979) have concluded that LPS is concentrated primarily in phagocytic vacuoles of the hepatic Kupffer cells whereas Willerson et al (1970) have concluded that LPS accumulates in the mitochondrial fraction of rodent liver and spleen. A central question in addressing the role of the liver in the pathophysiology of LPS shock is whether parenchymal cells are affected directly by LPS or whether metabolic changes observed in endotoxicosis are mediated through soluble products released from macrophages after phagocytosis of LPS (Berry 1977).

Elucidation of the molecular mechanism of action of a toxin can be achieved only by characterizing the active site of the toxin molecule, by defining the receptor, by determining the nature of the initial interaction between the toxin and its receptor, and by relating the initial events to the sequelae of cellular alterations that culminate in characteristic symptoms in susceptible animals. Although the molecular bases of the toxicity of gram-negative bacterial LPS is not resolved definitively, a substantial body of information now exists so that reasonable models can be formulated.

Lipid A represents the endotoxically active center of the LPS molecule whereas the polysaccharide portion serves as a solubilizing carrier for an otherwise insoluble lipid (Bradley 1979). Complexes of myristic acid with bovine serum albumin, myristic acid with concanavalin A, dimethyl myristamide with concanavalin A, and sphingosine with bovine serum albumin, possess endotoxic activity (Bradley 1976). Springer et al (1977) have isolated glycerophosphatides from platelets and leukocytes that combine with LPS. These observations are consistent with the proposition that the lipid A moiety binds to a lipid receptor or inserts into a lipoidal constituent of the cell (Bradley 1979). Macrophages release prostaglandins upon exposure to LPS or lipid A in vitro (Fischer et al 1977). Prostaglandins E_1, E_2, and $F_{2\alpha}$ can evoke hypothermia in mice and this action is suppressed by indometacin. Similarly, Halushka et al (1981) have found that acetylsalicylic acid significantly prolongs the survival of rats in endotoxin shock. Early-phase tolerance, which is elicited by daily injections for 4 days of a small dose of LPS, seems to result from the inability of Kupffer cells to release endogenous pyrogen upon subsequent challenge with LPS (Greer and Rietschel 1978). Early-phase tolerance may be based upon refractoriness of macrophages to release prostaglandins when provoked by LPS or lipid A. Prostaglandin E_2 stimulates the rate of intracellular protein degradation in rat muscle (Rodemann and Goldberg 1982). Prostaglandins may increase intracellular protein degradation by activating the lysosomal apparatus. The elevated lysosomal activity observed in animals and cells treated with LPS (McGivney and Bradley 1977, 1978) may be elicited by prostaglandins.

16. CONCLUDING REMARKS

Animals treated with lethal doses of LPS do not die until several hours have elapsed. This delay reflects the time required to (a) elicit a lethal effector, (b) accumulate a lethal amount of an effector, (c) consume or eliminate an essential vital factor, or (d) provoke degradation of an essential vital factor. Inhibition of protein synthesis in the intact animals renders them hyperreactive to LPS, indicating that little new protein synthesis is required for the lethal sequelae and that protein synthesis is probably palliative in coping with endotoxicosis. Inhibitors of protein synthesis do not act synergistically with LPS in test systems in vitro, corroborating the proposition that protein synthesis is needed to repair the injury precipitated by LPS. LPS acts directly on a wide range of cell types, leading to altered Ca^{2+} flux, release of hydrolytic enzymes and mediators, increased lysosomal enzyme activity, and impaired mitochondrial function. Endotoxin does not appear to kill cells by a direct cytotoxic effect, but indirectly through responses involving the released mediators. Death in intact animals results from dysregulation precipitated by the sudden release of mediators rather than by the action of specific toxins. Mediators acting in concert, for example, may produce severe hypoglycemia and an insulin-shock type of death. Still other mediators produce coagulopathies, leading to disseminated intravascular coagulation and fatal renal failure. The sequelae leading to death varies according to species, to route of administration, and to the immunologic status of the animal.

The following model for the cellular and molecular bases for the action of endotoxin is proposed: (a) LPS is taken into the cell by endocytosis, (b) LPS is translocated through the cytoplasm in an endocytic vacuole, (c) the endotoxic toxophore is transferred to a receptor in the mitochondrial membrane, (d) the mitochondrial proton gradient is destroyed, (e) ADP and NADH accumulate in the cytosol leading to enhanced glycolytic metabolism and lactic acid production, (f) permeability of the plasma membrane to Ca^{2+} is altered by LPS and the Ca^{2+} flux is further altered by mitochondrial injury, (g) peroxides and superoxides accumulate and oxygenate lipids producing prostaglandins, (h) these deleterious events enhance autophagy, activate lysosomes, and lead to release of lysosomal enzymes, and (i) intracellular proteins are oxidized and degraded, thereby providing precursors for new protein synthesis (Fig. 1). Newly synthesized proteins include endogenous pyrogen, lysosomal enzymes, and the proteins needed to repair cellular damage, for example plasma membrane proteins and mitochondrial proteins. These cellular alterations culminate in the characteristic symptoms of endotoxicosis: perturbed regulation of body temperature, biphasic changes in glucose and calcium levels, increased lactic acid production and lysosomal enzyme activities, disseminated intravascular coagulation, and severe circulatory hypotension (Bradley 1979, 1981). These host responses to LPS not only provide defenses

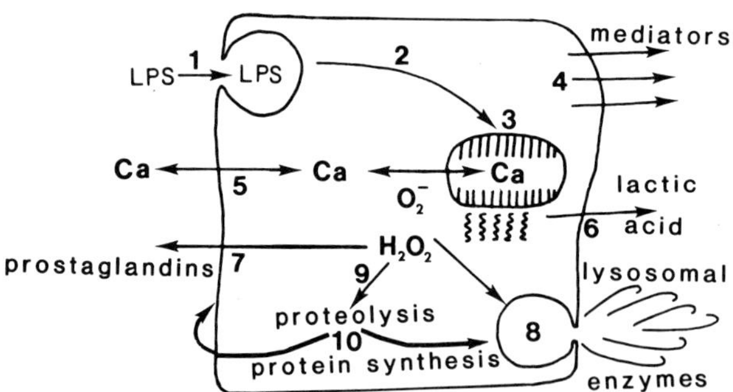

Fig. 1 *Cellular basis for the action of LPS. (1) Endocytosis; (2) translocation of LPS through the cytoplasm; (3) LPS toxophore binds to the mitochondrial membrane; (4) mediators are released during endocytosis; (5) Ca^{2+} flux is altered during endocytosis and by mitochondrial injury; (6) ADP and NADH accumulate in the cytosol thereby stimulating lactic acid production; (7) peroxides and superoxides oxygenate lipids, producing prostaglandins; (8) lysosomes are activated and lysosomal enzymes are released; (9) intracellular proteins are oxidized and then degraded; and (10) the amino acid pool is used to synthesize new proteins.*

against gram-negative bacterial infections but may also serve to protect the host from serious trauma. In trauma as well as in infection, extensive new protein synthesis is required even though intake of nutrients may be limited. The association between man and the colonic bacterial flora is so intimate that LPS has come to serve as an important natural immunomodulator.

REFERENCES

Allison AC, Davies P, Page RC (1973) Effects of endotoxin on macrophage and other lymphoreticular cells. *J. Infect. Dis. 128 Suppl.*, 212-219.

Apte RN, Hertogs ChF, Pluznik DH (1980) Regulation of lipopolysaccharide-induced granulopoiesis and macrophage formation by spleen cells. II. Macrophage-lymphocyte interactions in the process of generation of colony-stimulating factor. *J. Immunol. 124*, 1223-1229.

Ballard FJ (1978) Intracellular protein degradation. *Essays Biochem. 13*, 1-37.

Bates HM, Margolin S (1968) Endotoxin-induced inhibition of renal function in the mouse. *Proc. Soc. Exp. Biol. Med. 129*, 642-646.

Beeson PB (1947) Effect of reticuloendothelial blockade on immunity to the Shwartzman phenomenon. *Proc. Soc. Exp. Biol. Med. 64*, 146-149.

Beisel WR (1975) Metabolic responses to infection. *Annu. Rev. Med. 26*, 9-20.

Benacerraf B, Thorbecke GJ, Jacoby D (1959) Effect of zymosan on endotoxin toxicity in mice. *Proc. Soc. Exp. Biol. Med. 100*, 796-799.

Berry LJ (1964) Effect of endotoxins on the level of selected enzyme metabolites. In: Landy M, Braun W (Eds), *Bacterial Endotoxins*, pp 151-159. Rutgers University Press, New Brunswick, NJ.

Berry LJ (1977) Bacterial toxins. *CRC Crit. Rev. Toxicol. 5*, 239-318.

Berry LJ, Smythe DS (1964) Effects of bacterial endotoxins on metabolism. VII. Enzyme induction and cortisone protection. *J. Exp. Med. 120*, 721-732.

Bradley SG (1974) Enhanced toxicity for mice of combinations of *Escherichia coli* lipopolysaccharide with antitumor drugs. *Dev. Ind. Microbiol. 15*, 368-376.

Bradley SG (1976) Endotoxic activity of complexes of myristic acid and proteins. *Proc. Soc. Exp. Biol. Med. 151*, 267-270.

Bradley SG (1978) Alteration of the sensitivity of mice to bacterial endotoxin. In: Rosenberg P (Ed), *Toxins: Animal, Plant and Microbial*, pp 907-920. Pergamon Press, Oxford.

Bradley SG (1979) Cellular and molecular mechanism of action of bacterial endotoxins. *Annu. Rev. Microbiol. 33*, 67-94.

Bradley SG (1981) Direct action of bacterial endotoxin on cells, mitochondria, and lysosomes. In: Majde JA, Person RJ (Eds), *Pathophysiological Effects of Endotoxins at the Cellular Level*, pp 3-14. Alan R. Liss, New York.

Bradley SG, Bond JS (1975) Toxicity, clearance, and metabolic effects of pactamycin in combination with bacterial lipopolysaccharide. *Toxicol. Appl. Pharmacol. 31*, 208-221.

Bradley SG, Rose WC (1973) Enhanced toxicity for mice of pactamycin with bacterial endotoxin. *Antimicrob. Agents Chemother. 3*, 686-692.

Bradley SG, Adams AC, Smith MC (1975a) Potentiation of the toxicity of mithramycin by bacterial lipopolysaccharide. *Antimicrob. Agents Chemother. 7*, 322-327.

Bradley SG, Marecki NM, Bond JS, Munson AE, John DT (1975b) Enhanced cytotoxicity in mice of combinations of concanavalin A and selected antitumor drugs. *Adv. Exp. Med. Biol. 55*, 291-307.

Bradley SG, Munson AE, Dewey WL, Harris LS (1977) Enhanced susceptibility of mice to combinations of Δ^9-tetrahydrocannabinol and live or killed gram-negative bacteria. *Infect. Immun. 17*, 325-329.

Butcher RW, Sutherland EW (1962) Adenosine 3',5'-phosphate in biological materials. I. Purification and properties of cyclic 3',5'-nucleotide phosphodiesterase and the use of this enzyme to characterize adenosine 3',5'-phosphate in human urine. *J. Biol. Chem. 237*, 1244-1250.

Clowes GHA, O'Donnell TF, Blackburn GL, Maki TN (1976) Energy metabolism and proteolysis in traumatized and septic man. *Surg. Clin. North Am. 56*, 1169-1184.

Cook JA, Marconi EA, DiLuzio NR (1974) Lead, cadmium and endotoxin interaction: effect on mortality and hepatic function. *Toxicol. Appl. Pharmacol. 28*, 292-297.

Corrigan Jr JJ, Bell BM (1971) Comparison between the polymyxins and gentamicin in preventing endotoxin-induced coagulation and leukopenia. *Infect. Immun. 4*, 563-566.

Crafton CG, DiLuzio NR (1969) Relationship of reticuloendothelial functional activity to endotoxin lethality. *Am. J. Physiol. 217*, 736-742.

Cremer N, Watson DW (1957) Influence of stress on distribution of endotoxin in RES determined by fluorescein antibody technique. *Proc. Soc. Exp. Biol. Med. 95*, 510-513.

Cuffini A, Carlone NA, Forni G (1980) Bacterial phagocytosis by macrophages from lipopolysaccharide responder and nonresponder mouse strains. *Infect. Immun. 28*, 762-765.

Dewey WL, Martin BR, Harris LS (1976) Chronic effects of Δ^9-THC in animals: tolerance and biochemical changes. In: Braude MC, Szara S (Eds), *Pharmacology of Marihuana, Vol 2*, pp 585-594. Raven Press, New York.

Dobson EL, Kelly LS (1973) The combined stimulation of the reticuloendothelial system by estradiol and endotoxin. *J. Reticuloendothel. Soc. 13*, 61-69.

Doe WF, Henson PM (1979) Macrophage stimulation by bacterial lipopolysaccharides. III. Selective unresponsiveness of C3H/HeJ macrophages to the lipid A differentiation signals. *J. Immunol. 123*, 2304-2310.

Dowling JN, Feldman HA (1970) Quantitative biological assay of bacterial endotoxins. *Proc. Soc. Exp. Biol. Med. 134*, 861-864.

Evans R (1974) Specific and nonspecific activation of macrophages. In: Wagner W-H, Hahn H (Eds), *Activation of Macrophages*, pp 305-314. Elsevier Publishing Company, Amsterdam/New York.

Filkins JP (1979) Adrenergic blockage and glucoregulation in endotoxin shock. *Circ. Shock 6*, 99-107.

Filkins JP (1980) Endotoxin-enhanced secretion of macrophage insulin-like activity. *J. Reticuloendothel. Soc. 27*, 507-511.

Filkins JP, Cornell RP (1974) Depression of hepatic gluconeogenesis and the hypoglycemia of endotoxin shock. *Am. J. Physiol. 227*, 778-781.

Fischer H, Lohmann-Matthes ML, Peskar B, Suter D, Rietschel E, Weidemann M, Wekerle H (1977) Prostaglandin synthesis and phagocytosis-induced chemoluminescence in endotoxin-activated macrophages. *Eur. Surg. Res. 9*, 286-287.

Floersheim GL, Logara-Kalantzis A (1972) Synergistic toxicity of endotoxins with oxytetracycline and tetracycline. *Lancet 1*, 1126.

Floersheim GL, Szeszak JJ (1971) Poly I:Poly C and endotoxins share immunosuppressive properties and increase the toxicity of α-amanitin and hexobarbital. *Agents Actions 2*,

150-155.

Freudenberg MA, Bøg-Hansen TC, Back U, Galanos C (1980) Interaction of lipopolysaccharides with plasma high-density lipoprotein in rats. *Infect. Immun. 28*, 373-380.

Gabridge MG (1974) Synergistic effect of pactamycin and sparsomycin on *Mycoplasma* induced lethal toxicity of mice. *Antimicrob. Agents Chemother. 5*, 453-457.

Galanos C, Rietschel ETh, Westphal O, Kim YB, Watson DW (1972) Biological activities of lipid A complexed with bovine serum albumin. *Eur. J. Biochem. 31*, 230-233.

Garlick PJ, McNurlan M, Fern E, Tomkins A, Waterlow JC (1980) Stimulation of protein synthesis and breakdown by vaccination. *Br. Med. J. 281*, 263-265.

Glode LM, Jacques A, Mergenhagen SE, Rosenstreich DL (1977) Resistance of macrophages from C3H/HeJ mice to *in vitro* cytotoxic effects of endotoxin. *J. Immunol. 119*, 162-166.

Golub S, Gröschel D, Nowotny A (1968) Factors which affect the reticuloendothelial system uptake of bacterial endotoxins. *J. Reticuloendothel. Soc. 5*, 324-339.

Goodrum KJ, Berry LJ (1978) The effect of glucocorticoid antagonizing factor on hepatoma cells. *Proc. Soc. Exp. Biol. Med. 159*, 359-363.

Greer GG, Rietschel ETh (1978) Lipid A-induced tolerance and hyperreactivity to hypothermia in mice. *Infect. Immun. 19*, 357-368.

Haeseler F, Bodel P, Atkins E (1977) Characteristics of pyrogen production by isolated rabbit Kupffer cells *in vitro*. *J. Reticuloendothel. Soc. 22*, 569-581.

Halushka PV, Wise WC, Cook JA (1981) Protective effects of aspirin in endotoxic shock. *J. Pharmacol. Exp. Ther. 218*, 464-469.

Hershko A, Ciechanover A (1982) Mechanisms of intracellular protein breakdown. *Annu. Rev. Biochem. 51*, 335-364.

Janoff A, Weissmann G, Zweifach BW, Thomas L (1962) Pathogenesis of experimental shock. IV. Studies on lysosomes in normal and tolerant animals subjected to lethal trauma and endotoxemia. *J. Exp. Med. 116*, 451-466.

Karp RD, Bradley SG (1968) Synergistic toxicity of endotoxin with pactamycin and sparsomycin. *Proc. Soc. Exp. Biol. Med. 128*, 1075-1080.

Kim YB, Watson DW (1970) A purified group A streptococcal pyrogenic exotoxin. Physicochemical and biological properties including enhancement of susceptibility to endotoxin lethal shock. *J. Exp. Med. 131*, 611-628.

Knapp MR, Severinson-Gronowicz E, Schroeder J, Strober S (1979) Characterization of the spontaneous murine B cell leukemia (BCL$_1$). II. Tumor cell proliferation and IgM secretion after stimulation by LPS. *J. Immunol. 123*, 1000-1006.

Levine S, Sowinski R (1976) Endotoxin toxicity in rats is enhanced by tilorone. *Proc. Soc. Exp. Biol. Med. 151*, 329-332.

Loewe S (1957) Antagonisms and antagonists. *Pharmacol. Rev. 9*, 237-242.

Long CL, Birkhahn RH, Geiger JW, Betts JE, Schiller WR, Blakemore WS (1981) Urinary excretion of 3-methylhistidine: an assessment of muscle protein catabolism in adult normal subjects and during malnutrition, sepsis, and skeletal trauma. *Metabolism 30*, 765-776.

Maier RV, Mathison JC, Ulevitch RJ (1981) Interactions of bacterial lipopolysaccharides with tissue macrophages and plasma lipoproteins. In: Majde JA, Person RJ (Eds), *Pathophysiological Effects of Endotoxins at the Cellular Level*, pp 133-155. Alan R. Liss, New York.

Mantilla-Plata B, Harbison RD (1974) Effects of phenobarbital and SKF525A pretreatment, sex, liver, injury, and vehicle on Δ^9-tetrahydrocannabinol toxicity. *Toxicol. Appl. Phar-*

macol. 27, 123-130.

Mantilla-Plata B, Harbison RD (1976) Influence of alteration of tetrahydrocannabinol metabolism on tetrahydrocannabinol-induced teratogenesis. In: Braude MC, Szara S (Eds), *Pharmacology of Marihuana, Vol 2*, pp 773-742. Raven Press, New York.

Marecki NM (1974) Methylprednisolone protection against endotoxin in lead-sensitized mice. *Infect. Immun. 9*, 962-963.

Marecki NM, Bradley SG (1973) Enhanced toxicity for mice of combinations of bacterial endotoxin with antitumor drugs. *Antimicrob. Agents Chemother. 3*, 599-606.

Marecki NM, Bradley SG (1974) Enhanced toxicity for mice of 6-mercaptopurine with bacterial endotoxin. *Antimicrob. Agents Chemother. 5*, 413-419.

Marecki NM, Bradley SG, Munson AE, Drummond DC (1975) Effect of bacterial lipopolysaccharide, lipid A, and concanavalin A on lethality of 5-fluorouracil for mice. *Toxicol. Appl. Pharmacol. 31*, 83-92.

Martin BR, Montgomery J, Dewey WL, Harris LS (1978) Alterations in the pharmacodynamics of ^{3}H-Δ^9-THC in mice by bacterial endotoxin. *Drug Metab. Dispos. 6*, 282-287.

Mathison JC, Ulevitch RJ (1979) The clearance, tissue distribution, and cellular localization of intravenously injected lipopolysaccharide in rabbits. *J. Immunol. 123*, 2133-2143.

McCallum RE (1981) Hepatocyte-Kupffer cell interactions in the inhibition of hepatic gluconeogenesis by bacterial endotoxin. In: Majde JA, Person RJ (Eds), *Pathophysiological Effects of Endotoxins at the Cellular Level*, pp 99-113. Alan R. Liss, New York.

McGivney A, Bradley SG (1977) Enhanced lysosomal enzyme activity in mouse cells treated with bacterial endotoxin *in vitro*. *Proc. Soc. Exp. Biol. Med. 155*, 390-394.

McGivney A, Bradley SG (1978) Effects of bacterial endotoxin on lysosomal enzyme activities of human cells in continuous culture. *J. Reticuloendothel. Soc. 23*, 223-230.

McGivney A, Bradley SG (1979a) Effects of bacterial endotoxin on lysosomal and mitochondrial enzyme activities of established cell cultures. *J. Reticuloendothel. Soc. 26*, 307-316.

McGivney A, Bradley SG (1979b) Action of bacterial endotoxin and lipid A on mitochondrial enzyme activities of cells in culture and subcellular fractions. *Infect. Immun. 25*, 664-671.

McGivney A, Bradley SG (1980a) Action of bacterial lipopolysaccharide on the respiration of mouse liver mitochondria. *Infect. Immun. 27*, 102-106.

McGivney A, Bradley SG (1980b) Susceptibility of mitochondria from endotoxin-resistant mice to lipopolysaccharide. *Proc. Soc. Exp. Biol. Med. 163*, 56-59.

McMillan DE, Dewey WL, Harris LS (1971) Characteristics of tetrahydrocannabinol tolerance. *Ann. NY Acad. Sci. 91*, 83-99.

Mela L (1981) Direct and indirect effects of endotoxin on mitochondria function. In: Majde JA, Person RJ (Eds), *Pathophysiological Effects of Endotoxins at the Cellular Level*, pp 15-21. Alan R. Liss, Inc., New York.

Miller ML, Samuelson CO, Ward JR, Hiramoto RN (1979) Endotoxin toxicity in rats with 6-sulfanilamidoindazole arthritis. *Infect. Immun. 25*, 337-344.

Moeller GR, Terry L, Snyderman R (1978) The inflammatory response and resistance to endotoxin in mice. *J. Immunol. 120*, 116-123.

Mørland B (1979) Studies on selective induction of lysosomal enzyme activities in mouse peritoneal macrophages. *J. Reticuloendothel. Soc. 26*, 749-762.

Mørland B, Mørland J (1978) Selective induction of lysosomal enzyme activities in mouse peritoneal macrophages. *J. Reticuloendothel. Soc. 23*, 469-477.

Munson AE, Regelson W (1971) Sensitization to endotoxin by pyran copolymer. *Proc. Soc. Exp. Biol. Med. 137*, 553-557.

Munson AE, Drummond DC, Adams AC, Bradley SG (1976) Enhanced toxicity for mice of combinations of bacterial lipopolysaccharide and vincristine. *Antimicrob. Agents Chemother. 9*, 840-847.

Munson AE, Sanders VM, Bradley SG, Loveless SE, Harris LS (1978) Lethal interaction of bacterial lipopolysaccharide and naturally occurring cannabinoids. *J. Reticuloendothel. Soc. 24*, 647-655.

Nowakowski M, Edelson PJ, Bianco C (1980) Activation of C3H/HeJ macrophages by endotoxin. *J. Immunol. 125*, 2189-2194.

Nygren H, Dahlén G, Möller G (1979) Bacterial lipopolysaccharides bind selectively to lymphocytes from lipopolysaccharide high-responder mouse strains. *Scand. J. Immunol. 10*, 555-561.

Peavy DL, Shands Jr JW, Adler WH, Smith RT (1973) Mitogenicity of bacterial endotoxins: characterization of the mitogenic principle. *J. Immunol. 111*, 352-357.

Perry RP, Kelley DE (1979) Immunoglobulin messenger RNAs in murine cell lines that have characteristics of immature B lymphocytes. *Cell 18*, 1333-1339.

Phillips FS, Sternberg SS, Hamilton L, Clarke DA (1954) The toxic effects of 6-mercaptopurine and related compounds. *Ann. NY Acad. Sci. 60*, 283-296.

Pieroni RE, Levine L (1969) Enhancing effect of insulin on endotoxin lethality. *Experientia 25*, 507-508.

Pieroni RE, Broderick EJ, Bundeally A, Levine L (1970) A simple method for the quantitation of submicrogram amounts of bacterial endotoxin. *Proc. Soc. Exp. Biol. Med. 133*, 790-794.

Prior RL, Visek WJ (1973) Effects of modifying urea hydrolysis in acute nephrectomy on survival, ammonia in cecal contents and blood metabolites. *Proc. Soc. Exp. Biol. Med. 144*, 184-188.

Rodemann HP, Goldberg AL (1982) Arachidonic acid, prostaglandin E_2 and $F_{2\alpha}$ influence rates of protein turnover in skeletal and cardiac muscle. *J. Biol. Chem. 257*, 1632-1638.

Rose WC (1973) Interaction of bacterial toxins in the toxicity of chemotherapeutic agents. *CRC Crit. Rev. Toxicol. 3*, 159-209.

Rose WC, Bradley SG (1969) Retardation by methylprednisolone of the synergistic toxicity of endotoxin with sparsomycin or pactamycin. *Proc. Soc. Exp. Biol. Med. 132*, 729-731.

Rose WC, Bradley SG (1971) Enhanced toxicity for mice of combinations of antibiotics with *Escherichia coli* cells or *Salmonella typhosa* endotoxin. *Infect. Immun. 4*, 550-555.

Rose WC, Bradley SG, Lee IP (1972a) Enhanced toxicity for mice of vincristine and other chemotherapeutic agents with *Salmonella typhosa* endotoxin and *Pseudomonas aeruginosa. Antimicrob. Agents Chemother. 1*, 489-495.

Rose WC, Bradley SG, Lee IP (1972b) Potentiation of the toxicity of several antitumor agents by *Salmonella typhosa* endotoxin. *Toxicol. Appl. Pharmacol. 23*, 102-111.

Ruco LP, Meltzer MS (1978a) Defective tumoricidal capacity of macrophages from C3H/HeJ mice. *J. Immunol. 120*, 329-334.

Ruco LP, Meltzer MS (1978b) Macrophage activation for tumor cytotoxicity: development of macrophage cytotoxic activity requires completion of a sequence of short-lived intermediary reactions. *J. Immunol. 121*, 2035-2042.

Schlievert PM (1982) Enhancement of host susceptibility to lethal endotoxin shock by staphylococcal pyrogenic exotoxin type C. *Infect. Immun. 36*, 123-128.

Selye H, Tuckweber B, Bertok L (1966) Effect of lead acetate on the susceptibility of rats to bacterial endotoxins. *J. Bacteriol. 91*, 884-890.

Severinson-Gronowicz E, Doss C, Assisi F, Vitetta ES, Coffman RL, Strober S (1979a) Surface Ig isotypes on cells responding to lipopolysaccharide by IgM and IgG secretion. *J. Immunol. 123*, 2049-2056.

Severinson-Gronowicz E, Doss C, Schröder J (1979b) Activation to IgG secretion by lipopolysaccharide requires several proliferation cycles *J. Immunol. 123*, 2057-2062.

Seyberth HW, Schmidt-Gayk H, Hackenthal E (1972) Toxicity, clearance, and distribution of endotoxin in mice as influenced by actinomycin D, cycloheximide, α-amanitin and lead acetate. *Toxicon 10*, 491-500.

Shands Jr JW, Miller V, Martin H (1969) The hypoglycemic activity of endotoxin. I. Occurrence in animals hyperreactive to endotoxin. *Proc. Soc. Exp. Biol. Med. 130*, 413-417.

Siemens AJ, Kalant H (1975) Metabolism of Δ^1-tetrahydrocannabinol by the rat *in vivo*. *Biochem. Pharmacol. 24*, 755-762.

Silver IA (1981) Some effects of *Escherichia coli* endotoxin on cells in culture. In: Majde JA, Person RJ (Eds), *Pathophysiological Effects of Endotoxins at the Cellular Level*, pp 81-95. Alan R. Liss, Inc., New York.

Sipe JD, Vogel SN, Ryan JL, McAdam KPWJ, Rosenstreich DL (1979) Detection of a mediator derived from endotoxin-stimulated macrophages that induces the acute phase serum amyloid A response in mice. *J. Exp. Med. 150*, 597-606.

Skarnes RC (1966) The inactivation of endotoxin after interaction with certain proteins of normal serum. *Ann. NY Acad. Sci. 133*, 644-662.

Sone S, Fidler IJ (1980) Tumor cytotoxicity of rat alveolar macrophages activated *in vitro* by endotoxin. *J. Reticuloendothel. Soc. 27*, 269-279.

Springer GF, Adye JC, Mergenhagen SE, Rosenstreich DL (1977) Endotoxin interactions with phosphatides from human leukocytes and platelets, and with lymphoid cells of susceptible and resistant mice. In: Schlessinger D (Ed), *Microbiology 1977*, pp 326-329. American Society of Microbiology, Washington DC.

Suter E, Ullman GE, Hoffman RG (1958) Sensitivity of mice to endotoxin after vaccination with BCG (Bacillus Calmette-Guérin). *Proc. Soc. Exp. Biol. Med. 99*, 167-169.

Tagliabue A, McCoy JL, Heberman RB (1978) Refractoriness to migration inhibition factors of macrophages of LPS nonresponder mouse strains. *J. Immunol. 121*, 1223-1226.

Tavakoli H, Mela L (1982) Alterations of mitochondrial metabolism and protein concentrations in subacute septicemia. *Infect. Immun. 38*, 536-541.

Tavakoli H, Moon RJ (1981) *In vitro* interactions of endotoxin, chromatin, DNA, and steroid hormone receptor. In: Majde JA, Person RJ (Eds), *Pathophysiological Effects of Endotoxins at the Cellular Level*, pp 33-43. Alan R. Liss, Inc., New York.

Trejo RA, Loose LD, DiLuzio NR (1972) Influence of diethylstilbestrol (DES) on reticuloendothelial function, tissue distribution, and detoxification of *S. enteritidis* endotoxin. *J. Reticuloendothel. Soc. 11*, 88-97.

Tsuda M, Honjo T, Shimiza A, Mizuno D, Naton S (1980) Induced synthesis of immunoglobulin messenger RNA accompanies induction of immunoglobulin production in cultured mouse spleen cells. *J. Biochem. 84*, 1285-1290.

Ulevitch RJ, Johnston AR (1978) The modification of biophysical and endotoxic properties of bacterial lipopolysaccharides by serum. *J. Clin. Invest. 62*, 1313-1324.

Urbaschek B, Urbaschek R (1977) Some aspects of microcirculatory and metabolic changes in endotoxemia and in endotoxin tolerance. In: Schlessinger D (Ed), *Microbiology 1977*, pp 286-292. American Society of Microbiology, Washington DC.

Vogel NS, Rosenstreich DL (1979) Defective Fc receptor mediated phagocytosis in C3H/ HeJ macrophages. I. Correction by lymphokine induced stimulation. *J. Immunol. 123*, 2842-2850.

Walker RI, Snyder SL (1978) Endotoxemia in irradiated mice. In: Rosenberg P (Ed), *Toxins: Animal, Plant and Microbial*, p 933-946. Pergamon Press, Oxford.

Willerson JT, Trelstad RL, Pincus T, Levy SB, Wolff SM (1970) Subcellular localization of *Salmonella enteritidis* endotoxin in liver and spleen of mice and rats. *Infect. Immun. 1*, 440-445.

Wilson ME, Munkenbeck P, Morrison DC (1981) Influence of bacterial endotoxins on neutrophilic leukocytes: lack of a correlation between *in vivo* and *in vitro* responses. In: Majde JA, Person RJ (Eds), *Pathophysiological Effects of Endotoxins at the Cellular Level*, pp 157-172. Alan R. Liss, Inc., New York.

Wolfe RR, Elahi D, Spitzer J (1977) Glucose and lactate kinetics after endotoxin administration in dogs. *Am. J. Physiol. 232*, E180-E185.

Yarkoni E, Rapp HJ (1979) Influence of oil and Tween concentrations on enhanced endotoxin lethality in mice pretreated with emulsified trehalose-6,6′dimycolate (Cord Factor). *Infect. Immun. 24*, 571-572.

Yuan D, Tucker PW (1982) Effect of lipopolysaccharide stimulation on the transcription and translation of messenger (RNA) for cell surface immunoglobulin M. *J. Exp. Med. 156*, 962-974.

Zlydaszyk JC, Moon RJ (1976) Fate of ^{51}Cr-labeled lipopolysaccharide in tissue culture cells and livers of normal mice. *Infect. Immun. 14*, 100-105.

Handbook of Endotoxin, Vol. 3: Cellular Biology of Endotoxin
L.J. Berry, editor
© Elsevier Science Publishers B.V., 1985

CHAPTER 16

Interactions of bacterial lipopolysaccharides and plasma high density lipoproteins*

RICHARD J. ULEVITCH

1. INTRODUCTION

The host factors which regulate the interactions of bacterial lipopolysaccharides (LPS) with cells are not well defined, although serum proteins are presumed to play a major role in controlling these events. It is now appreciated that LPS binds to a variety of serum proteins and that this can result in modifications of the biophysical, immunologic or biologic properties of the LPS. Recently a role for high density lipoproteins (HDL) as carriers of LPS in blood has been described, and the purpose of this chapter is to consider experiments which have explored this newly identified function of HDL. To do this, the chapter will be divided into three sections which will review: (a) studies of the detoxification of LPS by serum which provided the initial evidence implicating lipoproteins in LPS–serum protein interactions; (b) studies on the role of high density lipoprotein in LPS–serum protein interactions and the mechanism of LPS-lipoprotein binding, and (c) studies of the biologic properties of the LPS-HDL complex.

Because the complexes of LPS and HDL remain stable in blood, are cleared slowly, and retain the capacity to activate certain key host mediation systems, this form of LPS may be of great importance in the pathogenesis of LPS-induced shock and disseminated intravascular coagulation. Therefore, studies of the mechanism of interaction of LPS and HDL, the structure of the complex and the biologic properties of this material may provide new insight into the mechanisms of LPS action. The study of the biology and chemistry of LPS–HDL interactions has received little attention until recently but is now becoming an area of active investigation. This chapter will consider all of the recent, published work related to this new facet of LPS research.

* The author appreciates the support of U.S.P.H.S. Grants AI-15136 and AI-00391 (RCDA), Army Contract No. C-2101.

2. LPS INTERACTIONS WITH SERUM PROTEINS

Since the early 1950s, the effect of serum on LPS has been an area of active research. From the initial inquiries a number of seemingly inconsistent findings emerged. On the one hand, it was noted that addition of serum to LPS could result in the inhibition of LPS activity, while on the other hand, the potentiation of certain biologic effects attributed to LPS was reported. These seemingly diverse findings depended on the experimental conditions and the source of LPS. The greatest efforts were devoted to defining the conditions for serum-induced detoxification of LPS, and from these early studies evolved the basis for experiments performed over twenty years later which have defined the mechanism of LPS interactions with HDL. Some of the early studies of LPS–serum interactions which suggested a role for lipoproteins will be reviewed in Section 2.1.

2.1. Detoxification of LPS by serum

Many of the initial studies of the effect of serum on LPS were performed to investigate the mechanism of LPS detoxification by serum (Rosen et al 1958; Skarnes et al 1958) and from these efforts the first suggestion that LPS interacts with lipoproteins was provided. Using antibody to human serum proteins, Skarnes (1966) reported results suggesting that in serum endotoxins may bind to two classes of lipoproteins; β- and α_1-lipoprotein. He suggested that this binding was associated with a loss of endotoxicity as well as alterations in the physical and immunochemical properties of the LPS. These findings were extended to include efforts to isolate a serum fraction that would alter the immunochemical and biologic properties of LPS. Upon fractionation of serum on DEAE-cellulose a protein peak rich in both albumin and α_1-lipoprotein (HDL) was isolated that after addition to LPS increased the diffusion rates of endotoxins in agar gels. The increased diffusion rate was suggested to result from enzymatic degradation of the LPS (Skarnes 1966), producing LPS of a much smaller particle size. This early work related to LPS detoxification by serum is being considered in detail in Chapter 3 and will not be reviewed further here.

In other studies Skarnes (1968, 1970), and Moreau and Skarnes (1973) described results suggesting that LPS-lipoprotein complexes formed in vivo in plasma. However, they did not provide direct biochemical evidence of complex formation between LPS and lipoproteins. Subsequently, Back et al (1976) demonstrated a change in the crossed immunoelectrophoretic patterns of LPS after exposure to serum and suggested that this change resulted from degradation of LPS by an arylesterase associated with HDL. However, the altered immunoelectrophoretic patterns of LPS observed by these investigators could have resulted from alternative processes, including LPS binding to plasma proteins, LPS disaggregation by serum proteins, LPS degradation by enzymes, or a combination of any of these phenomena. The relative importance of these various mechanisms was not evaluated by Back and his colleagues.

Despite these suggestions that LPS–HDL interactions may be involved in LPS modification by serum, neither the mechanism of LPS-lipoprotein binding nor the biochemical properties of the putative LPS-lipoprotein complex were explored until very recently. During the past five years, however, studies have been performed that have clarified much about the mechanism of LPS–lipoprotein interaction and these will be reviewed in Sections 2.2–2.3.

2.2. LPS–lipoprotein interactions: mechanism of interaction with high density lipoproteins

The use of equilibrium isopycnic density gradient ultracentrifugation of LPS/serum mixtures incorporating high-purity, radiolabeled LPS as a tracer provides a quantitative, analytical method to determine serum effects on LPS. This technique is particularly useful since most LPS preparations have buoyant densities in the range of 1.4 g/cm^3 and after binding to HDL the LPS is converted to a form with a density of less than 1.2 g/cm^3 (Munford et al 1981a, 1981b; Ulevitch et al 1978, 1979). The original application of this technique yielded the key observations which have led to a greater understanding of the mechanism of LPS binding to HDL, and ultracentrifugation continues to provide data for new insights into the mechanisms of LPS interactions with plasma proteins.

In a series of studies performed in our laboratory (Ulevitch et al 1978, 1979, 1981), we observed that after intravenous injection of LPS or after mixing LPS with serum in vitro, the buoyant density of the LPS decreased to less than 1.2 g/cm^3. Our initial observations in vivo were obtained in experiments with LPS from *Escherichia coli* O111:B4 which was radioiodinated (Ulevitch 1978) and then injected into rabbits (Ulevitch et al 1978). Blood samples were removed, the cells and plasma separated, and the LPS/plasma mixture subjected to equilibrium density gradient ultracentrifugation in CsCl. The data derived from isopycnic equilibrium density gradient analysis of LPS/plasma mixtures indicated that the LPS in plasma rapidly undergoes a marked change in hydrated buoyant density. The *E. coli* O111:B4 LPS is rapidly partitioned into at least two major forms with a density of approximately 1.4 g/cm^3 and a density of less than 1.2 g/cm^3. The LPS at d $\simeq$ 1.4 g/cm^3 is either removed from the blood or converted to the d < 1.2 g/cm^3 form, so that by 30 minutes post-injection more than 90% of the LPS in plasma is found at d < 1.2 g/cm^3. This form of LPS (d < 1.2 g/cm^3) remains in the plasma and disappears with a $T_{1/2}$ of more than 10 hours. This phenomenon of an in vivo change in density of LPS to <1.2 g/cm^3 was subsequently observed by Munford et al (1981a, 1982) with LPS from *Salmonella typhimurium* in studies performed in both the rat as well as the squirrel monkey.

Analysis of the mechanism of the change in density of LPS occurring in vivo was substantially aided by the observation that this density change occurs in vitro when LPS is mixed with either plasma or serum. Studies have now been described using LPS from a number of different bacterial sources isolated by different extraction procedures. These include LPS from *E. coli* O111:B4, *E. coli* O111:B4:55, *Sal-*

monella minnesota R595, *Salmonella typhimurium*, and *Salmonella abortus equi*, and with serum from several different species including man, rabbit, mouse and rat produce qualitatively similar changes in the density of LPS. In general, the addition of serum to LPS results in a time-dependent, small decrease in the density of the parent LPS accompanied by disaggregation of the LPS, followed by the appearance of LPS at d $<$ 1.2 g/cm^3. The rate of the density change of LPS to d $<$ 1.2 g/cm^3 depends on temperature and the concentration of LPS and it varies with the source of LPS and serum as well. For some LPS preparations, the presence of divalent ions, particularly Ca^{2+}, may inhibit the serum-induced reduction in LPS density, although this can be overcome by the addition of EDTA to serum or by prior treatment of LPS with electrodialysis (Galanos and Lüderitz 1976) or chelex resin treatment (Strain et al 1983; R.J. Ulevitch et al unpublished data). These treatments reduce the particle size of LPS by removal of tightly bound cationic substances and presumably facilitate binding to HDL in this way. On the basis of current data from several different laboratories, a mechanism to explain the serum-induced decrease in LPS density must include at least two steps as shown below. The evidence to support this mechanism will be discussed in detail in the following sections.

Step 1 (disaggregation): $\qquad\qquad\qquad$ LPS$_{(aggregated)}$ $\rightleftharpoons$ LPS$_{(disaggregated)}$

Step 2 (binding to HDL): $\qquad$ LPS$_{(disaggregated)}$ + HDL $\rightarrow$ LPS-HDL$_{d\,<\,1.2\,\text{g/cm}^3}$

In Section 2.2.3. the observations of Tobias and Ulevitch (1983) demonstrating that the binding of LPS to HDL may be controlled by protein(s) present in acute-phase serum will be considered; these new data underscore the potential importance of other serum proteins in regulating the binding of LPS to HDL and the complexity of these reactions.

2.2.1. Evidence for disaggregation of LPS in serum

A number of observations provide evidence that disaggregation of LPS is produced by components of lipid-free serum. If serum is depleted of lipids by extraction with butanol/diisopropyl ether (Cham and Knowles 1976), or by ultracentrifugation to remove lipoproteins (Munford et al 1981b), this lipid-depleted serum does not decrease the buoyant density of LPS to less than 1.2 g/cm^3. However, as shown by Ulevitch et al (1979) and Munford et al (1981b), exposure of LPS to delipidated serum results in a change in the sedimentation behavior of LPS in sucrose gradients which is consistent with a decrease in the particle size of LPS aggregates. Munford et al (1981b) also demonstrated that the presence of 1 mM Ca^{2+} in lipid-depleted serum reduced the extent of disaggregation while 2 mM Ca^{2+} inhibited disaggregation essentially completely. When LPS and delipidated serum were mixed prior to the addition of Ca^{2+}, then no effect on the disaggregation was observed. These data suggest that the disaggregation step is sensitive to divalent ions, but that once disaggregation occurs it is not easily reversible.

At the present time, little information is available concerning the identity of the serum proteins responsible for LPS disaggregation other than the observations that this component is present in lipid-depleted serum. It is unlikely that the disaggregating factors are bile salts although this possibility has not been rigorously excluded. Johnson et al (1977) reported the isolation of an α-globulin from human serum which disaggregates LPS. These data are consistent with prior observations of Skarnes (1966) which implicated an α-globulin in LPS disaggregation. Johnson et al (1977) reported that disaggregation of LPS by the α-globulin resulted in the partial loss of endotoxic activity when tested by Limulus lysate gelation and toxicity in actinomycin D-treated mice. The α-globulin is 4.5S, is apparently not a lipoprotein, and the ability of the factor to disaggregate LPS is not blocked by irreversible inhibitors of serine esterases. The role which this protein may play in the disaggregation of LPS in serum prior to binding of LPS to HDL needs to be explored further.

Of particular interest is the question of whether, in addition to disaggregation of LPS, transfer to disaggregated LPS to HDL is effected by a separate group of serum proteins. A possible role in this regard should be considered for phospholipid or glycolipid exchange proteins known to be present in serum (Ihm and Harmony 1980; Pownall et al 1982) and needs to be explored in future studies.

Other data which support the concept that LPS be disaggregated prior to binding to HDL derives from characterization of the LPS at $d < 1.2$ g/cm^3 when *S. minnesota* R595 LPS is mixed with plasma. The isolated LPS-HDL complex (see below) elutes from Sepharose 6B in a position equivalent to 310 kilodaltons which is comparable to the size of rabbit HDL and thus must consist of disaggregated LPS bound to an HDL particle. Finally, in our own studies (Ulevitch et al 1979) as well as the studies of Munford et al (1981b), it was demonstrated that addition of the parent LPS (in the absence of delipidated serum) to purified HDL did not result in reduction of the density of the LPS or the binding of LPS to HDL. Munford et al also showed that prior treatment of LPS with sodium deoxycholate markedly enhanced the association of the LPS with HDL (Munford et al 1981b). Since it is known that deoxycholate reduces the particle size of LPS by disaggregation, this finding adds additional strength to the contention that a reduction in the particle size of LPS is required for binding to HDL. It is possible that reduction in the particle size of LPS aggregates by removal of cations, which can be accomplished by a number of techniques including electrodialysis (Galanos and Lüderitz 1976), chelex resin treatment (Strain et al 1983) or a combination of these treatments, might facilitate the binding of LPS to HDL in the absence of other serum components. This question deserves further analysis.

2.2.2. *The role of HDL in alterations of LPS by serum*

The first direct evidence that HDL is involved in the reduction of LPS density to $d < 1.2$ g/cm^3 in serum was obtained in experiments reported by Ulevitch et al (1979). Delipidated serum was mixed with varying concentrations of purified lipoproteins: very low density lipoproteins (VLDL), low density lipoproteins (LDL) or HDL, and then this reconstituted serum was examined for the ability to reduce the buoyant density of LPS to less than 1.2 g/cm^3. Studies using LPS from either *S. minnesota* Re595 or *E. coli* O111:B4 showed that only HDL and not the other lipoproteins effectively reconstituted the chemically delipidated serum. Similarly, Munford et al (1981b) showed that purified rat HDL added back to lipoprotein-depleted rat serum reconstituted the lipid-deficient serum to reduce the density of LPS to less than 1.2 g/cm^3. Thus, these data support the hypothesis that HDL is involved in the decrease in buoyant density of LPS in serum but they do not provide direct evidence that LPS at d < 1.2 g/cm^3 results from the binding of LPS to HDL.

More compelling evidence that LPS at d < 1.2 g/cm^3 results from the binding of disaggregated LPS to HDL was obtained by Ulevitch et al (1981) using immunoaffinity chromatography. Antibody to *S. minnesota* Re595 LPS was immunopurified, insolubilized on Sepharose 4B (anti Re595-Sepharose 4B), and used to isolate LPS at d < 1.2 g/cm^3 from LPS/serum mixtures. Mixtures of Re595 LPS and rabbit serum were passed through the anti Re595-Sepharose 4B and after washing, the bound LPS and associated substances were eluted with 2.5 M potassium thiocyanate. The properties of the material found in the fraction eluted by potassium thiocyanate are summarized below.

This fraction contained LPS which had a d < 1.2 g/cm^3 when centrifuged in CsCl gradients, contained both protein and lipid components in addition to the LPS, displayed an apparent molecular weight of 310 kilodaltons when examined by molecular exclusion chromatography, and reacted with anti-Re595 LPS antibody in immunodiffusion gels although the diffusion rate was markedly greater than that of the parent LPS. Because this eluate contains LPS as well as HDL protein and lipid (see below), this will be referred to as the LPS-HDL complex. Thus the eluted fraction had all of the properties which would be expected from a complex of disaggregated LPS and HDL.

After treatment of the LPS-HDL complex with reducing agents under denaturing conditions, sodium dodecylsulfate polyacrylamide gel electrophoresis (SDS-PAGE) demonstrated that a polypeptide with an apparent molecular weight of approximately 25.7 kilodaltons is the major protein component. Evidence that this polypeptide is derived from HDL was obtained from amino acid analysis of the 25.7-kilodalton band which showed that the amino acid composition of this 25.7-kilodalton polypeptide of the LPS-HDL complex was identical to that of the apo AI present in rabbit HDL (Ulevitch et al 1981). The apo AI also has an apparent molecular weight of 25.7 kilodaltons in SDS-PAGE.

The lipid composition of the LPS-HDL complex isolated by immunoaffinity chromatography is shown in Table 1 and compared with that of rabbit HDL. These data indicate that the lipids of the potassium thiocyanate eluate were likely to be derived from HDL, although inspection of these data reveals several differences between the lipid composition of the LPS-HDL complex and that of bulk HDL. Most notable is the variation in the cholesterol ester/cholesterol ratios which de-

TABLE 1   *Composition of LPS-HDL complex isolated from Re595 LPS-rabbit serum mixture by immunoaffinity chromatography*

	Composition (% by weight)	
	LPS-HDL*	HDL**
Protein	53	54
Total lipid	47	46

	Composition (% by weight of total lipid content)	
Lipid class	LPS-HDL*	HDL**
Triglyceride	20	9
Cholesterol ester	15	36
Cholesterol	16	6
Phospholipid	46	46
Ceramide monohexoside	3	trace

	Composition (% by weight of total phospholipid content)	
Phospholipid class	LPS-HDL*	HDL**
Lysophosphatidyl choline	6	2
Sphingomyelin	9	6
Phosphatidyl choline	34	76
Phosphatidyl ethanolamine	40	8
Phosphatidyl serine Phosphatidyl inositol	11	7.2

* Prepared from a mixture of Re595 LPS and rabbit plasma (containing 20 mM EDTA); the lipid analysis was performed as described (Ulevitch et al 1981).
** Rabbit HDL was prepared from pooled normal rabbit serum and the lipid analysis was performed as described (Ulevitch et al 1981).

crease from 6/1 in HDL to less than 1 in the isolated LPS-HDL complex. When the phospholipid composition of the LPS-HDL complex was analyzed for individual phospholipid classes and compared with that of serum HDL, a striking difference was noted. These data shown in Table 1 demonstrate that the formation of an LPS-HDL complex results in a marked change in the phospholipid content of HDL. The phosphatidyl choline/phosphatidyl ethanolamine ratio of less than 1 observed in the LPS-HDL complex stands in marked contrast to the ratio of approximately 10 typically observed in HDL. The mechanism of this alteration in phospholipids of the LPS-HDL complex is not known at the present time and is a subject for future investigation.

Freudenberg et al (1980) have also confirmed a role for HDL in interactions of LPS in serum. With the use of crossed immunoelectrophoresis it was shown that the presence of either smooth or rough form LPS changed the electrophoretic mobility of HDL. This alteration in the charge of HDL results from binding of LPS to HDL, with complexes of smooth form LPS and HDL migrating slower, while those of rough form LPS and HDL migrate faster than HDL. These differences may result from the more negative charge of rough form LPS and suggest that the biological properties of LPS-HDL complexes formed with either rough or smooth forms of LPS may differ. In the same study these investigators demonstrated more rapid clearance kinetics of rough form LPS compared to smooth form LPS injected into rats under conditions where binding of LPS to HDL was expected to be rapid and complete. Whether this difference in kinetics results from dissimilar rates of binding of rough and smooth forms of LPS to HDL or from different rates of clearance of LPS-HDL complexes formed with rough and smooth forms of LPS remains to be determined.

2.2.3. Binding of LPS to HDL in acute-phase serum

The serum concentrations of a number of acute-phase proteins are known to increase dramatically as a result of infection, after exposure to nonspecific inflammatory substances such as silver nitrate or turpentine, and importantly following LPS injection. Because of the possibility that one or more of the acute-phase proteins might be involved in regulating the interactions of LPS with serum proteins or cells, we have recently initiated studies to examine the binding of LPS to HDL in acute-phase serum.

Consideration was first given to the possibility that one acute-phase serum protein, serum amyloid A (SAA), an apolipoprotein of HDL, might either influence the binding of LPS to SAA-containing HDL particles or the subsequent biologic activity of the LPS-HDL (SAA) complex (Hoffman and Bendit 1982; Tobias et al 1982). As noted in recently published studies by Tobias et al (1982), however, evidence to support this hypothesis has not yet been obtained although these studies are still being pursued.

As part of these studies, Tobias et al (1982) examined the dependence of the formation of the LPS-HDL complex on LPS concentration in acute-phase serum. Careful inspection of these data yielded a striking difference between normal and acute-phase serum. At concentrations of LPS between 1 and 50 µg/ml, the addition of LPS to acute-phase serum resulted in the formation of LPS complexes with a density of 1.3 g/cm^3 (termed complex 1.3) as well as the expected complex of LPS-HDL at d < 1.2 g/cm^3. In more recent studies, these observations were extended to include a comparison of the kinetics of binding of LPS to HDL in normal and acute-phase serum (Tobias and Ulevitch 1983). Upon mixing 10 µg LPS with normal rabbit serum at 37°C (in the presence of 20 mM EDTA), the half-life for LPS binding to HDL is typically 2–3 minutes. When the same experiment is done using acute-phase rabbit serum (collected 24 hours post-induction with a sub-

with these host components. Unfortunately, a systematic examination of these changes has not been completed, although a number of recent studies provide some insight into the nature of these effects. Some comparisons of the activity of the parent LPS and LPS bound to HDL may be made from our own data (Mathison and Ulevitch 1981; Ulevitch et al 1978, 1979). These data for *S. minnesota* Re595 LPS are summarized in Table 2 and show that the effect of LPS binding to HDL produces a broad spectrum of changes in LPS activity ranging from a minimal loss of activity to as much as a 1000-fold decrease in activity. The most dramatic effects are observed on activities associated with LPS in a highly aggregated state, such as activation of the classical complement pathway. LPS from the rough strain of *S. minnesota* Re595, normally an excellent activator of the classical complement pathway when bound to HDL, is at least 100-fold less active than the parent LPS in the ability to activate complement. As expected, effects which are known to be dependent on complement activation, such as transient thrombocytopenia observed after an intravenous injection in rabbits, are then inhibited in a similar manner. The ability of LPS to produce a pyrogenic response in rabbits is also markedly diminished as a result of binding of LPS to HDL. In contrast are the lesser effects of the formation of an LPS-HDL complex on a variety of other LPS-dependent activities including mitogenicity in murine B cells, macrophage activation, and the ability to produce hypotension, disseminated intravascular coagulation and death in experimental animals.

Unfortunately, for a number of reasons, it is not possible at the present time to develop a general hypothesis about the result of LPS binding to HDL on the biologic activity of LPS. The data of Table 2 have been obtained with only one type of LPS, and although similar data are available for other rough and smooth forms of LPS, the studies are limited. It is already known that rough and smooth forms of LPS form LPS-HDL complexes of different charge (Freudenberg et al 1980). How this might effect biologic activity is unknown and will require more

TABLE 2 *Comparison of some properties of parent LPS and LPS bound to HDL of S. minnesota*

	Re595 LPS	Re595 LPS–HDL	Relative activity LPS–HDL *vs* LPS
Pyrogenicity	50 ng	5 000 ng	1/100
Leukopenia	50 ng	50 000 ng	1/1 000
Immediate, transient thrombocytopenia	10 µg	> 500 µg	at least 1/50
CH_{50} reduction (50%)	1*	1 000*	1/1 000
Shock and DIC in rabbits	100–200 µg	50–200 µg	unchanged
LD_{50} adrenalectomized mice	45 ng	280 ng	1/6

* Relative amount of LPS needed to reduce rabbit CH_{50} level by 50%.

study. Another gap in our knowledge results from the observation that the binding of LPS to HDL, at least in the cases of LPS from *S. minnesota* Re595 or *E. coli* O111:B4, is saturable, so that the LPS/HDL ratio may vary depending on the initial LPS concentration. Many of the data of Table 2 have been obtained for LPS-HDL complexes formed at a single LPS concentration, and therefore the effect of varying the LPS/HDL ratio on the biologic activities of the LPS/HDL complex has not yet been fully evaluated.

Finally, since HDL is a complex mixture of lipoproteins defined by buoyant density properties and differing in both apolipoprotein and lipid content, the selective binding of LPS to a specific HDL subfraction may result in LPS-HDL complexes with unique properties. This is particularly relevant when considering the variety of apoproteins which may be present in HDL and are known to influence the interactions of HDL with cell membrane receptors (Nicoll et al 1980).

3.1. Effect of HDL on the clearance of LPS from blood

A dramatic result of the formation of the LPS-HDL complex is observed when the kinetics of clearance from blood and the specificities of the interactions of LPS-HDL complexes with cells are analyzed. The rate of clearance of LPS from blood has been shown to depend on several factors including the type of LPS (rough or smooth), the method of purification and subsequent treatment such as electrodialysis or chelex resin treatment, and the host components to which the LPS binds. Despite this complex set of variables, numerous studies have documented that the clearance kinetics of LPS are usually characterized by at least two very distinct phases: a rapid clearance with a $T_{1/2}$ of less than 10 minutes and a much slower phase with a $T_{1/2}$ of more than 5 hours (Braude et al 1955; Herring et al 1963; Mathison and Ulevitch 1979). Although in the past this was considered to result from selective removal of large aggregates of LPS by cells of the reticuloendothelial system, it is now appreciated that these two phases reflect the partitioning of LPS between HDL and non-HDL targets of LPS in blood. In recent studies with LPS-HDL complexes formed in vitro by adding *S. minnesota* Re595 LPS to rabbit serum, we showed that injection of these complexes resulted in clearance kinetics of LPS characterized by the absence of the rapid disappearance phase and that the clearance of LPS is described by a first-order decay curve with a $T_{1/2}$ of 15 hours.

In elegant experiments by Munford et al (1982), these findings were confirmed and extended. These investigators showed that the clearance rate and tissue uptake of LPS-HDL complexes formed with a smooth form LPS approximated that of serum HDL. These studies revealed that the LPS-HDL complex was actually cleared more slowly than HDL, and it was suggested that this occurred because of decreased interaction of the LPS-HDL complex with cell uptake systems for HDL. An alternative explanation is that the LPS bound to a subfraction of HDL which is cleared at a somewhat slower rate than that observed for total serum HDL. Despite this difference in clearance kinetics, the tissue distribution patterns of the LPS in LPS-HDL complexes and ^{125}I-HDL were similar. In the studies of Mathison and Ulevitch (1979, 1981) and Munford et al (1982), the effect of varying the LPS-HDL ratio by forming complexes at differing LPS input concentrations was unfortunately not evaluated.

In our own studies (Mathison and Ulevitch 1979, 1981) as well as the studies of Munford et al (1981a, 1982) it also was noted that LPS was bound at high concentrations in tissue which have specific uptake mechanisms for HDL. The best example of this is the adrenal cortex. Munford et al (1981a) demonstrated that uptake of LPS into adrenal tissues could be diminished by increasing plasma HDL levels 5-fold or by administering dexamethasone, while ACTH pretreatment enhanced LPS uptake into the adrenals. These latter hormonal treatments are known to modulate the number of HDL receptors on adrenocortical cells with ACTH increasing and dexamethasone decreasing the number of receptors. We have also recently shown that the LPS which is localized in the adrenal cortex is associated with lipid droplets in the cytoplasm of these cells (J.C. Mathison and R.J. Ulevitch, unpublished data). Thus these data strongly support the hypothesis that LPS, when bound to HDL, can enter cells by the same mechanisms which are responsible for the internalization of HDL. These findings serve to highlight the effect that formation of the LPS-HDL complex has on the interactions of LPS with cells. These data also provide a possible explanation for the observations of Jones and Carter (1955) that after intravenous injection, a fraction of ^{14}C-polysaccharides of *Klebsiella pneumoniae* (presumably containing the LPS of this bacteria) localized in the adrenal cortex of guinea pigs.

It is likely that binding of LPS to HDL slows down the rapid uptake of LPS into the fixed macrophages of the RES while facilitating the uptake into parenchymal cells such as hepatocytes. A good example of this may be seen in studies of the cellular localization of *S. minnesota* Re595 LPS in liver where two distinct findings have been reported. In our own studies in rabbits with *S. minnesota* Re595 LPS – where in vivo binding to HDL is limited because of the large aggregate form of this LPS – localization at 5 or 180 minutes after injection is exclusively in the liver macrophages (Mathison and Ulevitch 1979). In contrast, studies of Freudenberg et al (1982), using rats injected with Re595 LPS (which was treated by electrodialysis after extraction), reported the rapid localization of LPS in both liver macrophages and hepatocytes. This difference in findings may result from the increased ability of the electrodialyzed Re595 LPS, which has a much smaller particle size than the untreated parent Re595 LPS, to bind to HDL. However, since the studies of Freudenberg et al (1982) determined the presence of LPS in cells by immunofluorescence, the rates and extent of uptake into hepatocytes versus macrophages were not measured.

Although there was a large difference in the amount of LPS injected in these two studies, perhaps differences in the composition of HDL in rabbits and rats are of the greatest importance. In regard to this latter consideration, rat HDL has been shown to contain apolipoprotein E (Quarfordt et al 1978), whereas normal rabbit HDL contains little of this apoprotein (Chapman 1980). Since the presence of apolipoprotein E on HDL will direct this particle to cells with the high-affinity LDL receptor, LPS bound to apolipoprotein E containing HDL particles might be expected to accumulate in hepatocytes which are known to contain this receptor. This is a question for further analysis. Several excellent reviews of the mechanisms of uptake of lipoproteins by receptor-mediated endocytosis have recently appeared in which these subjects are discussed in detail (Brown et al 1979; Brown and Goldstein 1983; Goldstein and Brown 1977; Nicoll et al 1980).

3.2. Release of LPS from bacterial membranes in the serum and binding to HDL

Although it is generally agreed that many of the physiologic alterations occurring in patients with gram-negative bacteremia result from direct effects of LPS, there is a paucity of unambiguous data documenting the presence of LPS in blood or tissues during sepsis. This results from the lack of a quantitative, sensitive assay to measure the presence of LPS in biological fluids and especially in blood. A very limited number of studies have attempted to examine the question of circulating LPS in animal models of bacteremia (Rissing et al 1981), and the question of the concentration, physical and chemical form and association with host blood elements of LPS during sepsis or experimental models of sepsis remains unanswered. Therefore, despite all the information that is now available regarding the interactions of purified LPS with host elements, including the studies of the binding of LPS to HDL, virtually nothing is known about the in vivo release of LPS from bacterial membranes and interactions of this LPS with host cells and proteins.

Recently Munford et al (1982) developed a system using intact bacteria or isolated bacterial membranes which has yielded new and potentially very important information regarding the release and interactions of LPS with HDL. To accomplish this, these investigators used a mutant of *Salmonella typhimurium*, G-30, which incorporates ^{3}H-galactose almost exclusively into LPS and thus provides a specific, in situ biosynthetically labeled LPS. Using this source of ^{3}H-LPS, Munford et al prepared four LPS-containing fractions to examine the binding of LPS to HDL in both serum as well as in vivo. These fractions are phenol-extracted LPS, intact bacteria, bacterial outer membranes and outer membrane fragments. The results of this work indicated that neither HDL alone nor lipoprotein-depleted plasma alone produced an LPS-HDL complex with any of the four LPS-containing fractions. In contrast, addition of both HDL and lipoprotein-depleted plasma to the four fractions yielded the following data with respect to LPS-HDL binding: (a) with purified LPS more than 90% of the LPS bound to HDL, (b) with outer membrane fragments more than 50% of the LPS bound to HDL, (c) with outer membranes more than 20% of the LPS bound to HDL and (d) with intact bacteria approximately 10% of the LPS bound to HDL. These data demonstrate that release of LPS from bacterial membranes occurs in serum and that binding to HDL requires the presence of serum protein(s) in lipid-depleted serum. The investigators extended these findings to show that LPS released from to various sources of bacterial membrane which did associate with HDL was not pyrogenic and that it exhibited clearance kinetics and associated with cells in a manner which would be predicted from previous studies with LPS-HDL complexes formed with purified LPS. These data suggest that release of LPS from bacteria and subsequent binding of LPS to HDL may result during sepsis. It is likely that the formation of the LPS-HDL complex would occur by a mechanism which is similar, if not identical, to that described for the binding of purified LPS to HDL in serum.

The chemical composition of the bacterial membrane may influence the extent to which LPS is released and interacts with HDL. The non-LPS components of the bacterial membrane, such as lipids and proteins, may be modulated by multiple host factors including degradation of bacterial membranes by serum enzymes or complement-mediated reactions, and degradation of bacterial membranes internalized by phagocytic cells. In this latter regard the release of biologically active LPS from macrophages following ingestion of intact gram-negative bacteria has recently been described (Duncan and Morrison 1983). This may provide a mechanism for the continuous release of LPS from phagocytic cells during sepsis and allow for the binding of LPS to HDL.

ACKNOWLEDGEMENT

The author wishes to thank Ms. Elizabeth Goddard for her assistance in the preparation of this manuscript.

REFERENCES

Back U, Moller BB, Bog-Hansen TC (1976) Host resistance to lipopolysaccharides in the pathogenesis of multiple sclerosis and membrane proliferative glomerulonephritis. *Lancet 1*, 188-192.

Braude AI, Carey FJ, Zalesky M (1955) Studies with radioactive endotoxin. II. Correlation of physiologic effects with distribution of radioactivity in rabbits injected with lethal doses of *E. coli* endotoxin labeled with radioactive sodium chromate. *J. Clin. Invest. 34*, 858-866.

Brown MS, Goldstein JL (1983) Lipoprotein metabolism in the macrophage: implications for cholesterol deposition in atherosclerosis. *Annu. Rev. Biochem. 52*, 223-261.

Brown MS, Kovanen PT, Goldstein JL (1979) Receptor-mediated uptake of lipoprotein-cholesterol and its utilization for steroid synthesis in the adrenal cortex. *Recent Prog. Hormone Res. 35*, 215-257.

Cham BE, Knowles BR (1976) A solvent system for delipidation of plasma or serum without protein precipitation. *J. Lipid Res. 17*, 176-181.

Chapman MJ (1980) Animal lipoproteins: chemistry, structure and comparative aspects. *J. Lipid Res. 21*, 789-853.

Duncan Jr RL, Morrison DC (1983) Fate of endotoxic lipopolysaccharide (LPS) following ingestion of *E. coli* by murine macrophages in vitro. *Fed. Proc. 42*, 419.

Freudenberg MA, Bog-Hansen TC, Back U, Galanos C (1980) Interaction of lipopolysaccharides with plasma high density lipoproteins in rats. *Infect. Immun. 28*, 373-380.

Freudenberg MA, Freudenberg N, Galanos C (1982) Time course of cellular distribution of endotoxin in liver, lungs, and kidneys of rats. *Br. J. Exp. Pathol. 63*, 56-65.

Galanos C, Lüderitz O (1976) The role of the physical state of lipopolysaccharides in the interaction with complement. *Eur. J. Biochem. 65*, 403-408.

Goldstein JL, Brown MS (1977) The low-density lipoprotein pathway and its relation to atherosclerosis. *Annu. Rev. Biochem. 46*, 897-930.

Herring WB, Herion JC, Walker RJ, Palmer JG (1963) Distribution and clearance of circulating endotoxin. *J. Clin. Invest. 42*, 79-87.

Hoffman JS, Benditt EP (1982) Changes in high density lipoprotein content following endotoxin administration in the mouse. Formation of serum amyloid protein-rich subfractions. *J. Biol. Chem. 257*, 10510-10517.

Ihm J, Harmony JAK (1980) Simultaneous transfer of cholesterol ester and phospholipids by protein(s) isolated from lipoprotein-free plasma. *Biochem. Biophys. Res. Commun. 93*, 1114-1120.

Jann B, Reske K, Jann K (1975) Heterogeneity of lipopolysaccharides. Analysis of polysaccharide chain lengths by sodium dodecylsulfate-polyacrylamide gel electrophoresis. *Eur. J. Biochem. 60*, 235-246.

Johnson KW, Ward PA, Goralnick S, Osborne MJ (1977) Isolation from human serum of an inactivator of bacterial lipopolysaccharides. *Am. J. Pathol. 77*, 559-574.

Jones RS, Carter Y (1955) Incorporation in adrenal cortex of C^{14}labeled fractions of *Klebsiella pneumoniae. Proc. Soc. Exp. Biol. Med. 90*, 148-153.

Mathison JC, Ulevitch JR (1979) The clearance, distribution and cellular localization of intravenously injected lipopolysaccharide in rabbits. *J. Immunol. 123*, 2133-2143.

Mathison JC, Ulevitch RJ (1981) *In vivo* interaction of bacterial lipopolysaccharide (LPS) with rabbit platelets: modulation by C3 and high density lipoproteins. *J. Immunol. 126*, 1575-1580.

Moreau SC, Skarnes RC (1973) Host resistance to bacterial endotoxemia: mechanisms in endotoxin-tolerant animals. *J. Infect. Dis. 128*, (Suppl), S122-S133.

Munford RJ, Andersen JM, Dietschy JM (1981a) Sites of tissue binding and uptake in vivo of bacterial lipopolysaccharide-high density lipoprotein complexes. *J. Clin. Invest. 68*, 1503-1513.

Munford RJ, Hall CC, Dietschy JM (1981b) Binding of *Salmonella typhimurium* lipopolysaccharides to rat high density lipoproteins. *Infect. Immun. 34*, 835-843.

Munford RJ, Hall CC, Lipton JM, Dietschy JM (1982) Biological activity, lipoprotein-binding behavior, and *in vivo* deposition of extracted and native forms of *Salmonella typhimurium* lipopolysaccharides. *J. Clin. Invest. 70*, 877-888.

Nicoll A, Miller NE, Lewis B (1980) High-density lipoprotein metabolism. *Adv. Lipid Res. 17*, 53-106.

Pownall HJ, Hickson D, Gotto Jr AM, Massay JB (1982) Kinetics of spontaneous and plasma-stimulated sphingomyelin transfer. *Biochim. Biophys. Acta 712*, 169-176.

Quarfordt SH, Jain RS, Jakoi L, Robinson S, Shelburne F (1978) The heterogeneity of rat high density lipoproteins. *Biochem. Biophys. Res. Commun. 83*, 786-793.

Rissing JP, Buxton TB, Harris R, Moore Jr W (1981) Assessment of lipopolysaccharide and outer membrane of *Bacteroides fragilis* by an antibody-inhibition enzyme-linked immunosorbent assay in physiologic fluids and infected animals. *J. Lab. Clin. Med. 98*, 784-794.

Rosen FS, Skarnes RC, Landy M, Shear MJ (1958) Inactivation of endotoxin by a humoral component. III. Role of divalent cation and a dialyzable component. *J. Exp. Med. 108*,

701-711.

Rosner MR, Verret RC, Khorana HG (1979) The structure of lipopolysaccharide from *Escherichia coli* heptoseless mutant. III. Two fatty acyl amidases from Dictyostelium discoideum and their action on lipopolysaccharide derivatives. *J. Biol. Chem. 254*, 5926-5933.

Rudbach JA, Johnson AG (1964) Restoration of endotoxin activity following alteration by plasma. *Nature (London) 202*, 811-812.

Skarnes RC (1966) The inactivation of endotoxin at the interaction with certain proteins of normal serum. *Ann. N.Y. Acad. Sci. 133*, 644-662.

Skarnes RC (1968) *In vivo* interaction of endotoxin with a plasma lipoprotein having esterase activity. *J. Bacteriol. 95*, 2031-2034.

Skarnes RC (1970) Host defense against bacterial endotoxemia: mechanism in normal animals. *J. Exp. Med. 132*, 300-316.

Skarnes RC, Rosen FS, Shear MJ, Landy M (1958) Inactivation of endotoxin by a humoral component. II. Interaction of endotoxin with serum or plasma. *J. Exp. Med. 108*, 685-699.

Strain SM, Fesik SW, Armitage IM (1983) Characterization of lipopolysaccharide from a heptoseless mutant of *Escherichia coli* by Carbon 13 nuclear magnetic resonance. *J. Biol. Chem. 258*, 2906-2910.

Tobias PS, Ulevitch RJ (1983) Control of lipopolysaccharide-high density lipoprotein binding by an acute phase protein(s). *J. Immunol. 131*, 1913-1916.

Tobias PS, McAdam KPWJ, Ulevitch RJ (1982) Interactions of bacterial lipopolysaccharide with acute-phase rabbit serum and isolation of two forms of rabbit serum amyloid A. *J. Immunol. 128*, 1420-1427.

Ulevitch RJ (1978) The preparation and characterization of a radioiodinated bacterial lipopolysaccharide. *Immunochemistry 15*, 157-164.

Ulevitch RJ, Johnston AR (1978) The modification of the biophysical and endotoxic properties of bacterial lipopolysaccharide by serum. *J. Clin. Invest. 62*, 1313-1324.

Ulevitch RJ, Johnston AR, Weinstein DB (1979) New function for high density lipoproteins. Their participation in intravascular reactions of bacterial lipopolysaccharides. *J. Clin. Invest. 64*, 1516-1524.

Ulevitch RJ, Johnston AR, Weinstein DB (1981) New function for high density lipoproteins. Isolation and characterization of a bacterial lipopolysaccharide-high density lipoprotein complex formed in rabbit plasma. *J. Clin. Invest. 67*, 827-837.

Verett CR, Rosner MR, Khorana HG (1982) Fatty acylamidases from Dictyostelium discoideum that act on lipopolysaccharide and derivatives. I. Partial purification and properties. *J. Biol. Chem. 257*, 10222-10227.

Handbook of Endotoxin, Vol. 3: Cellular Biology of Endotoxin
L.J. Berry, editor
© Elsevier Science Publishers B.V., 1985

CHAPTER 17

Antitumor effects of endotoxins*

A. NOWOTNY

> *This review is dedicated to the memory of Dr. Murray Shear, who initiated and led extensive work in this field. His results formed the foundation of our present understanding of the mode of action of antitumor effects of gram-negative bacteria and their products.*

1. HISTORICAL INTRODUCTION

Reports correlating some infectious diseases with antitumor effects including spontaneous regressions can be found in the medical literature written several hundred years ago.

Hoffman wrote in 1675 on the beneficial effects of erysipelas infection on the outcome of other diseases: 'quin salubritatis plane notam quandoque fert: ac novi alios morbos erysipelate superveniente, fuisse feliciter sublata'. Deidier (1725) wrote detailed studies on a variety of tumors, including nonmalignant ones. He said that one cannot use the same treatment on all tumors. Some must even be left alone. He quotes Hippocrates who said that patients with tumors often live longer if they remain untreated than those to whom 'remedies' were applied (Aphorisms, Chapter six). Deidier points out that some tumors must be opened up as soon as possible, others must be treated so that suppuration will be induced. Talking about erysipelas, Deidier describes the symptoms of strong inflammation, where the blood supply is increased, the indurations are hot and the lymphatic

---

DEFINITIONS AND ABBREVIATIONS:
The enhancement of nonspecific resistance to subsequent tumor challenge is abbreviated in this text as tumor resistance or TUR. Hemorrhagic damage of established tumors is indicated as tumor hemorrhage or TUH.

Endotoxin, lipopolysaccharide and pyrogen are synonyms used to identify the biologically multipotent component of gram-negative bacterial cell walls. Instead of distinguishing among these, we use the term endotoxin exclusively throughout this chapter.

---

* Research presented in this chapter was supported by USPHS grants CA 24628, 16934, AI 05581 and 5-K3-A1-19, 487.

vessels are enlarged. The same author also stated that tumors of syphilitic patients
are cured more often than others (Deidier 1725). Dupré de Lisle (1774) described
several cancer case histories in his therapeutic manual in full detail. He was alo in
favor of prolonged and maintained suppuration of the lesions since this led in a
few cases to regression of the tumors.

Trnka (1783) studied the curative effects of the fever caused by malaria. There
is a reference in this outstanding treatise to a 63-year-old woman with bilateral
mastoid tumors whose condition was greatly improved by the fever. Astute Belgian
and French physicists observed that prostitutes infected with syphilis had a lower
frequency of cancer than the average population. Although this observation was
roundly critisized by contemporary savants, attempts were made to infect cancerous patients with syphilis. Alquie (1851) and Didot (1851) used 'syphilization' to
cure cancer. They either placed a compress soaked with 'pus bonum et laudabile'
from very active chancres on the tumor, or, more effectively, they transferred pus
from such lesions to the tumor by incision with a lancet. These treatments induced
extensive local inflammation, strong suppuration and other signs of syphilitic infection in the area. In twenty days all the cancerous growths regressed. Cicatrisation
of the tumors showed signs of healing. The syphilis, a lesser evil, was treated with
mercury preparations, and the patient was put on a fortifying diet. Unfortunately
the improvement was only transient. Six weeks later the patient was again in the
same or worse condition as at the beginning of the therapy and 'rien ne rappellait
les heureux espérances d'autrefois' (Alquie 1851). Didot (1851) assumed that the
mechanism of the beneficial effect was that the 'virus' of syphilis destroyed the
'virus' of cancer, since antagonism exists between the causative agents of these two
diseases. Ricord (1853) and several others failed to observe any beneficial effects
of the syphilization.

Deliberate induction of erysipelas for tumor treatment was first applied by
Busch, who reported in 1866 at the meeting of the Niederrheinische Gesellschaft
für Natur und Heilkunde in Bonn, Germany, that he often observed the regression
or even complete disappearance of lupus nodes on patients with nosocomially
acquired erysipelas. He described that the cells of the nodes disintegrated and
were resorbed, the ulceration was followed by the collapse of the nodes and later
by the healing of the skin. It was also stated in this report that the healing of the
nodes was only temporary, since after the erysipelas was over, the nodes started
to grow again. Even more interesting was his case history on the regression of a
skin sarcoma of a patient who had several attacks of erysipelas. The more severe
the infection was, the more obvious was the healing, which finally resulted in the
complete regression of the tumorous nodes. A few more cases were described by
Busch with similar consequences, but unfortunately the rapid resorption of the
large tumor masses caused considerable clinical complications and the patients
died either from these or from rapid regrowth of the tumors (Busch 1866).

A few years later the first therapeutic attempts were reported by Busch (1868)
where the patient with an inoperable lymphosarcoma was deliberately injured and
exposed to erysipelas. A white-hot coin-shaped iron was used to cauterize the

tumor and the patient was put into a certain hospital bed where so far every patient who had open wounds acquired erysipelas. Within one week unmistakable signs of this infection were evident, and two weeks later the size of the tumor was significantly reduced. Unfortunately, this patient developed the same shock symptoms as those observed previously: weak and irregular pulse, low body temperature, and colorless mucosa, which were attributed to the rapid disintegration and resorption of the large tumor. The patient was transferred to a room which was previously ventilated for two weeks, she received milk bouillon with egg and Hungarian wine and recovered from the shock in a week. Finally, several weeks later, the tumor grew again, and the patient could not be saved. Whether they tried this radical therapy mollified with bouillon and Hungarian wine again, was not reported.

The first reports of Busch were followed by several accounts of clinical investigators describing the consequences of tumor regression induced by erysipelas. Fehleisen (1882) identified the pathogenic microorganism as *Streptococcus erysipelatos*. He injected cultures of living bacteria into 7 patients with inoperable malignancies, and in 3 cases significant reduction of the tumor size was reported. Verneuil (1886) surveyed the relevant literature on the curative effects of erysipelas. He himself studied the symptoms of this disease, and as early as 1849 observed that infections with *Strep. erysipelatos* sometimes have remarkable curative effects on various diseases, including a few cases of malignancies. He faithfully reported completely negative effects also. Schwimmer (1887) described several reports where erysipelas eliminated other diseases, such as osteoporosis, sarcoma of the oral cavity, or syphilitic lesions. Less effective was erysipelas on lupus, psoriasis or lymphomas. Schwimmer strongly emphasizes that erysipelas is not a high-mortality disease. In his clinic in Hungary, in the year 1886, all 49 cases recovered from it in spite of the fact that some had wide-spread and severe infections at the time of admission.

Bruns (1888) confirmed the observations of Busch and Fehleisen and discussed at great length the curative effects of erysipelas in certain malignant diseases. This report describes the permanent cure of 3 out of 5 cases. Bruns assumed that the high fever, elicited by the bacterial infection, resulted in selective destruction of malignant cells. Lassar (1891) tried to avoid the consequences of deliberate infection with live bacteria, used sterile culture filtrate of *Strep. erysipelatos*, but failed to elicit clinically manifested regression of tumors.

The first animal experiments were designed and executed by Spronck (1892). He killed the *Strep. erysipelatos* bacteria to avoid virulent infection and concluded that direct effect of the bacteria on the tumor can be ruled out as a mechanism of antitumor action. As oncogenic agents he suspected parasites or other microorganisms to be involved and assumed that injection of bacterial products have a suppressive effect on the proliferation of these tumorigenic agents.

Following the work of Fehleisen (1882), Coley started out by infecting cancer patients with live *Strep. erysipelatos (pyogenes)* together with an additional microorganism, *Serratia marcescens (Bacillus prodigiosus)* which he often observed to be present in infections leading to 'spontaneously' occurring tumor regression (Coley 1893).

The 'mixed bacterial toxin' preparation was developed which contained sterile culture supernate from the two microorganisms (Coley 1898). During the following decades, more than a thousand patients were subjected to treatment with 'Coley's toxin', and in a few cases definitive curative effects were reported. Particularly responsive were lymphosarcomas. The disease-free survival time described in numerous publications varied considerably, the longest being 70 years. These remarkable results were collected and presented in a series of monographs published by the Cancer Research Institute in New York (Fowler 1969a, 1969b, 1970; Nauts et al 1946; Nauts 1973, 1975, 1980). Table 1 taken from one of these publications gives only a partial account of the results.

Several other hospitals and laboratories used the mixed bacterial toxin for therapeutic as well as experimental purposes. Although many applications did not result in beneficial effects on the host, one can conclude that at least transient regressions were observed in a few cases, indicating that Coley's toxin has antitumor effects, but that its effectiveness is limited to certain tumor types and its potency also shows individual variations among patients with the same types of tumors.

The experimental work was continued along several lines. One of these was the search for other bacteria and bacterial products with antitumor effects. Another was the establishment of animal tumor models that were used to test the efficiency of bacterial preparations in damaging the neoplastic growth. A number of laboratories attempted to elucidate the possible mechanisms of the anti-tumor effect.

Beebe and Tracy (1907) used the same toxin combination as Coley, but observed that *Bacterium coli commune* suspension was equally effective on lymphosarcomas. On the other hand, *Staphylococcus aureus* was without effect. These authors also attempted to isolate the TUH-inducing component of *B. prodigiosus* (*S. marcescens*) cells. The bacteria were grown and autolysed at 38 °C for two weeks, and passed through a porous ceramic filter (Berkefeld) which retained the cells but let through bacterial product capable of producing highly toxic effects in animals. The toxin could be precipitated from the filtrate by alcohol. Uhlenhuth et al (1910) and Beck (1911) injected bacterial preparations intratumorally to sarcoma-bearing rats and mice. They obtained results similar to those of Beebe and Tracy. Gratia and Linz (1931) used the culture filtrate of coliform bacteria to induce hemorrhages in the transplanted liposarcomas of guinea pigs. The lesion induced by two injections of culture filtrates resembled the Sanarelli-Shwartzman reactions. Shwartzman and Michailovsky (1932) followed these examples but injected meningococcal cell filtrate into murine Sarcoma 180. Duran-Reynals (1933) took *Escherichia coli* filtrates and studied the effects of these on a variety of tumors in mice and rats. He concluded that the TUH induced by the filtrate is not elicited by mechanisms which are similar to the local Shwartzman reaction. In the same year Apitz (1933) stated that the two mechanisms are similar since both affect the blood vessels, but he also assumed a direct action of the filtrate on the malignant cell.

Shear, who contributed the most to our present understanding of the antitumoral effect of various bacterial products, first injected meningococcal culture filtrate intravenously into mice bearing Sarcoma 180. He observed only marginal effects on tumor growth. Later, Shear and coworkers isolated a TUH-producing fraction

TABLE 1 *5-year survival of 897 patients with various types of tumors treated with coley toxins (mixed bacterial vaccines)*

Type of tumor	Total no. of cases	5-year survival			
		inoperable		operable	
		No./total	%	No./total	%
Bone tumors					
Ewing's sarcoma	114	11/52	21	18/62	29
Osteogenic sarcoma	162	3/23	13	43/139	31
Reticular cell sarcoma	72	9/49	18	13/23	57
Multiple myeloma	12	4/8	50	2/4	50
Giant cell tumor	57	15/19	79	33/38	87
Soft tissue sarcomas					
Lymphosarcoma	86	42/86	49	–	–
Hodgkin's disease	15	10/15	67	–	–
Other soft tissue sarcomas	188	78/138	57	36/50	73
Gynecological tumors					
Breast cancer	33	13/20	65	13/13	100
Ovarian cancer	16	10/15	67	1/1	(100)
Cervical carcinoma	3	2/3	67	–	–
Uterine sarcoma	11	8/11	73	–	–
Other tumors					
Testicular cancer*	64	14/43*	34	15/21	71
Malignant melanoma	31	10/17	60	10/14	71
Colorectal cancer	13	5/11	46	2/2	(100)
Renal cancer (adult)	8	3/7	43	1/1	(100)
Renal cancer (Wilms' tumor)	3	–	–	1/3	33
Neuroblastoma	9	1/6	17	2/3	67
TOTAL	897	238/523	46	190/374	51

* Including 16 terminal cases.

We are most thankful to Mrs. Helen Coley Nauts for permitting the reproduction of this summary table.

from *E. coli* filtrate (Shear and Andervont 1936; Shear 1936) but the most active among such preparations was obtained from *B. prodigiosus* culture supernatants (Shear and Turner 1943). This was the beginning of systematic studies on the chemical characterization of the gram-negative bacterial components with tumor-destructive activity and on the possible mechanisms by which this effect is achieved.

While this work was in progress, other laboratories tested a large number of bacterial isolates for TUH-inducing activity. As already mentioned above meningococcal cell washings could cause TUH in Sarcoma 180. Carminati (1933) used a *Lactobacillus* strain on murine tumor but could not demonstrate significant effects. Duran-Reynals (1935) used *Bacillus coli* and a human typhoid bacillus on various tumors of mice. Shwartzman (1935) used *E. typhosa* and reported that while antibodies of the bacterium could neutralize the toxic properties of the filtrate they did not diminish its TUH effect on murine sarcoma. Later, he used the culture filtrates of *B. coli*, *Bacillus enteritidis*, *E. typhosa*, *Staph. aureus*, *Strep. pyogenes*, tubercle bacilli and meningococci. The most active in TUH were the gram-negative ones, but all filtrates were equally toxic. The conclusion was that TUH and toxic activities are elicited by separate chemicals (Shwartzman 1936). Shapiro (1940) found that it is the polysaccharide moiety of the *Salmonella enteritidis* bacteria which affects transplantable tumor cells in vitro. Finally, Zahl and coworkers (1945) listed 35 gram-negative bacteria, which contained TUH-active components and they reported that none of the tested 32 strains of *Streptococci* showed effects on experimental tumors. At that time it was generally accepted that gram-negative bacterial products are the most active in TUH, while gram-positive strains have no comparable cell constituents.

While this research was being conducted, the era of radio and chemotherapy began. Because of the severe side effects and the limited benefits, the use of 'mixed bacterial toxins' was almost completely discontinued. Simultaneously, the once fervent research using experimental tumor models to test bacterial products also declined. In the years between 1950 and 1970 only a few laboratories continued such work, and this is reflected in the greatly reduced number of publications on the subject in these twenty years.

Before we close this incomplete survey we would like to refer to one of the latest therapeutic approaches. 118 years after Alquie's use of syphilis infection to cure malignancies, Mathé and associates developed a procedure in combination with chemotherapy to prolong the remission of patients with acute lymphoblastic leukemia (Mathé 1969). Twenty cuts were made on the skin, each 5 cm long, arranged in a square; into this was poured 2 ml of a suspension containing 150 mg of living *Bacillus Calmette Guérin* (BCG) bacteria. Many reports on the effects of this therapeutic approach followed, several of these describing significantly prolonged survival times while others published no effects on various types of malignancies. Further research on experimental models as well as on clinical application of BCG therapy is being conducted in several institutions.

The circle appears to be closed. The return to infection with live bacteria to prolong the life of cancer patients appears as an approach which is hardly acceptable in our world of dazzling achievements in technology and biology. Still, it would be unscholarly and impetuous to dismiss the significance of these early studies which established beyond doubt that some bacterial infections can induce at least a transient regression of certain neoplasms. It was the more recent BCG treatment which renewed the interest in nonspecific therapy with biological response modifiers to boost the resistance of patients against their malignancies. We gained considerable insight during these years in the chemical nature of bacterial products, and through the progresses in oncology, immunology, pharmacology and related basic sciences we are undeniably better equipped to elucidate the molecular events which lead to enhanced resistance to the uncontrollable growth of neoplasia.

The aim of this chapter is to survey the most salient features of these achievements, with emphasis on the chemical entities involved and particularly on the possible mechanisms through which antitumor effects can be achieved.

2. IDENTIFICATION OF THE CHEMICAL ENTITIES ELICITING ANTITUMOR EFFECTS

The two microorganisms used to produce the mixed bacterial toxin preparation of Coley were *Streptococcus pyogenes* and *Serratia marcescens*. The classification of streptococci was established by the work of Schottmüller, Brown, Lancefield and others during the first half of this century. The cell wall structure was studied by many laboratories and it was reported that it has endotoxic properties (Robertson and Schwab 1969). The bacteria synthesize and release a large number of extracellular products, hemolysins, toxins, kinases, DNAase, hyaluronidase, which were all studied in great detail regarding their biochemical functions, and immunological and pathogenetic roles. Relatively little has been done to identify the products of streptococci which may be involved in antitumor actions in spite of the quite promising results discussed earlier. For a brief review of these results see Ginsburg (1970). More recently Japanese scientists developed a commercial preparation, called Picibanil extracted from a mutant strain of Group A hemolytic *Streptococcus*. Kondo et al (1982) tested its activity in the usual parameters before it was used for clinical trials. The results showed significantly increased survival times if Picibanil was used in combination with chemotherapy. The preparation caused fever and chills when given intravenously, and local bleeding and ulceration when given intradermally. The most advantageous route of injection was intratumoral. Unfortunately, no chemical analytical data of the preparations are given in the review of Kondo and associates.

The other microorganism in Coley's toxin was *S. marcescens*. The fundamental work of Shear and associates identified a polysaccharide-rich preparation as a TUH inducer isolated from this gram-negative strain (Hartwell et al 1941, 1943a, 1943b; Kahler 1943; Shear and Andervont 1936; Shear 1941). This compound contained about 2.2% nitrogen and 1% phosphorus. Phospholipid was also detected (16%), which could be precipitated by hydrolysis with 1.5 N HCl at 100 °C. Digestion with trypsin did not effect TUH activity, hydrolysis with 0.1 N NaOH destroyed it. Hydrolysis with 1.5 N HCl liberated 60% reducing sugar and an insoluble precipitate (probably lipid A) which contained both nitrogen and phosphorus (Hartwell et al 1943b).

Regarding the homogeneity of the 'Shear's polysaccharide', ultracentrifugal and electrophoretic analyses revealed the presence of minor components ($\sim$10%) in all preparations. The molecular weight (MW) of the major component was calculated from the sedimentation data and was found to be around 8×10^6 (Kahler et al 1943). It was assumed that the TUH activity resides in this macromolecule.

It is evident today that Shear's polysaccharide was an endotoxin preparation. Zahl established a relationship between the TUH-active component and the endotoxins of gram-negative bacteria (Zahl et al 1942). Several publications followed which described the findings of various laboratories regarding the optimal conditions of the endotoxin-induced TUH and provided some insight into the possible mode of action of this phenomenon (Creech et al 1948a, 1948b, 1949; O'Malley et al 1962a, 1962b; Rathgeb and Sylven 1954a, 1954b; Zahl et al 1945; Zahl 1950). Unfortunately, all this work was carried out with rather crude endotoxin preparations, and therefore it is not possible to draw any conclusions regarding the chemical nature of the TUH-inducing active site(s).

Systematic studies of the subunits of the endotoxin structure were initiated by Boivin et al (1933) and continued by Westphal and Lüderitz (1954). Mihich et al claimed that the 'lipid A' part of the endotoxin complex is responsible for TUH (Mihich et al 1961). Claims were made at the same time about the 'lipid A' precipitate as the representative of the toxic site of the endotoxin, and these claims were later substantiated by numerous investigators. The most convincing data came from experiments using rough mutants of some gram-negative bacteria.

Smooth and rough variants of some gram-negative bacterial strains were compared for their carbohydrate content as early as 1926, and it was found that smooth to rough conversion is accompanied by partial or complete loss of the polysaccharide moiety of their antigenic complex (Reimann 1926; White 1927, 1928, 1929). Boivin and Mesrobeanu (1937) compared the toxicity of their trichloroacetic acid (TCA)-extracted O antigens obtained from smooth and rough strains of *Salmonella typhimurium* and found that the smooth endotoxic O antigen was ten times more toxic than the rough mutant endotoxin. Using *Shigella paradysenteriae* strains, a 5-fold difference was found. Mackenzie et al (1940) reached the same conclusion. Zahl and associates (1945) compared both toxicity and TUH activity of Boivin-type extracts from smooth and rough *Salmonella choleraesuis* and *Hemophilus influenzae* strains. They reported that TUH activity decreased as a result of decreasing carbohydrate content and this went parallel with the toxicity of the preparations.

The use of rough mutants of *Salmonella minnesota* 1114 strain by Lüderitz et al (1966) resulted in a breakthrough in the structural studies of the polysaccharide moiety of the endotoxic lipopolysaccharide. Dr. O. Lüderitz kindly provided our laboratories with the same mutant endotoxins and we used these for the comparison of their biological activities in a number of parameters (Kasai and Nowotny 1967; Tripodi and Nowotny 1965, 1966). We found that only minor differences can be established in their toxicity when the preparations are properly solubilized. In the induction of enhanced nonspecific resistance, their potency was indistinguishable. Kim and Watson reached similar conclusions (1967). The TUH activity was measured on Sarcoma 37 and it was found that the heptose-less Re mutant R595 of *Salmonella minnesota* was at least as active as smooth endotoxin. The conclusion of these experiments was that the O-antigenic polysaccharide moiety is not required for TUH induction (Nowotny et al 1971).

Continuation of these studies applied Re mutants purified by thin layer chromatography (TLC) from 5 different gram-negative strains to measure their capacity to increase TUR to TA3-Ha tumor. It was found that not all Re mutants are equally active in TUR, although their toxicities were very similar (Ng et al 1976). Accordingly, we assumed that toxicity and antitumor activity must have different structural requirements. While toxicity seems to be dependent upon the presence of certain fatty acids in a specific arrangement, TUR activity might be dependent upon the presence of other substituents (Ng et al 1976). It was found during the same studies that the chemical compositions of the lipid moieties of the TLC-purified Re mutant endotoxins are also significantly different. The molar ratios of hexosamine:phosphorus:3-deoxy-D-manno-2-octulosonic acid (KDO): fatty acids demonstrated this point.

Rietschel et al (1977) showed the existence of unusual lipid A structures in some uncommon bacterial strains. Continuation of these studies revealed that TUH activities of these preparations were also significantly diverse (Westphal et al 1980). Ribi and coworkers took toxic *S. minnesota* R595 Re mutant endotoxin, subjected it to very mild acidic hydrolysis and obtained a preparation which was nontoxic but active in the induction of tumor regression. These authors established with analytical chemical measurements, carried out on samples purified by TLC, that acidic hydrolysis cleaved one of the two phosphate functional groups present on the R595 mutant subunits. This treatment was sufficient to render the mutant endotoxin nontoxic, but it maintained its tumor-regressive potency (Ribi et al 1983). These results resemble earlier findings, where CH_3OK treatment was used to reduce toxicity of TCA-extracted preparations from *S. marcescens*. The detoxified 'endotoxoid' samples maintained their TUH effect (Nowotny et al 1971) and could also enhance TUR against TA3-Ha challenge (Nowotny et al 1975a; Yang-Ko et al 1973). It is interesting to note that CH_3OK treatment cleaves O-ester linkages, including phosphate functional groups (Nowotny 1964).

According to these findings, one is inclined to conclude that all endotoxic activities reside in or near the lipid moiety of the structure. This conclusion may be valid for the induction of TUH but may not apply to TUR induction. The first set of evidence comes from experiments in which TUR enhancement was achieved with gram-positive bacterial cells or with their products, which did not induce hemorrhagic necrosis in established tumors. These cells do not contain endotoxin, although some of them can elicit febrile response. Their mechanism for TUR enhancement seems to be related to the activation of the immune defenses, including both the specific and the nonspecific arms of it.

Additional evidence was generated by inducing TUR with gram-negative cell products which were free of the lipid moiety. It is known that mild acidic hydrolysis of endotoxins which precipitates the lipids (Boivin et al 1933; Westphal and Lüderitz 1954) cleaves water-soluble components. Among these are polysaccharides and oligosaccharides, together with glycopeptides and a small percent of phosphorylated breakdown products. The carboxylic acid content of this preparation is less than 1%. Due to its high polysaccharide content we designated this preparation as PS. We found that this nontoxic and lipid A-free PS preparation was active as an inducer of colony stimulating factor (CSF), and as an immune adjuvant both in vitro and in vivo. It could protect against lethal irradiation, and when combined with BCG infection, it could generate TUR (Nowotny et al 1975b; Behling and Nowotny 1977; Chang et al 1974; Frank et al 1977; Nowotny et al 1975b; Nowotny 1977). PS given to Sarcoma 37-bearing mice did not induce TUH.

Another preparation could be obtained by subjecting whole lyophylized bacterial cells to mild, 0.2 N acetic acid hydrolysis, as described by White (1929). The supernatant contained no detectable carboxylic acids, no phosphorus, approximately 60–70% carbohydrates, 5–8% hexosamines and 7–12% amino acids. This preparation, called White-type polysaccharide (WPS), was nontoxic, active in CSF induction, and in radiation protection (Urbaschek 1980), and could prolong the survival of L1210 leukemia-injected mice. It was also active as an immune adjuvant and as an activator of osteoclasts (Rothman et al 1985; Sanavi et al 1982).

A comparison of dose–response relationships indicated that some of these activities were considerably less than the activities of endotoxins. These were adjuvancy, CSF induction or radiation protection by PS. Some WPS preparations were as potent as the fully toxic endotoxin extracts of the same bacterium in adjuvant or in CSF assays, or in the prolongation of L1210 survival time. The lack of toxicity and their chemical composition makes it unlikely that they contain residual lipid A, but we cannot exclude the possibility at this time that some of the many split products present in a PS or WPS preparation derived from the lipid moiety of the endotoxins (Rothman et al 1985). Some of these breakdown products may still carry those structural elements of the intact endotoxin which are the essential molecular effectors of reactions leading to the above biological consequences. Only careful isolation and accurate chemical structural analysis of these components can clarify their origin.

While discussing the composition of PS or WPS, we referred to the obvious heterogeneity of such preparations. Before closing this part of the review, we should also elaborate briefly on the heterogeneity of the endotoxin preparations. It is quite important to realize that the terms endotoxin or lipopolysaccharide do not describe a single molecular species. They identify multicomponent systems since rapidly accumulating evidence shows that even the most acclaimed purified preparations consist of several constituents (Chester and Meadow 1975; DiRienzo and MacLeod 1978; Garfield et al 1976; Goldman and Leive 1980; Jann et al 1975; Jarrell and Kropinsky 1977; Kabir 1976; Le Dur et al 1980; Leive 1980; Morrison and Leive 1975; Nowotny 1966, 1971; Nowotny et al 1966; Palva and Mäkelä 1980; Russell and Johnson 1975; Van der Zwan et al 1978; and several others).

There are two types of heterogeneities which we know about at this time. One is simple contamination of the endotoxin preparations by other bacterial cell components. These are coextracted and cannot be separated from endotoxin by the conventional methods of purification such as precipitation with alcohol and sedimentation in an ultracentrifuge. High resolution methods such as column chromatography or gel electrophoresis readily revealed the presence of these bacterial cell components in all preparations so far studied.

The other type of heterogeneity is an intrinsic property of the macromolecules with endotoxic activity. It was first reported by separating Re mutant endotoxins on thin layers of silicic acid (Chen et al 1973, 1975; Chen and Nowotny 1974; Kasai and Nowotny 1967; Ng et al 1974, 1976; Nowotny et al 1967). The results showed that these mutant endotoxins consist of chromatographically distinct molecules. Molar ratios indicated that relatively minor differences exist in their compositions which might explain their diverse chromatographic behavior, while they only have a slight effect on their endotoxicity. Greater differences were found by measuring their beneficial effects such as enhancement of nonspecific resistance to infections or to transplantable tumors (Ng et al 1976). Multiple bands were detected by other laboratories separating isotope-labeled 'smooth' endotoxins by sodium dodecylsulfate-polyacrylamide gel electrophoresis (SDS-PAGE). This heterogeneity was attributed to different degrees of polymerization of the repeating units in the polysaccharide moiety (Goldman and Leive 1980; Palva and Mäkelä 1980; Tsai and Frash 1982). For a recent review on the heterogeneity of endotoxins the reader is referred to Volume 1 of this series (Chapter 10).

At any rate, attempts to assign a biological activity to one component in a heterogeneous mixture of diverse molecules is not possible without the isolation of the component in pure form and without the presentation of convincing data regarding the activity of the isolated component in the biological assay system in question. More frequently than we wish, purification of a biologically active mixture leads to reduction or loss of the activity. Such observations we have reported earlier (Nowotny 1971). This is particularly applicable to activities which are the summations of various mechanisms triggered by diverse effector molecules acting on multiple targets. These mechanisms also may be interdependent, and therefore, if one of the effector molecules is removed during purification, the final activity might be reduced or completely absent. The enhancement of nonspecific resistance to virulent bacteria or malignant cells by endotoxin may fall into this category.

In conclusion, it appears to be well documented that toxicity of the preparation is not a prerequisite for the TUR or TUH effects. This was shown by a number of independent laboratories using either detoxified endotoxins or nontoxic bacterial cell wall components. Generally speaking, the elicitation of the various biological effects of endotoxins appear to require different structural features (Johnson and Nowotny 1964; LeGarrec et al 1981; Nowotny 1963, 1964; Nowotny et al 1971; Nowotny and Butler 1980; Ribi et al 1982, 1983; Schultz et al 1978; Tanamoto et al 1978, 1979, 1980, 1982; Yang-Ko et al 1973). For reviews on this topic see Nowotny (1969, 1983) or Sultzer (1971).

It is also quite clear that not only endotoxins, but also other components of the gram-negative as well as gram-positive bacteria can enhance the host's resistance to experimental tumors. Furthermore, TUH and TUR seem to be the end result of different mechanisms, and the molecules which can trigger and/or amplify these mechanisms may have different chemical characteristics. Totally different mechanisms may be required to induce TUH or TUR. It is most likely that the active sites of the endotoxins which elicit TUH or enhance TUR also differ both in their chemical compositions and in their biological effectiveness. Toxic endotoxins on their attenuated varieties are the prerequisites for TUH induction, while TUR enhancement can be achieved, not only by nontoxic remnants of endotoxins, but also by some constituents of gram-negative and gram-positive bacteria unrelated to endotoxin.

3. POSSIBLE MECHANISMS

3.1. Direct effect of endotoxin on neoplastic cells

Seligman et al (1948) labeled the Shear's polysaccharide preparation with iodine isotope and followed its fate in tumor-bearing mice. The accumulation of the radioactivity was not significantly different in the necrotizing tumors from that in the other organs. It was concluded that the mechanism of TUH does not involve direct action of endotoxin on the tumors. Hager et al (1969) arrived at the same conclusion. They used several endotoxins on a variety of tumor cells but observed no direct cytotoxicity in vitro in spite of the fact that significant protection of the tumor-inoculated mice could be achieved by repeated injection of some endotoxins.

We studied the fate of endotoxin in normal and in Sarcoma 37-bearing mice. The detection of the injected *S. marcescens* endotoxin was carried out by O-specific rabbit antiserum labeled with fluorescein. The quick accumulation of endotoxin was clearly manifested in those organs which are rich in cells of the reticuloendothelial system, but not in the tumor itself (Nowotny et al 1971). Bara et al (1973a, 1973b) used another type of endotoxin, the lipophylic mutant glycolipid from *S. minnesota* R595, and measured its insertion into the cell membrane of normal as well as SV40-transformed fibroblasts. A significantly higher intake of the labeled glycolipid by the SV40-transformed cells was evident. The same laboratory reported that the major effect of glycolipid insertion was a reduction of the growth rate of transformed cells without the same effect on normal cells (Brailovsky et al 1973). Nigam (1975) pointed out that the shorter the polysaccharide chain of a rough mutant endotoxin, the greater its antitumor effect as determined by lengthened survival time of mice injected with Ehrlich ascites tumor. This is in agreement with our finding that the R595 heptose-less endotoxic glycolipid from *S. minnesota* Re mutant which induced TUH of Sarcoma 37 was more potent than endotoxins from wild strains (Nowotny et al 1971). Regarding the possible method of enhancing tumor resistance by these glycolipids, Nigam (1975) suggested increased immune response to the tumor cells. Incubation of TA3-Ha cells for 7 hours with endotoxin from *S. marcescens* did not cause any detectable change in the cell viability, and the cells maintained their tumorigenicity (Yang and Nowotny 1974).

On the basis of the findings in these surveyed experimental systems, a direct cytotoxic effect by endotoxin on malignant cells may be ruled out. Whether this conclusion is generally applicable to all types of endotoxins and tumors remains to be seen. What is not questionable is the fact that endotoxins have a great affinity to eukaryotic cell membranes.

This has been observed by several authors. Keogh et al (1948) reported the absorption of bacterial polysaccharides to erythrocytes. Neter and coworkers developed the very useful passive hemagglutination and passive hemolysis methods for the quantitation of antibodies to bacteria, which was based on the above phenomenon (Neter et al 1952). Endotoxins may be fixed by erythrocytes under extreme conditions in vivo (Buxton 1959; Springer and Horton 1964). Springer and coworkers isolated endotoxin receptors from erythrocytes (Springer et al 1974) and from leukocytes and platelets (Springer and Adye 1975). Shands (1973) did electron microscopic observations and described how endotoxin particles associated with cell membranes by 'edge attachment'. McLaughlin et al (1978) expanded the findings of Nigam and associates by determining the enhanced binding of rough mutant endotoxins to guinea pig line 10 hepatocarcinoma cells, as compared to the binding of wild-type endotoxins.

Much less investigated are the consequences of the endotoxin–membrane interaction. Weissman and Thomas (1962), and Bona et al (1971) assumed that endotoxin–membrane interaction destabilizes lysosomal membranes. This reaction may lead to excessive spill of lysosomal hydrolases and consequently to tissue damage. It has been assumed (Bara et al 1973a, 1973b; Schuster et al 1970; Shands 1973, and several others) that interaction between the endotoxin and the cell membrane involves nonpolar interaction between the lipid moiety of the endotoxin and bimolecular lipid layers of the eurkaryotic cell. Isotope-labeled endotoxin was used by several authors to study the mechanism of its uptake by various cells. Davies et al (1978) observed a membrane reorganization due to interaction with *E. coli* endotoxin. They also described a cyclic fluctuation between absorption and desorption of labeled endotoxin. Bona et al (1969; Bona 1971, 1973) studied the pinocytosis of endotoxin by macrophages and observed a strong adsorption within the first 15 minutes. A full uptake was completed in 24 hours. Important for our discussion is the observation that the endotoxin within the macrophage maintains its Shwartzman reactivity. Furthermore, the macrophages could transfer their endotoxin content to adhering cells (Bona et al 1972). It remains to be investigated whether such transfer of 'processed' endotoxin to neoplastic cells can occur.

Braun (1973) assumed that endotoxin–membrane interaction introduces changes in the cyclic AMP levels of the tumor membrane which might alter cell–cell interactions such as the immune response to the tumor, which in turn would affect the proliferation of the malignant cell. Braun did not exclude the possibility that elevated cAMP levels could stop the proliferation of the tumor cell itself. The untimely death of Braun prevented the generation of sufficient data to test these hypotheses, but other laboratories continued investigations along the same lines. Wolff and Cook (1975) reported that *Salmonella typhosa*, *E. coli* and *S. marcescens* endotoxins stimulate adenylate cyclase activity. Y-1 cells have receptors for endotoxin which are different from ACTH receptors on the cell membrane. Wolff and Cook proposed that the endotoxin–membrane interaction induces conformational changes in the latter, resulting in adenylcyclase activation. Jones (1977) and Kilpatric-Smith et al (1981) studied the effect of endotoxin on the energy metabolism of tumor cells in vitro. Their conclusion was that profound changes are caused in the mitochondrial metabolism by endotoxin. Tatsumi et al (1979) described that leukemia cells showed enhanced DNA synthesis due to exposure to endotoxin. Malignant HeLa cells were exposed to smooth and rough endotoxins from *S. typhimurium* strains by Kihlström (1980). He observed that all endotoxins tested not only reacted rapidly with the HeLa cells but the interaction caused membrane changes indicated by the increased reactivity of the cells with *S. typhimurium* bacterial cells. The HeLa cells retained their viability during the process.

One of the most obvious consequences one can assume here is the altered immunogenicity of the malignant cell. This may be due to membrane rearrangement, resulting in better exposure of cryptic immunogenic sites and in an improved immunogenic expression of neoplastic membrane components. Increased steroidogenesis (Wolff and Cook 1975) may significantly contribute to this. Another possible consequence is the acquisition of new immunogenicity by the malignant (or normal) cells through the insertion of endotoxin into the membrane. Experimental insertion of endotoxin or lipid A into synthetic liposomal membranes resulted in immunogenic particles which reacted with the corresponding immunoglobulins and the complement components.

Summarizing these findings, it is evident that endotoxin–tumor cell interactions induce profound changes in some tumor types which do not appear to have immediate toxic effects on the cell. These changes are manifested in enzyme activity levels, in the energy metabolism and in a few isolated cases in biosynthetic activities, quite similar to the effects of endotoxins on some normal cell types. With the methods available to us, no direct cytotoxic effects can be determined, but this does not eliminate the possibility that some of these changes will alter the cellular mechanisms of the tumor to a level at which a slow but irreversible deterioration of the vital functions can set in. Another consequence of endotoxin–tumor cell interaction might be a changed immunogenic expression of the tumor cell. This may consist of enhanced immunogenicity of tumor-specific membrane components, or of newly acquired immunogenicity through the insertion of endotoxin into the cell membrane. Finally, we have to remember, that endotoxins admixed to immunogens greatly enhance the immune response even to the weakest immune stimuli. If malignant membranes have a significantly enhanced affinity to endotoxins, several of the theoretical possibilities listed here may constitute important factors in the antitumor effects. (For a further discussion of the interaction of endotoxin with cell membranes, see Chapter 2.)

3.2. Effects on the vascular system

Sanarelli (1924) wrote a report on a very significant observation resulting from repeated intravenous injection of culture filtrate from *Vibrio cholerae*. The aim of these experiments was to establish a model system to study the pathogenesis of cholera infection. When guinea pigs were injected repeatedly, the intestinal walls and the mucosal surfaces in these animals began to hemorrhage extensively. Shwartzman (1928) continued these experiments. A local skin test was developed where the first injection of culture filtrate, or endotoxin, is given intradermally (this is the 'preparing' injection) followed 24 hours later by an intravenous injection of the same preparation ('provoking' injection). The first injection causes local inflammation, erythema, and swelling of the tissues, changes in the microcirculation, formation of disseminated intravascular coagulation and greatly enhanced vascular permeability. The second injection causes rupture of the capillaries. Because of extensive thrombosis in the area, anoxia and necrosis of the skin occurs. The mechanism, including the participation of factors of the blood clotting system and the role of platelets as well as other leukocytes, has been extensively investigated, as reviewed by Taichman (1971). (See also Chapter 16.)

Gratia and Linz (1931) induced local hemorrhages in the transplanted sarcoma of guinea pigs by a single injection of *B. coli* filtrate. Autopsy showed hemorrhage in the tumors, but no such changes were seen in normal tissues. Shwartzman and Michailovsky (1932) extended these experiments to Sarcoma 180-bearing mice. Single injection of meningococcal culture filtrate caused identical results as described by Gratia and Linz (1931). A considerable percentage of the animals died following the injection of the culture filtrate, but several of the survivors were cured indefinitely. In every group a few mice showed hemorrhagic necrosis of the tumor, but nonnecrotized residual tumors grew again and killed the animals.

Within a few years several laboratories reported similar findings (Andervont 1936; Apitz 1933; Duran-Reynals 1935; Fogg 1936; Gerber and Bernheim 1938; Jacobi 1936; Shear and Andervont 1936), which were described as the Shwartzman reaction of tumors. The efforts to elucidate the possible mechanism soon led to the hypothesis that bacterial endotoxins are the effector substance and the microvasculature is the target. Youngner and Algire (1949) described the sequence of circulatory events in the tumor starting with a slowdown of the blood flow, agglutination of erythrocytes, stasis and occlusion of the capillaries within the tumor as well as in adjacent normal striated muscle. The same authors established that artificial obstruction of the blood flow to the tumor for as short as 60 minutes caused hemorrhage in the tumors, which became visible first after 24 hours. While this phenomenon was reproducible in all experiments using L sarcoma, the mammary adenocarcinoma C3HBA of the same strain of mice did not show hemorrhage or other signs of damage. As a result of these as well as of other experiments, Algire and coworkers concluded that prolonged hypotension causes TUH but only in some sensitive tumors (Algire and Legallais 1951; Algire et al 1947, 1952).

Dachy (1967) reviewed the sensitivity of normal and pathologically altered tissues to the hemorrhagic effect of endotoxin. Zweifach (1964) studied the vascular effects of endotoxin in normal tissues. The essence of the findings is that while endotoxin induced rapid changes in the mircocirculation in the skin of rabbits, no comparable effect was noticeable in the wall of the mesentery. Again, a different response was obtained when the wall of the small intestine was injected. Zweifach assumed that the sensitivity of the capillary walls may be different in these organs, although no evidence of damage to the endothelial wall or basement membrane of the capillaries and of the venules could be established. Hinshaw (1964) described that intravenous endotoxin injection elicits a series of reactions which involve blood constituents and tissue factors. Histamine is liberated, and vasoactive amines other than histamine are released, thus inducing profound changes not only in the blood pressure but also in vascular permeability. The effect of endotoxin on the microcirculation was most impressively demonstrated by Urbaschek et al (Urbaschek 1972; Urbaschek and Urbaschek 1976; Urbaschek et al 1967).

The differences found by Zweifach or Hinshaw in the reactivity of the endothelial layers of various organs to endotoxin very well demonstrate the difference which might exist in the responsiveness between normal and neoplastic vasculature to endotoxin. Cater et al (1962, 1963, 1964, 1966) reported that the vessels of tumors are particularly sensitive to amines and to the mediators of inflammatory reactions. This was quite skillfully demonstrated by Oswald and Cater (1969), who followed the injection of endotoxin into tumor-bearing rats 2 hours later by an intravenous injection of india ink. Heavy deposition of the ink was visible between the endothelial cells indicating enhanced vascular permeability. Particularly heavy was the staining of the blood vessels in the tumor. According to Oswald and Cater (1968), the changes in the tumor caused by endotoxin are the result of effects on the vessels in the tumor. The venules in the tumor are particularly precarious and the rapidly growing tumors are especially vulnerable to anoxia. Anoxia will lead to more inflammation, thus amplifying the endotoxin-induced exudate formation, adherence of platelets and granulocytes to the endothelial layers, and the spread of thrombosis in the general area.

Contreras and Bale (1968) exposed Walker 256 rat tumor to 1500 R radiation to sensitize it to a sublethal dose of *S. marcescens* endotoxin. The quick necrosis observed was attributed to enhanced release of coagulation-producing substances into the circulation; these products were trapped in the tumor, thereby causing stasis and destruction of the malignant tissue. Accordingly, the vasculature of the tumor can be rendered even more sensitive to endotoxin by preceding irradiation. McGrath and Stewart (1969) described that the endothelial layers of normal blood vessels are severely damaged after endotoxin injection. The most prominent features were distorted, vacuolized or destroyed nuclei. The role of granulocytes in the endotoxin-induced vascular damage was studied by Gaynor (1973). It was found that the neutropenia induced by nitrogen mustard did not protect the rabbits from endotoxin-induced epithelial damage.

It is evident from the accumulated experimental data that not all tumors respond to endotoxin, but those that do appear to be more sensitive to its action, just as normal tissue blood vessels after the 'preparative' injection in the Sanarelli-Shwartzman phenomenon. There is little doubt that the above vascular changes are the key events leading to TUH.

3.3. Enhancement of antitumor immune response

Since the early work of Foley (1952), Prehn and Main (1957) and many other investigators, it is firmly established that most tumors have detectable immunogenicity in syngeneic or in autochtonous hosts. Immune response of human patients against their own malignant cells could also be manifested in a considerable number of cases (Herberman 1974, 1977; McCoy et al 1975, 1978). It is reasonable to assume that endotoxins, the very potent immune enhancers, may act as immune adjuvants, thus potentiating the rather weak immunogenicity of the tumor cells. This may be achieved either by passive attachment to the cell surface, or by stimulation of the host's immune response mechanisms. The first possibility has already been discussed in Section 3.1. Quite a few publications reported findings which may be the consequences of the adjuvant action of endotoxin, potentiating the immune reactions against the tumor.

3.3.1. Humoral immune response

The adjuvant effect of endotoxin on the humoral immune response to the tumor may have a profound effect on the growth pattern of tumors. The presence of antitumor antibodies in the circulation of tumor-bearing patients as well as experimental animals has been demonstrated in the past by numerous laboratories, and the references given here are only a few out of many relevant publications (Hartman et al 1974; Hellström et al 1968; Lewis et al 1969, 1976; Morton et al 1968; Muna et al 1969; Old et al 1966). Kaliss and Bryant (1958) described that it is possible to generate antisera of which the immunoglobulin components will react with the tumor in a specific fashion but will do no harm to it (Kaliss 1962). As a matter of fact, the tumors will grow even better under the protective coat of these antibodies than without them. This phenomenon was called 'immune enhancement' and its possible mechanism was thoroughly studied by Möller (1963a, 1963b, 1963c) and several others. Other authors described the host-protective effect of syngeneic antitumor sera and endotoxin against viable tumor challenge but only if these were introduced during the first 6 days after the challenge (Shinoda et al 1977a, 1977b).

Probably the most important, and from a therapeutic point of view the most promising use of antitumor antibodies can be found through the mechanisms of antibody-dependent cell-mediated cytotoxicity (ADCC). The antitumor immunoglobulins react with the tumor cell while their Fc portion is available for interaction with cells which have Fc receptors. The most important among these cells appears to be the macrophage since it has endotoxin-inducible tumoricidal activity. The ADCC phenomenon and its activation by endotoxin will be discussed later in Section 3.4.2. in connection with macrophage actions on tumor targets.

It is also well established that some oncogenic viruses and some tumors are strongly immunosuppressive (Bendinelli 1968; Bendinelli et al 1985; Dent 1972; Friedman and Ceglowski 1971; Friedman and Kately 1975; Kamo et al 1975a, 1975b; Salaman 1969). Compensation for such immunosuppressive effect is possible by endotoxin application or by post-endotoxin sera (Butler et al 1979, 1980; Butler and Friedman 1979). The immunosuppressive effect of some chemotherapeutic drugs could also be eliminated by the proper combination of the drug with endotoxin or its nontoxic split products (Gaal and Nowotny 1979). It is feasible that one of the contributing factors in the endotoxin-induced TUR enhancement lies in the immune restorative effect of the endotoxin.

The role of humoral immune response in the antitumor defense of the host is obviously important, but the details of the mechanism remain unclear. Tumor growth enhancement as well as cytotoxic effects were reported in using certain tumor lines in selected hosts before or after treatment with antitumor immunoglobulins. The effect of endotoxin on the humoral immune response may also be enhancing or suppressive, and this appears to be a time- and dose-dependent phenomenon (Behling and Nowotny 1977, 1980, 1982). It is obvious that the role of endotoxin-induced alterations in the humoral response to tumors, and the role of these alterations in the antitumor defenses of the host are quite complex, and they have been much less investigated in the last two decades than the role of cell-mediated immunity to malignant cells in tumor development.

The results of the ADCC system and the more recent availability of monoclonal antibodies to some tumor membrane constituents will undoubtedly refocus research interest on the consequences of tumor–immunoglobulin interactions. It remains to be seen how endotoxin, the potent enhancer of humoral immune responses, can contribute to these.

3.3.2. *Cell-mediated immunity*

Endotoxins were shown to be potent enhancers of B cell mitogenesis (Andersson et al 1972; Bona et al 1976; Gery et al 1972; Gormus et al 1974; Peavy et al 1970, 1978; Sredni and Roseszajn 1978). Many laboratories uniformly confirmed these findings. The use of rough mutant endotoxins, where the polysaccharide moiety is deficient or absent but the lipid moiety appears to be complete, showed that it is the lipid moiety which is responsible for the B cell mitogenesis (Andersson et al 1973; Chiller et al 1973; Rosenstreich et al 1973). We synthesized N-pamitoyl-Dglycosamine and found that this very simple model of one of the characteristic structural elements in the lipid A structure could also induce B cell mitogenesis, provided that this glycolipid is properly dispersed (Rosenstreich et al 1974).

Nonspecific release of hemolysins from nonimmunized mouse spleen cells by endotoxin-containing typhoid vaccine or by phytohemagglutinin (PHA) was reported by Hege and Cole (1967). The observation was confirmed and further extended by Andersson et al (1972) and Nakano et al (1975). It is concluded that activation of B cells by endotoxin induces the release of immunoglobulins, capable of lysing sheep red blood cells in the presence of complement in spite of the fact that the mice were not immunized with sheep red blood cells.

We studied the possible role of endotoxin-activated B cells in the antitumor defense and observed that in vivo pretreatment of mice with endotoxin enhances their resistance to the allogeneic TA3-Ha transplantable tumors (Grohsman and Nowotny 1972). Continuation of these studies led to the analysis of the mechanisms with particular emphasis on the cells involved. The resistance could be transferred by spleen cells of normal endotoxin-injected mice (Yang and Nowotny 1974; Yang-Ko et al 1973). We found that the resistance of the low-responder C3H/HeJ mice could not be enhanced to this tumor challenge by endotoxin injection. Other authors established a B cell deficiency in these mice. On the other hand, we found that the T cell-deficient athymic Balb/c nude mice showed a good TUR enhancement as a result of intravenous endotoxin injection (Nowotny et al 1975a; Nowotny and Butler 1980). Our conclusion was that B cell–endotoxin interaction is part of the mechanisms which led to enhanced TUR in the above system (for a review of these results, see Butler 1983).

Some laboratories reached similar conclusions. Ralph et al (1974) found that B cell mitogens, such as endotoxin or dextran sulfate, can stimulate TUR in lymphosarcoma transplants. Saluk (1983) and Berendt and Saluk (1976) reported that endotoxin-induced TUR is mediated by a humoral factor. Continued studies by the same group (Berendt et al 1978a, 1978b) established that TUR requires adhering peritoneal exudate cells (probably macrophages) and a radiosensitive lymphocyte population (probably B cells). Bober et al (1976a, 1976b) reported that chronic administration of endotoxin stimulated the growth of oil-induced plasma cell tumors, but 1–5 ng endotoxin given before MOPC 315 plasma cell challenge protected a significant percentage of the animals. Somewhat similar results were obtained by Kedar et al (1978) who found that small quantities of mycobacterial preparations enhanced cytotoxicity while a dose that was 10 times greater suppressed it.

Prager et al (1975) stimulated immune response to modified lymphoma cells by endotoxin. A high dose of endotoxin given together with the iodoacetamide-treated tumor cells caused the 70–75% rejection of a subsequent challenge by a viable tumor. Prager and coworkers demonstrated specific antibodies in the sera of the surviving mice, and they concluded that endotoxin-induced B cell activation produced this effective humoral antitumor immune response. Cell-mediated immunity could also be demonstrated in tumor-resistant mice (Prager et al 1976).

Probably the most intensively pursued studies of cell-mediated destruction of tumor cells in the last 10–15 years focused on the role of T cells. Tripodi et al (1970) inhibited the establishment of Sarcoma 180 in mice and assumed that endotoxin did not act on the tumor but shifted the host response to a mechanism which led to the rejection of the tumor inocula by delayed hypersensitivity-type response. For a while it was thought that endotoxins do not influence T cell reactions either directly or indirectly, but accumulating evidence made this assumption untenable. Suppression of graft-versus-host (GVH) reactions by endotoxin or by its detoxified derivatives was demonstrated and thoroughly studied by Liacopoulos et al (1967a, 1967b; Liacopoulos 1983). Endotoxin given to the cell donor before transplantation greatly reduced the GVH. When endotoxin was given after cell transfer, potentiation of the GVH reaction became evident. Skopinska and coworkers (Skopinska and Oluwasanni 1972; Skopinska et al 1977; Skopinska-Rozewska 1983) described prolongation of graft survival when the donor was treated with endotoxin before the organ transplantation. Truitt et al (1978; Truitt 1983) suppressed graft rejection with endotoxins while they maintained the antileukemic effect of graft transfer.

The indirect action of endotoxin on T cells was observed by several laboratories. One of these examples is the potentiation of concanavalin A (Con A) responsiveness by endotoxin (Ozato et al 1975; Schmidtke and Najarian 1975). Scheid et al (1973), and Adorini and coworkers (1976) observed accelerated maturation of T cells induced by endotoxin. McGhee et al (1980) assumed that T cell activation is a critical event in the adjuvant effector mechanism of endotoxin.

Bona et al (1976) studied the consequences of endotoxin–lymphocyte interactions. They found that both B and T cells have endotoxin receptors but only B cells will undergo mitogenic transformation. Considerable differences exist between T and B cells regarding their endotoxin receptors and in the fate of cell-bound endotoxin. While B cells quickly internalized the labeled endotoxin, it remained on the surfaces of T cells for 48 hours. The internalization is the event which leads to enhanced metabolism and blast transformation. The T cells do not show such changes but their affinity to endotoxin is evident. What other consequences this might have and whether suppressor and helper cells respond to this interaction in different ways is not known at this time. Grant and Alexander (1974) studied the specific and nonspecific cytotoxic action of T lymphocytes in C57Bl mice against lymphoma. The effect of endotoxin on macrophages and lymphocytes by direct as well as mediated action was discussed in this paper.

Cytotoxicity induction in mixed leukocyte culture (MLC) was studied by several laboratories. Narayanan et al (1978) and Narayanan and Sundharadas (1978) obtained an 8–10-fold enhancement of cytotoxicity when endotoxin was added to the MLC. The cytotoxic cells were specific to the sensitizing cell preparations and endotoxin was acting on the responding cells without requiring the presence of adherent cell populations. Vallera et al (1980a, 1980b) studied the mechanism of cytotoxic T cell generation in MLC. They observed that endotoxin added to the culture reduced the number of cytotoxic cells and showed that endotoxin–B cell interaction generated a factor which had a profound effect on the T cell subpopulation(s) involved in cell-mediated cytotoxicity. Berendt et al (1978c) investigated the causes of responsiveness/unresponsiveness of certain tumor types to endotoxin treatment, a problem brought up by Shear and his coworkers decades ago. The conclusion was that only those experimental tumors which were sufficiently immunogenic to the host to begin with will undergo a complete regression after hemorrhagic necrosis. This immunogenicity was T-dependent as shown by lack of tumor regression when the animals were immunosuppressed by whole body irradiation and thymectomy. The same laboratory believed that an endotoxin-induced humoral component, the tumor necrotizing factor, causes only TUH but not tumor regression (Berendt et al 1978c).

Before closing this section, one has to mention the discussion of less-defined cell types which may have a very significant effect on the antitumor effects induced by endotoxin. The 'natural killer' (NK) cells were detected in the last decade, and their characterization as well as their mode of action is the subject of several current very intensive studies (Herberman 1982; Herberman and Holden 1978). Recently attempts were made to activate NK cells by endotoxin or other microbial preparations. Gangemi et al (1980) used the low responder C3H/HeJ mice to induce NK activation by endotoxin, *Corynebacterium parvum* or by virus. The normal responder control strain was C3H/HeN. Endotoxin was activating NK cells only in C3H/HeN mice, but not in the low responder strain. All other preparations were active in both strains. Peritoneal macrophages of C3H/HeJ were not activated by *C. parvum* either, but in spite of this, both strains showed enhanced NK cytotoxicity induced by *C. parvum*. This finding indicates that macrophage and NK activations are the consequences of separate mechanisms. Chow et al (1981) studied the genetics of NK and naturally occurring antitumor immunoglobulin induction. Both were increased by endotoxin administration which indicates that certain aspects of the regulation may be common for both, but differences were also observed which argued against the notion that NK activity is due to passively acquired tumor-specific immunoglobulins.

In collaboration with A.G. Johnson we tested NK cell activation by endotoxin using human leukemia cells as targets of NK cytotoxicity. The preliminary results we obtained can be summarized as follows: (a) both wild-type and rough mutant endotoxins could enhance NK cytotoxicity, and (b) reduced but still significant NK activation was measurable by the injection of split products of endotoxins, in which the lipid moiety has been destroyed (Nowotny et al 1982).

Summarizing the action of endotoxin on lymphatic cells, it is evident that not only B cells, but T and NK cells as well can be affected in various ways by endotoxin. Most probably we are confronted here with both direct action on some cell types as well as with cellular responses due to the release of mediators triggered by endotoxin. What the involvement of cell type subsets in these direct and indirect actions is, is largely unknown at this time, but there are indications that the effect of endotoxin on these is manifested at different time intervals after the injection of endotoxin (Behling and Nowotny 1982). Nothing final can be said regarding the identification of the cells which can execute tumor target killing in the endotoxin-induced antitumor models. All options must be kept open in spite of the fact that recent results, started by the work of Alexander and Evans (1971), put a much greater emphasis on the role of macrophages than any other cell types. These findings are discussed below.

3.4. The role of macrophages

3.4.1. *Macrophage activation for antitumor action*

As far as we know, Evans and Alexander (1970) were the first to describe nonspecific activation of isolated macrophages by endotoxin, which rendered them cytotoxic to tumor cells. An avalanche of publications followed, almost all confirming this observation both in animal models as well as in human macrophages. In the following review of these papers, we will discuss the mechanism of activation, the humoral and cellular factors contributing to the cytotoxicity, the role of antitarget antibodies as well as other humoral components in the endotoxin induced macrophage-mediated killing of tumor cells.

Evans and Alexander (1970) found that macrophages from mice immunized with syngeneic lymphoma cells reduced the growth rate of lymphoma cells in vitro. While this reaction was immunologically specific, there were nonspecific ways to induce macrophage cytotoxicity. Double stranded RNA or endotoxin was capable of inducing similar effects. Alexander and Evans (1971) extended these studies to other lymphomas induced by methylcholanthrane. Regarding the mechanism of activation, it was found that an exposure of macrophages to endotoxin of as little as 30 minutes was sufficient to activate the cells. Supernatants of these activated macrophages did not contain cytotoxic substances for lymphoma targets. As far as the lytic action is concerned, they established that contact between the activated macrophages and the target cell had to be continuous for at least 24 hours, otherwise the tumor cells would recover and show no reduction of tumorigenicity. Alexander and Evans (1971) assumed the existence of endotoxin receptors on macrophage surfaces through which the activation took place, and hypothesized that the activated macrophage developed a new receptor on its surface which is needed for the contact with the target cell. They assumed that this latter receptor is lost after successful target cell destruction since the macrophages were not able to kill any of the freshly added lymphoma cells. Regarding the relevance of these findings to TUH, Alexander and Evans (1971) wrote that while the effect of endotoxin on the vasculature within the solid tumors is unquestionably the leading mechanism in TUH, additional reactions are needed to destroy the tumor cells. This is provided by the activated macrophages. Further details on the above points were published by Alexander in a series of articles (Alexander 1976a, 1976b; Alexander et al 1976).

Several laboratories initiated studies aimed at elucidating the mechanism of macrophage activation. The major question was whether the activation occurs by direct action of endotoxin on macrophage, or endotoxin acts on other cells which

410

then results in the generation of macrophage-activating factors. Bruley-Rosset et al (1976a) activated isolated macrophages in vitro and rendered them cytotoxic by endotoxin. The interpretation of the findings was that this agent exerted a direct action on the macrophage. Doe and Henson (1978) and Doe et al (1978) arrived at the same conclusion using several endotoxin preparations. They also found, in agreement with Alexander and Evans (1971), that the lipid A moiety of the endotoxin is the site which activates the macrophage. Doe and Henson (1979) found later that not only the lipid A but also an isolated protein-containing fragment of the endotoxin complex (lipid A-associated protein) can also induce macrophage activation. This was shown by using macrophages from the low-responder C3H/HeJ mice. These macrophages could not be induced to lyse tumor targets by 'protein free' endotoxin or lipid A, but became fully reactive after exposure to this protein-rich, nontoxic fraction obtained from conventional endotoxin preparations. Although Schubert and David (1980) studied pinocytosis instead of cytotoxicity as a measure of macrophage activation, they also found that direct action of endotoxin on macrophages is sufficient to trigger activation.

In contrast to these reports, numerous investigators concluded that macrophage activation by endotoxin is the result of mediators (macrophage activating factor or MAF) generated and released by other cells. Bona et al (1972) described that macrophages and lymphocytes adhere to each other and form a cellular island which cooperates in the generation of the immune response, including the processing and the transfer of immunogens. Allison et al (1973) found that such cooperation is required between macrophages, and B and T lymphocytes. Evans and Alexander (1976) provided evidence that products of activated T cells can render macrophages cytotoxic to tumors. Berendt and Saluk (1976) suggested that cell–cell interaction results in the production of a cytotoxic humoral component. Wrigley and Saluk (1981) co-cultured normal macrophages with lymphocytes stimulated in vivo by endotoxin and observed significantly enhanced phagocytosis of erythrocytes as compared with macrophage alone. Vogel et al (1982) showed that T cells regulate the in vitro sensitivity of macrophages to endotoxin. Panke et al (1982) used T cells and macrophages from normal and from BCG-sensitized hamsters. Their thorough studies showed that macrophages from BCG-sensitized donors can be rendered highly cytotoxic to tumors. Pulmonary macrophages from nonsensitized hamsters did not respond to endotoxin unless they were co-cultured for 3 hours with T cells from sensitized animals. It is notable that endotoxin concentrations of nanograms per milliliter were sufficient to induce cytotoxicity.

Extensive literature deals with the role various mediators may play in the activation of macrophages. Bruley-Rosset et al (1976a, 1976b) found that MAF is the lymphokine which can enhance the cytotoxic potential of macrophages. Sone et al (1980) rendered alveolar macrophages tumoricidal by MAF which was encapsulated into multilamellar liposomes. The MAF-activating potential became several thousand times higher by this procedure. Ruco and Meltzer (1977) generated lymphokines by the exposure of BCG immune spleen cells to purified protein derivative (PPD). The same authors noted that such activation quickly diminished and became undetectable after 20 hours, that it required two activating stimuli (Ruco and Meltzer 1978a) and the completion of a sequence of short-lived intermediary reactions (Ruco and Meltzer 1978b). One of these is provided by lymphokines (Meltzer et al 1980), the effect of which is synergistically augmented by endotoxin. The sequence of these effects on the macrophage is critical for effective cytotoxicity (Meltzer and Wohl 1979; Meltzer 1981). The authors divided the reaction sequence into three phases. The first is the precursor differentiation which involves the recruitment of blood-derived immature macrophages and their maturation into lymphokine-responsive cells. The second is reaction with the lymphokines which primes the macrophages for the third step when endotoxin renders them cytotoxic for tumor targets (Meltzer et al 1982).

Taramelli et al (1980) elaborated a microcytotoxicity test based on a similar scheme. MAF-containing supernatant samples were used to 'prime' proteose-peptone-induced macrophages. These cells showed a low tumoricidal activity but when as little as 10 ng/ml endotoxin was added, the cytolysis was greatly enhanced. Hansen et al (1982) found that this 'triggering' effect of endotoxin is evident only if protease-peptone-induced macrophages are used, indicating that these cells are better prepared for the lymphokine-endotoxin action sequence than macrophages obtained by other means. When the macrophages came from BCG-sensitized mice, they already had a high level of tumor cytotoxicity, and this could not be further amplified by lymphokine and endotoxin. Pace and Russell (1981) used lymphokine-rich supernatant of Con A-stimulated spleen cells to induce tumor-cell killing. They reported that when all reagents and media were free of endotoxin, no cytotoxicity was seen, but when the endotoxin contamination (determined by the Limulus assay) was greater than 0.125 ng/ml, the lymphokine significantly induced target cell lysis. The assay developed on the basis of this observation made use of endotoxin-free solutions and glassware, to which a constant 3 ng/ml endotoxin concentration was added. Trace amounts of lymphokines could be detected when such samples were added to the above system.

It only follows that every experiment should use endotoxin-free reagents and glassware if the aim is to see the level of induced macrophage activity (Martin et al 1978; Olsson et al 1980; Pace et al 1981). The glassware can be rendered pyrogen-free if heated for 3 hours at 180 °C. Most plastics can be treated overnight at room temperature or at 37 °C with 1 N KOH in anhydrous methanol. Very thorough rinsing with pyrogen-free double distilled water must follow. Inorganic salts or dextrose-containing solutions must be prepared in pyrogen-free sterile distilled water or filtered through several layers of activated charcoal to remove the endotoxin content. Sera or other fluids cannot be made pyrogen-free without absorbing or denaturing important components.

Regulators of the macrophage tumoricidal capacity were also found. Chapman and Hibbs (1977) isolated a high MW serum lipoprotein which inhibited and a low MW lipoprotein which promoted macrophage activation. Local environment has an important influence on the macrophage cytotoxicity (Hibbs et al 1977). Conditions of endotoxin-induced macrophage activation were also studied by the same laboratory (Chapman et al 1979; Hibbs et al 1979). Drysdale and Shin (1981) investigated the role of prostaglandins in the induction of macrophage activation by endotoxin or by interferon. The findings demonstrated the existence of a complex regulatory system which controls the cytotoxic function of macrophages.

The results which described antimacrophage effects of endotoxins are somewhat paradoxical. Shands et al (1974) separated peritoneal exudate cells into adhering and nonadhering populations. While the latter gave a mitogenic response when incubated with endotoxin, the adhering cells showed no mitosis but manifested cytotoxic effects. Ralph and Nakoinz (1977a) investigated the reaction of four established macrophage lines to various types of immunopotentiation, among them to BCG and endotoxin. The growth of all four macrophage lines was inhibited by one or another of the immunostimulators. Endotoxin was inhibitory on one of the four lines. Ralph and Nakoinz reported that growth inhibition did not interfere with differentiated functions of the macrophages since neither lysozyme nor CSF production was inhibited by endotoxin. Vogel et al (1979) showed that endotoxin suppresses macrophage phagocytic activity. Hammerström and Unsgaard (1979) found that long exposure to *E. coli* endotoxin interferes adversely with some functions of human macrophages such as phagocytosis and intracellular digestion of yeast cells, antitumor effect and the number of lysosomes. Endotoxin at a concentration of 10 µg/ml and 72-hour exposure was toxic to the monocytes. A highly malignant macrophage line M5076 did not show any division when it was treated with endotoxin. Interestingly enough, the nondividing macrophage maintained its viability and functioned both as a phagocytic and as a tumoricidal cell (Talmadge et al 1982).

The above diversities in macrophage response to endotoxin clearly depend on the macrophage under study, and on its origin and maintenance (Hansen et al 1982; Lee et al 1981; Panke et al 1982; Ralph and Nakoinz 1977a, 1977b; Russell et al 1977; Weinberg 1981). Another very important factor may be the quality and the quantity of the endotoxin used to activate the macrophages.

It is of particular interest to know how macrophages behave within the tumor and whether they respond to endotoxin-induced tumor-cell killing as normal tissue or circulating macrophages. Woodward and Daynes (1978) isolated macrophages from murine tumors induced by ultraviolet irradiation. Spleen cells of mice treated with ultraviolet irradiation could be rendered cytotoxic to the tumor cells in vitro, but this reaction required the presence of tumor-derived macrophages. Furthermore, these cytotoxic spleen cells were also tumoricidal to benzpyrene-induced syngeneic fibrosarcoma cells. An important role has been assigned in these activities to the resident macrophages. Taniyama and Holden (1979) obtained macrophages from progressing and from regressing murine sarcoma virus-induced tumors. There was no difference in early tumors, but after 50–65 days the progressing tumor macrophage had considerably lower cytotoxicity. Endotoxin could augment cytotoxicity in those macrophages where cytolytic activity preexisted, but did not show any effect on cells which were inactive. Peri et al (1981) concluded that the few macrophages residing in the human ascites or solid ovarian carcinomas probably do not have any beneficial effect to the host since they were less cytotoxic than normal macrophages. In this connection, one should also point out that the existence of a suppressor of macrophage activities was assumed by Kurashige and Mitsuhashi (1973) and it was thought to be produced by the growing tumor itself. A low MW inhibitor of macrophage tumoricidal activity was isolated from Balb/c mice tumor tissue by Cheung et al (1979). Nathan (1982), assuming that the cytotoxic effect of the activated macrophage is due to the secretion of oxygen intermediates, hypothesized that tumors produce anti-oxydants in self-defense (Arrick et al 1982).

3.4.2. *Mechanisms of the tumoricidal effect of endotoxin-activated macrophages*

Allison (1978) summarized the extensive work done in his as well as in other laboratories on the mechanisms and consequences of macrophage activation. Enzyme secretion, complement activation, interferon production, release of endogenous pyrogen, CSF, thromboplastin formation, plasminogen activation and synthesis of prostaglandins were particularly emphasized in this review. According to Allison, C3a generation seems to play a key role in the mechanism of the tumor target destruction. This is achieved by an endotoxin-induced enzyme release from the macrophage which cleaves C3 (Schorlemmer et al 1977) into C3a and C3b. C3b attached to the complement receptor on the macrophage is a powerful activator and C3a is a tumoricidal factor (Allison 1978). No further confirmation of this assumption has been published. Attachment of the macrophage to the target cell seems to be a prerequisite of cytolytic action, according to Piessens (1978), and Adams and Marino (1980). The increased adhesion of macrophages to surfaces is the basis of the methods for their separation from nonadhering lymphocytes and other cells. This spreading and adherence was particularly evident in the endotoxin-induced tumoricidal macrophages as described by Saluk et al (1981). The nature of this binding and the molecular mechanism of it is quite unclear. Piessens (1978) as well as Marino and Adams (1982) assumed that lymphokines are augmenting the adhesion.

The requirement of close cell–cell interaction for cytolysis is substantiated by the mechanism of the antibody-dependent cell-mediated cytotoxicity (ADCC) phenomenon. Tumor-specific antibodies react with the target cell and through their Fc region with the macrophage which has recognition sites for Fc. Ralph and Nakoinz (1977b) lysed lymphoma cells (EL4) by incubating them with either H-2 allosera or with anti-thy 1.2 in the presence of various murine macrophage or macrophage-related tumor lines. Normal nonimmune sera were without effect. This lysis could be enhanced by 76 to 86% if the macrophages were stimulated with endotoxin or PPD of mycobacteria. Ralph and Nakoinz pointed out that the various macrophage lines do not behave in an identical fashion in these assays. Continuation of these studies dealt with such differences among the malignant macrophage lines on the one hand, and other tumors and normal macrophages on the other (Ralph et al 1978). The most salient among the dissimilarities are phagocytic capacities and sensitivity to immunostimulating agents. One of these macrophage lines was inactive in the cytolysis of B or T lymphocytic tumor cells, others showed only moderate spontaneous cytotoxicity, but this could be greatly enhanced by target-specific antibody, endotoxin or phorbol myristate acetate. Pretreatment of the macrophage lines with endotoxin did not trigger nonspecific killing, but such pretreatment very much enhanced ADCC (Ralph and Nakoinz 1981; Ralph et al 1982).

Glaser et al (1976) used normal rat spleen cells and activated them in vitro with endotoxin or PHA. The cytolytic activity of the cells was measured on Gross virus-induced lymphoma cells. The PHA-triggered cytotoxicity was dependent upon the presence of T cells, endotoxin required the presence of macrophages. Mitogenic concentration of the two agents also enhanced ADCC. Shinoda et al (1977a) used syngeneic immune serum on tumors in passive immunotherapy. It was found that such antibodies can lyse tumor cells in cooperation with activated macrophages. The results showed that intraperitoneal injection of antiserum suppressed the growth of intraperitoneally injected tumors but did not affect already established ones. Continuation of this work led to the use of immunopotentiators such as endotoxin, in addition to the passive immunotherapy. This combination synergistically enhanced the survival rate, even in those mice which had an established ascites tumor (Shinoda et al 1977b). Sera of the cured mice was investigated for cytotoxicity in three systems: in complement-mediated cytotoxicity, in lymphocyte-mediated cytotoxicity, in lymphocyte– and macrophage-mediated reactions. The ADCC was the most effective when activated macrophages were used (Saito et al 1978). Lymphocytes or the complement-mediated lytic effects were considerably lower. It is important to note that the macrophage-ADCC levels remained high for 160 days after the cure.

Another school of thought emphasizes the release of cytotoxic substances by the endotoxin-activated macrophages. Reed and Lucas (1975) investigated the time span of the activated stage in vitro. Tumoricidal activity reached its maximum 8–16 hours after attachment and became undetectable in 36 to 48 hours. No endotoxin was added to this system initially. When it was added to the culture which showed no cytotoxicity anymore, the activity was restored. Reed and Lucas also found in the supernatant of the adhering macrophages a soluble component which was cytotoxic to tumor cells. Fractionation of the supernatant by Sephadex chromatography revealed that this toxin is not identical with lymphotoxin. Independently from Reed and Lucas, Currie and Basham (1975) isolated a cytotoxic agent which was lytic only to sarcoma cells but did not influence the growth of normal cells. Induction of the release of their cytotoxin was achieved by endotoxin from *Salmonella typhosa* bacteria. Berendt and Saluk (1976) exposed peritoneal exudate cells to endotoxin and observed the release of soluble factors which inhibited tumor growth in vivo. Young et al (1980) speculated that inhibition of tumor metastasis development is achieved by a component that inhibits tumor cell migration. They isolated it from endotoxin-activated macrophage supernatant. Interestingly enough, this soluble factor inhibited the migration of all four tumor lines tested but had no such effect on normal lymphoblasts, lymphocytes or macrophages.

The existence of this soluble cytotoxin seems to be well established. Its induction, the kinetics of the activation as well as the mode of action were studied in detail by Reidarson et al (1982a, 1982b). This toxin is labile and it is released first in culture for a short period of time if the macrophages obtained from tumor-inoculated mice form a monolayer. Addition of tumor target cells (but not normal cells) triggers the release again. Poly I:poly C oligonucleotide or endotoxin can act similarly on the macrophage monolayer. Fractionation resulted in two major peaks: one has a MW 140 000 to 160 000 (alpha toxin), the other is smaller, its MW is approximately 60 000 daltons (beta toxin). Quite interestingly, normal resident peritoneal macrophages could not be induced to release these cytotoxins. When the macrophages were obtained after an intraperitoneal thioglycolate injection, good yield of the cytotoxin was obtained after stimulation with endotoxin or oligonucleotide. Reidarson and coworkers showed, in agreement with other investigators, that the toxin lyses only neoplastic cells. The continuation of these studies focused on the lytic mechanism. It was established that the cytotoxin binds readily to neoplastic cells, whether they be allogeneic or syngeneic murine tumors, but did not bind to human K562 tumor cells. When the target cells (NS-1) and the macrophages were separated by a Millipore membrane and the macrophages were stimulated, destruction of the allogeneic target took place. Unstimulated macrophage did not exert any effect across the membrane. This highly interesting finding clearly indicated that direct contact between the macrophage and its target cell is not required in the above system. Finally, the same laboratory established that the cytotoxin is probably a protease. This conclusion was reached as a number of protease inhibitors clearly abolished the cytotoxic activity (Reidarson et al 1982b).

Pabst and Johnston (1980) studied the production of superoxide anion in the endotoxin activated macrophage. On the basis of earlier studies of the macrophage-induced killing of microorganisms (Johnston 1978; Johnston et al 1975), they assumed that this event is involved in the tumoricidal reaction. The results showed that both endotoxin and muramyl dipeptide (MDP) probably act directly on the macrophage since admixture of nonadhering cells to the culture did not enhance the cytotoxicity. The major event in the activation process was the generation of peroxide anions. Pabst and Johnston assumed that previous in vitro exposure of macrophages to endotoxin primes them to generate these toxic oxygen metabolites when they come into contact with tumor cells. Nathan reviewed this field and studied the possible role of these oxygen metabolites in the effector function of activated macrophages (Nathan 1982). The conclusion was that the ability of macrophages to secrete toxic oxygen metabolites correlates with their capability to induce nonphagocytic destruction of tumor cells. Nathan brought up the possibility that tumor cells minimize the oxidation caused injury by antioxidant defenses, and suggested that the oxidation-reduction cycle of glutathion is the major self-defending pathway of the tumor cells (Arrick et al 1982; Nathan 1982).

According to Meltzer et al (1982), and Adams and Marino (1981) the tumoricidal action goes through various phases. The first is the activation of the macrophage which consists of a series of events leading to a morphologically but particularly functionally altered effector cell. The target cell destruction is believed to require a close cell-to-cell contact, although the recent results of Reidarson et al (1982b) strongly suggest reinvestigation of this particular point. The cell killing appears to be nonphagocytic and it is caused by toxic substances released by the macrophage. Among these, proteolytic enzymes and toxic oxygen metabolites were studied in detail. The simultaneous presence of both is not excluded.

After surveying the overwhelming amount of data on the possible role of macrophages in the antitumor effect of endotoxin, there seems to be little doubt today that macrophage activation and all its possible consequences play pivotal roles in the endotoxin-induced TUR. Enhanced tumoricidal effect may be further amplified if tumor-specific immunoglobulins are present and calls for cellular cooperation in this action may be sent out by the turned-on macrophage in the form of soluble mediators (some of which are discussed in the next paragraphs). Unquestionably, at the moment this seems to be the most promising lead to follow and we are only at the beginning of this work. This is the area where we can expect the clarification of some of the mechanisms. Even if the use of endotoxins will not result in more than the possibility to study and understand some of the molecular events leading to the generation of tumoricidal macrophages in a few experimental systems, the work was indeed well invested.

3.5. Non-antibody humoral factors in the antitumor reaction mechanisms

3.5.1. Humoral factors influencing cell proliferation and/or differentiation

The injection of endotoxin induces the release of a number of humoral factors. Passive transfer of these can trigger or mediate several beneficial reactions known to be induced by endotoxins. At the same time these humoral factors do not elicit any of the well-known harmful effects of endotoxin injections. The very important fact that several endotoxic reactions may be mediated by humoral factors was first recognized by Berry (1971), who extensively studied the effect of endotoxin on various regulations of metabolic processes. The endotoxin-induced humoral mediators were reviewed by Berry (1978).

Proliferation of certain cell types of the murine bone marrow could be initiated by various conditioned media (Pluznik and Sachs 1965). MacNeill (1970), Metcalf (1971), and Quesenberry et al (1972) showed that the injection of various antigen-containing preparations, including endotoxins, can induce the release of a factor into the sera of mice, which, if added to normal bone marrow cell suspensions in semi-solid agar, stimulates the formation of colonies. The condition of the generation and some of the molecular characteristics of this colony stimulating factor (CSF) were published by Stanley et al (1975) and Burgess et al (1977). We found that those endotoxin preparations, or their split products, which were highly active in the enhancement of TUR against the allogeneic TA3-Ha adenocarcinoma of mice were also particularly active in CSF induction (Butler and Nowotny 1976). More detailed studies revealed that the post-endotoxin sera of normal mice, rich in CSF, could passively transfer TUR. Particularly active were the post-endotoxin sera of BCG-infected mice both in CSF induction and in TUR enhancement (Butler et al 1978). These sera were not cytotoxic to TA3-Ha cells in vitro. We assumed that either the CSF activity contributes to enhanced TUR, or if CSF and TUR are elicited by two independent humoral factors, their generation seems to require identical experimental conditions (Nowotny and Butler 1980). These experiments confirmed the earlier conclusions we reached when we saw no direct effect of endotoxin on tumor cells and assumed that humoral factors might have been involved in the antitumor effects elicited by endotoxins (Nowotny et al 1971).

Hinterberger and associates (1979) studied CSF induction in leukemic patients. Pharmaceutical endotoxin preparations were injected into normal individuals and into acute as well as into chronic myeloid leukemia patients. Normal individuals showed a sharp increase in CSF production, but most of the leukemic patients showed no CSF activity in response to a fever-producing dose of endotoxin. More details are given in an up-to-date review of this work by Hinterberger et al (1983).

Several laboratories reported that CSF-rich sera can induce the differentiation of leukemic cells. Yamamoto et al (1981) used products of gram-negative as well as gram-positive cells, synthetic polynucleotide or MDP derivatives to elicit the production of a so-called differentiating factor (DF). Mouse myeloid leukemia cells responded to post-endotoxin sera. Similarly active were BCG or *Nocardia rubra* cell walls in DF induction. MDP, or cell walls from *C. parvum* were inactive. Metcalf (1982a, 1982b) used mouse myelomonocytic leukemia cells (WEHI3B) and studied their differentiation induced by post-endotoxin sera. It was reported that athymic nude mice did respond with DF production to endotoxin while C3H/HeJ mice did not. Partly purified DF irreversibly modified some of the newly synthesized daughter chromatids of myeloid leukemic cells during the stem cell suppression and differentiation.

3.5.2.   *Tumor necrotizing factor*

O'Malley et al (1962b) studied the mechanism of TUH, described and extensively investigated by Shear and associates. O'Malley and coworkers reported that *S. marcescens* endotoxin injection induces the appearance of a TUH-causing serum component in normal mice.

The mechanism of TUH induced by endotoxin was also studied by the investigators of the Memorial Sloan-Kettering Cancer Center. Kassel et al (1973) found that serum can mediate leukemia cell destruction in AKR mice. Mamaril et al (1974) reported that BCG-infected and endotoxin-injected mouse liver cell microsomes release large quantities of NADase into the circulation. Neither BCG infection nor endotoxin injection alone elicited such changes. Green et al (1974) induced TUH in Meth A sarcoma by post-BCG–endotoxin sera. Carswell et al (1975) published the first extensive description of the serum component which induced TUH and named it tumor necrosis factor of TNF. Helson and coworkers (1975) found that TNF is cytotoxic for human melanoma cells in vitro. Green et al (1976a, 1977) replaced BCG with *C. parvum* and isolated TNF from the liver cells of *C. parvum*–endotoxin-injected mice. Green et al (1976b) partly purified the TNF, and described it as follows: its MW is 150 000, it migrates with globulins in the electrophoretic field and it consists of at least four subunits. Hoffman et al (1976) found that TNF could replace T helper cells in the induction of immune response to T-dependent immunogens in nude mice. Shah et al (1977) studied the possible relationship between CSF and TNF. Hoffmann et al (1978) further extended these studies and compared the action of TNF with endotoxin in a number of immune reactions. They found that TNF is not a B cell mitogen but it promotes maturation of B cells. It inhibits the response of T and B cells to mitogens and it does not induce polyclonal B cell activation. TNF appears to be a macrophage product. One of the observations was of major importance, namely that TNF has no cytotoxicity on normal cells. Later publications on TNF described it as a glycoprotein with 70 000 MW (Oettgen et al 1980). Other laboratories confirmed and extended most of these findings (Männel et al 1979, 1980a, 1980b, 1980c). Ruff and Gifford (1980) purified and characterized rabbit TNF. Prince et al (1981) induced TNF in the ascites fluid of Sarcoma-bearing mice by endotoxin treatment. Primi et al (1977) described that post-endotoxin sera of mice is cytotoxic to autologous spleen cells and therefore not exclusively tumor-specific. Kull and Cuatrecasas (1981) developed a sensitive and quantitative in vitro assay to measure the cytotoxicity of the tumor necrotic serum (TNS). They separated several fractions by gel filtration and found that cytotoxicity is present in more than one peak, and the MW varies from 100 000 to 250 000. A smaller fraction (MW 50 000) was also found. Two of these cytotoxic fractions did not induce tumor necrosis. This activity was found in a peak with approximately 160 000 MW. The cytotoxic activity could not be inhibited by protease inhibitors, but certain proteolytic treatments or periodate oxydation of the cytotoxic fractions abolished their activity.

The TNS had remarkable activities in a number of other systems. We already discussed that it had high levels of CSF and that it could also prevent the take of TA3-Ha adenocarcinoma grown in ascites form (Butler et al 1978; Nowotny and Butler 1980). Parant (1980) found that TNS could protect mice against challenge with viable *Klebsiella* and *Listeria* microorganisms. We found that TNS is a potent immune stimulator (Butler et al 1979, 1980; Friedman et al 1980), and macrophages were identified as producers of this immune adjuvant factor (Butler et al 1979).

The pharmacodynamics of the TUH and of the TNF-release was studied by Kuper et al (1982) and Bloksma et al (1982a, 1982b). Histological investigation of Meth A sarcomas after endotoxin injection revealed vasodilation, diffuse hemorrhage and degenerating tumor cells. Adrenoceptor antagonists such as phenoxybenzamine, an alpha-blocker, and propranolol, a beta-blocker, caused a reduction of the vasodilation a few hours after endotoxin, but enhanced it 24 hours later. Both agents enhanced tumor size by 24 hours if they were given alone, but only phenoxybenzamine stimulated tumor growth for a prolonged period of time.

Phenoxybenzamine reduced the TUH effect of endotoxin. Both adrenoceptor antagonists as well as epinephrine diminished the release of TNF. Interferon production was inhibited by the α-adrenoceptor blockers but not by propranolol. It was assumed that the endotoxin-induced release of antitumor humoral components is controlled by either the α-adrenoceptor, or by serotonin or by choline receptors or by a combination of these components. Bloksma and colleagues also suggested that the growth of the tumor in the untreated as well as in the endotoxin-injected mice is under α-adrenergic control and TUH might be mediated by the release of adrenal catecholamines.

3.5.3. Interferon

Endotoxin is also known to release interferon as described by Ho (1964), and Youngner and Stinebring (1965). Ho (1983) reviewed the conditions of endotoxin-induced interferon release and described the characterization of this type of interferon both from qualitative as well as from quantitative points of view.

The effect of interferon on cell division, on the immune response and particularly on tumors was recognized by various laboratories, but it was usually attributed to the presence of impurities in the interferon preparation. Stewart et al (1971) reported first on these 'non-antiviral functions' of interferon. The evidence of interferon-induced inhibition of malignant cell multiplication came from the laboratories of Gresser et al (1969; Gresser 1972). May other laboratories confirmed the findings using various tumor target cells. Interferon also has an indirect effect on cell lysis since it may potentiate the cytotoxicity of sensitized lymphocytes. Skurkovich et al (1976) found that interferon can render the target cell surfaces more susceptible to cytotoxic antibodies. Gresser and Tovey (1978), and Stewart (1979) reviewed the mode of action of interferon and found that changes on the target cell surface are among the first to take place. Ng et al (1983) studied these surface changes in great detail and found that they occur simultaneously with a slowdown of the multiplication rate of the cells. The killing mechanism per se is not completely understood. In connection with this we should refer to the results of Kato et al (1979a, 1979b, 1980). These authors pointed out that not only interferon but also a cytotoxic factor is generated by endotoxin in BCG-sensitized mice. This 'cytotoxin' is clearly distinguishable from interferon by the condition of its generation, its resistance to heat, and sensitivity to low pH.

It is quite clear that interferon, CSF, TNF, interleukin 1 or 2 and other humoral factors released after endotoxin injection are only a few of the several serum components which are detectable in the post-endotoxin sera. Bradfield et al (1980) observed elevated levels of transaminase and transferase enzymes in sera of endotoxin injected mice. Gorecka-Tisera et al (1981) found elevated serum lysozyme levels. Kawakami et al (1982) described that endotoxin injection releases a lipoprotein lipase inhibitor from peritoneal exudate cells. Abd-El-Fatah et al (1976) detected seventeen plasma protein level changes by two dimensional immunoelectrophoresis. Several of these showed an increase of more than 100% after the injection of endotoxin into rats.

Concluding the discussion on the role of mediators, we can say that the most lasting impression is caused, not so much by the number of the mediators released after endotoxin treatment, but by the variety of functions these mediators can fulfill. Some of these molecular messengers of intercellular communication are known, but the discovery of many more should be expected. It might be superfluous to emphasize the importance of these post-endotoxin serum factors, but one has to stress that the use of impure preparations makes it impossible to assign cytotoxic or other effects to any one of the components present in them. Without purification of homogenous molecular species from these sera, we will remain woefully ignorant about the identity of the active components in it which cause or contribute to the antitumor effect induced by endotoxin.

3.6. Endotoxin in combination with other biological response modifiers

The combined use of agents is not new in this field. The mixed bacterial toxin of Coley is one of the first examples of such combinations. Later, in the 1920s, bacterial products were shown to potentiate each other's effects in a synergistic fashion. Sanarelli's classical experiments (1924) opened up a new field of studies. Bordet (1931) infected guinea pigs with BCG and then produced extensive intestinal mucosal hemorrhage by injecting them with an endotoxin-containing preparation. Gratia and Linz (1931), whose pioneering work in tumor therapy with bacterial products was discussed earlier, sensitized with *Bacillus anthracis* vaccine and induced the Sanarelli phenomenon with *E. coli* culture filtrate. Freund (1934) took tuberculous guinea pigs and observed a greatly increased skin reactivity (local Shwartzman) to endotoxins. Suter et al (1958) reported up to 1000-fold increase in the lethal effect of endotoxin preparations when the mice were hypersensitized with BCG infection. Suter also replaced BCG infection by treatment with cord factor obtained from mycobacteria and reported that 5 μg of this substance was almost as active as the infection with BCG to hypersensitize mice to endotoxin (Suter 1964a). The mechanism of this sensitization was also studied (Suter and Kirsanow 1961; Suter and Munoz 1963; Suter 1964b).

Carswell et al applied the same phenomenon in 1975 to induce the TNF which we discussed earlier. Ribi and associates (1975) caused hemorrhagic necrosis of regressing line 10 hepatocarcinoma in guinea pigs by intratumoral injection of endotoxin and a purified fraction of cord factor, called P3. Numerous publications on this observation followed, with detailed descriptions of the preparations and of the optimal conditions for their use (Ribi et al 1976a, 1976b, 1979, 1983; Richman et al 1980).

Glaser et al (1976) augmented in vitro cytotoxicity of normal rat spleen cells against syngeneic Gross virus-induced lymphoma by 24-hour incubation with mitogenic concentrations of endotoxin and PHA. The cytolytic reaction did not involve the release of toxic substances but it required the presence of T cells.

Our laboratory studied the prevention of tumor take (TUR). A variety of immunostimulators were tested alone and in combination to measure their ability to enhance the nonspecific resistance to a challenge with viable inocula of TA3-Ha tumor. Toxic endotoxin as well as nontoxic, partly hydrolyzed products of these (PS) were used alone and in combination with BCG infection, killed BCG, P3 preparation of Ribi, cord factor, synthetic glycolipid adjuvants, levamisole and MDP. Great differences were found among these agents and the combinations thereof in their capacity to induce TUR, but it became evident that TUR can be enhanced in an additive or sometimes synergistic fashion by using combinations of certain agents which presumably stimulate different cell preparations of the host's defenses (Butler and Nowotny 1979). Other laboratories combined endotoxin with mycobacterial sulfolipid and obtained results identical with the combination of endotoxin and cord factor, but without the toxic effect of the latter (Yarkoni et al 1979a, 1979b). The combined use of endotoxin and lentinan was very effective against Ehrlich carcinoma and syngeneic mammary carcinoma MM46 in C3H/He mice (Abe et al 1982).

Hammerström (1979) induced cytolytic activity in human monocytes against a human tumor cell line (NHIK 3025) by mediators. These were obtained from the culture supernatant of human lymphocytes exposed to *C. parvum*. This cytotoxicity was enhanced by the addition of endotoxin. It is interesting to note that human monocytes could not be rendered cytotoxic by direct exposure to endotoxin. Several other laboratories reported that human macrophages or monocytes could be stimulated to cytotoxicity to various tumor targets (Cameron and Churchill 1980; Schacter et al 1981; Sone et al 1982). The explanation for this discrepancy may lie in the composition of the effector cell population, or in the sensitivity of the target cells to cytolytic action. The nature and purity of the endotoxins used is another potential source for diverse results.

Interaction between chemotherapeutic drugs and endotoxins were also studied in our laboratory. Five drugs were used alone and in combination with each other as well as with endotoxin and PS to test their effect on the immune response of mice to sheep red blood cells. It was found that some of the drugs can act, not only as immune suppressants, but as immune enhancers if the time interval between their injection and the immunization is varied. Proper timing of certain combinations resulted in an immunopotentiation which was 23 times greater. Furthermore, we found that the combined use of drugs and endotoxin or its nontoxic PS derivative could compensate for some of the immunosuppressive effect. With the use of endotoxin–drug combinations striking immune adjuvant effect was achieved by the use of vincristine and endotoxin, or vincristine and PS (Gaal and Nowotny 1979).

Dye and North (1980) combined cyclophosphamide and endotoxin and observed the effect of this on the growth of SA-1 spindle cell sarcoma in ascites form. Cyclophosphamide alone reduced the number of tumor cells by 90%, but did not affect the number of macrophages in the peritoneal cavity. The remaining SA-1 cells, in spite of the presence of an overwhelming number of macrophages, continued to grow and finally killed the host. On the other hand, when endotoxin was given after treatment with cyclophosphamide, the ascites tumor regressed completely. It was assumed that endotoxin generated and antitumor immunity through its action as an immune enhancer. The possible activation of macrophages by endotoxin was not studied. Yamamura et al (1980) used DBA/2J mice bearing syngeneic mammary adenocarcinoma in ascites form. The treatment was melphalan and endotoxin, which did not only result in regression but even in permanent cures in some of the mice. Optimal treatment was with endotoxin (or PHA) given to tumor-bearing mice 3 days before melphalan injection. The assumptive evaluation of the results stated that the collaboration between immune-specific cytotoxic cells and the drug resulted in the elimination of the tumor.

We are, or at least we all should be, keenly aware of the multiplicity and extreme complexity of the possible mechanisms which might lead to the control of tumor growth or tumor elimination. It is clear that the various possible mechanisms elicited by endotoxin alone run different pathways, enhancing some while inhibiting others, but certainly not mobilizing all available ways and means necessary to eradicate neoplasms. It is only logical that the use of several effective biological response modifiers which will trigger multiple mechanisms is expected to be more efficient than the use of a single stimulator of the antitumor defenses.

3.7. Tumor growth enhancement by endotoxin

The enhancement of tumor growth was observed by a number of investigators, including ourselves. Henderson (1968) made an unusual experiment, where Balb/c mammary tumor single cell suspension was 'sprayed…onto the broad subcutaneous expanses of mice of the same inbred stock'. Boivin-type endotoxin exerted an adjuvatant influence on the tumor growth by making the host's connective tissue expanses more susceptible to tumor cell implantation.

Cruse et al (1971) reported that endotoxin tolerance, achieved by repeated injection of endotoxin, prolonged the survival of tumor allografts in mice. Strausser and Bober (1972) injected 10 ng endotoxin 3 times per week for 60 days starting the day after the inoculation of Moloney sarcoma virus (MSV) tumor. Accelerated tumorous death was recorded. When the dose was raised to 100 or 1000 ng, the percentage of survivors and the regression rate of tumors were increased. Bober et al (1976b) extended these studies on the growth rate of mineral oil-induced plasmacytomas. Endotoxin in doses of nanograms increased the incidence of plasma cell tumor development. Platica and Hollander (1978) measured the endotoxin content of the peritoneal fluid of mineral oil-injected mice and found that the injection causes endotoxin accumulation in the peritoneal cavity. They assumed that endotoxin plays a role in the pathogenesis by promoting plasmacytoma development. Finally, the work of Kearney and Harrop (1980) should be mentioned, who found that the subcutaneous growth of methylcholanthrene-induced fibrosarcomas was promoted when 0.2 µg endotoxin was injected intraperitoneally 1 day before tumor challenge. On the other hand, when the same injection was given 6 days before the fibrosarcoma cells, definite TUR effect was seen.

It was observed that negative effects are often elicited by improper timing and/or overdose of endotoxin injections. Amounts of endotoxin in the order of nanograms activate macrophages, while doses of micrograms per milliliter are toxic for them. The adjuvant effect is best manifested when endotoxin is given together with the immunogen; if injected 2–5 days before, immunosuppression may be the result (Franzl and McMaster 1968; Kind and Johnson 1959; McMaster and Franzl 1968; Nakano et al 1976). Kimball et al (1968) observed an enhanced sensitivity of mice to infectious agents shortly after endotoxin injection which gradually passed into enhanced resistance. A well-recognizable oscillation between positive and negative effects was detected by us when groups of mice were given single endotoxin injections and their response tested at various time intervals afterwards. These responses included resistance to lethal irradiation, immune adjuvant effect to sheep red blood cells measured by the Jerne plaque assay and by rosette formation and resistance to a variety of tumor challenges (Behling and Nowotny 1977, 1978, 1980; Nowotny 1977; Nowotny et al 1982). We observed that the timing necessary to elicit the TUR effect was different for TA3-Ha and for Lewis lung of L1210 tumors. The timing required to obtain negative effects were also dissimilar in the above three tumors. Optimal times for TUR induction were either 1–3 days or 13–16 days before TA3-Ha challenge (Nowotny 1977). The Lewis lung carcinoma in C57B16 mice could be retarded in growth when endotoxin was given 5–7 days or 1–2 days before challenge. The L1210 tumor could be slowed down only slightly in CBA/2J mice and only when the intraperitoneal injection was given 1 day after 10^3 L1210 cells (Nowotny et al 1982, and unpublished data). At other time intervals endotoxin had either no effect or enhanced the tumor growth.

It is difficult to explain why the hosts require different endotoxin pretreatment times to reach the peak of their TUR capacity when they are challenged with various tumors. The site of the tumor growth and of the mouse strain used may affect the time needed for full TUR development. It is more difficult to explain how the oscillation phenomenon can develop after a single endotoxin injection. Over-compensation for the disturbed homeostatic equilibrium, differences in the activation times needed for some cellular and/or humoral mechanisms, synchrony of the maturation and cell cycles are only hypothetical possibilities (Behling and Nowotny 1982). The true mechanisms remain to be elucidated.

4. CONCLUSION

Some bacterial infections can induce reactions which harm neoplastic tissues, and we now have a few preparations isolated from these microorganisms with which similar effects can be observed. Bacterial endotoxins are among these products and their effectiveness in a few experimental tumor systems is well proven.

Regarding the mechanisms through which the antitumor effect of endotoxin is executed, several possibilities have to be considered, such as direct action on certain tumor cells, effect on the precarious vasculature of the rapidly developing malignant tumors, activation of macrophages and lymphocytes in an immunologically nonspecific way, and release of similarly nonspecific humoral factors. Since many tumors were shown to be immunogenic, we have to consider that endotoxin, one of the most potent immunostimulators, will enhance the receptiveness of the immune system in those experimental models where pretreatment is efficient, or it may strengthen the faltering immune defenses when progressive tumor growth and their products seem to overcome the limited capacity of the immune system.

It is much more difficult, if not impossible to point out the most dominant mechanism through which endotoxin exerts its antitumor effect. Probably several of these play key roles. If one of them is absent, the others cannot intensify their functions to fill the void and if they are interdependent, no mechanism will serve the host.

One of the most likely reasons why one cannot define a generally applicable reaction mechanism is that there is no single one which would be applicable to all tumors, to all endotoxins, in every possible host species. It is common knowledge that cancer is not one disease, and also that among morphologically and clinically similar malignancies, great differences can be observed in the course of the disease. It is well documented that differences exist between the cells of primary and metastatic lesions in the same host, and that heterogeneity exists even among the cells of a single tumor (Owens et al 1982). Many of the seemingly contradictory findings we reviewed in this chapter are due to the fact that the many possible variables were not identical from experiment to experiment. The location of the tumor obviously influences its accessibility to the effector mechanism. The stage of tumor development, the tumor load of the host, impairment or enhancement of the nonspecific or specific resistance by subclinical infections may be factors which determine success or failure of the experiment.

Similarly inconsistent are the chemical compositions of the conventional endotoxin preparations. It is known and has to be kept in mind that not all bacterial endotoxins will have the same potency in inducing TUH or TUR. Results obtained by the use of one endotoxin may not be reproducible when one uses an endotoxin from another gram-negative strain. With the endotoxin of any one strain, simple electrophoretic analyses will show that two types of heterogeneities exist: one is due to incomplete purification, the other to the intrinsic heterogeneity of the assembled macromolecules. Both of these are obvious sources of deviations in experimental results, and with the exception of a very few laboratories, no attention has been paid to this crucial point. Neglect of the problem of heterogeneity will not make it go away, it will only continue to generate poorly reproducible results. It is a must to work with preparations which were tested for chromatographic purity and even then one should not expect identical results comparing endotoxins from different bacterial strains.

After having said that, one finds it quite remarkable how many different experimental or clinical trials showed the antitumor effects of endotoxins. These indicate that there must be shared features among some tumors as well as among most endotoxins. That being the case, we are allowed the assumption that the course of certain mechanisms follow similar pathways. The most investigated mechanism in the last years appears to be the macrophage activation and its consequences, but under no conditions should we forego those possible modes of action which became less popular in recent years.

Considering the multiform nature of spontaneously occurring malignancies mentioned above, we have to realize that there never will be a single remedy which will cure all forms of neoplasia, unless we find a dominant and uniformly shared feature among them which will be amenable to our extrinsic interference. It only follows that we probably will never cure cancer by endotoxin alone. What seems to be a more realistic aim is that by the use of pure endotoxins or other bacterial cell products we will gain a better insight into some of the molecular actions which have the potential to induce antitumor effects. Should that be only one of the many pathways which we will some day fully understand and govern, we will be able to improve the fighting chances of those who need our help the most.

ACKNOWLEDGEMENT

I wish to thank the many students and associates of mine, whose dedicated work generated the results summarized in this review.

REFERENCES

Abd-El-Fattah M, Scherer R, Ruhenstroth-Bauer G (1976) Application of quantitative two-dimensional immunoelectrophoresis in the study of the acute-phase reaction following injection of the lipid A component of bacterial lipopolysaccharides in rats. *J. Mol. Med. 1*, 211-221.

Abe S, Yoshioka O, Masuko Y, Tsubouchi J, Kohno M, Nakajima H, Yamazaki M, Mizono D (1982) Combination antitumor therapy with lentinan and bacterial lipopolysaccharide against murine tumors. *Gann 73*, 91-96.

Adams DO, Marino PA (1981) Evidence for a multistep mechanism of cytolysis by BCG-activated macrophages: the interrelationship between the capacity for cytolysis, target binding, and secretion of cytolytic factor. *J. Immunol. 126*, 981-987.

Adorini L, Ruco L, Uccini S, Soravito de Franceschi G, Baroni CD, Doria G (1976) Biological effects of *Escherichia coli* lipopolysaccharide (LPS) in vivo. II. Selection in the mouse thymus of PHA– and CON A-responsive cells. *Immunology 31*, 225-232.

Alexander P (1976a) Functions of macrophage in malignant disease. *Annu. Rev. Med. 27*, 207-224.

Alexander P (1976b) Macrophages and tumors. *Schweiz. Med. Wochenschr. 106*, 1345-1350.

Alexander P, Evans R (1971) Endotoxin and double stranded RNA render macrophages cytotoxic. *Nature New Biol. 232*, 76-78.

Alexander P, Eccles SA, Gauci CLL (1976) Significance of macrophages in human and experimental tumors. *Ann. N.Y. Acad. Sci. 276*, 124-133.

Algire GE, Legallais FY (1951) Vascular reactions of normal and malignant tissues *in vivo*. IV. The effect of peripheral hypotension on transplanted tumors. *J. Natl Cancer Inst. 12*, 399-421.

Algire GE, Legallais FY, Park HD (1947) Vascular reactions of normal and malignant tissues *in vivo*. II. The vascular reactions of normal and neoplastic tissues of mice to a bacterial polysaccharide from *Serratia marcescens* (*Bacillus prodigiosus*) culture filtrates. *J. Natl Cancer Inst. 8*, 53-62.

Algire GE, Legallais FY, Anderson BF (1952) Vascular reactions of the normal and malignant tissues *in vivo*. V. The role of hypotension in the action of a bacterial polysaccharide on tumors. *J. Natl Cancer Inst. 12*, 1279-1295.

Allison AC (1978) Macrophage activation and nonspecific immunity. *Int. Rev. Exp. Pathol. 18*, 303-346.

Allison AC, Davies P, Page RC (1973) Effects of endotoxin on macrophages and other lymphoreticular cells. *J. Infect. Dis. 128*, S212-219.

Alquié (1851) Inoculation de la syphilis au cancer. *Gaz. Hôp. 24*, 546-548.

Andersson J, Sjöberg O, Möller G (1972) Induction of immunoglobulin and antibody synthesis *in vitro* by lipopolysaccharides. *Eur. J. Immunol. 2*, 349-353.

Andersson J, Melchers F, Galanos C, Lüderitz O (1973) The mitogenic effect of lipopolysaccharide on bone marrow-derived mouse lymphocytes; lipid A as the mitogenic part of the molecule. *J. Exp. Med. 137*, 943-953.

Andervont HB (1936) The reaction of mice and various mouse tumors to the injection of bacterial products. *Am. J. Cancer 27*, 77-83.

Apitz K (1933) Über Blutungsreaktionen am Impfcarcinom der Maus. *Z. Krebsforsch. 40*, 50-70.

Arrick BA, Nathan CF, Griffith OW, Cohn ZA (1982) Glutathione depletion sensitizes tumor cells to oxydative cytolysis. *J. Biol. Chem. 257*, 1231-1237.

Bara J, Lallier R, Brailovsky C, Nigam VN (1973a) Fixation of a *Salmonella minnesota* R-form glycolipid on the membrane of normal and transformed rat-embryo fibroblasts. *Eur. J. Biochem. 35*, 489-494.

Bara J, Lallier R, Trudel M, Brailovsky C, Nigam VN (1973b) Molecular models on the insertion of a *Salmonella minnesota* R-form glycolipid into the cell membrane of normal and transformed cells. *Eur. J. Biochem. 35*, 495-498.

Beck M (1911) Versuche über Mausekrebs. *Z. Krebsforsch. 10*, 149-154.

Beebe SP, Tracy M (1907) The treatment of experimental tumors with bacterial toxins. *J. Am. Med. Assoc. 49*, 1493-1498.

Behling UH, Nowotny A (1977) Immune adjuvancy of lipopolysaccharide and a nontoxic hydrolytic product demonstrating oscillating effects with time. *J. Immunol. 118*, 1905-1907.

Behling UH, Nowotny A (1978) Long-term adjuvant effect of bacterial endotoxins in prevention and restoration of radiation-caused immunosuppression. *Proc. Soc. Exp. Biol. Med. 157*, 348-353.

Behling UH, Nowotny A (1980) Cyclic changes of positive and negative effect of single endotoxin injection. In: Agarwal MK (Ed), *Bacterial Endotoxins and Host Response*, pp 11-26. Elsevier/North-Holland Biomedical Press, Amsterdam.

Behling UH, Nowotny A (1982) Bacterial endotoxins as modulators of specific and nonspecific immunity. In: Hiernaux J, DeLisi C (Eds), *Regulatory Implication of Oscillatory Dynamics in the Immune Response* Chemical Rubber Company Press, Cleveland, Ohio.

Bendinelli M (1968) Haemolytic plaque formation by mouse peritoneal cells and the effect on it of Friend virus infection. *Immunology 14*, 837-850.

Bendinelli M, Matteuci D, Friedman H (1985) Retrovirus-induced acquired immunodeficiencies. *Adv. Cancer Res.*, in press.

Berendt MJ, Saluk PH (1976) Tumor inhibition in mice by LPS-induced peritoneal cells and an induced soluble factor. *Infect. Immun. 14*, 965-969.

Berendt MJ, Mezrow GF, Saluk PH (1978a) Requirement for a radio-sensitive lymphoid cell in the generation of lipopolysaccharide-induced rejection of a murine tumor allograft. *Infect. Immunity 21*, 1033-1035.

Berendt MJ, North RJ, Kirstein DP (1978b) The immunological basis of endotoxin-induced tumor regression. Requirement for T-cell-mediated immunity. *J. Exp. Med. 148*, 1550-1559.

Berendt MJ, North RJ, Kirstein DP (1978c) The immunological basis of endotoxin-induced tumor regression. Requirement for a pre-existing state of concomitant anti-tumor immunity. *J. Exp. Med. 148*, 1560-1569.

Berry LJ (1971) Metabolic effects of bacterial endotoxins. In: Kadis S, Weinbaum G, Ajl SJ (Eds), *Microbial Toxins Vol 5: Bacterial Endotoxins*, pp 165-208. Academic Press, New York.

Berry LJ (1978) The mediation of endotoxemic effects. In: Rosenberg P (Ed), *Toxins: Animal, Plant and Microbial*, pp 869-887. Pergamon Press, Oxford-New York.

Bloksma N, Hofhuis FM, Willers JMN (1982a) Effect of adrenoceptor blockade on hemorrhagic necrosis of Meth A sarcomata induced by endotoxin or tumor necrosis serum. *Immunopharmacology 4*, 163-171.

Bloksma N, Hofhuis FM, Benaisse-Trouw B, Willers JMN (1982b) Endotoxin induced release of tumor necrosis factor and interferon *in vivo* is inhibited by prior adrenoceptor blockade. *Cancer Immun. Immunother. 14*, 41-45.

Bober LA, Kranepool MJ, Hollander VP (1976a) Inhibitory effect of endotoxin on growth of plasma-cell tumor. *Cancer Res. 36*, 927-929.

Bober LA, Kranepool MJ, Bojko C, Steiner G, Hollander VP (1976b) Endotoxin enhancement of plasma cell tumor development in mice given injections of mineral oil. *Cancer Res. 36*, 1947-1949.

Boivin A, Mesrobeanu L (1937) Recherches sur les toxines des bacilles dysentériques. Sur l'existence d'un principe toxique thermolabile et neurotrope dans les corps bactériens du bacille de Shiga. *C. R. Soc. Biol. 126*, 222-225.

Boivin A, Mesrobeanu J, Mesrobeanu L (1933) Extraction d'un complexe toxique et antigénique à partir du bacille d'Aertrycke. *C. R. Soc. Biol. 114*, 307-310.

Bona CA (1971) La pinocytose de toxines bactériennes par les macrophages et les implications immunologiques. *Arch. Biol. (Liège) 82*, 323-392.

Bona C (1973) Fate of endotoxin in macrophages: biological and ultrastructural aspects. *J. Infect. Dis. 128*, S74-81.

Bona C, Anteunis A, Robineaux R, Astesano A (1969) Etude radioautographique ultrastructurale du transfert in vitro de l'ARN immunogène produit par les crophages qui ont capté l'endotoxine de Salmonella typhimurium 'S'. *C. R. Acad. Sci. D 269*, 1145-1147.

Bona C, Chedid L, Lamensans A (1971) In vitro attachment of radioactive endotoxins to lysosomes. *Infect. Immun. 4*, 532-536.

Bona C, Anteunis A, Robineaux R, Astesano A (1972) Transfer of antigenic macromolecules from macrophages to lymphocytes. I. Autoradiographic and quantitative study of (14C)endotoxin and (125)haemocyanin transfer. *Immunology 23*, 799-816.

Bona C, Juy D, Truffa-Bachi P, Kaplan GJ (1976) Binding, capping and internalization of lipopolysaccharide in thymic and non-thymic lymphocytes of mouse – biological and autoradiographic study. *J. Microsc. Biol. Cell 25*, 47-56.

Bordet P (1931) Contribution à l'étude de l'allergie non-spécifique. *C. R. Soc. Biol. (Paris)*

106, 1251-1254.

Bradfield JWB, Whitmarsh-Everiss T, Palmer DB, Payne R, Symes MO (1980) Hyperphagocytosis and the effect of lipopolysaccharide injection in tumour-bearing mice. *Br. J. Cancer 42*, 900-907.

Brailovsky C, Trudel M, Lallier R, Nigam V (1973) Growth of normal and transformed rat embryo fibroblasts: effect of glycolipids from *Salmonella minnesota* R mutants. *J. Cell Biol. 57*, 124-132.

Braun W (1973) Immunologic and antineoplastic effects of endotoxin: role of membranes and mediation by cyclic adenosine-3′,5′-monophosphate. *J. Infect. Dis. 128*, S188-S197.

Bruley-Rosset M, Florentin I, Khalil AM, Mathé G (1976a) Nonspecific macrophage activation by systemic adjuvants. Evaluation by lysosomal enzyme and *in vitro* tumoricidal activities. *Int. Arch. Allergy Appl. Immunol. 51*, 594-607.

Bruley-Rosset M, Florentin I, Mathé G (1976b) In vivo and in vitro macrophage activation by systemic adjuvants. *Agents Actions 6*, 251-255.

Bruns P (1888) Die Heilwirkung des Erysipelas auf Geschwülste. *Beitr. Klin. Chir. 3*, 443-466.

Burgess AW, Camakaris J, Metcalf D (1977) Purification and properties of colony-stimulating factor from mouse lung-conditioned medium. *J. Biol. Chem. 252*, 1998-2003.

Busch W (1866) in 'Verhandlungen ärztlicher Gesellschaften'. *Berl. Klin. Wochenschr. 3*, 245-246.

Busch W (1868) In: 'Verhandlungen ärtzlicher Gesellschaften'. *Berl. Klin. Wochenschr. 5*, 137-138.

Butler RC (1983) Enhancement of nonspecific resistance to tumor by endotoxin. In: Nowotny A (Ed), *Beneficial Effects of Endotoxin*, pp 497-512. Plenum Publishing Company, New York.

Butler RC, Friedman H (1979) Leukemia virus-induced immunosuppression: reversal by subcellular factors. *Ann. N.Y. Acad Sci. 332*, 446-450.

Butler RC, Nowotny A (1976) Colony stimulating factor (CSF) containing serum has antitumor effect. *IRCS Med. Sci. 4*, 206.

Butler RC, Nowotny A (1979) Combined immunostimulation in the prevention of tumor take in mice using endotoxins, their derivatives and other immune adjuvants. *Cancer Immunol. Immunother. 6*, 255-262.

Butler RC, Abdelnoor AM, Nowotny A (1978) Bone marrow colony-stimulating factor and tumor resistance-enhancing activity of postendotoxin mouse sera. *Proc. Natl Acad. Sci. USA 75*, 2893-2896,

Butler RC, Nowotny A, Friedman H (1979) Macrophage factors that enhance the antibody response. *Ann. N.Y. Acad. Sci. 332*, 564-578.

Butler RC, Friedman H, Nowotny A (1980) Restoration of depressed antibody responses of leukemic splenocytes treated with LPS-induced factors. *Adv. Exp. Med. Biol. 121A*, 315-322.

Buxton A (1959) The in vivo sensitization of avian erythrocytes with *Salmonella gallinarum* polysaccharide. *Immunology 2*, 203-210.

Cameron DJ, Churchill WH (1980) Cytotoxicity of human macrophages for tumor cells: enhancement by bacterial lipopolysaccharides (LPS). *J. Immunol. 124*, 708-712.

Carminati V (1933) Influence of lactic acid bacilli on mouse cancer (in Italian). *Boll. Ist. Sieroter. Milan. 12*, 205-220.

Carswell EA, Old LJ, Kassel RL, Green S, Fiore N, Williamson B (1975) An endotoxin-induced serum factor that causes necrosis of tumors. *Proc. Natl Acad. Sci. USA 72*, 3666-

3670.

Cater DB, Grigson CMB, Watkinson DA (1962) Changes of oxygen tension in tumors induced by vasoconstrictor and vasodilator drugs. *Acta Radiol. 58*, 401-434.

Cater DB, Schoeniger EL, Watkinson DA (1963) Effect of breathing high pressure oxygen upon tissue oxygen tension in rat and mouse tumours. *Acta Radiol. Ther. Phys. Biol. 1*, 233-252.

Cater DB, Petrie A, Watkinson DA (1964) Effect of 5-hydroxytryptamine and cyproheptadine on tumour blood flow. Estimation by rate of cooling after microwave diathermy. *Acta Radiol. Ther. Phys. Biol. 3*, 109-128.

Cater DB, Adair HM, Grove CA (1966) Effects of vasomotor drugs and 'mediators' of the inflammatory reaction upon the oxygen tension of tumors and tumor blood-flow. *Br. J. Cancer 20*, 504-525.

Chang H, Thompson JJ, Nowotny A (1974) Release of colony stimulating factor (CSF) by non-endotoxic breakdown products of bacterial lipopolysaccharides. *Immunol. Commun. 3*, 401-409.

Chapman HA Jr, Hibbs JB Jr (1977) Modulation of macrophage tumoricidal capability by components of normal serum: a central role for lipid. *Science 197*, 282-285.

Chapman HA Jr, Vavrin Z, Hibbs JB Jr (1979) Modulation of plasminogen activator secretion by activated macrophages: influence of serum factors and correlation with tumoricidal potential. *J. Reticuloendothel Soc. 26*, 21-30.

Chen CH, Nowotny A (1974) Direct determination of molar ratios of various chemical constituents in endotoxic glycolipids in silicic acid scrapings from thin-layer chromatographic plates. *J. Chromatogr. 97*, 39-45.

Chen CH, Johnson AG, Kasai N, Key BA, Levin J, Nowotny A (1973) Heterogeneity and biological activity of endotoxic glycolipid from *Salmonella minnesota* R595. *J. Infec. Dis. 128*, S43-51.

Chen CH, Chang CM, Nowotny AM, Nowotny A (1975) Rapid biological and chemical analyses of bacterial endotoxins separated by preparative thinlayer chromatography. *Anal. Biochem. 63*, 183-194.

Chester IR, Meadow PM (1975) Heterogeneity of lipopolysaccharide from *Pseudomonas aeruginosa*. *Eur. J. Biochem. 58*, 273-282.

Cheung HT, Cantarow WD, Sundharadas G (1979) Tumoricidal activity of macrophages induced by lipopolysaccharide and its inhibition by a low molecular weight factor extracted from tumors. *J. Reticuloendothel. Soc. 26*, 21-30.

Chiller JM, Skidmore BJ, Morrison DC, Weigle WO (1973) Relationship of the structure of bacterial lipopolysaccharide to its function in mitogenesis and adjuvanticity. *Proc. Natl Acad. Sci. USA 70*, 2129-2133.

Chow DA, Wolosin LB, Greenberg AH (1981) Genetics, regulation, and specificity of murine natural antitumor antibodies and natural killer cells. *J. Natl Cancer Inst. 67*, 445-453.

Coley WB (1893) The treatment of malignant tumors by repeated inoculations of erysipelas; with a report of ten original cases. *Am. J. Med. Sci. 105*, 487-511.

Coley WB (1898) The treatment of inoperable sarcoma with the mixed toxins of erysipelas and *Bacillus prodigiosus*; immediate and final results in 140 cases. *J. Am. Med. Assoc. 31*, 389-395; 456-465.

Contreras Ma De Los Angeles, Bale WF (1968) Endotoxin, epinephrine, and ellagic acid effects on the radiation-sensitized Walker 256 rat carcinosarcoma. *Radiat. Res. 36*, 166-179.

Creech HJ, Hamilton MA, Diller IC (1948a) Comparative studies of the immunological, toxic and tumor-necrotizing properties of polysaccharides from *Serratia marcescens* (*Bacillus prodigiosus*). *Cancer Res. 8*, 318-329.

Creech HJ, Hamilton MA, Nishimura ET, Hankwitz RF Jr (1948b) The influence of antibody-containing fractions on the lethal and tumornecrotizing actions of polysaccharides from *Serratia marcescens* (*Bacillus prodigiosus*) *Cancer Res. 8*, 330-336.

Creech HJ, Hankwitz RF Jr, Wharton DRA (1949) Further studies of the immunological properties of polysaccharides from *Serratia marcescens* (*Bacillus prodigiosus*) I. The effects of passive and active immunization of the lethal activity of the polysaccharides. *Cancer Res. 9*, 150-157.

Cruse JM, Forbes JT, Gillespie GY, Lewis GK (1971) Effect of endotoxin tolerance on tumor allograft enhancement in mice. *Am. J. Pathol. 62*, A-65.

Currie GA, Basham C (1975) Activated macrophages release a factor which lyses malignant cells but not normal cells. *J. Exp. Med. 142*, 1600-1605.

Dachy A (1967) De la sensibilité à l'action hémorragipare des endotoxines. *Rev. Immunol. (Paris) 31*, 133-292.

Davies M, Stewart-Tull DES, Jackson DM (1978) The binding of lipopolysaccharide from *Escherichia coli* to mammalian cell membranes and its effect on liposomes. *Biochim. Biophys. Acta 508*, 260-276,

Deidier A (1725) *Dissertation Médecinal et Chirugical sur les Tumeurs.* C.-M. d'Houry, Paris.

Dent PB (1972) Immunodepression by oncogenic viruses. *Prog. Med. Virol. 14*, 1-35.

Didot A (1851) Essai sur la prophylaxie du cancer par la syphilization artificielle. *Bull. Acad. R. Med. Belg. 11*, 100-172.

DiRienzo JM, MacLeod RA (1978) Composition of the fractions separated by polyacrylamide gel electrophoresis of the LPS of a marine bacterium. *J. Bacteriol. 136*, 158-167.

Doe WF, Henson PM (1978) Macrophage stimulation by bacterial lipopolysaccharides. I. Cytolytic effect on tumor target cells. *J. Exp. Med. 148*, 544-555.

Doe WF, Henson PM (1979) Macrophage stimulation by bacterial LPS. III. Selective unresponsiveness of C3H/HeJ macrophages to the lipid A differentiation signal. *J. Immunol. 123*, 2304-2310.

Doe WF, Yang ST, Morrison DC, Betz SJ, Henson PM (1978) Macrophage stimulation by bacterial lipopolysaccharides. II. Evidence for differentiation signals delivered by lipid A and by a protein rich fraction of lipopolysaccharides. *J. Exp. Med. 148*, 557-566.

Drysdale BE, Shin HS (1981) Activation of macrophages for tumor cell cytotoxicity: identification of indomethacin sensitive and insensitive pathways. *J. Immunol. 127*, 760-765.

Dupré de Lisle (1774) *Traité, sur le Vice Cancereux.* Couturier, Paris.

Duran-Reynals F (1933) Reaction of transplantable and spontaneous tumors to blood-carried bacterial toxin in animals unsusceptible to the Shwartzman phenomenon. *Proc. Soc. Exp. Biol. Med. 31*, 341-344.

Duran-Reynals F (1935) Reaction of spontaneous mouse carcinoma to blood-carried bacterial toxins. *Proc. Soc. Exp. Biol. Med. 32*, 1517-1521.

Dye ES, North RJ (1980) Macrophage accumulation in murine ascites tumors. I. Cytoxan-induced dominance of macrophages over tumor cells and the anti-tumor effect of endotoxin. *J. Immunol. 125*, 1650-1657.

Evans R, Alexander P (1970) Cooperation of immune lymphoid cells with macrophages in tumour immunity. *Nature (London) 228*, 620-622.

Evans R, Alexander P (1976) Mechanism of extracellular killing of nucleated mammalian cells by macrophages. In: Nelson DS (Ed), *Immunobiology of the Macrophage*, pp 536-576. Academic Press, New York.

Fehleisen F (1882) Uber die Züchtung der Erysipelkokken auf künstlichem Nährboden und die Ubertragbarkeit auf den Menschen. *Dtsch. Med. Wochenschr. 8*, 553-554.

Fogg LC (1936) Effect of certain bacterial products upon the growth of mouse tumor. *Public Health Rep. 51*, 56-64.

Foley EJ (1952) Immunity of C3H mice to lymphosarcoma 6C3H-Ed following regression of the implanted tumor. *Proc. Soc. Exp. Biol. Med. 80*, 675-677.

Fowler GA (1969a) Enhancement of natural resistance to malignant melanoma with special reference to the beneficial effects of concurrent infections or bacterial toxin therapy. End results in 50 cases. *N.Y. Cancer Res. Inst. Monogr. 9*.

Fowler GA (1969b) Apparently beneficial effects of acute bacterial infections, inflammation or bacterial toxin therapy on cancer of the colon or rectum. *N.Y. Cancer Res. Inst. Monogr. 10*.

Fowler GA, Nauts HC (1970) The apparently beneficial effects of concurrent infections, inflammation, fever, and of bacterial toxin therapy on neuroblastoma. *N.Y. Cancer Res. Inst. Monogr. 11*.

Frank S, Specter S, Nowotny A, Friedman H (1977) Immunocyte stimulation in vitro by nontoxic bacterial lipopolysaccharide derivatives. *J. Immunol. 119*, 855-869.

Franzl RE, McMaster PD (1968) The primary immune response in mice. I. The enhancement and suppression of hemolysin production by a bacterial endotoxin. *J. Exp. Med. 127*, 1087-1107.

Freund J (1934) Hemorrhages in tuberculous guinea pigs at the site of injection of irritans following intravascular injections of injurious substances (Shwartzman phenomenon). *J. Exp. Med. 60*, 669-685.

Friedman H, Ceglowski WS (1971) Immunosuppression by tumor viruses. *Progr. Immunol. 1*, 815.

Friedman H, Kateley JR (1975) Lymphocyte surface receptors and leukemia virus-induced immunosuppression. *Am. J. Clin. Pathol. 63*, 735-747.

Friedman H, Spector S, Butler RC, Nowotny A (1980) Humoral factors in the adjuvant effect of lipopolysaccharides. In: Schlessinger D (Ed), *Microbiology – 1980*, pp 44-48. American Society for Microbiology, Washington D.C.

Gaal D, Nowotny A (1979) Immune enhancement by tumor therapeutic drugs and endotoxin. *Cancer Immunol. Immunother. 6*, 9-15.

Gangemi JD, Ghaffar A, Trauger RL, Sigel MM (1980) Natural killer cell activation in lipopolysaccharide-responsive and -nonresponsive mice by viral and bacterial agents. *J. Reticuloendothel. Soc. 27*, 525-533.

Garfield DJ, Rebers RA, Heddleston KL (1976) Immunogenic and toxic properties of a purified lipopolysaccharide-protein complex from *Pasteurella multocida. Infect. Immun. 14*, 990-999.

Gaynor E (1973) The role of granulocytes in endotoxin-induced vascular injury. *Blood 61*, 797-807.

Gerber IE, Bernheim AI (1938) Morphologic study of the reactivity of mouse sarcoma 180 to bacterial filtrates. *Arch. Pathol. 26*, 971-983.

Gery I, Kruger GJ, Spiesel SZ (1972) Reactions of thymus-deprived mice and karyotypic analysis of dividing cells in mice bearing T_6T_6-thymus grafts. *J. Immunol. 108*, 1088-1091.

Ginsburg I (1970) Streptolysin S. In: Montie TC, Kadis S, Ajl SJ (Eds), *Microbial Toxins Vol 3: Bacterial Protein Toxins*, pp 100-171. Academic Press, New York.

Glaser M, Djeu JY, Kirchner H, Herberman RB (1976) Augmentation of cell-mediated cytotoxicity against syngeneic Gross virus-induced lymphoma in rats by phytohemagglutinin and endotoxin. *J. Immunol. 116*, 1512-1519.

Goldman RC, Leive L (1980) Heterogeneity of antigenic-side-chain length in lipopolysaccharide from *Escherichia coli* O111 and *Salmonella typhimurium* LT2. *Eur. J. Biochem. 107*, 145-153.

Gorecka-Tisera A, Proctor JW, Yamamura Y, Harnaha J, Meinert K (1981) Dose, route, and time dependence of serum lysozyme and antitumor activity following administration of glucan, *Corynebacterium parvum*, pyran, or lipopolysaccharide to mice. *J. Natl Cancer Inst. 67*, 911-915.

Gormus BJ, Crandall RB, Shands JW Jr (1974) Endotoxin-stimulated spleen cells: mitogenesis, the occurrence of the C3 receptor and the production of immunoglobulin. *J. Immunol. 112*, 770-775.

Grant CK, Alexander P (1974) Nonspecific cytotoxicity of spleen cells and the specific cytotoxic action of thymus-derived lymphocytes *in vitro. Cell. Immunol. 14*, 46-51.

Gratia A, Linz R (1931) Le phénomène de Shwartzman dans le sarcome du cobaye. *C. R. Soc. Biol. (Paris) 108*, 427-428.

Green S, Carswell E, Old LJ, Fiore N, Mamaril F, Dobrjansky A, Schwartz MK (1974) Mechanism of endotoxin-induced tumor hemorrhagic necrosis. *Proc. Am. Ass. Cancer Res. 15*, 139.

Green S, Dobrjansky A, Chiasson MA, Carswell EA, Helson L, Schwartz MK, Old LJ (1976a) Necrosis of meth A tumors by a factor from liver of *C. parvum* (CP)-endotoxin treated mice. *Proc. Am. Ass. Cancer Res. 17*, 84.

Green S, Dobrjansky A, Carswell EA, Kassel RL, Old LJ, Fiore N, Schwartz MK (1976b) Partial purification of a serum factor that causes necrosis of tumors. *Proc. Natl Acad. Sci. USA 73*, 381-385.

Green S, Dobrjansky A, Chiasson MA, Carswell E, Schwartz MK, Old LJ (1977) *Corynebacterium parvum* as the priming agent in the production of tumor necrosis factor in the mouse. *J. Natl Cancer Inst. 59*, 1519-1522.

Gresser I (1972) Antitumor effects of interferon. *Adv. Cancer Res. 16*, 97-140.

Gresser I, Tovey MG (1978) Antitumor effects of interferon. *Biochim. Biophys. Acta 516*, 231-247.

Gresser I, Bourali C, Levy JP (1969) Increased survival in mice inoculated with tumor cells and treated with interferon preparations. *Proc. Natl Acad. Sci. USA 63*, 51-56.

Grohsman J, Nowotny A (1972) The immune recognition of TA3 tumors, its facilitation by endotoxin, and abrogation by ascites fluid. *J. Immunol. 109*, 1090-1095.

Hager EB, Bailey GL, Hampers CL, Merrill JP (1969) Immunity against the Ehrlich ascites carcinoma produced by bacterial endotoxins. *Transplant. Proc. 1*, 124-125.

Hammerström J (1979) In vitro influence of endotoxin on human mononuclear phagocyte structure and function. 2. Enhancement of the expression of cytostatic and cytolytic activity of normal and lymphokine-activated monocytes. *Acta Pathol. Microbiol. Scand. Sect. C, 87*, 1-9.

Hammerström J, Unsgaard G (1979) In vitro influence of endotoxin on human mononuclear phagocyte structure and function. I. Depression of protein synthesis, phagocytosis of *Candida albicans* and induction of cytostatic activity. *Acta Pathol. Microbiol. Scand. Sect. C, 87*, 384-389.

Hansen JG, Bennedsen J, Rhodes JM, Larsen SO (1982) In vitro studies on the tumour cytotoxicity of normal, stimulated and immunologically activated mouse macrophages.

Acta Pathol. Microbiol. Immunol. Scand. Sect. C 90, 27-32.

Hartman D, Proctor JW, Lewis MG, Lyons H (1974) In vitro interactions between antitumour antibodies and anti-antibodies in malignancy. *Lancet 2*, 1481-1483.

Hartwell JL, Shear MJ, Kahler H, Turner FC (1941) Properties of the fraction of *B. prodigiosus* which produced hemorrhage in mouse sarcomas. *Cancer Res. 1*, 741.

Hartwell JL, Shear MJ, Adams JR Jr (1943a) Chemical treatment of tumors. VII. Nature of the hemorrhage-producing fraction from *Serratia marcescens* (*Bacillus prodigiosus*). *J. Natl Cancer Inst. 4*, 107-122.

Hartwell JL, Shear MJ, Adams JR Jr (1943b) Nature of the bacterial polysaccharide which produces hemorrhage in mouse sarcoma. *Cancer Res. 3*, 122.

Hege JS, Cole LJ (1967) Antibody plaque-forming cells in unsensitized mice. Specificity and response to neonatal thymectomy, X-irradiation and PHA. *J. Immunol. 99*, 61-70.

Hellström I, Hellström KE, Pierce GE, Yang PJ (1968) Cellular and humoral immunity to different types of human neoplasms. *Nature (London) 220*, 1352-1354.

Helson L, Green S, Carswell E, Old LJ (1975) Effect of tumour necrosis factor on cultured human melanoma cells. *Nature (London) 258*, 731-732.

Henderson JS (1968) Endotoxin as adjuvator to the transplantation of a mouse mammary tumor. *J. Exp. Med. 128*, 1363-1375.

Herberman RB (1974) Delayed hypersensitivity skin reactions to antigens on human tumors. *Cancer 34*, 1469-1473.

Herberman RB (1977) Immunogenicity of tumor antigens. *Biochim. Biophys. Acta 473*, 93-119.

Herberman RB (1982) Overview on NK cells and possible mechanisms for their cytotoxic activity. *Adv. Exp. Med. Biol. 146*, 337-351.

Herberman RB, Holden HT (1978) Natural cell-mediated immunity. *Adv. Cancer Res. 27*, 305-377.

Hibbs JB Jr, Taintor RR, Chapman HA Jr, Weinberg JB (1977) Macrophage tumor killing: influence of the local environment. *Science 197*, 279-282.

Hibbs JB, Weinberg JB, Chapman HA (1979) Modulation of the tumoricidal function of activated macrophages by bacterial endotoxin and mammalian macrophage activation factor(s). *Adv. Exp. Med. Biol. 121B*, 433-453.

Hinshaw LB (1964) The release of vasoactive agents by endotoxin. In: Landy M, Braun W (Eds), *Bacterial Endotoxins*, pp 118-125. Rutgers University Press. New Brunswick, NJ.

Hinterberger W, Paukovits WR, Mittermayer K, Wallner U (1979) Endotoxin-induced colony stimulating activity in normal and myeloid leukaemic subjects. *Scand. J. Haematol. 22*, 280-287.

Hinterberger W, Graninger W, Paukovits WR (1983) Colony stimulating activity induction in patients with acute leukemia. In: Nowotny A (Ed), *Beneficial Effects of Endotoxin*, pp 415-432. Plenum Publishing Company, New York.

Ho M (1964) Interferon-like viral inhibitor in rabbits after intravenous administration of endotoxin. *Science 146*, 1472-1474.

Ho M (1983) Induction of interferon by endotoxin. In: Nowotny A (Ed), *Beneficial Effects of Endotoxin*, pp 381-395. Plenum Publishing Company, New York.

Hoffmann F (1675) *Opera Omnia*, p 100.

Hoffmann MK, Green S, Old LJ, Oettgen HF (1976) Serum containing endotoxin-induced tumour necrosis factor substitutes for helper T cells. *Nature London 263*, 416-417.

Hoffmann MK, Oettgen HF, Old LJ, Mittler RS, Hämmerling U (1978) Induction and immunological properties of tumor necrosis factor. *J. Reticuloendothel. Soc. 23*, 307-318.

Jacobi M (1936) The effect of the Shwartzman reaction with bacterial filtrate on transplantable tumors in animals. *Am. J. Cancer 26*, 770-774.

Jann B, Reske K, Jann K (1975) Heterogeneity of lipopolysaccharides. Analysis of polysaccharide chain lengths by sodium dedecylsulfate-polyacryl-amide gel electrophoresis. *Eur. J. Biochem. 60*, 239-246.

Jarrell K, Kropinski AM (1977) The chemical composition of the LPS from *Pseudomonas aeruginosa* strain PAO and a spontaneously derived rough mutant. *Microbios 19*, 103-116.

Johnson AG, Nowotny A (1964) Relationship of structure to function in bacterial O-antigens. III. Biological properties of endotoxoids. *J. Bacteriol. 87*, 809-814.

Johnston RB Jr (1978) Oxygen metabolism and the microbicidal activity of macrophages. *Fed. Proc. 37*, 2759-2764.

Johnston RB, Keele BB Jr, Misra H-P, Lehmeyer JE, Webb LS, Baehner RL, Rajagopalan KV (1975) The role of superoxide anion generation in phagocytic bactericidal activity. Studies with normal and chronic granulomatous disease leucocytes. *J. Clin. Invest. 55*, 1357-1372.

Jones GRN (1977) Cancer: energy uncoupling as a primary event in endotoxin-induced necrosis of the S180 sarcoma in situ. *Biochem. Soc. Trans. 5*, 214-216.

Kabir S (1976) Electrophoretic studies of smooth and rough Salmonella lipopolysaccharides on polyacrylamide gel. *Microbios Let. 1*, 79-81.

Kahler H (1943) Chemical treatment of tumors. VIII. Ultracentrifugal and electrophoretic analysis of the hemorrhage-producing fraction from *Serratia marcescens* (*Bacillus prodigiosus*) culture filtrate. *J. Natl Cancer Inst. 4*, 123-129.

Kaliss N (1962) The elements of immunologic enhancement: a consideration of mechanisms. *Ann. N.Y. Acad. Sci. 101*, 64-79.

Kaliss N, Bryant BR (1958) Factors determining homograft destruction and immunological enhancement in mice receiving successive tumor inocula. *J. Natl Cancer Inst. 20*, 691-704.

Kamo I, Kateley J, Friedman H (1975a) Mastocyloma-induced suppression of in vitro antibody formation. *Proc. Soc. Exp. Biol. Med. 148*, 883-886.

Kamo I, Patel C, Kateley J, Friedman H (1975b) Immunosuppression induced in vitro by mastocytoma tumor cells and cell-free extracts. *J. Immunol. 114*, 1749-1756.

Kasai N, Nowotny A (1967) Endotoxic glycolipid from a heptose-less mutant of Salmonella minnesota. *J. Bacteriol. 94*, 1824-1836.

Kassel RL, Old LJ, Carswell EA, Fiore NC, Hardy WD Jr (1973) Serum-mediated leukemia cell destruction in AKR mice. *J. Exp. Med. 138*, 925-938.

Kato N, Nakashima I, Ohta M, Naito S, Kojima T (1979a) Interferon and cytotoxic factor (cytotoxin) released in the blood of mice infected with Mycobacterium bovis BCG. I. Enhanced production of interferon and appearance of cytotoxin stimulated by capsular polysaccharide of *Klebsiella pneumoniae* or lipopolysaccharide. *Microbiol. Immunol. 23*, 383-394.

Kato N, Nakashima I, Ohta M, Naito S (1979b) Interferon and cytotoxic factor (cytotoxin) released in the blood of mice infected with Mycobacterium bovis BCG. II. Influence of time after BCG inoculation on production of interferon and cytotoxin by capsular polysaccharide of Klebsiella pneumoniae or by bacterial lipopolysaccharide and on hyperreactivity to their lethal effects. *Microbiol. Immunol. 23*, 395-402.

Kato N, Nakashima I, Ohta M, Nagase F, Yokochi T, Naito S (1980) Interferon and cytotoxic factor (cytotoxin) released in the blood of mice infected with Mycobacterium bovis BCG. III. Interferon and cytotoxin induced by the specific antigen as compared with those induced by bacterial lipopolysaccharide. *Microbiol. Immunol. 24*, 1043-1051.

Kawakami M, Pekala PH, Lane MD, Cerami A (1982) Lipoprotein lipase suppression in 3T3-L1 cells by an endotoxin-induced mediator from exudate cells. *Proc. Natl Acad. Sci. USA 79*, 912-916.

Kearney R, Harrop P (1980) Potentiation of tumor growth by endotoxin in serum from syngeneic tumor-bearing mice. *Br. J. Cancer 42*, 559-567.

Kedar E, Nahas F, Unger E, Weiss DW (1978) In vitro induction of cell-mediated immunity to murine leukemia cells. III. Effect of methanol extraction residue fraction of BCG on the generation of cytotoxic lymphocytes against leukemia. *J. Natl Cancer Inst. 60*, 1097-1106.

Keogh EV, North EA, Warburton MF (1948) Adsorption of bacterial polysaccharides to erythrocytes. *Nature (London) 161*, 687-688.

Kihlström E (1980) The effects of lipopolysaccharides on the association of *Salmonella typhimurium* with HeLa cells. *Scand. J. Infect. Dis. 24*, 141-143.

Kilpatric-Smith L, Erecinska M, Silver IA (1981) Early cellular responses in vitro to endotoxin administration. *Circ. Shock 8*, 585-600.

Kim YB, Watson DW (1967) Biologically active endotoxins from Salmonella mutants deficient in O- and R-polysaccharides and heptose. *J. Bacteriol. 94*, 1320-1326.

Kimball HR, Williams TW, Wolff SM (1968) Effect of bacterial endotoxin on experimental fungal infections. *J. Immunol. 100*, 24-33.

Kind P, Johnson AG (1959) Studies on the adjuvant action of bacterial endotoxins on antibody formation. *J. Immunol. 82*, 415-427.

Kondo M, Kato H, Yoshikawa T, Katano M, Torisu M (1982) Clinical evaluation of anticancer activity of a streptococcal preparation OK-432 (Picibanil). In: Jeljaszewicz J, Pulverer G, Roszkowski W (Eds), *Bacteria and Cancer*, pp 415-434. Academic Press, London.

Kull FC Jr, Cuatrecasas P (1981) Preliminary characterization of the tumor cell cytotoxin in tumor necrosis serum. *J. Immunol. 126*, 1279-1283.

Kuper CF, Bloksma N, Hofbuis FM, Bruyntjes JP, Willers JM (1982) Influence of adrenoceptor blockade on endotoxin-induced histopathological changes in murine Meth A sarcoma. *Int. J. Immunopharmacol. 4*, 49-55.

Kurashige S, Mitsuhashi S (1973) Macrophage activities in sarcoma 180-bearing mice and EL4-bearing mice. *Gann 73*, 85-90.

Lassar O (1891) Zur Erysipelimpfung. *Dtsch. Med. Wochenschr. 17*, 898-899.

LeDur A, Chaby R, Szabo L (1980) Isolation of two protein-free and chemically different lipopolysaccharides from *Bordetella pertussis* phenol-extracted endotoxin. *J. Bacteriol. 143*, 78-88.

Lee KC, Wong M, McIntyre D (1981) Characterization of macrophage subpopulations responsive to activation by endotoxin and lymphokines. *J. Immunol. 126*, 2474-2479.

LeGarrec Y, Martin A, Collet B, Toujas L, Pilet C (1981) Antitumor properties of chemically detoxified killed *Brucella abortus* organisms. *Cancer Immunol. Immunother. 11*, 63-68.

Leive L (1977) Domains involving nonrandom distribution of lipopolysaccharide in outer membrane of *Escherichia coli*. *Proc. Natl Acad. Sci. USA 74*, 5065-5068.

Lewis MG, Ikonopison RL, Nairn RC (1969) Tumor-specific antibodies in human malignant melanoma and their relationship to extent of disease. *Br. Med. J. 3*, 547-552.

Lewis MG, Hartman D, Jerry LM (1976) Antibodies and anti-antibodies in human malignancy: expression of deranged immune regulation. *Ann. N.Y. Acad. Sci. 2*, 316-327.

Liacopoulos P (1983) Effect of endotoxin on graft-vs-host reactions. In: Nowotny A (Ed), *Beneficial Effects of Endotoxins*, pp 433-450. Plenum Publishing Company, New York.

Liacopoulos P, Merchant B, Harrel BE (1967a) Effect of donor immunization with somatic polysaccharides on the graft-versus-host reactivity of transfered donor splenocytes. *Proc. Soc. Exp. Biol. Med. 125*, 958-962.

Liacopoulos P, Merchant B, Harrel BE (1967b) Inhibition of the graft-versus-host reaction by pretreatment of donors with various antigens. *Transplantation 5*, 1423-1435.

Lüderitz O, Galanos C, Risse HJ, Ruschman E, Schlecht S, Schmidt G, Schulte-Holthausen H, Wheat R, Westphal O (1966) Structural relationships of *Salmonella* O and R antigens. *Ann. N.Y. Acad. Sci. 133*, 349-374.

MacKenzie GM, Pike RM, Swinney RE (1940) Virulence of Salmonella typhimurium. II. Studies of the polysaccharide antigens of virulent and avirulent strains. *J. Bacteriol. 40*, 197-214.

Mamaril FP, Liu RY, Schwartz MK, Old LJ, Green S (1974) Release of NADase into circulation of BCG-infected mice after treatment with endotoxin. *Fed. Proc. 33*, 1308.

Männel DN, Rosenstreich DL, Mergenhagen SE (1979) Mechanism of LPS-induced tumor necrosis: Requirement for LPS-sensitive lymphoreticular cells. *Infect. Immun. 24*, 573-576.

Männel DN, Meltzer MS, Mergenhagen SE (1980a) Generation and characterization of a LPS-induced and serum-derived cytotoxic factor for tumor cells. *Infect. Immun. 28*, 204-211.

Männel DN, Moore RN, Mergenhagen SE (1980b) Macrophages as a source of tumoricidal activity (Tumor-Necrotizing Factor). *Infect. Immun. 30*, 523-530.

Männel DN, Farrar JJ, Mergenhagen SE (1980c) Separation of a serum-derived tumoricidal factor from a helper factor for plaqueforming cells. *J. Immunol. 124*, 1106-1110.

Marino PA, Adams DO (1982) The capacity of activated murine macrophages for augmented binding of neoplastic cells: analysis of induction by lymphokine containing MAF and kinetics of the reaction. *J. Immunol. 128*, 2816-2823.

Martin F, Martin M, Jeannin JF, Lagneau A (1978) Rat macrophage-mediated toxicity to cancer cells; effect of endotoxins and endotoxin inhibitors contained in culture media. *Eur. J. Immunol. 8*, 607-611.

Mathé G (1969) Acive immunotherapy for acute lymphoblastic leukemia. *Lancet, 1*, 697-699.

McCoy JL, Jerome LF, Dean JH, Perlin E, Oldham RK, Char DH, Cohen MH, Felix EL, Herberman RB (1975) Inhibition of leukocyte migration by tumor-associated antigens in soluble extracts of human malignant melanoma. *J. Natl Cancer Inst. 55*, 19-23.

McCoy JL, Dean JH, Cannon GB, Jerome LJ, Alford TC, Parks WP, Gilden RV, Oroszlan ST, Herberman RB (1978) Leukocyte migration inhibition and lymphocyte blastogenesis responses in breast carcinoma patients to mouse mammary tumor virus and to virion gp52 antigen and Rauscher murine leukemia virus-Kirsten sarcoma virus gp69/71 antigen. *J. Natl Cancer Inst. 60*, 1259-1267.

McGhee JR, Michalek SM, Barb JL, Kiyono H, Rosenstreich DL, Mergenhagen SE (1980) Lipopolysaccharide regulation of the immune response: cellular basis of adjuvancy and suppression. In: Schlessinger D (Ed), *Microbiology – 1980*, pp 49-54.

McGrath MJ, Stewart GJ (1969) The effects of endotoxin on vascular endothelium. *J. Exp. Med. 129*, 833-839.

McLaughlin CA, Bickel WD, Kyle JS, Ribi E (1978) Synergistic tumor regressive activity observed following treatment of line-10 hepatocellular carcinomas with deproteinized

BCG cell walls and mutant *Salmonella typhimurium* glycolipid. *Cancer Immunol. Immunother. 5*, 45-52.

McMaster PD, Franzl RE (1968) The primary immune response in mice. II. Cellular responses of lymphoid tissue accompanying the enhancement or complete suppression of antibody formation by a bacterial endotoxin. *J. Exp. Med. 127*, 1109-1125.

McNeill TA (1970) Antigenic stimulation of bone marrow colony forming cells. III. Effect in vivo. *Immunology 18*, 61-72.

Meltzer MS (1981) Macrophage activation for tumor cytotoxicity: characterization of priming and trigger signals during lymphokine activation. *J. Immunol. 127*, 1-183.

Meltzer MS, Wohl LM (1979) Synergistic interaction of lymphokines (LK) and bacterial lipopolysaccharides (LPS) for macrophage tumor cytotoxicity and secretion of prostaglandin E2 (PGE2). *Fed. Proc. 38*, 3719.

Meltzer MS, Ruco LP, Leonard EJ (1980) Macrophage activation of tumor cytotoxicity: mechanism of macrophage activation by lymphokines. *Adv. Exp. Med. Biol. 121*, 381-398.

Meltzer MS, Occhionero M, Ruco IP (1982) Macrophage activation for tumor cytotoxicity: regulatory mechanisms for induction and control of cytotoxic activity. *Fed. Proc. 41*, 2198-2205.

Metcalf D (1971) Acute antigen-induced elevation of serum colony stimulating factor (CSF) levels. *Immunology 21*, 427-436.

Metcalf D (1982a) Sources and biology of regulatory factors active on mouse myeloid leukemic cells. *J. Cell Physiol. Suppl. 1*, 175-183.

Metcalf D (1982b) Regulator-induced suppression of myelomonocytic leukemic cells: clonal analysis of early cellular events. *Int. J. Cancer 30*, 203-210.

Mihich E, Westphal O, Lüderitz O, Neter E (1961) The tumor necrotizing effect of lipoid A component of Escherichia coli endotoxin. *Proc. Soc. Exp. Biol. Med. 107*, 816-819.

Morrison DC, Leive L (1975) Fractions of lipopolysaccharide from E. coli U111-B4 prepared by 2 extraction procedures. *J. Biol. Chem. 250*, 2911-2919.

Morton DL, Malmgren RA, Holmes EC, Ketcham AS (1968) Demonstration of antibodies against human malignant melanoma by immunofluorescence. *Surgery 64*, 233-240.

Möller G (1963a) Studies on the mechanism of immunological enhancement of tumor homografts. I. Specificity of immunological enhancement. *J. Natl Cancer Inst. 30*, 1153-1176.

Möller G (1963b) Studies on the mechanism of immunological enhancement of tumor homografts. II. Effect of isoantibodies on various tumor cells. *J. Natl Cancer Inst. 30*, 1177-1204.

Möller G (1963c) Studies on the mechanism of immunological enhancement of tumor homografts. III. Interaction between humoral isoantibodies and immune lymphoid cells. *J. Natl Cancer Inst. 30*, 1205-1226.

Muna NM, Marcus S, Smart C (1969) Detection by immunofluorescence of antibodies specific for human malignant melanoma cells. *Cancer 23*, 88-93.

Nakano M, Uchiyama T, Tanabe MJ, Saito K (1975) Nonspecific elicitation of antibody-forming cells in the mouse spleen by bacterial lipopolysaccharide. *Jpn. J. Microbiol. 19*, 141-148.

Nakano M, Tanabe JM, Saito T, Shimizu T (1976) Immunosuppressive effect of bacterial lipopolysaccharide on antibody response. *Jpn. J. Microbiol. 20*, 53-58.

Narayanan PR, Sundharadas G (1978) Differential effects of polyadenylic:polyuridylic acid and lipopolysaccharide on the generation of cytotoxic T lymphocytes. *J. Exp. Med. 147*,

1355-1361.

Narayanan PR, Kloehn DB, Sundharadas G (1978) Immune response to alloantigens in vitro, amplification of the development of cytotoxic T lymphocytes by lipopolysaccharide and polyadenylic:polyuridylic acid. *J. Immunol. 121*, 2502-2508.

Nathan CF (1982) Secretion of oxygen intermediates: role in effector functions of activated macrophages. *Fed. Proc. 41*, 2206-2211.

Nauts HC (1973) Enhancement of natural resistance to renal cancer: beneficial effects of concurrent infections and immunotherapy with bacterial vaccines. *N.Y. Cancer Res. Inst. Monogr. 12.*

Nauts HC (1975) Giant cell tumor of bone: end results following immunotherapy (Coley toxins) alone or combined with surgery and/or radiation (66 cases), or with concurrent infection (4 cases). *N.Y. Cancer Res. Inst. Monogr. 4 (2nd ed).*

Nauts HC (1980) The beneficial effects of bacterial infections on host resistance to cancer. End results in 449 cases. *N.Y. Cancer Res. Inst. Monogr. 8 (2nd ed).*

Nauts HC, Swift WE, Coley BL (1946) The treatment of malignant tumors by bacterial toxins as developed by the late William B. Coley, M.D., reviewed in the light of modern research. *Cancer Res. 6*, 205-216.

Neter E, Bertram LF, Arbesman CF (1952) Demonstration of *Escherichia coli* 055 and 0111 antigens by means of hemagglutination test. *Proc. Soc. Exp. Biol. Med. 79*, 255-257.

Ng AK, Chang CM, Chen CH, Nowotny A (1974) Comparison of the chemical structure and biological activities of the glycolipids of *Salmonella minnesota* R595 and *Salmonella typhimurium* SL1102. *Infect. Immun. 10*, 938-947.

Ng A-K, Butler C, Chen CH, Nowotny A (1976) Relationship of structure to function in bacterial endotoxins IX. Differences in the lipid moiety of endotoxic glycolipids. *J. Bacteriol. 126*, 511-519.

Ng A-K, Imai K, Pellegrino MA, Vitiello A, Indiveri F, Wilson BS, Ferrone S (1983) Modulation of immune lysis of tumor cells by interferon. In: Nowotny A (Ed), *Pathological Membranes*, pp 313-339. Plenum Publishing Company, New York.

Nigam VN (1975) Effect of core lipopolysaccharides from *Salmonella minnesota* R mutants on the survival times of mice bearing Ehrlich tumor. *Cancer Res. 35*, 628-633.

Nowotny A (1963) Endotoxoid preparations. *Nature (London) 197*, 721-722.

Nowotny A (1964) Chemical detoxification of bacterial endotoxins. In: Landy M, Braun W (Eds), *Bacterial Endotoxins*, pp 29-37. Institute of Microbiology, Rutgers University Press, New Brunswick, NJ.

Nowotny A (1966) Heterogeneity of endotoxic bacterial lipopolysaccharides revealed by ion exchange column chromatography. *Nature (London) 210*, 278-280.

Nowotny A (1969) Molecular aspects of endotoxic reactions. *Bacteriol. Rev. 33*, 72-98.

Nowotny A (1971) Relationship of structure and biological activity of bacterial endotoxins. *Naturwissenschaften 58*, 397-409.

Nowotny A (1977) Relation of structure to function in bacterial endotoxins. In: Schlessinger D (Ed), *Microbiology – 1977*, pp 247-252. American Society for Microbiology, Washington D.C.

Nowotny A (1983) In search of the active sites in endotoxins. In: Nowotny A (Ed), *Beneficial Effects of Endotoxin*, pp 1-55. Plenum Publishing Company, New York.

Nowotny A, Butler RC (1980) Studies on the endotoxin induced tumor resistance. *Adv. Exp. Med. Biol. 121B*, 455-469.

Nowotny A, Cundy K, Neale N, Nowotny AM, Radvany R, Thomas S, Tripodi D (1966) Relation of structure to function in bacterial O-antigens. IV. Fractionation of the components. *Ann. N.Y. Acad. Sci. 133*, 586-603.

Nowotny A, Kasai N, Tripodi D (1967) The use of mutants for the study of the relationship between structure and function in endotoxins. In: *Structure et Effets Biologiques de Produits Bactériens Provenant de Bacilles Gram-Negatifs*, pp 79-94. International Colloquium, CNRS, Paris.

Nowotny A, Golub S, Key B (1971) Fate and effect of endotoxin derivatives in tumor-bearing mice. *Proc. Soc. Exp. Biol. Med. 136*, 66-69.

Nowotny A, Chang CM, Chen CH, Grohsman J, Liang H, Ng AK, Rote N Jr, Thompson JJ, Yang-Ko C (1975a) Endotoxin-induced tumor resistance. In: Urbaschek B, Urbaschek R, Neter E (Eds), *Gram Negative Bacterial Infections*, pp 210-213. Springer Verlag, New York-Wien.

Nowotny A, Behling UH, Chang HL (1975b) Relation of structure to function in bacterial endotoxins. VIII. Biological activities in a polysaccharide-rich fraction. *J. Immunol. 115*, 199-203.

Nowotny A, Abdelnoor A, Behling UH, Butler RC, Johnson AG, Nowotny AM (1982) Molecular aspects of the adjuvant and other beneficial effects of endotoxins. In: Yamamura Y, Kotani S, Azuma I, Koda A, Shiba T (Eds), *Immunomodulation by Microbial Products and Related Synthetic Compounds*, pp 103-116. Excerpta Medica, Amsterdam-Oxford-Princeton.

Oettgen HF, Carswell EA, Kassel RL, Fiore N, Williamson B, Hoffmann MK, Haranaka K, Old LJ (1980) Endotoxin-induced tumor necrosis factor. *Recent Results Cancer Res. 75*, 207-212.

Old LJ, Boyse EA, Oettgen HF, DeHarven E, Geering G, Williamson B, Clifford P (1966) Precipitating antibody in human serum to an antigen present in cultured Burkitt's lymphoma cells. *Proc. Soc. Natl Acad. Sci. USA 56*, 1699-1704.

Olsson NO, Jeannin JF, Martin F (1980) Modulation of macrophage-mediated toxicity to cancer cells in vitro by liposomes and endotoxins. *C. R. Soc. Biol. (Paris) 174*, 1067-1071.

O'Malley WE, Achinstein B, Shear MJ (1962a) Action of bacterial polysaccharide on tumors. I. Reduced growth of Sarcoma 37, and induced refractoriness, by *Serratia marcescens* polysaccharide. *J. Natl Cancer Inst. 29*, 1161-1168.

O'Malley WE, Achinstein B, Shear MJ (1962b) Action of bacterial polysaccharide on tumors. II. Damage of Sarcoma 37 by serum of mice treated with *Serratia marcescens* polysaccharide, and induced tolerance. *J. Natl Cancer Inst. 29*, 1169-1175.

Oswald NTA, Cater DB (1969) Effect of endotoxin from *Serratia marcescens* on the permeability of vessels in hepatomas and carrageenin granulomas of rats. *Br. J. Exp. Pathol. 50*, 84-96.

Owens AH Jr, Coffey DS, Baylin SB (Eds) (1982) *Tumor Cell Heterogeneity. Origins and Implications*. Bristol Myers Cancer Symposia Vol 4. Academic Press, New York-London.

Ozato K, Adler WH, Ebert JD (1975) Synergism of bacterial lipopolysaccharides and Concanavalin A in the activation of thymic lymphocytes. *Cell. Immunol. 17*, 532-541.

Pabst MJ, Johnston RB Jr (1980) Increased production of superoxide anion by macrophages exposed in vitro to muramyl dipeptide or lipopolysaccharide. *J. Exp. Med. 151*, 101-114.

Pace JL, Russell SW (1981) Activation of mouse macrophages for tumor-cell killing. I. Quantitative-analysis of interactions between lymphokine and LPS. *J. Immunol. 126*, 1863-1867.

Pace JL, Taffet SM, Russell SW (1981) The effect of endotoxin in eliciting agents on the activation of mouse macrophages for tur cell killing. *J. Reticuloendothel. Soc. 30*, 15-21.

Palva ET, Mäkelä PH (1980) Lipopolysaccharide heterogeneity in Salmonella typhimurium analyzed by sodium dodecyl sulfate/polyacrylamide gel electrophoresis. *Eur. J. Biochem. 107*, 137-143.

Panke ES, Zwilling BS, Somers SD, Campolito LB, Packer BJ (1982) The effects of lipopolysaccharide, BCG-immune T lymphocytes, and lymphokines on generation of tumoricidal pulmonary macrophages in Syrian hamster. *Exp. Lung Res. 3*, 81-90.

Parant M (1980) Antimicrobial resistance enhancing activity of tumor necrosis serum factor induced by endotoxin in BCG treated mice. *Recent Results Cancer Res. 75*, 213-219.

Peavy DL, Adler WH, Smith RT (1970) The mitogenic effects of endotoxin and staphylococcal enterotoxin B on mouse spleen cells and human peripheral lymphocytes. *J. Immunol. 105*, 1453-1458.

Peavy DL, Baughn RE, Musher DM (1978) Mitogenic activity of bacterial LPS in vivo: morphological and functional characterization of responding cells. *Infect. Immun. 19*, 71-78.

Peri G, Polentar N, Sessa C, Mangioni C, Mantovani A (1981) Tumoricidal activity of macrophages isolated from human ascitic and solid ovarian carcinomas – augmentation by interferon, lymphokines and endotoxin. *Int. J. Cancer 28*, 143-152.

Piessens WF (1978) Increased binding of tumor cells by macrophages activated in vitro with lymphocyte mediators. *Cell. Immunol. 35*, 303-317.

Platica M, Hollander V (1978) Role of lipopolysaccharide in the production of plasma cell tumors in mice given mineral oil injections. *Cancer Res. 38*, 703-705.

Pluznik DH, Sachs L (1965) The cloning of normal 'mast' cells in tissue culture. *J. Cell. Comp. Physiol. 66*, 319-324.

Prager MD, Ludden CM, Mandy WJ, Allison JP, Kitto GB (1975) Endotoxin-stimulated immune response to modified lymphoma cells. *J. Natl Cancer Inst. 75*, 773-775.

Prager MD, Gordon WC, Baechtel FS (1976) Immunogenicity of modified tumor cells in syngenic hosts. *Ann. N.Y. Acad. Sci. 276*, 61-74.

Prehn RT, Main JM (1957) Immunity to methylcholantrene induced sarcoma. *J. Natl Cancer Inst. 18*, 769 778.

Primi D, Smith CIE, Hammarström L, Möller G (1977) Sera from lipopolysaccharide (LPS)-injected mice exhibit complement-dependent cytotoxicity against syngeneic and autologous spleen cells. *Cell. Immunol. 32*, 252-262.

Prince HE, Anderson WL, Tomasi TB (1981) An endotoxin-elicited tur necrosis-like factor in the ascitic fluid of tumor-bearing mice. *Fed. Proc. 40*, 5230.

Quesenberry P, Morley A, Stohlman F, Rickard K, Howard D, Smith M (1972) Effect of endotoxin on granulopoiesis and colony-stimulating factor. *N. Engl. J. Med. 286*, 227.

Ralph P, Nakoinz I (1977a) Direct toxic effects of immunopotentiators on monocytic, myelomonocytic, and histiocytic or macrophage tumor cells in culture. *Cancer Res. 37*, 546-550.

Ralph P, Nakoinz I (1977b) Antibody-dependent killing of erythrocyte and tumor targets by macrophage-related cell lines: enhancement by PPD and LPS. *J. Immunol. 119*, 950-954.

Ralph P, Nakoinz I (1981) Differences in antibody-dependent cellular cytotoxicity and activated killing of tumor cells by macrophage cell lines. *Cancer Res. 41*, 3546-3550.

Ralph P, Nakoinz I, Raschke WC (1974) Lymphosarcoma cell growth is selectively inhibited by B lymphocyte mitogens: LPS, dextran sulfate and PPD. *Biochem. Biophys. Res. Commun. 61*, 1268-1275.

Ralph P, Nakoinz I, Broxmeyer HE, Schrader S (1978) Immunologic functions and *in vitro* activation of cultured macrophage tumor lines. *Natl Cancer Inst. Monogr. 48*, 303-310.

Ralph P, Williams N, Nakoinz I, Jackson H, Watson JD (1982) Distinct signals for antibody-dependent and nonspecific killing of tumor targets mediated by macrophages. *J. Immunol. 129*, 427-432.

Rathgeb P, Sylven B (1954a) Fractionation studies on the tumor-necrotizing agent from Serratia marcescens (Shear's Polysaccharide). *J. Natl Cancer Inst. 14*, 1099-1108.

Rathgeb P, Sylven B (1954b) Chemical studies on the tumor-necrotizing agent from Serratia marcescens (Shear's Polysaccharide). *J. Natl Cancer Inst. 14*, 1109-1117.

Reed WP, Lucas ZJ (1975) Cytotoxic activity of lymphocytes. V. Role of soluble toxin in macrophage-inhibited cultures of tumor cells. *J. Immunol. 115*, 395-404.

Reidarson TH, Levy WE, Klostergaard J, Granger GA (1982a) Inducible macrophage cytotoxins I. Biokinetics of activation and release in vitro. *J. Natl Cancer Inst. 69*, 878-887.

Reidarson TH, Granger GA, Klostergaard J (1982b) Inducible macrophage cytotoxins. II. Tumor lysis mechanism involving target cell-binding proteases. *J. Natl Cancer Inst. 69*, 889-894.

Reimann HA (1926) Serological relationships of type-specific and degraded Pneumococci. *J. Exp. Med. 43*, 107-117.

Ribi EE, Granger DL, Milner KC, Strain SM (1975) Tumor regression caused by endotoxins and mycobacterial fractions. *J. Natl Cancer Inst. 55*, 1253-1257.

Ribi E, Milner KC, Granger DL, Kelly MT, Yamamoto K, Brehmer W, Parker R, Smith RF, Strain SM (1976a) Immunotherapy with nonviable microbial components. *Ann. N.Y. Acad. Sci. 227*, 228-238.

Ribi E, Takayama K, Milner K, Gray GR, Goren M, Parker R, McLaughlin Ch, Kelly M (1976b) Regression of tumors by an endotoxin combined with trehalose mycolates of differing structure. *Cancer Immunol. Immunother. 1*, 265-270.

Ribi E, Parker R, Strain SM, Mizuno Y, Nowotny A, Von Eschen KB, Cantrell JL, McLaughlin Ch, Hwang KM, Goren MB (1979) Peptides as requirement for immunotherapy of the guinea-pig Line-10 tumor with endotoxins. *Cancer Immunol. Immunother. 7*, 43-58.

Ribi E, Amano K, Cantrell J, Schwartzman S, Parker R, Takayama K (1982) Preparation and antitumor activity of nontoxic lipid A. *Cancer Immunol. Immunother. 12*, 91-96.

Ribi E, Cantrell JL, Takayama K, Amano K-i (1983) Enhancement of antitumor resistance by mycobacterial products and endotoxin. In: Nowotny A (Ed), *The Beneficial Effects of Endotoxin*, pp 529-554. Plenum Publishing Company, New York.

Ricord Ph (1853) *Letters on Syphilis* (translated by D.D. Slade) D. Clapp, Boston.

Richman SP, Chism VT, Murphy SG (1980) Effect of microbial fractions and vehicle on survival of mice bearing Lewis lung carcinoma. *Cancer Treat. Rep. 64*, 1329-1333.

Rietschel ET, Hase S, King M-T, Redmond J, Lehmann V (1977) Chemical structure of lipid A. In: Schlessinger D (Ed), *Microbiology - 1977*, pp 262-280. American Society for Microbiology, Washington D.C.

Robertson JS, Schwab JH (1961) Endotoxic properties associated with cell walls of group A streptococci. *J. Infect. Dis. 108*, 25-34.

Rosenstreich DL, Nowotny A, Chused T, Mergenhagen SE (1973) In vitro transformation of mouse bone-marrow derived (B) lymphocytes induced by the lipid component of endotoxin. *Infect. Immun. 8*, 406-411.

Rosenstreich DL, Asselineau J, Mergenhagen SE, Nowotny A (1974) A synthetic glycolipid with B-cell mitogenic activity. *J. Exp. Med. 140*, 1404-1409.

Rothman J, Johnson AG, Friedman H, Kovats E, Pham PH, Sanavi F, Nowotny AM, Nowotny A (1985) Biological effects of White-type polysaccharides of gram-negative bacteria. *J. Biol. Response Modif. 4*, 169-184.

Ruco LP, Meltzer MS (1977) Macrophage activation for tumor cytotoxicity: induction of tumoricidal macrophages by supernatants of PPD-stimulated bacillus Calmette-Guérin-immune spleen cell cultures. *J. Immunol. 119*, 889-896.

Ruco LP, Meltzer MS (1978a) Macrophage activation for tumor cytotoxicity: development of macrophage cytotoxic activity requires completion of a sequence of short-lived intermediary reactions. *J. Immunol. 121*, 2035-2042.

Ruco LP, Meltzer MS (1978b) Macrophage activation for tumor cytotoxicity: tumoricidal activity by macrophages from C3H/HeJ mice requires at least two activation stimuli. *Cell. Immunol. 41*, 35-51.

Ruff MR, Gifford GE (1980) Purification and physico-chemical characterization of rabbit tumor necrosis factor. *J. Immunol. 125*, 1671-1677.

Russell RRB, Johnson KG (1975) SDS-polyacrylamide gel electrophoresis of lipopolysaccharides. *Can. J. Microbiol. 21*, 2013-2018.

Russell SW, Doe WF, McIntosh AT (1977) Functional characterization of a stable, non-cytolytic stage of macrophage activation in tumors. *J. Exp. Med. 146*, 1511-1520.

Saito M, Yamazaki M, Mizuno D (1978) Immune response of mice in immunotherapy of tumors with syngeneic antitumor serum plus lipopolysaccharide. *Gann 69*, 331-337.

Salaman M (1969) Immunodepression by viruses. *Antibiot. Chemother. (Basel) 15*, 393.

Saluk PH (1983) Induced enhanced resistance to transplantable tumours in mice. In: Nowotny A (Ed), *Beneficial Effects of Endotoxins*, pp 513-517. Plenum Publishing Company, New York.

Saluk PH, Edmundowicz S, Wrigley DM, Tabor DR (1981) Coexpression of enhanced spreading, ingestion, and inhibition of tumor cell proliferation by macrophages from lipopolysaccharide-injected mice. *Cancer Immunol. Immunother. 11*, 159-164.

Sanarelli G (1924) De la pathogénie du choléra. Le choléra expérimental. *Ann. Inst. Pasteur. 38*, 11-72.

Sanavi F, Behling UH, Nejman G, Kováts E, Nowotny A (1982) In vitro bone resorption by endotoxic and non-toxic lipopolysaccharide (LPS) preparations. *J. Dental Res. 61*, 338.

Schacter B, Kleinhen ME, Edmonds K, Ellner JJ (1981) Spontaneous cytotoxicity of human peripheral blood mononuclear cells for the lymphoblastoid cell line CCRF-CEM-augmentation by bacterial lipopolysaccharide. *Clin. Exp. Immunol. 46*, 640-648.

Scheid MP, Hoffmann MK, Komuro K, Hämmerling U, Abott J, Boyse EA, Cohen GH, Hooper JA, Schulof RS, Goldstein AL (1973) Differentiation of T cells induced by preparations from thymus and by nonthymic agents. The determined state of the precursor cell. *J. Exp. Med. 138*, 1027-1032.

Schmidtke JR, Najarian JS (1975) Synergistic effects on DNA synthesis of phytohemagglutinin or Concanavalin A and lipopolysaccharide in human peripheral blood lymphocytes. *J. Immunol. 114*, 742-746.

Schorlemmer HV, Ferluga J, Allison AC (1977) Interaction of macrophages and complements in the pathogenesis of chronic inflammation. In: Willoughby DA, Giroud JP, Velo GP (Eds), *Perspectives in Inflammation*, pp 218-248. University Park Press, Baltimore, MD.

Schubert RD, David JR (1980) Stimulation of guinea pig macrophage pinocytosis by lipopolysaccharide (LPS): evidence that LPS acts directly on the macrophage. *Cell.*

Immunol. 55, 166-173.

Schultz RM, Chirigos MA, Pavlidis NA, Youngner JS (1978) Macrophage activation and antitur activity of a *Brucella abortus* ether extract, Bru-Pel. *Cancer Treat. Rep. 62*, 1937-1941.

Schuster BG, Palmer RF, Aronson RS (1970) The effect of endotoxin on thin lipid bilayer membranes. *J. Membr. Biol. 3*, 67-72.

Schwimmer E (1887) Az orbáncz gyógyhatása egyes kóralakoknál (The curative effect of erysipelas on some diseases, in Hungarian). *Gyógyászat 22*, 353-355.

Seligman AM, Shear MJ, Leiter J, Sweet B (1948) Chemical alteration of polysaccharide from Serratia marcescens. I. Tumor-necrotizing polysaccharide tagged with radioactive iodine. *J. Natl Cancer Inst. 9*, 13-18.

Shah RG, Green S, Moore MAS (1977) The effect of *Corynebacterium parvum* endotoxin treated mouse serum on granulocyte-macrophage colony growth. *Blood 50*, 160.

Shands JW Jr (1973) Affinity of endotoxin for membranes. *J. Infect. Dis. 128*, S197-201.

Shands JW Jr, Peavy DL, Gormus BJ, McGraw J (1974) In vitro and in vivo effects of endotoxin on mouse peritoneal cells. *Infect. Immun. 9*, 106-112.

Shapiro CJ (1940) The effect of toxic carbohydrate complex from *S. enteritidis* on transplantable rat tumors in tissue culture. *Am. J. Hyg. 31*, 114-126.

Shear MJ (1936) Chemical treatment of tumors. IV. Properties of hemorrhage-producing fraction of *B. coli* filtrate. *Proc. Soc. Exp. Biol. Med. 34*, 325-326.

Shear MJ (1941) Effect of a concentrate from *B. prodigiosus* filtrate on subcutaneous primary induced mouse tumors. *Cancer Res. 1*, 731-732.

Shear MJ, Andervont HB (1936) Chemical treatment of tumors. III. Separation of hemorrhage-producing fraction of *B. coli* filtrate. *Proc. Soc. Exp. Biol. Med. 34*, 323-325.

Shear MJ, Turner FC (1943) Chemical treatment of tumors. V. Isolation of the hemorrhage-producing fraction from *Serratia marcescens* (*Bacillus prodigiosus*) culture filtrate. *J. Natl Cancer Inst. 4*, 81-97.

Shinoda H, Yamazaki M, Mizuno D (1977a) Protective effect of syngeneic antitumor serum on mammary tumor in a syngeneic host. *Gann 68*, 561-565.

Shinoda H, Yamazaki M, Mizuno D (1977b) Passive immunotherapy of established tumors with syngeneic antitumor serum in combination with immunopotentiators. *Gann 68*, 567-571.

Shwartzman G (1928) Studies of *Bacillus typhosus* toxic substances. I. The phenomenon of local skin reactivity to *B. typhosus* culture filtrate. *J. Exp. Med. 48*, 247-268.

Shwartzman G (1935) Hemorrhagic necrosis and regression of Sarcoma 180. *Science 82*, 201.

Shwartzman G (1936) Reactivity of malignant neoplasms to bacterial filtrates. II. Relation of mortality to hemorrhagic necrosis and regression elicited by certain bacterial filtrates. *Arch. Pathol. 21*, 509-524.

Shwartzman G, Michailovsky N (1932) Phenomenon of local skin reactivity to bacterial filtrates in the treatment of mouse sarcoma 180. *Proc. Soc. Exp. Biol. Med. 29*, 737-741.

Skopinska E, Oluwasanni JO (1972) Effect of Escherichia coli endotoxin on skin allografts in mice. *Folia Biol. (Prague) 18*, 270-285.

Skopinska E, Nowaczyk M, Gorski A, Glyda J (1977) Studies on MIF production following transplantation of an allogeneic kidney in the rabbit. *Immunol. Pol. 2*, 153-169.

Skopinska-Rozewska E (1983) The effect of endotoxins on skin and kidney transplantation reactions. In: Nowotny A (Ed), *Beneficial Effects of Endotoxins*, pp 451-473. Plenum Publishing Company, New York.

Skurkovich SV, Klinova EG, Eremkina EI (1976) Stimulation of humoral immunity to surface antigens of leukemic cells. *Bull. Exp. Biol. Med. 81*, 429-439.

Sone S, Poste G, Fidler IJ (1980) Rat alveolar macrophages are susceptible to activation by free and liposome-encapsulated lymphokines. *J. Immunol. 124*, 2197-2202.

Sone S, Moriguchi S, Shimizu E, Ogushi F, Tsubura E (1982) In vitro generation of tumoricidal properties in human alveolar macrophages following interaction with endotoxin. *Cancer Res. 42*, 2227-2231.

Springer GF, Adye JC (1975) Endotoxin-binding substances from human leukocytes and platelets. *Infect. Immun. 12*, 978-9.

Springer GF, Horton RE (1964) Erythrocyte sensitization by blood group-specific bacterial antigens. *J. Gen. Physiol. 37*, 1229-1250.

Springer GF, Adye JC, Bezkorovainy A, Jirgenson B (1974) Properties and activity of the lipopolysaccharide-receptor from human erythrocytes. *Biochemistry 13*, 1379-1389.

Spronck CHH (1892) Tumeurs malignes et maladies infectieuses. *Ann. Inst. Pasteur 6*, 683-707.

Sredni B, Rosenszajn LA (1978) Relation between structure of bacterial LPS and stimulation of B-lymphocytes into colony formation. *J. Cell. Physiol. 96*, 53-61.

Stanley ER, Hansen G, Woodcock J, Metcalf D (1975) Colony stimulating factor and the regulation of granulopoiesis and macrophage production. *Fed. Proc. 34*, 2272-2278.

Stewart WE (1979) *The Interferon System*. Springer Verlag, Vienna-New York.

Stewart WE, Gosser LB, Lockart RZ Jr (1971) Priming: a non-anti-viral function of interferon. *J. Virol. 7*, 792-801.

Strausser HR, Bober LA (1972) Inhibition of tumor growth and survival of aged mice inoculated with Moloney tumor transplants and treated with endotoxin. *Cancer Res. 32*, 3156-3159.

Sultzer BM (1971) Chemical modification of endotoxin and inactivation of its biological properties. In: Kadis S, Weinbaum G, Ajl SJ (Eds), *Microbial Toxins Vol 5: Bacterial Endotoxins*, pp 91-126. Academic Press, New York.

Suter E (1964a) Hyperreactivity to endotoxin in infection. In: Landy M, Braun W (Eds), *Bacterial Endotoxins*, pp 435-447. Rutgers University Press, New Brunswick, NJ.

Suter E (1964b) Hyperreactivity to endotoxin after infection with BCG. Studies on its distinguishing properties. *J. Immunol. 92*, 49-54.

Suter E, Kirsanow EM (1961) Hyperreactivity to endotoxin in mice infected with mycobacteria. Induction and elicitation of the reactions. *Immunology 4*, 354-356.

Suter E, Munoz JJ (1963) Effect of BCG infection on sensitivity of mice to histamine, serotonine and passively induced anaphylaxis. *Proc. Soc. Exp. Biol. Med. 114*, 211-213.

Suter E, Ullman GE, Hoffman RG (1958) Sensitivity of mice to endotoxin after vaccination with BCG. *Proc. Soc. Exp. Biol. Med. 99*, 167-169.

Taichman NS (1971) The local Shwartzman reaction. In: Movat HZ (Ed), *Inflammation, Immunity and Hypersensitivity*, pp 479-525. Harper and Row, New York.

Talmadge JE, Donova PA, Hart TR (1982) Inhibition of cellular division of a murine macrophage tumor by macrophage-activating agents. *Cancer Res. 42*, 1850-1855.

Tanamoto K-i, Homma JY (1982) Essential regions of the lipopolysaccharide of Pseudomonas aeruginosa responsible for pyrogenicity and activation of the proclotting enzyme of horseshoe crabs. Comparison with antitumor, interferon-inducing and adjuvant activities. *J. Biochem. 91*, 741-74.

Tanamoto K-i, Abe C, Homma Y, Kuretani K, Hoshi A, Kojima Y (1978) A compound possessing antitumor and interferon-inducing activities derived from the common antigen (OEP) of *Pseudomonas aeruginosa*. *J. Biochem. (Tokyo) 83*, 711-718.

Tanamoto K-i, Abe C, Homma JY, Kojima Y (1979) Regions of the lipopolysaccharide of *Pseudomonas aeruginosa* essential for antitumor and interferon-inducing activities. *Eur. J. Biochem. 97*, 623-629.

Tanamoto K-i, Cho Y, Abe C, Homma JY, Kojima Y (1980) Regions of the lipopolysaccharide of *Pseudomonas aeruginosa* essential for antitumor, interferon-inducing and adjuvant activities. *Jpn. J. Med. Sci. Biol. 33*, 34-35.

Taniyama T, Holden HT (1979) Cytolytic activity of macrophages isolated from primary murine sarcoma virus (MSV)-induced tumors. *Int. J. Cancer 24*, 151-160.

Taramelli D, Holden HT, Varesio L (1980) Endotoxin requirement for macrophage activation by lymphokines in a rapid microcytotoxicity assay. *J. Immunol. Methods 37*, 225-232.

Tatsumi E, Domae N, Takiuchi Y, Sawada H, Shirakawa S, Uchino H (1979) Lipopolysaccharide responsiveness of malignant lymphoid cells in a patient with hairy cell leukemia. *Blood 54*, 524-529.

Tripodi D, Nowotny A (1965) Relation of structure to function in bacterial O-antigens V. Nature of active sites in endotoxin lipopolysaccharides of Serratia marcescens. *Information Exchange Group No. 5, Scientific memo #67.*

Tripodi D, Nowotny A (1966) Relation of structure to function in bacterial O-antigens. V. Nature of active sites in endotoxic lipopolysaccharides of Serratia marcescens. *Ann. N.Y. Acad. Sci. 133*, 604-621.

Tripodi D, Hollennbeck L, Pollack W (1970) The effect of endotoxin on the implantation of a mouse sarcoma. *Int. Arch. Allergy Appl. Immunol. 37*, 575-585.

Trnka W (1783) *Historia Febris Hecticae*. Vindobonae, prostat apud Rud. Graefferum.

Truitt RL (1983) Adoptive immunotherapy of leukemia by endotoxin-induced modulation of the graft-versus-host reaction. In: Nowotny A (Ed), *Beneficial Effects of Endotoxins*, pp 475-496. Plenum Publishing Company, New York.

Truitt RL, Rose WC, Rimm AA, Bortin MM (1978) Graft-versus-leukemia VIII. Selective reduction in anti-host reactivity without loss of antileukemic reactivity by treatment of donor mice with lipopolysaccharide. *Exp. Hematol. 6*, 488-498.

Tsai C-M, Frasch CE (1982) A sensitive silver stain for detecting lipopolysaccharides in polyacrylamide gels. *Anal. Biochem. 119*, 115-119.

Uhlenhuth, Haendel, Steffenhagen (1910) Beobachtungen über Immunität bei Rattensarkom. *Z. Immunitaetsforsch. Exp. Ther. 6*, 654-664.

Urbaschek B (1972) Die Vitalmikroskopie in der interdisziplinären medizinischen Forschung. *Med. Welt 23*, 1513-1516.

Urbaschek R (1980) Effects of bacterial products on granulopoiesis. *Adv. Exp. Med. Biol. 121B*, 51.

Urbaschek B, Urbaschek R (1977) Some aspects of microcirculatory and metabolic changes in endotoxemia and in endotoxin tolerance. In: Schlessinger D (Ed), *Microbiology – 1977*, pp 286-292. American Society for Microbiology, Washington D.C.

Urbaschek B, Versteyl R, Götte D (1967) Zur Hemmung der Endotoxinwirkung an der Mikrozirkulation durch ein Glucofuranosid-Derivat. *Klin. Wochenschr. 45*, 955-956.

Vallera DA, Pflugfelder U, Schmidtke JR (1980a) Lipopolysaccharide induced immunomodulation of the generation of cell-mediated cytotoxicity. I. Suppression of the development of cytotoxic lymphocytes. *J. Immunol. 124*, 635-640.

Vallera DA, Gamble CE, Schmidtke JR (1980b) Lipopolysaccharide-induced immunomodulation of the generation of cell-mediated cytotoxicity. II. Evidence for the involvement of a regulatory B lymphocyte. *J. Immunol. 124*, 641-649.

Van Der Zwan JC, Dankert J, DeVries K, Orie NGM, Kauffman HF (1978) Purification of specific precipitinogen and extraction of endotoxin from *Haemophilus influenzae. J. Clin. Pathol. 31*, 370-377.

Verneuil (1886) De l'inoculation de l'érysipèle comme moyen curatif. *Union Méd. 41*, 217-221.

Vogel SN, Marshall ST, Rosenstreich DL (1979) Analysis of the effects of lipopolysaccharide on macrophages: differential phagocytic responses of C3H/HeN and C3H/HeJ macrophages in vitro. *Infect. Immun. 25*, 328-336.

Vogel SN, Weedon LL, Wahl LM, Rosenstreich DL (1982) BCG-induced enhancement of endotoxin sensitivity in C3H/HeJ mice. 2. T cell modulation of macrophage sensitivity to LPS in vitro. *Immunobiology 160*, 479-493.

Weinberg JB (1981) In vivo modulation of macrophage tumoricidal activity: enhanced tumor cell killing by peritoneal macrophages from mice given injections of sodium periodate. *J. Natl Cancer Inst. 66*, 529-533.

Weissmann G, Thomas L (1962) Studies on lysosomes (The effects of endotoxin, endotoxin tolerance, and cortisone on the release of acid hydrolases from granular fraction of rabbit liver). *J. Exp. Med. 116*, 433-450.

Westphal O, Lüderitz O (1954) Chemische Erforschung von Lipopolysacchariden gram negativer Bakterien. *Angew. Chem. 66*, 407-417.

Westphal O, Lüderitz O, Rietschel ET, Galanos C (1980) Bacterial lipopolysaccharide and its lipid A component: some historical and some current aspects. *Biochem. Soc. Trans. 9*, 191-195.

White PB (1927) On the relation of the alcohol-soluble constituents of bacteria to their spontaneous agglutination. *J. Pathol. Bacteriol. 30*, 113-130.

White PB (1928) Further notes on spontaneous agglutination of bacteria. *J. Pathol. Bacteriol. 31*, 423-433.

White PB (1929) Notes on intestinal bacilli with special reference to smooth and rough races. *J. Pathol. Bacteriol. 32*, 85-94.

Wolff J, Cook GH (1975) Endotoxic lipopolysaccharides stimulate steroidogenesis and adenylate cyclase in adrenal tumor cells. *Biochim. Biophys. Acta 413*, 291-297.

Woodward JG, Daynes RA (1978) Cell-mediated immune response to syngeneic UV induced tumors. I. The presence of tumor associated macrophages and their possible role in the in vitro generation of cytotoxic lymphocytes. *Cell. Immunol. 41*, 304-319.

Wrigley DM, Saluk PH (1981) Two populations of lipopolysaccharide-stimulated lymphocytes cooperate to induce C3B-mediated phagocytosis by macrophages. *Fed. Proc. 40*, 4629.

Yamamoto Y, Tomida M, Hozumi M, Azuma I (1981) Enhancement by immunostimulants of the production by mouse spleen cells of factor(s) stimulating differentiation of mouse myeloid leukemic cells. *Gann 72*, 828-833.

Yamamura Y, Proctor JW, Fischer BC, Harnaha JB, Mahvi TA (1980) Collaboration between specific anti-tumor immunity and chemotherapeutic agents. *Int. J. Cancer 25*, 417-423.

Yang C, Nowotny A (1974) Effect of endotoxin on tumor resistance in mice. *Infect. Immun. 9*, 95-100.

Yang-Ko C, Grohsman J, Rote N Jr, Nowotny A (1973) Non-specific resistance induced in allogeneic mice to the transplantable TA3-Ha tumor by endotoxin and its derivatives. In: Daikos GK (Ed), *Progress in Chemotherapy*, pp 193-197. Proceedings of the VIII International Congress on Chemotherapy, Athens.

Yarkoni E, Goren MB, Rapp HJ (1979a) Regression of a transplanted guinea pig hepatoma after intralesional injection of an emulsified mixture of endotoxin and nycobacterial sulfolipid. *Infect. Immun. 24*, 357-362.

Yarkoni E, Rapp HJ, Polonsky J, Varenne J, Lederer E (1979b) Regression of a murine fibrosarcoma after intralesional injection of a synthetic C39 glycolipid related to cord factor. *Infect. Immunity 26*, 462-466.

Young MR, Cheung HT, Sundharadas G (1980) Selective inhibition of tumor cell migration by culture supernatants derived from normal and lipopolysaccharid activated macrophages. *J. Reticuloendothel. Soc. 27*, 143-150.

Youngner JS, Algire GH (1949) The effect of vascular occlusion on transplanted tumors. *J. Natl Cancer Inst. 10*, 565-572.

Youngner JS, Stinebring WR (1965) Interferon appearance stimulated by endotoxin, bacteria or viruses in mice pretreated with *Escherichia coli*, endotoxin or infected with *Mycobacterium tuberculosis. Nature (London) 108*, 456-458.

Zahl PA (1950) Action of bacterial toxins on tumors. VIII. Factors in their use for cancer therapy. *J. Natl Cancer Inst. 11*, 279-288.

Zahl PA, Hutner SH, Spitz S, Sugiura K, Cooper FS (1942) The actions of bacterial toxins on tumors. I. Relationship of the tumor hemorrhagic agent to the endotoxin antigens of gram-negative bacteria. *Am. J. Hyg. 36*, 224-242.

Zahl PA, Starr MP, Hutner SH (1945) Effect of bacterial toxins on turs VII. Tumor-hemorrhage factor in bacteria. *Am. J. Hyg. 41*, 41-56.

Zweifach BW (1964) Vascular effects of bacterial endotoxin. In: Landy M, Braun W (Eds), *Bacterial Endotoxins*, pp 110-117. Institute of Microbiology, Rutgers State University, New Brunswick, NJ.

Subject index